**Petroleum
Production
Engineering**

Petroleum Production Engineering

A Computer-Assisted Approach

Boyun Guo, Ph.D.
University of Louisiana at Lafayette

William C. Lyons, Ph.D.
New Mexico Institute of Mining and Technology

Ali Ghalambor, Ph.D.
University of Louisiana at Lafayette

AMSTERDAM • BOSTON • HEIDELBERG • LONDON • NEW YORK • OXFORD
PARIS • SAN DIEGO • SAN FRANCISCO • SINGAPORE • SYDNEY • TOKYO

Gulf Professional Publishing is an imprint of Elsevier

Gulf Professional Publishing is an imprint of Elsevier
30 Corporate Drive, Suite 400, Burlington, MA 01803, USA
Linacre House, Jordan Hill, Oxford OX2 8DP, UK

Copyright © 2007, Elsevier Inc. All rights reserved.

No part of this publication may be reproduced, stored in a retrieval system, or transmitted in any form or by any means, electronic, mechanical, photocopying, recording, or otherwise, without the prior written permission of the publisher.

Permissions may be sought directly from Elsevier s Science & Technology Rights Department in Oxford, UK: phone: (+44) 1865 843830, fax: (+44) 1865 853333, E-mail: permissions@elsevier.com. You may also complete your request on-line via the Elsevier homepage (http://elsevier.com), by selecting "Support & Contact" then "Copyright and Permission" and then "Obtaining Permissions."

∞ Recognizing the importance of preserving what has been written, Elsevier prints its books on acid-free paper whenever possible.

Library of Congress Cataloging-in-Publication Data
Application submitted

British Library Cataloguing-in-Publication Data
A catalogue record for this book is available from the British Library.

ISBN 13: 978-0-7506-8270-1
ISBN 10: 0-7506-8270-1

For information on all Gulf Professional Publishing publications visit our Web site at www.books.elsevier.com

07 08 09 10 11 12 10 9 8 7 6 5 4 3 2 1

Printed in the United States of America

Working together to grow
libraries in developing countries

www.elsevier.com | www.bookaid.org | www.sabre.org

ELSEVIER   BOOK AID International   Sabre Foundation

Dedication

This book is dedicated to my wife Huimei Wang for her understanding and encouragement that were as responsible as the experience and knowledge that have been inscribed herein.

–Boyun Guo

Contents

Preface ix
List of Symbols xi
List of Tables xv
List of Figures xvii

PART I PETROLEUM PRODUCTION ENGINEERING FUNDAMENTALS

1 Petroleum Production System 1/3
1.1 Introduction 1/4
1.2 Reservoir 1/4
1.3 Well 1/5
1.4 Separator 1/8
1.5 Pump 1/9
1.6 Gas Compressor 1/10
1.7 Pipelines 1/11
1.8 Safety Control System 1/11
1.9 Unit Systems 1/17
Summary 1/17
References 1/17
Problems 1/17

2 Properties of Oil and Natural Gas 2/19
2.1 Introduction 2/20
2.2 Properties of Oil 2/20
2.3 Properties of Natural Gas 2/21
Summary 2/26
References 2/26
Problems 2/26

3 Reservoir Deliverability 3/29
3.1 Introduction 3/30
3.2 Flow Regimes 3/30
3.3 Inflow Performance Relationship 3/32
3.4 Construction of IPR Curves Using Test Points 3/35
3.5 Composite IPR of Stratified Reservoirs 3/37
3.6 Future IPR 3/39
Summary 3/42
References 3/42
Problems 3/43

4 Wellbore Performance 4/45
4.1 Introduction 4/46
4.2 Single-Phase Liquid Flow 4/46
4.3 Multiphase Flow in Oil Wells 4/48
4.4 Single-Phase Gas Flow 4/53
4.5 Mist Flow in Gas Wells 4/56
Summary 4/56
References 4/57
Problems 4/57

5 Choke Performance 5/59
5.1 Introduction 5/60
5.2 Sonic and Subsonic Flow 5/60
5.3 Single-Phase Liquid Flow 5/60
5.4 Single-Phase Gas Flow 5/60
5.5 Multiphase Flow 5/63
Summary 5/66
References 5/66
Problems 5/66

6 Well Deliverability 6/69
6.1 Introduction 6/70
6.2 Nodal Analysis 6/70
6.3 Deliverability of Multilateral Well 6/79
Summary 6/84
References 6/85
Problems 6/85

7 Forecast of Well Production 7/87
7.1 Introduction 7/88
7.2 Oil Production during Transient Flow Period 7/88
7.3 Oil Production during Pseudo-Steady Flow Period 7/88
7.4 Gas Production during Transient Flow Period 7/92
7.5 Gas Production during Pseudo-Steady Flow Period 7/92
Summary 7/94
References 7/94
Problems 7/95

8 Production Decline Analysis 8/97
8.1 Introduction 8/98
8.2 Exponential Decline 8/98
8.3 Harmonic Decline 8/100
8.4 Hyperbolic Decline 8/100
8.5 Model Identification 8/100
8.6 Determination of Model Parameters 8/101
8.7 Illustrative Examples 8/101
Summary 8/104
References 8/104
Problems 8/104

PART II EQUIPMENT DESIGN AND SELECTION

9 Well Tubing 9/109
9.1 Introduction 9/110
9.2 Strength of Tubing 9/110
9.3 Tubing Design 9/111
Summary 9/114
References 9/114
Problems 9/114

10 Separation Systems 10/117
10.1 Introduction 10/118
10.2 Separation System 10/118
10.3 Dehydration System 10/125
Summary 10/132
References 10/132
Problems 10/132

11 Transportation Systems 11/133
11.1 Introduction 11/134
11.2 Pumps 11/134
11.3 Compressors 11/136
11.4 Pipelines 11/143
Summary 11/156
References 11/157
Problems 11/157

PART III ARTIFICIAL LIFT METHODS

12 Sucker Rod Pumping 12/161
12.1 Introduction 12/162
12.2 Pumping System 12/162

12.3	Polished Rod Motion	12/165
12.4	Load to the Pumping Unit	12/168
12.5	Pump Deliverability and Power Requirements	12/170
12.6	Procedure for Pumping Unit Selection	12/172
12.7	Principles of Pump Performance Analysis	12/174
Summary		12/179
References		12/179
Problems		12/179

13	**Gas Lift**	**13/181**
13.1	Introduction	13/182
13.2	Gas Lift System	13/182
13.3	Evaluation of Gas Lift Potential	13/183
13.4	Gas Lift Gas Compression Requirements	13/185
13.5	Selection of Gas Lift Valves	13/192
13.6	Special Issues in Intermittent-Flow Gas Lift	13/201
13.7	Design of Gas Lift Installations	13/203
Summary		13/205
References		13/205
Problems		13/205

14	**Other Artificial Lift Methods**	**14/207**
14.1	Introduction	14/208
14.2	Electrical Submersible Pump	14/208
14.3	Hydraulic Piston Pumping	14/211
14.4	Progressive Cavity Pumping	14/213
14.5	Plunger Lift	14/215
14.6	Hydraulic Jet Pumping	14/220
Summary		14/222
References		14/222
Problems		14/223

PART IV PRODUCTION ENHANCEMENT

15	**Well Problem Identification**	**15/227**
15.1	Introduction	15/228
15.2	Low Productivity	15/228
15.3	Excessive Gas Production	15/231
15.4	Excessive Water Production	15/231
15.5	Liquid Loading of Gas Wells	15/231
Summary		15/241
References		15/241
Problems		15/242

16	**Matrix Acidizing**	**16/243**
16.1	Introduction	16/244
16.2	Acid–Rock Interaction	16/244
16.3	Sandstone Acidizing Design	16/244
16.4	Carbonate Acidizing Design	16/247
Summary		16/248
References		16/248
Problems		16/249

17	**Hydraulic Fracturing**	**17/251**
17.1	Introduction	17/252
17.2	Formation Fracturing Pressure	17/252
17.3	Fracture Geometry	17/254
17.4	Productivity of Fractured Wells	17/256
17.5	Hydraulic Fracturing Design	17/258
17.6	Post-Frac Evaluation	17/262
Summary		17/264
References		17/264
Problems		17/265

18	**Production Optimization**	**18/267**
18.1	Introduction	18/268
18.2	Naturally Flowing Well	18/268
18.3	Gas-Lifted Well	18/268
18.4	Sucker Rod–Pumped Well	18/269
18.5	Separator	18/270
18.6	Pipeline Network	18/272
18.7	Gas-Lift Facility	18/275
18.8	Oil and Gas Production Fields	18/276
18.9	Discounted Revenue	18/279
Summary		18/279
References		18/279
Problems		18/280

Appendix A	Unit Conversion Factors	282
Appendix B	The Minimum Performance Properties of API Tubing	283
Index		285

Preface

The advances in the digital computing technology in the last decade have revolutionized the petroleum industry. Using the modern computer technologies, today's petroleum production engineers work much more efficiently than ever before in their daily activities, including analyzing and optimizing the performance of their existing production systems and designing new production systems. During several years of teaching the production engineering courses in academia and in the industry, the authors realized that there is a need for a textbook that reflects the current practice of what the modern production engineers do. Currently available books fail to provide adequate information about how the engineering principles are applied to solving petroleum production engineering problems with modern computer technologies. These facts motivated the authors to write this new book.

This book is written primarily for production engineers and college students of senior level as well as graduate level. It is not authors' intention to simply duplicate general information that can be found from other books. This book gathers authors' experiences gained through years of teaching courses of petroleum production engineering in universities and in the petroleum industry. The mission of the book is to provide production engineers a handy guideline to designing, analyzing, and optimizing petroleum production systems. The original manuscript of this book has been used as a textbook for college students of undergraduate and graduate levels in Petroleum Engineering.

This book was intended to cover the full scope of petroleum production engineering. Following the sequence of oil and gas production process, this book presents its contents in eighteen chapters covered in four parts.

Part I contains eight chapters covering petroleum production engineering fundamentals as the first course for the entry-level production engineers and undergraduate students. Chapter 1 presents an introduction to the petroleum production system. Chapter 2 documents properties of oil and natural gases that are essential for designing and analysing oil and gas production systems. Chapters 3 through 6 cover in detail the performance of oil and gas wells. Chapter 7 presents techniques used to forecast well production for economics analysis. Chapter 8 describes empirical models for production decline analysis.

Part II includes three chapters presenting principles and rules of designing and selecting the main components of petroleum production systems. These chapters are also written for entry-level production engineers and undergraduate students. Chapter 9 addresses tubing design. Chapter 10 presents rule of thumbs for selecting components in separation and dehydration systems. Chapter 11 details principles of selecting liquid pumps, gas compressors, and pipelines for oil and gas transportation.

Part III consists of three chapters introducing artificial lift methods as the second course for the entry-level production engineers and undergraduate students. Chapter 12 presents an introduction to the sucker rod pumping system and its design procedure. Chapter 13 describes briefly gas lift method. Chapter 14 provides an over view of other artificial lift methods and design procedures.

Part IV is composed of four chapters addressing production enhancement techniques. They are designed for production engineers with some experience and graduate students. Chapter 15 describes how to identify well problems. Chapter 16 deals with designing acidizing jobs. Chapter 17 provides a guideline to hydraulic fracturing and job evaluation techniques. Chapter 18 presents some relevant information on production optimisation techniques.

Since the substance of this book is virtually boundless in depth, knowing what to omit was the greatest difficulty with its editing. The authors believe that it requires many books to describe the foundation of knowledge in petroleum production engineering. To counter any deficiency that might arise from the limitations of space, the book provides a reference list of books and papers at the end of each chapter so that readers should experience little difficulty in pursuing each topic beyond the presented scope.

Regarding presentation, this book focuses on presenting and illustrating engineering principles used for designing and analyzing petroleum production systems rather than in-depth theories. Derivation of mathematical models is beyond the scope of this book, except for some special topics. Applications of the principles are illustrated by solving example problems. While the solutions to some simple problems not involving iterative procedures are demonstrated with stepwise calculations, complicated problems are solved with computer spreadsheet programs. The programs can be downloaded from the publisher's website (*http://books.elsevier.com/companions/ 9780750682701*). The combination of the book and the computer programs provides a perfect tool kit to petroleum production engineers for performing their daily work in a most efficient manner. All the computer programs were written in spreadsheet form in MS Excel that is available in most computer platforms in the petroleum industry. These spreadsheets are accurate and very easy to use. Although the U.S. field units are used in the companion book, options of using U.S. field units and SI units are provided in the spreadsheet programs.

This book is based on numerous documents including reports and papers accumulated through years of work in the University of Louisiana at Lafayette and the New Mexico Institute of Mining and Technology. The authors are grateful to the universities for permissions of publishing the materials. Special thanks go to the Chevron and American Petroleum Institute (API) for providing Chevron Professorship and API Professorship in Petroleum Engineering throughout editing of this book. Our thanks are due to Mr. Kai Sun of Baker Oil Tools, who made a thorough review and editing of this book. The authors also thank Malone Mitchell III of Riata Energy for he and his company's continued support of our efforts to develop new petroleum engineering text and professional books for the continuing education and training of the industry's vital engineers. On the basis of the collective experiences of authors and reviewer, we expect this book to be of value to the production engineers in the petroleum industry.

Dr. Boyun Guo
Chevron Endowed Professor in Petroleum Engineering
University of Louisiana at Lafayette
June 10, 2006

List of Symbols

A	area, ft^2	f_{Ri}	flow performance function of the curvic section of lateral i
A_b	total effective bellows area, in.^{2}	f_{sl}	slug factor, 0.5 to 0.6
A_{eng}	net cross-sectional area of engine piston, in.^{2}	G	shear modulus, psia
A_{fb}	total firebox surface area, ft^2	g	gravitational acceleration, 32.17 ft/s^2
A'_i	inner area of tubing sleeve, in.^{2}	G_b	pressure gradient below the pump, psi/ft
A'_o	outer area of tubing sleeve, in.^{2}	g_c	unit conversion factor, 32.17 lb$_m$–ft/lb$_f$–s^2
A_p	valve seat area, gross plunger cross-sectional area, or inner area of packer, in.^{2}	G_{fd}	design unloading gradient, psi/ft
		G_i	initial gas-in-place, scf
A_{pump}	net cross-sectional area of pump piston, in.^{2}	G_p	cumulative gas production, scf
A_r	cross-sectional area of rods, in.^{2}	G_p^1	cumulative gas production per stb of oil at the beginning of the interval, scf
A_t	tubing inner cross-sectional area, in.^{2}		
$^o API$	API gravity of stock tank oil	G_s	static (dead liquid) gradient, psi/ft
B	formation volume factor of fluid, rb/stb	G_2	mass flux at downstream, lbm/ft^2/sec
b	constant 1.5×10^{-5} in SI units	GLR_{fm}	formation oil GLR, scf/stb
B_o	formation volume factor of oil, rb/stb	GLR_{inj}	injection GLR, scf/stb
B_w	formation volume factor of water, rb/bbl	GLR_{min}	minimum required GLR for plunger lift, scf/bbl
C_A	drainage area shape factor		
C_a	weight fraction of acid in the acid solution	$GLR_{opt,o}$	optimum GLR at operating flow rate, scf/stb
C_c	choke flow coefficient	GOR	producing gas-oil ratio, scf/stb
C_D	choke discharge coefficient	GWR	glycol to water ratio, gal TEG/lb$_m$ H$_2$O
C_g	correction factor for gas-specific gravity	H	depth to the average fluid level in the annulus, ft, or dimensionless head
C_i	productivity coefficient of lateral i		
C_l	clearance, fraction	h	reservoir thickness, ft, or pumping head, ft
C_m	mineral content, volume fraction	h_f	fracture height, ft
C_s	structure unbalance, lbs	HP	required input power, hp
C_t	correction factor for operating temperature	Hp_{MM}	required theoretical compression power, hp/MMcfd
c_t	total compressibility, psi^{-1}		
C_p	specific heat of gas at constant pressure, lbf-ft/lbm-R	H_t	total heat load on reboiler, Btu/h
		Δh	depth increment, ft
$\bar{C}_p$	specific heat under constant pressure evaluated at cooler	ΔHp_m	mechanical power losses, hp
		∇_{hi}	pressure gradient in the vertical section of lateral i, psi/ft
C_{wi}	water content of inlet gas, lb$_m$ H$_2$O/MMscf		
D	outer diameter, in., or depth, ft, or non-Darcy flow coefficient, d/Mscf, or molecular diffusion coefficient, m^2/s	J	productivity of fractured well, stb/d-psi
		J_i	productivity index of lateral i.
		J_o	productivity of non-fractured well, stb/d-psi
d	diameter, in.	K	empirical factor, or characteristic length for gas flow in tubing, ft
d_1	upstream pipe diameter, in.		
d_2	choke diameter, in.	k	permeability of undamaged formation, md, or specific heat ratio
d_b	barrel inside diameter, in.		
D_{ci}	inner diameter of casing, in.	k_f	fracture permeability, md
d_f	fractal dimension constant 1.6	k_H	the average horizontal permeability, md
D_h	hydraulic diameter, in.	k_h	the average horizontal permeability, md
D_H	hydraulic diameter, ft	k_i	liquid/vapor equilibrium ratio of compound i
D_i	inner diameter of tubing, in.	k_p	a constant
D_o	outer diameter, in.	k_{ro}	the relative permeability to oil
d_p	plunger outside diameter, in.	k_V	vertical permeability, md
D_{pump}	minimum pump depth, ft	L	length, ft, or tubing inner capacity, ft/bbl
D_r	length of rod string, ft	L_g	length of gas distribution line, mile
E	rotor/stator eccentricity, in., or Young's modulus, psi	L_N	net lift, ft
		L_p	length of plunger, in.
E_v	volumetric efficiency, fraction	M	total mass associated with 1 stb of oil
e_v	correction factor	M_2	mass flow rate at down stream, lbm/sec
e_p	efficiency	MW_a	molecular weight of acid
F_b	axial load, lb$_f$	MW_m	molecular weight of mineral
F_{CD}	fracture conductivity, dimensionless	N	pump speed, spm, or rotary speed, rpm
F_F	fanning friction factor	n	number of layers, or polytropic exponent for gas
F_{gs}	modified Foss and Gaul slippage factor		
f_{hi}	flow performance function of the vertical section of lateral i	N_{Ac}	acid capillary number, dimensionless
		N_{Cmax}	maximum number of cycles per day
f_{Li}	inflow performance function of the horizontal section of lateral i	n_G	number of lb-mole of gas
		N_i	initial oil in place in the well drainage area, stb
f_M	Darcy-Wiesbach (Moody) friction factor	n_i	productivity exponent of lateral i
F_{pump}	pump friction-induced pressure loss, psia		

Symbol	Description
n_L	number of mole of fluid in the liquid phase
N_{max}	maximum pump speed, spm
n_p	number of pitches of stator
N_p^1	cumulative oil production per stb of oil in place at the beginning of the interval
$N_{p,n}^f$	forcasted annual cumulative production of fractured well for year n
$N_{p,n}^{nf}$	predicted annual cumulative production of nonfractured well for year n
$N_{p,n}^{no}$	predicted annual cumulative production of non-optimized well for year n
$N_{p,n}^{op}$	forcasted annual cumulative production of optimized system for year n
N_{Re}	Reunolds number
N_s	number of compression stages required
N_{st}	number of separation stages -1
n_V	number of mole of fluid in the vapor phase
N_w	number of wells
$\Delta N_{p,n}$	predicted annual incremental cumulative production for year n
P	pressure, lb/ft^2
p	pressure, psia
p_b	base pressure, psia
p_{bd}	formation breakdown pressure, psia
P_c	casing pressure, psig
p_c	critical pressure, psia, or required casing pressure, psia, or the collapse pressure with no axial load, psia
p_{cc}	the collapse pressure corrected for axial load, psia
P_{cd2}	design injection pressure at valve 2, psig
P_{Cmin}	required minimum casing pressure, psia
$p_{c,s}$	casing pressure at surface, psia
$p_{c,v}$	casing pressure at valve depth, psia
P_d	pressure in the dome, psig
p_d	final discharge pressure, psia
$p_{eng,d}$	engine discharge pressure, psia
$p_{eng,i}$	pressure at engine inlet, psia
p_f	frictional pressure loss in the power fluid injection tubing, psi
P_h	hydraulic power, hp
p_h	hydrostatic pressure of the power fluid at pump depth, psia
p_{hf}	wellhead flowing pressure, psia
p_{hf_i}	flowing pressure at the top of lateral i, psia
p_L	pressure at the inlet of gas distribution line, psia
p_i	initial reservoir pressure, psia, or pressure in tubing, psia, or pressure at stage i, psia
p_{kd1}	kick-off pressure opposite the first valve, psia
p_{kf_i}	flowing pressure at the kick-out-point of lateral i, psia
p_L	pressure at the inlet of the gas distribution line, psia
P_{lf}	flowing liquid gradient, psi/bbl slug
P_{lh}	hydrostatic liquid gradient, psi/bbl slug
p_{Lmax}	maximum line pressure, psia
p_o	pressure in the annulus, psia
p_{out}	output pressure of the compression station, psia
P_p	W_p/A_t, psia
p_p	pore pressure, psi
p_{pc}	pseudocritical pressure, psia
$p_{pump,i}$	pump intake pressure, psia
$p_{pump,d}$	pump discharge pressure, psia
P_r	pitch length of rotor, ft
p_r	pseudoreduced pressure
P_s	pitch length of stator, ft, or shaft power, ft–lb$_f$/sec
p_s	surface operating pressure, psia, or suction pressure, psia, or stock-tank pressure, psia
p_{sc}	standard pressure, 14.7 psia
p_{sh}	slug hydrostatic pressure, psia
p_{si}	surface injection pressure, psia
$p_{suction}$	suction pressure of pump, psia
P_t	tubing pressure, psia
p_{tf}	flowing tubing head pressure, psig
p_{up}	pressure upstream the choke, psia
P_{vc}	valve closing pressure, psig
P_{vo}	valve opening pressure, psig
p_{wh}	upstream (wellhead) pressure, psia
p_{wf}	flowing bottom hole pressure, psia
p_{wf_i}	the average flowing bottom-lateral pressure in lateral i, psia
p_{wfo}	dynamic bottom hole pressure because of cross-flow between, psia
p_{wf}^c	critical bottom hole pressure maintained during the production decline, psia
p_{up}	upstream pressure at choke, psia
P_1	pressure at point 1 or inlet, lb$_f$/ft^2
P_2	pressure at point 2 or outlet, lb$_f$/ft^2
p_1	upstream/inlet/suction pressure, psia
p_2	downstream/outlet/discharge pressure, psia
$\bar{p}$	average reservoir pressure, psia
$\bar{p}_f$	reservoir pressure in a future time, psia
$\bar{p}_0$	average reservoir pressure at decline time zero, psia
$\bar{p}_t$	average reservoir pressure at decline time t, psia
ΔP	pressure drop, lb$_f$/ft^2
Δp	pressure increment, psi
δp	head rating developed into an elementary cavity, psi
Δp_f	frictional pressure drop, psia
Δp_h	hydrostatic pressure drop, psia
$\Delta p_{i\,avg}$	the average pressure change in the tubing, psi
$\Delta p_{o\,avg}$	the average pressure change in the annulus, psi
Δp_{sf}	safety pressure margin, 200 to 500 psi
Δp_v	pressure differential across the operating valve (orifice), psi
Q	volumetric flow rate
q	volumetric flow rate
Q_c	pump displacement, bbl/day
q_{eng}	flow rate of power fluid, bbl/day
Q_G	gas production rate, Mscf/day
q_G	glycol circulation rate, gal/hr
q_g	gas production rate, scf/d
$q_{g,inj}$	the lift gas injection rate (scf/day) available to the well
q_{gM}	gas flow rate, Mscf/d
$q_{g,total}$	total output gas flow rate of the compression station, scf/day
q_h	injection rate per unit thickness of formation, m^3/sec-m
q_i	flow rate from/into layer i, or pumping rate, bpm
$q_{i,max}$	maximum injection rate, bbl/min
q_L	liquid capacity, bbl/day
Q_o	oil production rate, bbl/day
q_o	oil production rate, bbl/d
q_{pump}	flow rate of the produced fluid in the pump, bbl/day
Q_s	leak rate, bbl/day, or solid production rate, ft^3/day
q_s	gas capacity of contactor for standard gas (0.7 specific gravity) at standard temperature (100 °F), MMscfd, or sand production rate, ft^3/day
q_{sc}	gas flow rate, Mscf/d
q_{st}	gas capacity at standard conditions, MMscfd
q_{total}	total liquid flow rate, bbl/day
Q_w	water production rate, bbl/day
q_w	water production rate, bbl/d

LIST OF SYMBOLS

q_{wh}	flow rate at wellhead, stb/day
R	producing gas-liquid ratio, Mcf/bbl, or dimensionless nozzle area, or area ratio A_p/A_b, or the radius of fracture, ft, or gas constant, 10.73 ft^3-psia/lbmol-R
r	distance between the mass center of counterweights and the crank shaft, ft or cylinder compression ratio
r_a	radius of acid treatment, ft
R_c	radius of hole curvature, in.
r_e	drainage radius, ft
r_{eH}	radius of drainage area, ft
R_p	pressure ratio
R_s	solution gas oil ratio, scf/stb
r_w	radius of wellbore, ft
r_{wh}	desired radius of wormhole penetration, m
R^2	A_o/A_i
∇_{Ri}	vertical pressure gradient in the curvic section of lateral i, psi/ft
S	skin factor, or choke size, $1/64$ in.
S_A	axial stress at any point in the tubing string, psi
S_f	specific gravity of fluid in tubing, water = 1, or safety factor
S_g	specific gravity of gas, air = 1
S_o	specific gravity of produced oil, fresh water = 1
S_s	specific gravity of produced solid, fresh water = 1
S_t	equivalent pressure caused by spring tension, psig
S_w	specific gravity of produced water, fresh water = 1
T	temperature, °R
t	temperature, °F, or time, hour, or retention time, min
T_{av}	average temperature, °R
T_{avg}	average temperature in tubing, °F
T_b	base temperature, °R, or boiling point, °R
T_c	critical temperature, °R
T_{ci}	critical temperature of component i, °R
T_d	temperature at valve depth, °R
TF_1	maximum upstroke torque factor
TF_2	maximum downstroke torque factor
T_m	mechanical resistant torque, lb$_f$-ft
t_r	retention time ≈ 5.0 min
T_{sc}	standard temperature, 520 °R
T_{up}	upstream temperature, °R
T_v	viscosity resistant torque, lb$_f$-ft
T_1	suction temperature of the gas, °R
$\bar{T}$	average temperature, °R
u	fluid velocity, ft/s
u_m	mixture velocity, ft/s
u_{SL}	superficial velocity of liquid phase, ft/s
u_{SG}	superficial velocity of gas phase, ft/s
V	volume of the pipe segment, ft^3
v	superficial gas velocity based on total cross-sectional area A, ft/s
V_a	the required minimum acid volume, ft^3
V_{fg}	plunger falling velocity in gas, ft/min
V_{fl}	plunger falling velocity in liquid, ft/min
V_g	required gas per cycle, Mscf
V_{gas}	gas volume in standard condition, scf
V_{G1}	gas specific volume at upstream, ft^3/lbm
V_{G2}	gas specific volume at downstream, ft^3/lbm
V_h	required acid volume per unit thickness of formation, m^3/m
V_L	specific volume of liquid phase, ft^3/mol-lb, or volume of liquid phase in the pipe segment, ft^3, or liquid settling volume, bbl, or liquid specific volume at upstream, ft^3/lbm
V_m	volume of mixture associated with 1 stb of oil, ft^3, or volume of minerals to be removed, ft^3
V_0	pump displacement, ft^3
V_P	initial pore volume, ft^3
V_r	plunger rising velocity, ft/min
V_{res}	oil volume in reservoir condition, rb
V_s	required settling volume in separator, gal
V_{slug}	slug volume, bbl
V_{st}	oil volume in stock tank condition, stb
V_t	$A_t(D - V_{slug}L)$, gas volume in tubing, Mcf
V_{Vsc}	specific volume of vapor phase under standard condition, scf/mol-lb
V_1	inlet velocity of fluid to be compressed, ft/sec
V_2	outlet velocity of compressed fluid, ft/sec
v_1	specific volume at inlet, ft^3/lb
v_2	specific volume at outlet, ft^3/lb
w	fracture width, ft, or theoretical shaft work required to compress the gas, ft-lb$_f$/lb$_m$
W_{air}	weight of tubing in air, lb/ft
W_c	total weight of counterweights, lbs
W_f	weight of fluid, lbs
W_{fi}	weight of fluid inside tubing, lb/ft
W_{fo}	weight of fluid displaced by tubing, lb/ft
WOR	producing water-oil ratio, bbl/stb
W_p	plunger weight, lb$_f$
W_s	mechanical shaft work into the system, ft-lbs per lb of fluid
w_w	fracture width at wellbore, in.
$\bar{w}$	average width, in.
X	volumetric dissolving power of acid solution, ft^3 mineral/ ft^3 solution
x_f	fracture half-length, ft
x_i	mole fraction of compound i in the liquid phase
x_1	free gas quality at upstream, mass fraction
y_a	actual pressure ratio
y_c	critical pressure ratio
y_i	mole fraction of compound i in the vapor phase
y_L	liquid hold up, fraction
Z	gas compressibility factor in average tubing condition
z	gas compressibility factor
z_b	gas deviation factor at T_b and p_b
z_d	gas deviation factor at discharge of cylinder, or gas compressibility factor at valve depth condition
z_s	gas deviation factor at suction of the cylinder
z_1	compressibility factor at suction conditions
$\bar{z}$	the average gas compressibility factor
ΔZ	elevation increase, ft

Greek Symbols

α	Biot's poroelastic constant, approximately 0.7
β	gravimetric dissolving power of acid solution, lb$_m$ mineral/lb$_m$ solution
ε'	pipe wall roughness, in.
ϕ	porosity, fraction
η	pump efficiency
γ	1.78 = Euler's constant
γ_a	acid specific gravity, water = 1.0
γ_g	gas-specific gravity, air = 1
γ_L	specific gravity of production fluid, water = 1
γ_m	mineral specific gravity, water = 1.0
γ_o	oil specific gravity, water = 1
γ_{oST}	specific gravity of stock-tank oil, water = 1
γ_S	specific weight of steel (490 lb/ft^3)
γ_s	specific gravity of produced solid, water = 1
γ_w	specific gravity of produced water, fresh water = 1
μ	viscosity
μ_a	viscosity of acid solution, cp
μ_{od}	viscosity of dead oil, cp

μ_f	viscosity of the effluent at the inlet temperature, cp	ρ_a	density of acid, lb_m/ft^3
μ_G	gas viscosity, cp	ρ_{air}	density of air, lb_m/ft^3
μ_g	gas viscosity at in-situ temperature and pressure, cp	ρ_G	in-situ gas density, lb_m/ft^3
		ρ_L	liquid density, lb_m/ft^3
μ_L	liquid viscosity, cp	ρ_m	density of mineral, lb_m/ft^3
μ_o	viscosity of oil, cp	ρ_{m2}	mixture density at downstream, lbm/ft^3
μ_s	viscosity of the effluent at the surface temperature, cp	$\rho_{o,st}$	density of stock tank oil, lb_m/ft^3
		ρ_w	density of fresh water, $62.4 \, lb_m/ft^3$
ν	Poison's ratio	ρ_{wh}	density of fluid at wellhead, lb_m/ft^3
ν_a	stoichiometry number of acid	ρ_i	density of fluid from/into layer i, lb_m/ft^3
ν_m	stoichiometry number of mineral	$\bar{\rho}$	average mixture density (specific weight), lb_f/ft^3
ν_{pf}	viscosity of power fluid, centistokes		
θ	inclination angle, deg., or dip angle from horizontal direction, deg.	σ	liquid-gas interfacial tension, dyne/cm
		σ_1	axial principal stress, psi,
		σ_2	tangential principal stress, psi
ρ	fluid density lb_m/ft^3	σ_3	radial principal stress, psi
ρ_1	mixture density at top of tubing segment, lb_f/ft^3	σ_b	bending stress, psi
		σ_v	overburden stress, psi
ρ_2	mixture density at bottom of segment, lb_f/ft^3	σ'_v	effective vertical stress, psi

List of Tables

Table 2.1: Result Given by the Spreadsheet Program *OilProperties.xls*
Table 2.2: Results Given by the Spreadsheet Program *MixingRule.xls*
Table 2.3: Results Given by the Spreadsheet *Carr-Kobayashi-Burrows-GasViscosity.xls*
Table 2.4: Results Given by the Spreadsheet Program *Brill.Beggs.Z.xls*
Table 2.5: Results Given by the Spreadsheet Program *Hall.Yarborogh.z.xls*
Table 3.1: Summary of Test Points for Nine Oil Layers
Table 3.2: Comparison of Commingled and Layer-Grouped Productions
Table 4.1: Result Given by *Poettmann-Carpenter BHP.xls* for Example Problem 4.2
Table 4.2: Result Given by *Guo.GhalamborBHP.xls* for Example Problem 4.3
Table 4.3: Result Given by *HagedornBrown Correlation.xls* for Example Problem 4.4
Table 4.4: Spreadsheet Average TZ.xls for the Data Input and Results Sections
Table 4.5: Appearance of the Spreadsheet *Cullender.Smith.xls* for the Data Input and Results Sections
Table 5.1: Solution Given by the Spreadsheet Program *GasUpChokePressure.xls*
Table 5.2: Solution Given by the Spreadsheet Program *GasDownChokePressure.xls*
Table 5.3: A Summary of C, m and n Values Given by Different Researchers
Table 5.4: An Example Calculation with Sachdeva's Choke Model
Table 6.1: Result Given by *BottomHoleNodalGas.xls* for Example Problem 6.1
Table 6.2: Result Given by *BottomHoleNodalOil-PC.xls* for Example Problem 6.2
Table 6.3: Result Given by *BottomHoleNodaloil-GG.xls.* for Example of Problem 6.2
Table 6.4: Solution Given by *BottomHoleNodalOil-HB.xls*
Table 6.5: Solution Given by *WellheadNodalGas-SonicFlow.xls.*
Table 6.6: Solution Given by *WellheadNodalOil-PC.xls*
Table 6.7: Solution Given by *WellheadNodalOil-GG.xls*
Table 6.8: Solution Given by *WellheadNodalOil-HB.xls.*
Table 6.9: Solution Given by *MultilateralGasWell Deliverability (Radial-Flow IPR).xls*
Table 6.10: Data Input and Result Sections of the Spreadsheet *MultilateralOilWell Deliverability.xls*
Table 7.1: Sroduction Forecast Given by *Transient ProductionForecast.xls*
Table 7.2: Production Forecast for Example Problem 7.2
Table 7.3: Oil Production Forecast for $N = 1$
Table 7.4: Gas Production Forecast for $N = 1$
Table 7.5: Production schedule forecast
Table 7.6: Result of Production Forecast for Example Problem 7.4
Table 8.1: Production Data for Example Problem 8.2
Table 8.2: Production Data for Example Problem 8.3
Table 8.3: Production Data for Example Problem 8.4
Table 9.1: API Tubing Tensile Requirements
Table 10.1: K-Values Used for Selecting Separators
Table 10.2: Retention Time Required Under Various Separation Conditions
Table 10.3: Settling Volumes of Standard Vertical High-Pressure Separators
Table 10.4: Settling Volumes of Standard Vertical Low-Pressure Separators
Table 10.5: Settling Volumes of Standard Horizontal High-Pressure Separators
Table 10.6: Settling Volumes of Standard Horizontal Low-Pressure Separators
Table 10.7: Settling Volumes of Standard Spherical High-Pressure Separators
Table 10.8: Settling Volumes of Standard Spherical Low-Pressure Separators (125 psi)
Table 10.9: Temperature Correction Factors for Trayed Glycol Contactors
Table 10.10: Specific Gravity Correction Factors for Trayed Glycol Contactors
Table 10.11: Temperature Correction Factors for Packed Glycol Contactors
Table 10.12: Specific Gravity Correction Factors for Packed Glycol Contactors
Table 11.1: Typical Values of Pipeline Efficiency Factors
Table 11.2: Design and Hydrostatic Pressure Definitions and Usage Factors for Oil Lines
Table 11.3: Design and Hydrostatic Pressure Definitions and Usage Factors for Gas Lines
Table 11.4: Thermal Conductivities of Materials Used in Pipeline Insulation
Table 11.5: Typical Performance of Insulated Pipelines
Table 11.6: Base Data for Pipeline Insulation Design
Table 11.7: Calculated Total Heat Losses for the Insulated Pipelines (kW)
Table 12.1: Conventional Pumping Unit API Geometry Dimensions
Table 12.2: Solution Given by Computer Program *SuckerRodPumpingLoad.xls*
Table 12.3: Solution Given by *SuckerRodPumping Flowrate&Power.xls*
Table 12.4: Design Data for API Sucker Rod Pumping Units
Table 13.1: Result Given by Computer Program *CompressorPressure.xls*
Table 13.2: Result Given by Computer Program *ReciprocatingCompressorPower.xls* for the First Stage Compression
Table 13.3: Result Given by the Computer Program *CentrifugalCompressorPower.xls*
Table 13.4: R Values for Otis Speedmaster Valves
Table 13.5: Summary of Results for Example Problem 13.7
Table 14.1: Result Given by the Computer Spreadsheet *ESPdesign.xls*
Table 14.2: Solution Given by *HydraulicPiston Pump.xls*
Table 14.3: Summary of Calculated Parameters
Table 14.4: Solution Given by Spreadsheet Program *PlungerLift.xls*

Table 15.1:	Basic Parameter Values for Example Problem 15.1	Table 17.2:	Summary of Some Commercial Fracturing Models
Table 15.2:	Result Given by the Spreadsheet Program *GasWellLoading.xls*	Table 17.3:	Calculated Slurry Concentration
		Table 18.1:	Flash Calculation with Standing's Method for k_i Values
Table 16.1:	Primary Chemical Reactions in Acid Treatments	Table 18.2:	Solution to Example Problem 18.3 Given by the Spreadsheet *LoopedLines.xls*
Table 16.2:	Recommended Acid Type and Strength for Sandstone Acidizing	Table 18.3:	Gas Lift Performance Data for Well A and Well B
Table 16.3:	Recommended Acid Type and Strength for Carbonate Acidizing	Table 18.4:	Assignments of Different Available Lift Gas Injection Rates to Well A and Well B
Table 17.1:	Features of Fracture Geometry Models		

List of Figures

Figure 1.1: A sketch of a petroleum production system.
Figure 1.2: A typical hydrocarbon phase diagram.
Figure 1.3: A sketch of a water-drive reservoir.
Figure 1.4: A sketch of a gas-cap drive reservoir.
Figure 1.5: A sketch of a dissolved-gas drive reservoir.
Figure 1.6: A sketch of a typical flowing oil well.
Figure 1.7: A sketch of a wellhead.
Figure 1.8: A sketch of a casing head.
Figure 1.9: A sketch of a tubing head.
Figure 1.10: A sketch of a "Christmas tree."
Figure 1.11: Sketch of a surface valve.
Figure 1.12: A sketch of a wellhead choke.
Figure 1.13: Conventional horizontal separator.
Figure 1.14: Double action piston pump.
Figure 1.15: Elements of a typical reciprocating compressor.
Figure 1.16: Uses of offshore pipelines.
Figure 1.17: Safety device symbols.
Figure 1.18: Safety system designs for surface wellhead flowlines.
Figure 1.19: Safety system designs for underwater wellhead flowlines.
Figure 1.20: Safety system design for pressure vessel.
Figure 1.21: Safety system design for pipeline pumps.
Figure 1.22: Safety system design for other pumps.
Figure 3.1: A sketch of a radial flow reservoir model: (a) lateral view, (b) top view.
Figure 3.2: A sketch of a reservoir with a constant-pressure boundary.
Figure 3.3: A sketch of a reservoir with no-flow boundaries.
Figure 3.4: (a) Shape factors for various closed drainage areas with low-aspect ratios. (b) Shape factors for closed drainage areas with high-aspect ratios.
Figure 3.5: A typical IPR curve for an oil well.
Figure 3.6: Transient IPR curve for Example Problem 3.1.
Figure 3.7: Steady-state IPR curve for Example Problem 3.1.
Figure 3.8: Pseudo–steady-state IPR curve for Example Problem 3.1.
Figure 3.9: IPR curve for Example Problem 3.2.
Figure 3.10: Generalized Vogel IPR model for partial two-phase reservoirs.
Figure 3.11: IPR curve for Example Problem 3.3.
Figure 3.12: IPR curves for Example Problem 3.4, Well A.
Figure 3.13: IPR curves for Example Problem 3.4, Well B.
Figure 3.14: IPR curves for Example Problem 3.5.
Figure 3.15: IPR curves of individual layers.
Figure 3.16: Composite IPR curve for all the layers open to flow.
Figure 3.17: Composite IPR curve for Group 2 (Layers B4, C1, and C2).
Figure 3.18: Composite IPR curve for Group 3 (Layers B1, A4, and A5).
Figure 3.19: IPR curves for Example Problem 3.6.
Figure 3.20: IPR curves for Example Problem 3.7.
Figure 4.1: Flow along a tubing string.
Figure 4.2: Darcy–Wiesbach friction factor diagram.
Figure 4.3: Flow regimes in gas-liquid flow.
Figure 4.4: Pressure traverse given by *Hagedorn BrownCorreltion.xls* for Example.
Figure 4.5: Calculated tubing pressure profile for Example Problem 4.5.
Figure 5.1: A typical choke performance curve.
Figure 5.2: Choke flow coefficient for nozzle-type chokes.
Figure 5.3: Choke flow coefficient for orifice-type chokes.
Figure 6.1: Nodal analysis for Example Problem 6.1.
Figure 6.2: Nodal analysis for Example Problem 6.4.
Figure 6.3: Nodal analysis for Example Problem 6.5.
Figure 6.4: Nodal analysis for Example Problem 6.6.
Figure 6.5: Nodal analysis for Example Problem 6.8.
Figure 6.6: Schematic of a multilateral well trajectory.
Figure 6.7: Nomenclature of a multilateral well.
Figure 7.1: Nodal analysis plot for Example Problem 7.1.
Figure 7.2: Production forecast for Example Problem 7.2.
Figure 7.3: Nodal analysis plot for Example Problem 7.2.
Figure 7.4: Production forecast for Example Problem 7.2
Figure 7.3: Production forecast for Example Problem 7.3.
Figure 7.4: Result of production forecast for Example Problem 7.4.
Figure 8.1: A semilog plot of q versus t indicating an exponential decline.
Figure 8.2: A plot of N_p versus q indicating an exponential decline.
Figure 8.3: A plot of $\log(q)$ versus $\log(t)$ indicating a harmonic decline.
Figure 8.4: A plot of N_p versus $\log(q)$ indicating a harmonic decline.
Figure 8.5: A plot of relative decline rate versus production rate.
Figure 8.6: Procedure for determining a- and b-values.
Figure 8.7: A plot of $\log(q)$ versus t showing an exponential decline.
Figure 8.8: Relative decline rate plot showing exponential decline.
Figure 8.9: Projected production rate by an exponential decline model.
Figure 8.10: Relative decline rate plot showing harmonic decline.
Figure 8.11: Projected production rate by a harmonic decline model.
Figure 8.12: Relative decline rate plot showing hyperbolic decline.
Figure 8.13: Relative decline rate plot showing hyperbolic decline.
Figure 8.14: Projected production rate by a hyperbolic decline model.
Figure 9.1: A simple uniaxial test of a metal specimen.
Figure 9.2: Effect of tension stress on tangential stress.
Figure 9.3: Tubing–packer relation.
Figure 9.4: Ballooning and buckling effects.
Figure 10.1: A typical vertical separator.
Figure 10.2: A typical horizontal separator.
Figure 10.3: A typical horizontal double-tube separator.
Figure 10.4: A typical horizontal three-phase separator.

LIST OF FIGURES

Figure 10.5:	A typical spherical low-pressure separator.
Figure 10.6:	Water content of natural gases.
Figure 10.7:	Flow diagram of a typical solid desiccant dehydration plant.
Figure 10.8:	Flow diagram of a typical glycol dehydrator.
Figure 10.9:	Gas capacity of vertical inlet scrubbers based on 0.7-specific gravity at 100 °F.
Figure 10.10:	Gas capacity for trayed glycol contactors based on 0.7-specific gravity at 100 °F.
Figure 10.11:	Gas capacity for packed glycol contactors based on 0.7-specific gravity at 100 °F.
Figure 10.12:	The required minimum height of packing of a packed contactor, or the minimum number of trays of a trayed contactor.
Figure 11.1:	Double-action stroke in a duplex pump.
Figure 11.2:	Single-action stroke in a triplex pump.
Figure 11.3:	Elements of a typical reciprocating compressor.
Figure 11.4:	Cross-section of a centrifugal compressor.
Figure 11.5:	Basic pressure–volume diagram.
Figure 11.6:	Flow diagram of a two-stage compression unit.
Figure 11.7:	Fuel consumption of prime movers using three types of fuel.
Figure 11.8:	Fuel consumption of prime movers using natural gas as fuel.
Figure 11.9:	Effect of elevation on prime mover power.
Figure 11.10:	Darcy–Wiesbach friction factor chart.
Figure 11.11:	Stresses generated by internal pressure p in a thin-wall pipe, $D/t > 20$.
Figure 11.12:	Stresses generated by internal pressure p in a thick-wall pipe, $D/t < 20$.
Figure 11.13:	Calculated temperature profiles with a polyethylene layer of 0.0254 M (1 in.).
Figure 11.14:	Calculated steady-flow temperature profiles with polyethylene layers of various thicknesses.
Figure 11.15:	Calculated temperature profiles with a polypropylene layer of 0.0254 M (1 in.).
Figure 11.16:	Calculated steady-flow temperature profiles with polypropylene layers of various thicknesses.
Figure 11.17:	Calculated temperature profiles with a polyurethane layer of 0.0254 M (1 in.).
Figure 11.18:	Calculated steady-flow temperature profiles with polyurethane layers of four thicknesses.
Figure 12.1:	A diagrammatic drawing of a sucker rod pumping system.
Figure 12.2:	Sketch of three types of pumping units: (a) conventional unit; (b) Lufkin Mark II unit; (c) air-balanced unit.
Figure 12.3:	The pumping cycle: (a) plunger moving down, near the bottom of the stroke; (b) plunger moving up, near the bottom of the stroke; (c) plunger moving up, near the top of the stroke; (d) plunger moving down, near the top of the stroke.
Figure 12.4:	Two types of plunger pumps.
Figure 12.5:	Polished rod motion for (a) conventional pumping unit and (b) air-balanced unit.
Figure 12.6:	Definitions of conventional pumping unit API geometry dimensions.
Figure 12.7:	Approximate motion of connection point between pitman arm and walking beam.
Figure 12.8:	Sucker rod pumping unit selection chart.
Figure 12.9:	A sketch of pump dynagraph.
Figure 12.10:	Pump dynagraph cards: (a) ideal card, (b) gas compression on down-stroke, (c) gas expansion on upstroke, (d) fluid pound, (e) vibration due to fluid pound, (f) gas lock.
Figure 12.11:	Surface Dynamometer Card: (a) ideal card (stretch and contraction), (b) ideal card (acceleration), (c) three typical cards.
Figure 12.12:	Strain-gage–type dynamometer chart.
Figure 12.13:	Surface to down hole cards derived from surface dynamometer card.
Figure 13.1:	Configuration of a typical gas lift well.
Figure 13.2:	A simplified flow diagram of a closed rotary gas lift system for single intermittent well.
Figure 13.3:	A sketch of continuous gas lift.
Figure 13.4:	Pressure relationship in a continuous gas lift.
Figure 13.5:	System analysis plot given by *GasLift Potential.xls* for the unlimited gas injection case.
Figure 13.6:	System analysis plot given by *GasLift Potential.xls* for the limited gas injection case.
Figure 13.7:	Well unloading sequence.
Figure 13.8:	Flow characteristics of orifice-type valves.
Figure 13.9:	Unbalanced bellow valve at its closed condition.
Figure 13.10:	Unbalanced bellow valve at its open condition.
Figure 13.11:	Flow characteristics of unbalanced valves.
Figure 13.12:	A sketch of a balanced pressure valve.
Figure 13.13:	A sketch of a pilot valve.
Figure 13.14:	A sketch of a throttling pressure valve.
Figure 13.15:	A sketch of a fluid-operated valve.
Figure 13.16:	A sketch of a differential valve.
Figure 13.17:	A sketch of combination valve.
Figure 13.18:	A flow diagram to illustrate procedure of valve spacing.
Figure 13.19:	Illustrative plot of BHP of an intermittent flow.
Figure 13.20:	Intermittent flow gradient at mid-point of tubing.
Figure 13.21:	Example Problem 13.8 schematic and BHP build.up for slug flow.
Figure 13.22:	Three types of gas lift installations.
Figure 13.23:	Sketch of a standard two-packer chamber.
Figure 13.24:	A sketch of an insert chamber.
Figure 13.25:	A sketch of a reserve flow chamber.
Figure 14.1:	A sketch of an ESP installation.
Figure 14.2:	An internal schematic of centrifugal pump.
Figure 14.3:	A sketch of a multistage centrifugal pump.
Figure 14.4:	A typical ESP characteristic chart.
Figure 14.5:	A sketch of a hydraulic piston pump.
Figure 14.6:	Sketch of a PCP system.
Figure 14.7:	Rotor and stator geometry of PCP.
Figure 14.8:	Four flow regimes commonly encountered in gas wells.
Figure 14.9:	A sketch of a plunger lift system.
Figure 14.10:	Sketch of a hydraulic jet pump installation.
Figure 14.11:	Working principle of a hydraulic jet pump.
Figure 14.12:	Example jet pump performance chart.
Figure 15.1:	Temperature and spinner flowmeter-derived production profile.
Figure 15.2:	Notations for a horizontal wellbore.

Figure 15.3:	Measured bottom-hole pressures and oil production rates during a pressure drawdown test.	Figure 17.4:	Concept of effective stress between grains.
Figure 15.4:	Log-log diagnostic plot of test data.	Figure 17.5:	The KGD fracture geometry.
Figure 15.5:	Semi-log plot for vertical radial flow analysis.	Figure 17.6:	The PKN fracture geometry.
Figure 15.6:	Square-root time plot for pseudo-linear flow analysis.	Figure 17.7:	Relationship between fracture conductivity and equivalent skin factor.
Figure 15.7:	Semi-log plot for horizontal pseudo-radial flow analysis.	Figure 17.8:	Relationship between fracture conductivity and equivalent skin factor.
Figure 15.8:	Match between measured and model calculated pressure data.	Figure 17.9:	Effect of fracture closure stress on proppant pack permeability.
Figure 15.9:	Gas production due to channeling behind the casing.	Figure 17.10:	Iteration procedure for injection time calculation.
Figure 15.10:	Gas production due to preferential flow through high-permeability zones.	Figure 17.11:	Calculated slurry concentration.
Figure 15.11:	Gas production due to gas coning.	Figure 17.12:	Bottom-hole pressure match with three-dimensional fracturing model PropFRAC.
Figure 15.12:	Temperature and noise logs identifying gas channeling behind casing.	Figure 17.13:	Four flow regimes that can occur in hydraulically fractured reservoirs.
Figure 15.13:	Temperature and fluid density logs identifying a gas entry zone.	Figure 18.1:	Comparison of oil well inflow performance relationship (IPR) curves before and after stimulation.
Figure 15.14:	Water production due to channeling behind the casing.	Figure 18.2:	A typical tubing performance curve.
Figure 15.15:	Preferential water flow through high-permeability zones.	Figure 18.3:	A typical gas lift performance curve of a low-productivity well.
Figure 15.16:	Water production due to water coning.	Figure 18.4:	Theoretical load cycle for elastic sucker rods.
Figure 15.17:	Prefracture and postfracture temperature logs identifying fracture height.	Figure 18.5:	Actual load cycle of a normal sucker rod.
Figure 15.18:	Spinner flowmeter log identifying a watered zone at bottom.	Figure 18.6:	Dimensional parameters of a dynamometer card.
Figure 15.19:	Calculated minimum flow rates with Turner et al.'s model and test flow rates.	Figure 18.7:	A dynamometer card indicating synchronous pumping speeds.
Figure 15.20:	The minimum flow rates given by Guo et al.'s model and the test flow rates.	Figure 18.8:	A dynamometer card indicating gas lock.
Figure 16.1:	Typical acid response curves.	Figure 18.9:	Sketch of (a) series pipeline and (b) parallel pipeline.
Figure 16.2:	Wormholes created by acid dissolution of limestone.	Figure 18.10:	Sketch of a looped pipeline.
Figure 17.1:	Schematic to show the equipment layout in hydraulic fracturing treatments of oil and gas wells.	Figure 18.11:	Effects of looped line and pipe diameter ratio on the increase of gas flow rate.
Figure 17.2:	A schematic to show the procedure of hydraulic fracturing treatments of oil and gas wells.	Figure 18.12:	A typical gas lift performance curve of a high-productivity well.
		Figure 18.13:	Schematics of two hierarchical networks.
Figure 17.3:	Overburden formation of a hydrocarbon reservoir.	Figure 18.14:	An example of a nonhierarchical network.

Part I: Petroleum Production Engineering Fundamentals

The upstream of the petroleum industry involves itself in the business of oil and gas exploration and production (E & P) activities. While the exploration activities find oil and gas reserves, the production activities deliver oil and gas to the downstream of the industry (i.e., processing plants). The petroleum production is definitely the heart of the petroleum industry.

Petroleum production engineering is that part of petroleum engineering that attempts to maximize oil and gas production in a cost-effective manner. To achieve this objective, production engineers need to have a thorough understanding of the petroleum production systems with which they work. To perform their job correctly, production engineers should have solid background and sound knowledge about the properties of fluids they produce and working principles of all the major components of producing wells and surface facilities. This part of the book provides graduating production engineers with fundamentals of petroleum production engineering. Materials are presented in the following eight chapters:

Chapter 1 Petroleum Production System 1/3
Chapter 2 Properties of Oil and Natural Gas 2/19
Chapter 3 Reservoir Deliverability 3/29
Chapter 4 Wellbore Performance 4/45
Chapter 5 Choke Performance 5/59
Chapter 6 Well Deliverability 6/69
Chapter 7 Forecast of Well Production 7/87
Chapter 8 Production Decline Analysis 8/97

1 Petroleum Production System

Contents
1.1 Introduction 1/4
1.2 Reservoir 1/4
1.3 Well 1/5
1.4 Separator 1/8
1.5 Pump 1/9
1.6 Gas Compressor 1/10
1.7 Pipelines 1/11
1.8 Safety Control System 1/11
1.9 Unit Systems 1/17
Summary 1/17
References 1/17
Problems 1/17

1.1 Introduction

The role of a production engineer is to maximize oil and gas production in a cost-effective manner. Familiarization and understanding of oil and gas production systems are essential to the engineers. This chapter provides graduating production engineers with some basic knowledge about production systems. More engineering principles are discussed in the later chapters.

As shown in Fig. 1.1, a complete oil or gas production system consists of a reservoir, well, flowline, separators, pumps, and transportation pipelines. The reservoir supplies wellbore with crude oil or gas. The well provides a path for the production fluid to flow from bottom hole to surface and offers a means to control the fluid production rate. The flowline leads the produced fluid to separators. The separators remove gas and water from the crude oil. Pumps and compressors are used to transport oil and gas through pipelines to sales points.

1.2 Reservoir

Hydrocarbon accumulations in geological traps can be classified as reservoir, field, and pool. A "reservoir" is a porous and permeable underground formation containing an individual bank of hydrocarbons confined by impermeable rock or water barriers and is characterized by a single natural pressure system. A "field" is an area that consists of one or more reservoirs all related to the same structural feature. A "pool" contains one or more reservoirs in isolated structures.

Depending on the initial reservoir condition in the phase diagram (Fig. 1.2), hydrocarbon accumulations are classified as oil, gas condensate, and gas reservoirs. An oil that is at a pressure above its bubble-point pressure is called an "undersaturated oil" because it can dissolve more gas at the given temperature. An oil that is at its bubble-point pressure is called a "saturated oil" because it can dissolve no more gas at the given temperature. Single (liquid)-phase flow prevails in an undersaturated oil reservoir, whereas two-phase (liquid oil and free gas) flow exists in a saturated oil reservoir.

Wells in the same reservoir can fall into categories of oil, condensate, and gas wells depending on the producing gas–oil ratio (GOR). Gas wells are wells with producing GOR being greater than 100,000 scf/stb; condensate wells are those with producing GOR being less than 100,000 scf/stb but greater than 5,000 scf/stb; and wells with producing GOR being less than 5,000 scf/stb are classified as oil wells.

Oil reservoirs can be classified on the basis of boundary type, which determines driving mechanism, and which are as follows:

- Water-drive reservoir
- Gas-cap drive reservoir
- Dissolved-gas drive reservoir

In water-drive reservoirs, the oil zone is connected by a continuous path to the surface groundwater system (aquifer). The pressure caused by the "column" of water to the surface forces the oil (and gas) to the top of the reservoir against the impermeable barrier that restricts the oil and gas (the trap boundary). This pressure will force the oil and gas toward the wellbore. With the same oil production, reservoir pressure will be maintained longer (relative to other mechanisms of drive) when there is an active water drive. Edge-water drive reservoir is the most preferable type of reservoir compared to bottom-water drive. The reservoir pressure can remain at its initial value above bubble-point pressure so that single-phase liquid flow exists in the reservoir for maximum well productivity. A steady-state flow condition can prevail in a edge-water drive reservoir for a long time before water breakthrough into the well. Bottom-water drive reservoir (Fig. 1.3) is less preferable because of water-coning problems that can affect oil production economics due to water treatment and disposal issues.

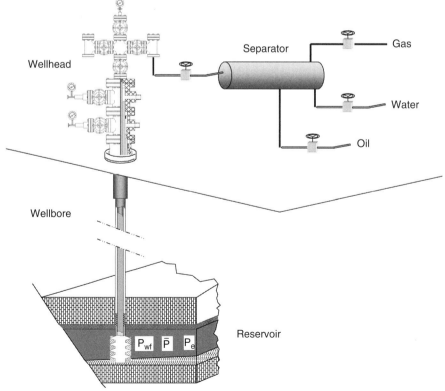

Figure 1.1 A sketch of a petroleum production system.

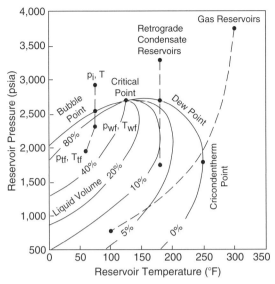

Figure 1.2 A typical hydrocarbon phase diagram.

In a gas-cap drive reservoir, gas-cap drive is the drive mechanism where the gas in the reservoir has come out of solution and rises to the top of the reservoir to form a gas cap (Fig. 1.4). Thus, the oil below the gas cap can be produced. If the gas in the gas cap is taken out of the reservoir early in the production process, the reservoir pressure will decrease rapidly. Sometimes an oil reservoir is subjected to both water and gas-cap drive.

A dissolved-gas drive reservoir (Fig. 1.5) is also called a "solution-gas drive reservoir" and "volumetric reservoir." The oil reservoir has a fixed oil volume surrounded by no-flow boundaries (faults or pinch-outs). Dissolved-gas drive is the drive mechanism where the reservoir gas is held in solution in the oil (and water). The reservoir gas is actually in a liquid form in a dissolved solution with the liquids (at atmospheric conditions) from the reservoir. Compared to the water- and gas-drive reservoirs, expansion of solution (dissolved) gas in the oil provides a weak driving mechanism in a volumetric reservoir. In the regions where the oil pressure drops to below the bubble-point pressure, gas escapes from the oil and oil–gas two-phase flow exists. To improve oil recovery in the solution-gas reservoir, early pressure maintenance is usually preferred.

1.3 Well

Oil and gas wells are drilled like an upside-down telescope. The large-diameter borehole section is at the top of the well. Each section is cased to the surface, or a liner is placed in the well that laps over the last casing in the well. Each casing or liner is cemented into the well (usually up to at least where the cement overlaps the previous cement job).

The last casing in the well is the production casing (or production liner). Once the production casing has been cemented into the well, the production tubing is run into the well. Usually a packer is used near the bottom of the tubing to isolate the annulus between the outside of the tubing and the inside of the casing. Thus, the produced fluids are forced to move out of the perforation into the bottom of the well and then into the inside of the tubing. Packers can be actuated by either mechanical or hydraulic mechanisms. The production tubing is often (particularly during initial well flow) provided with a bottom-hole choke to control the initial well flow (i.e., to restrict overproduction and loss of reservoir pressure).

Figure 1.6 shows a typical flowing oil well, defined as a well producing solely because of the natural pressure of the reservoir. It is composed of casings, tubing, packers, down-hole chokes (optional), wellhead, Christmas tree, and surface chokes.

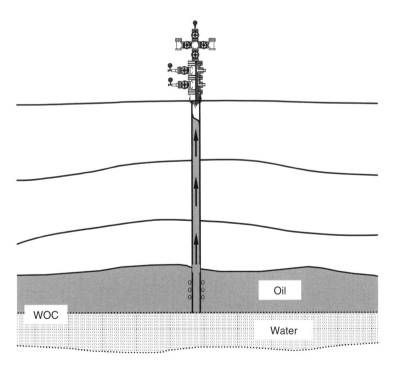

Figure 1.3 A sketch of a water-drive reservoir.

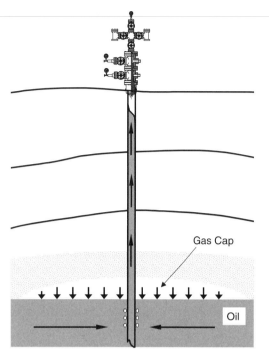

Figure 1.4 A sketch of a gas-cap drive reservoir.

Most wells produce oil through tubing strings, mainly because a tubing string provides good sealing performance and allows the use of gas expansion to lift oil. The American Petroleum Institute (API) defines tubing size using nominal diameter and weight (per foot). The nominal diameter is based on the internal diameter of the tubing body. The weight of tubing determines the tubing outer diameter. Steel grades of tubing are designated H-40, J-55, C-75, L-80, N-80, C-90, and P-105, where the digits represent the minimum yield strength in 1,000 psi. The minimum performance properties of tubing are given in Chapter 9 and Appendix B.

The "wellhead" is defined as the surface equipment set below the master valve. As we can see in Fig. 1.7, it includes casing heads and a tubing head. The casing head (lowermost) is threaded onto the surface casing. This can also be a flanged or studded connection. A "casing head" is a mechanical assembly used for hanging a casing string (Fig. 1.8). Depending on casing programs in well drilling, several casing heads can be installed during well construction. The casing head has a bowl that supports the casing hanger. This casing hanger is threaded onto the top of the production casing (or uses friction grips to hold the casing). As in the case of the production tubing, the production casing is landed in tension so that the casing hanger actually supports the production casing (down to the freeze point). In a similar manner, the intermediate casing(s) are supported by their respective casing hangers (and bowls). All of these casing head arrangements are supported by the surface casing, which is in compression and cemented to the surface. A well completed with three casing strings has two casing heads. The uppermost casing head supports the production casing. The lowermost casing head sits on the surface casing (threaded to the top of the surface casing).

Most flowing wells are produced through a string of tubing run inside the production casing string. At the surface, the tubing is supported by the tubing head (i.e., the tubing head is used for hanging tubing string on the production casing head [Fig. 1.9]). The tubing head supports the tubing string at the surface (this tubing is landed on the tubing head so that it is in tension all the way down to the packer).

The equipment at the top of the producing wellhead is called a "Christmas tree" (Fig. 1.10) and it is used to control flow. The "Christmas tree" is installed above the tubing head. An "adaptor" is a piece of equipment used to join the two. The "Christmas tree" may have one flow outlet (a tee) or two flow outlets (a cross). The master valve is installed below the tee or cross. To replace a master valve, the tubing must be plugged. A Christmas tree consists of a main valve, wing valves, and a needle valve. These valves are used for closing the well when needed. At the top of the tee structure (on the top of the "Christmas tree"), there is a pressure gauge that indicates the pressure in the tubing.

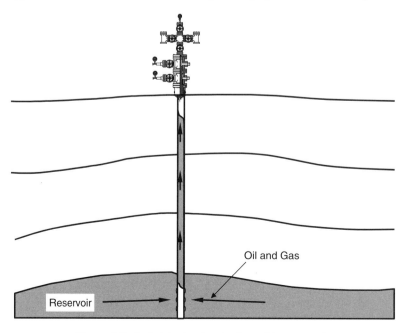

Figure 1.5 A sketch of a dissolved-gas drive reservoir.

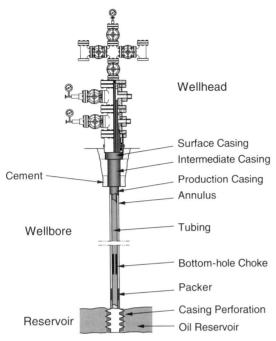

Figure 1.6 A sketch of a typical flowing oil well.

The wing valves and their gauges allow access (for pressure measurements and gas or liquid flow) to the annulus spaces (Fig. 1.11).

"Surface choke" (i.e., a restriction in the flowline) is a piece of equipment used to control the flow rate (Fig. 1.12). In most flowing wells, the oil production rate is altered by adjusting the choke size. The choke causes back-pressure in the line. The back-pressure (caused by the chokes or other restrictions in the flowline) increases the bottom-hole flowing pressure. Increasing the bottom-hole flowing pressure decreases the pressure drop from the reservoir to the wellbore (pressure drawdown). Thus, increasing the back-pressure in the wellbore decreases the flow rate from the reservoir.

In some wells, chokes are installed in the lower section of tubing strings. This choke arrangement reduces wellhead pressure and enhances oil production rate as a result of gas expansion in the tubing string. For gas wells, use of down-hole chokes minimizes the gas hydrate problem in the well stream. A major disadvantage of using down-hole chokes is that replacing a choke is costly.

Certain procedures must be followed to open or close a well. Before opening, check all the surface equipment such as safety valves, fittings, and so on. The burner of a line heater must be lit before the well is opened. This is necessary because the pressure drop across a choke cools the fluid and may cause gas hydrates or paraffin to deposit out. A gas burner keeps the involved fluid (usually water) hot. Fluid from the well is carried through a coil of piping. The choke is installed in the heater. Well fluid is heated both before and after it flows through the choke. The upstream heating helps melt any solids that may be present in the producing fluid. The downstream heating prevents hydrates and paraffins from forming at the choke.

Surface vessels should be open and clear before the well is allowed to flow. All valves that are in the master valve and other downstream valves are closed. Then follow the following procedure to open a well:

1. The operator barely opens the master valve (just a crack), and escaping fluid makes a hissing sound. When the fluid no longer hisses through the valve, the pressure has been equalized, and then the master valve is opened wide.
2. If there are no oil leaks, the operator cracks the next downstream valve that is closed. Usually, this will be

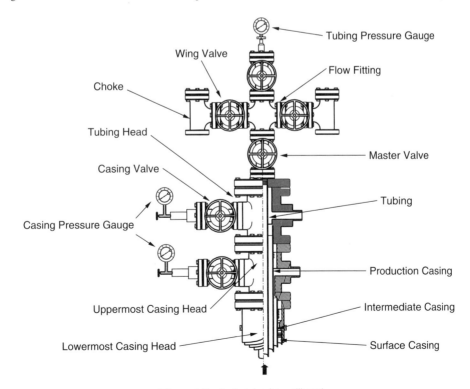

Figure 1.7 A sketch of a wellhead.

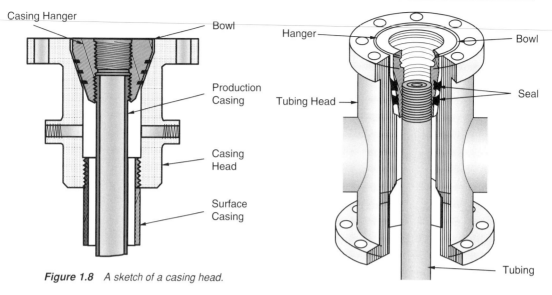

Figure 1.8 A sketch of a casing head.

Figure 1.9 A sketch of a tubing head.

either the second (backup) master valve or a wing valve. Again, when the hissing sound stops, the valve is opened wide.
3. The operator opens the other downstream valves the same way.
4. To read the tubing pressure gauge, the operator must open the needle valve at the top of the Christmas tree. After reading and recording the pressure, the operator may close the valve again to protect the gauge.

The procedure for "shutting-in" a well is the opposite of the procedure for opening a well. In shutting-in the well, the master valve is closed last. Valves are closed rather rapidly to avoid wearing of the valve (to prevent erosion). At least two valves must be closed.

1.4 Separator

The fluids produced from oil wells are normally complex mixtures of hundreds of different compounds. A typical oil well stream is a high-velocity, turbulent, constantly expanding mixture of gases and hydrocarbon liquids, intimately mixed with water vapor, free water, and sometimes solids. The well stream should be processed as soon as possible after bringing them to the surface. Separators are used for the purpose.

Three types of separators are generally available from manufacturers: horizontal, vertical, and spherical separators. Horizontal separators are further classified into

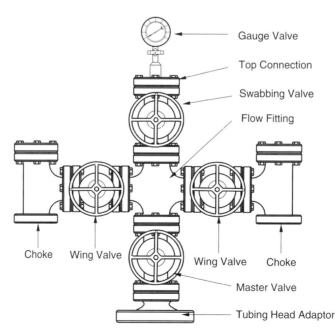

Figure 1.10 A sketch of a "Christmas tree."

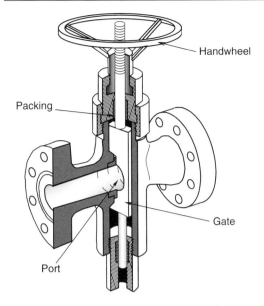

Figure 1.11 A sketch of a surface valve.

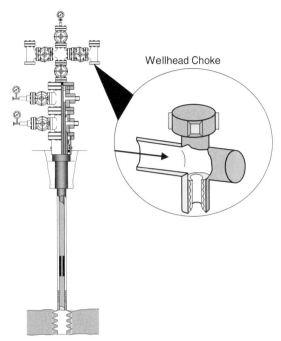

Figure 1.12 A sketch of a wellhead choke.

two categories: single tube and double tube. Each type of separator has specific advantages and limitations. Selection of separator type is based on several factors including characteristics of production steam to be treated, floor space availability at the facility site, transportation, and cost.

Horizontal separators (Fig. 1.13) are usually the first choice because of their low costs. Horizontal separators are almost widely used for high-GOR well streams, foaming well streams, or liquid-from-liquid separation. They have much greater gas–liquid interface because of a large, long, baffled gas-separation section. Horizontal separators are easier to skid-mount and service and require less piping for field connections. Individual separators can be stacked easily into stage-separation assemblies to minimize space requirements. In horizontal separators, gas flows horizontally while liquid droplets fall toward the liquid surface. The moisture gas flows in the baffle surface and forms a liquid film that is drained away to the liquid section of the separator. The baffles need to be longer than the distance of liquid trajectory travel. The liquid-level control placement is more critical in a horizontal separator than in a vertical separator because of limited surge space.

Vertical separators are often used to treat low to intermediate GOR well streams and streams with relatively large slugs of liquid. They handle greater slugs of liquid without carryover to the gas outlet, and the action of the liquid-level control is not as critical. Vertical separators occupy less floor space, which is important for facility sites such as those on offshore platforms where space is limited. Because of the large vertical distance between the liquid level and the gas outlet, the chance for liquid to re-vaporize into the gas phase is limited. However, because of the natural upward flow of gas in a vertical separator against the falling droplets of liquid, adequate separator diameter is required. Vertical separators are more costly to fabricate and ship in skid-mounted assemblies.

Spherical separators offer an inexpensive and compact means of separation arrangement. Because of their compact configurations, these types of separators have a very limited surge space and liquid-settling section. Also, the placement and action of the liquid-level control in this type of separator is more critical.

Chapter 10 provides more details on separators and dehydrators.

1.5 Pump

After separation, oil is transported through pipelines to the sales points. Reciprocating piston pumps are used to provide mechanical energy required for the transportation. There are two types of piston strokes, the single-action

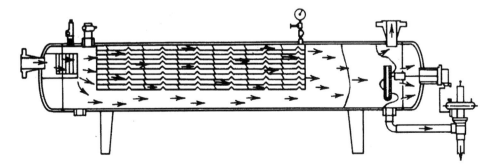

Figure 1.13 Conventional horizontal separator. (Courtesy Petroleum Extension Services.)

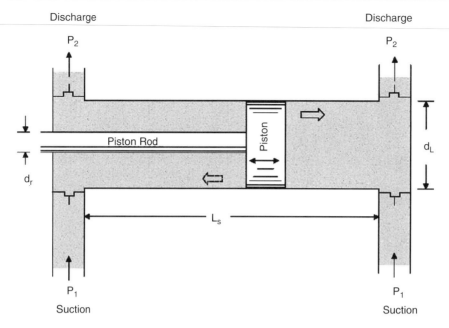

Figure 1.14 Double-action piston pump.

piston stroke and the double-action piston stroke. The double-action stroke is used for duplex (two pistons) pumps. The single-action stroke is used for pumps with three pistons or greater (e.g., triplex pump). Figure 1.14 shows how a duplex pump works. More information about pumps is presented in Chapter 11.

1.6 Gas Compressor

Compressors are used for providing gas pressure required to transport gas with pipelines and to lift oil in gas-lift operations. The compressors used in today's natural gas production industry fall into two distinct types: reciprocating and rotary compressors. Reciprocating compressors are most commonly used in the natural gas industry. They are built for practically all pressures and volumetric capacities.

As shown in Fig. 1.15, reciprocating compressors have more moving parts and, therefore, lower mechanical efficiencies than rotary compressors. Each cylinder assembly of a reciprocating compressor consists of a piston, cylinder, cylinder heads, suction and discharge valves, and other parts necessary to convert rotary motion to reciprocation motion. A reciprocating compressor is designed for a certain range of compression ratios through the selection of proper piston displacement and clearance volume within the cylinder. This clearance volume can be either fixed or variable, depending on the extent of the operation range and the percent of load variation desired. A typical reciprocating compressor can deliver a volumetric gas flow rate up to 30,000 cubic feet per minute (cfm) at a discharge pressure up to 10,000 psig.

Rotary compressors are divided into two classes: the centrifugal compressor and the rotary blower. A centrifu-

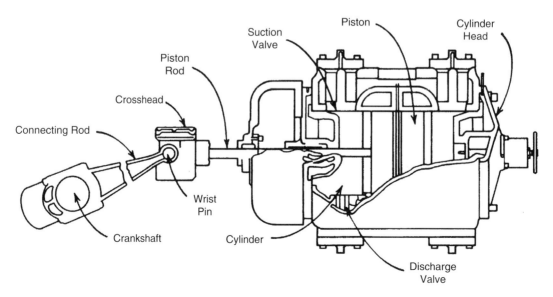

Figure 1.15 Elements of a typical reciprocating compressor. (Courtesy Petroleum Extension Services.)

gal compressor consists of a housing with flow passages, a rotating shaft on which the impeller is mounted, bearings, and seals to prevent gas from escaping along the shaft. Centrifugal compressors have few moving parts because only the impeller and shaft rotate. Thus, its efficiency is high and lubrication oil consumption and maintenance costs are low. Cooling water is normally unnecessary because of lower compression ratio and less friction loss. Compression rates of centrifugal compressors are lower because of the absence of positive displacement. Centrifugal compressors compress gas using centrifugal force. In this type of compressor, work is done on the gas by an impeller. Gas is then discharged at a high velocity into a diffuser where the velocity is reduced and its kinetic energy is converted to static pressure. Unlike reciprocating compressors, all this is done without confinement and physical squeezing. Centrifugal compressors with relatively unrestricted passages and continuous flow are inherently high-capacity, low-pressure ratio machines that adapt easily to series arrangements within a station. In this way, each compressor is required to develop only part of the station compression ratio. Typically, the volume is more than 100,000 cfm and discharge pressure is up to 100 psig. More information about different types of compressors is provided in Chapter 11.

1.7 Pipelines

The first pipeline was built in the United States in 1859 to transport crude oil (Wolbert, 1952). Through the one and half century of pipeline operating practice, the petroleum industry has proven that pipelines are by far the most economical means of large-scale overland transportation for crude oil, natural gas, and their products, clearly superior to rail and truck transportation over competing routes, given large quantities to be moved on a regular basis. Transporting petroleum fluids with pipelines is a continuous and reliable operation. Pipelines have demonstrated an ability to adapt to a wide variety of environments including remote areas and hostile environments. With very minor exceptions, largely due to local peculiarities, most refineries are served by one or more pipelines, because of their superior flexibility to the alternatives.

Figure 1.16 shows applications of pipelines in offshore operations. It indicates flowlines transporting oil and/or gas from satellite subsea wells to subsea manifolds, flowlines transporting oil and/or gas from subsea manifolds to production facility platforms, infield flowlines transporting oil and/or gas from between production facility platforms, and export pipelines transporting oil and/or gas from production facility platforms to shore.

The pipelines are sized to handle the expected pressure and fluid flow. To ensure desired flow rate of product, pipeline size varies significantly from project to project. To contain the pressures, wall thicknesses of the pipelines range from $\frac{3}{8}$ inch to $1\frac{1}{2}$ inch. More information about pipelines is provided in Chapter 11.

1.8 Safety Control System

The purpose of safety systems is to protect personnel, the environment, and the facility. The major objective of the safety system is to prevent the release of hydrocarbons

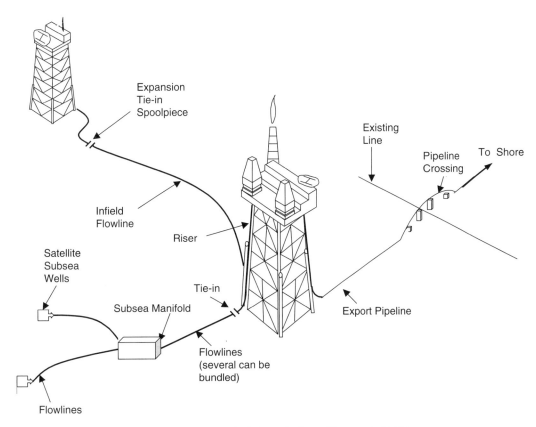

Figure 1.16 Uses of offshore pipelines. (Guo et al., 2005.)

from the process and to minimize the adverse effects of such releases if they occur. This can be achieved by the following:

1. Preventing undesirable events
2. Shutting-in the process
3. Recovering released fluids
4. Preventing ignition

The modes of safety system operation include

1. Automatic monitoring by sensors
2. Automatic protective action
3. Emergency shutdown

Protection concepts and safety analysis are based on undesirable events, which include

A. Overpressure caused by
   1. Increased input flow due to upstream flow-control device failure
   2. Decreased output flow due to blockage
   3. Heating of closed system
B. Leak caused by
   1. Corrosion
   2. Erosion
   3. Mechanical failure due to temperature change, overpressure and underpressure, and external impact force
C. Liquid overflow caused by
   1. Increased input flow due to upstream flow-control device failure
   2. Decreased output flow due to blockage in the liquid discharge
D. Gas blow-by caused by
   1. Increased input flow due to upstream flow-control device failure
   2. Decreased output flow due to blockage in the gas discharge
E. Underpressure caused by
   1. Outlet flow-control device (e.g., choke) failure
   2. Inlet blockage
   3. Cooling of closed system
F. Excess temperature caused by
   1. Overfueling of burner
   2. External fire
   3. Spark emission

Figure 1.17 presents some symbols used in safety system design. Some API-recommended safety devices are shown in Figs. 1.18 through 1.22.

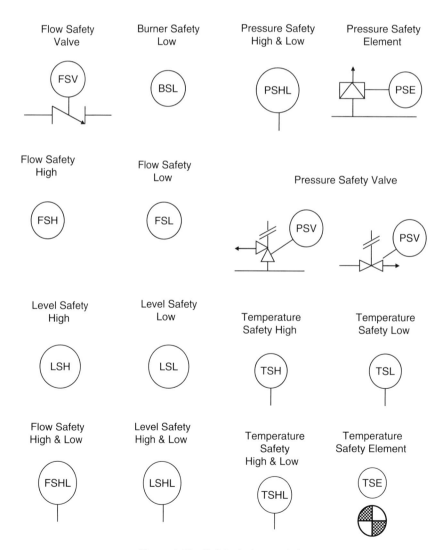

Figure 1.17 Safety device symbols.

PETROLEUM PRODUCTION SYSTEM 1/13

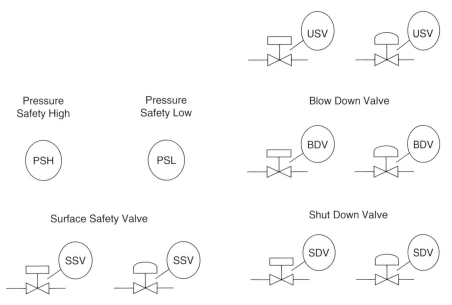

Figure 1.17 (Continued)

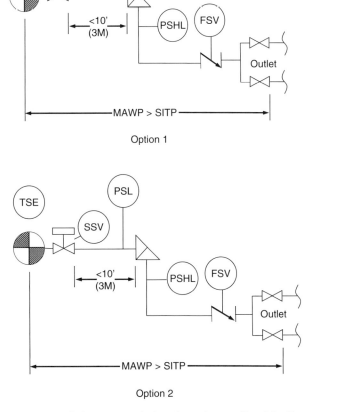

Figure 1.18 Safety system designs for surface wellhead flowlines.

1/14 PETROLEUM PRODUCTION ENGINEERING FUNDAMENTALS

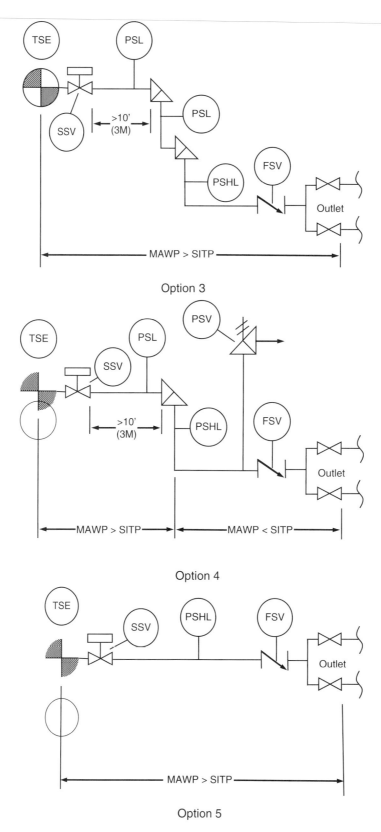

Option 3

Option 4

Option 5

Figure 1.18 (Continued)

PETROLEUM PRODUCTION SYSTEM 1/15

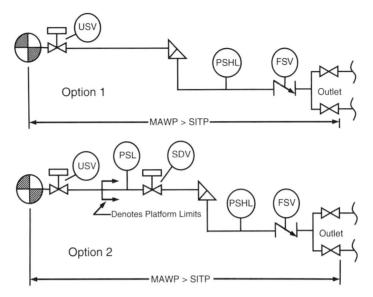

Figure 1.19 Safety system designs for underwater wellhead flowlines.

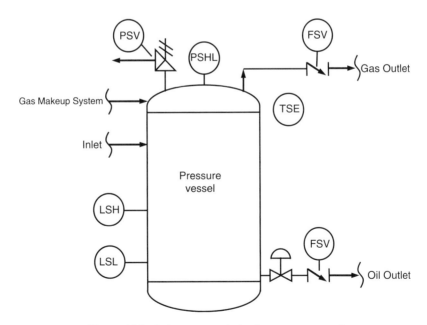

Figure 1.20 Safety system design for pressure vessel.

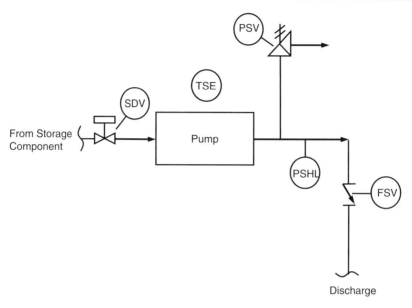

Figure 1.21 Safety system design for pipeline pumps.

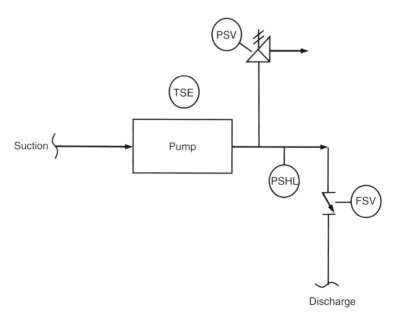

Figure 1.22 Safety system design for other pumps.

1.9 Unit Systems

This book uses U.S. oil field units in the text. However, the computer spreadsheet programs associated with this book were developed in both U.S. oil field units and S.I. units. Conversion factors between these two unit systems are presented in Appendix A.

Summary

This chapter provided a brief introduction to the components in the petroleum production system. Working principles, especially flow performances, of the components are described in later chapters.

References

American Petroleum Institute. "Bulletin on Performance Properties of Casing, Tubing, and Drill Pipe," 20th edition. Washington, DC: American Petroleum Institute. API Bulletin 5C2, May 31, 1987.

American Petroleum Institute. "Recommended Practice for Analysis, Design, Installation, and Testing of Basic Surface Safety Systems for Offshore Production Platforms," 20th edition. Washington, DC: American Petroleum Institute. API Bulletin 14C, May 31, 1987.

GUO, B. and GHALAMBOR A., *Natural Gas Engineering Handbook*. Houston: Gulf Publishing Company, 2005.

GUO, B., SONG, S., CHACKO, J., and GHALAMBOR A., *Offshore Pipelines*. Amsterdam: Elsevier, 2005.

SIVALLS, C.R. "Fundamentals of Oil and Gas Separation." Proceedings of the Gas Conditioning Conference, University of Oklahoma, Norman, Oklahoma, 1977.

WOLBERT, G.S., *American Pipelines*, University of Oklahoma Press, Norman (1952), p. 5.

Problems

1.1 Explain why a water-drive oil reservoir is usually an unsaturated oil reservoir.

1.2 What are the benefits and disadvantages of using down-hole chokes over wellhead chokes?

1.3 What is the role of an oil production engineer?

1.4 Is the tubing nominal diameter closer to tubing outside diameter or tubing inside diameter?

1.5 What do the digits in the tubing specification represent?

1.6 What is a wellhead choke used for?

1.7 What are the separators and pumps used for in the oil production operations?

1.8 Name three applications of pipelines.

1.9 What is the temperature safety element used for?

2 Properties of Oil and Natural Gas

Contents
2.1 Introduction 2/20
2.2 Properties of Oil 2/20
2.3 Properties of Natural Gas 2/21
Summary 2/26
References 2/26
Problems 2/26

2.1 Introduction

Properties of crude oil and natural gas are fundamental for designing and analyzing oil and gas production systems in petroleum engineering. This chapter presents definitions of these fluid properties and some means of obtaining these property values other than experimental measurements. Applications of the fluid properties appear in subsequent chapters.

2.2 Properties of Oil

Oil properties include solution gas–oil ratio (GOR), density, formation volume factor, viscosity, and compressibility. The latter four properties are interrelated through solution GOR.

2.2.1 Solution Gas–Oil Ratio

"Solution GOR" is defined as the amount of gas (in standard condition) that will dissolve in unit volume of oil when both are taken down to the reservoir at the prevailing pressure and temperature; that is,

$$R_s = \frac{V_{gas}}{V_{oil}}, \quad (2.1)$$

where

R_s = solution GOR (in scf/stb)
V_{gas} = gas volume in standard condition (scf)
V_{oil} = oil volume in stock tank condition (stb)

The "standard condition" is defined as 14.7 psia and 60 °F in most states in the United States. At a given reservoir temperature, solution GOR remains constant at pressures above bubble-point pressure. It drops as pressure decreases in the pressure range below the bubble-point pressure.

Solution GOR is measured in PTV laboratories. Empirical correlations are also available based on data from PVT labs. One of the correlations is,

$$R_s = \gamma_g \left[\frac{p}{18} \frac{10^{0.0125(°API)}}{10^{0.00091t}} \right]^{1.2048} \quad (2.2)$$

where γ_g and $°API$ are defined in the latter sections, and p and t are pressure and temperature in psia and °F, respectively.

Solution GOR factor is often used for volumetric oil and gas calculations in reservoir engineering. It is also used as a base parameter for estimating other fluid properties such as density of oil.

2.2.2 Density of Oil

"Density of oil" is defined as the mass of oil per unit volume, or lb_m/ft^3 in U.S. Field unit. It is widely used in hydraulics calculations (e.g., wellbore and pipeline performance calculations [see Chapters 4 and 11]).

Because of gas content, density of oil is pressure dependent. The density of oil at standard condition (stock tank oil) is evaluated by API gravity. The relationship between the density of stock tank oil and API gravity is given through the following relations:

$$°API = \frac{141.5}{\gamma_o} - 131.5 \quad (2.3)$$

and

$$\gamma_o = \frac{\rho_{o,st}}{\rho_w}, \quad (2.4)$$

where

$°API$ = API gravity of stock tank oil
γ_o = specific gravity of stock tank oil, 1 for freshwater
$\rho_{o,st}$ = density of stock tank oil, lb_m/ft^3
ρ_w = density of freshwater, 62.4 lb_m/ft^3

The density of oil at elevated pressures and temperatures can be estimated on empirical correlations developed by a number of investigators. Ahmed (1989) gives a summary of correlations. Engineers should select and validate the correlations carefully with measurements before adopting any correlations.

Standing (1981) proposed a correlation for estimating the oil formation volume factor as a function of solution GOR, specific gravity of stock tank oil, specific gravity of solution gas, and temperature. By coupling the mathematical definition of the oil formation volume factor with Standing's correlation, Ahmed (1989) presented the following expression for the density of oil:

$$\rho_o = \frac{62.4\gamma_o + 0.0136 R_s \gamma_g}{0.972 + 0.000147 \left[R_s \sqrt{\frac{\gamma_g}{\gamma_o}} + 1.25t \right]^{1.175}}, \quad (2.5)$$

where

t = temperature, °F
γ_g = specific gravity of gas, 1 for air.

2.2.3 Formation Volume Factor of Oil

"Formation volume factor of oil" is defined as the volume occupied in the reservoir at the prevailing pressure and temperature by volume of oil in stock tank, plus its dissolved gas; that is,

$$B_o = \frac{V_{res}}{V_{st}}, \quad (2.6)$$

where

B_o = formation volume factor of oil (rb/stb)
V_{res} = oil volume in reservoir condition (rb)
V_{st} = oil volume in stock tank condition (stb)

Formation volume factor of oil is always greater than unity because oil dissolves more gas in reservoir condition than in stock tank condition. At a given reservoir temperature, oil formation volume factor remains nearly constant at pressures above bubble-point pressure. It drops as pressure decreases in the pressure range below the bubble-point pressure.

Formation volume factor of oil is measured in PTV labs. Numerous empirical correlations are available based on data from PVT labs. One of the correlations is

$$B_o = 0.9759 + 0.00012 \left[R_s \sqrt{\frac{\gamma_g}{\gamma_o}} + 1.25t \right]^{1.2}. \quad (2.7)$$

Formation volume factor of oil is often used for oil volumetric calculations and well-inflow calculations. It is also used as a base parameter for estimating other fluid properties.

2.2.4 Viscosity of Oil

"Viscosity" is an empirical parameter used for describing the resistance to flow of fluid. The viscosity of oil is of interest in well-inflow and hydraulics calculations in oil production engineering. While the viscosity of oil can be measured in PVT labs, it is often estimated using empirical correlations developed by a number of investigators including Beal (1946), Beggs and Robinson (1975), Standing (1981), Glaso (1985), Khan (1987), and Ahmed (1989). A summary of these correlations is given by Ahmed (1989). Engineers should select and validate a correlation with measurements before it is used. Standing's (1981) correlation for dead oil is expressed as

$$\mu_{od} = \left(0.32 + \frac{1.8 \times 10^7}{API^{4.53}} \right) \left(\frac{360}{t + 200} \right)^A, \quad (2.8)$$

where

$$A = 10^{(0.43 + \frac{8.33}{API})} \quad (2.9)$$

and

μ_{od} = viscosity of dead oil (cp).

Standing's (1981) correlation for saturated crude oil is expressed as

$$\mu_{ob} = 10^a \mu_{od}^b, \quad (2.10)$$

where μ_{ob} = viscosity of saturated crude oil in cp and

$$a = R_s(2.2 \times 10^{-7} R_s - 7.4 \times 10^{-4}), \quad (2.11)$$

$$b = \frac{0.68}{10^c} + \frac{0.25}{10^d} + \frac{0.062}{10^e}, \quad (2.12)$$

$$c = 8.62 \times 10^{-5} R_s, \quad (2.13)$$

$$d = 1.10 \times 10^{-3} R_s, \quad (2.14)$$

and

$$e = 3.74 \times 10^{-3} R_s, \quad (2.15)$$

Standing's (1981) correlation for unsaturated crude oil is expressed as

$$\mu_o = \mu_{ob} + 0.001(p - p_b)(0.024\mu_{ob}^{1.6} + 0.38\mu_{ob}^{0.56}). \quad (2.16)$$

2.2.5 Oil Compressibility

"Oil compressibility" is defined as

$$c_o = -\frac{1}{V}\left(\frac{\partial V}{\partial p}\right)_T, \quad (2.17)$$

where T and V are temperature and volume, respectively. Oil compressibility is measured from PVT labs. It is often used in modeling well-inflow performance and reservoir simulation.

Example Problem 2.1 The solution GOR of a crude oil is 600 scf/stb at 4,475 psia and 140 °F. Given the following PVT data, estimate density and viscosity of the crude oil at the pressure and temperature:

Bubble-point pressure: 2,745 psia
Oil gravity: 35 °API
Gas-specific gravity: 0.77 air = 1

Solution Example Problem 2.1 can quickly solved using the spreadsheet program *OilProperties.xls* where Standing's correlation for oil viscosity was coded. The input and output of the program is shown in Table 2.1.

2.3 Properties of Natural Gas

Gas properties include gas-specific gravity, gas pseudo-critical pressure and temperature, gas viscosity, gas

Table 2.1 Result Given by the Spreadsheet Program OilProperties.xls

OilProperties.xls
Description: This spreadsheet calculates density and viscosity of a crude oil.
Instruction: (1) Click a unit-box to choose a unit system; (2) update parameter values in the Input data section; (3) view result in the Solution section and charts.

Input data	U.S. Field units	SI units
Pressure (p):	4,475 psia	
Temperature (t):	140 °F	
Bubble point pressure (p_b):	2,745 psia	
Stock tank oil gravity:	35 °API	
Solution gas oil ratio (R_s):	600 scf/stb	
Gas specific gravity (γ_g):	0.77 air = 1	
Solution		
$\gamma_o = \dfrac{141.5}{°API + 131.5}$	= 0.8498 H$_2$O = 1	
$\rho_o = \dfrac{62.4\gamma_o + 0.0136 R_s \gamma_g}{0.972 + 0.000147\left[R_s\sqrt{\frac{\gamma_g}{\gamma_o}} + 1.25t\right]^{1.175}}$	= 44.90 lb$_m$/ft^3	
$A = 10^{(0.43 + 8.33/API)}$	= 4.6559	
$\mu_{od} = \left(0.32 + \dfrac{1.8 \times 10^7}{API^{4.53}}\right)\left(\dfrac{360}{t + 200}\right)^A$	= 2.7956 cp	
$a = R_s(2.2 \times 10^{-7} R_s - 7.4 \times 10^{-4})$	= −0.3648	
$c = 8.62 \times 10^{-5} R_s$	= 0.0517	
$d = 1.10 \times 10^{-3} R_s$	= 0.6600	
$e = 3.74 \times 10^{-3} R_s$	= 2.2440	
$b = \dfrac{0.68}{10^c} + \dfrac{0.25}{10^d} + \dfrac{0.062}{10^e}$	= 0.6587	
$\mu_{ob} = 10^a \mu_{od}^b$	= 0.8498 cp	0.0008 Pa-s
$\mu_o = \mu_{ob} + 0.001(p - p_b)(0.024\mu_{ob}^{1.6} + 0.38\mu_{ob}^{0.56})$	= 1.4819 cp	0.0015 Pa-s

compressibility factor, gas density, gas formation volume factor, and gas compressibility. The first two are composition dependent. The latter four are pressure dependent.

2.3.1 Specific Gravity of Gas

"Specific gravity gas" is defined as the ratio of the apparent molecular weight of the gas to that of air. The molecular weight of air is usually taken as equal to 28.97 (~79% nitrogen and 21% oxygen). Therefore, the gas-specific gravity is

$$\gamma_g = \frac{MW_a}{28.97}, \quad (2.18)$$

where MW_a is the apparent molecular weight of gas, which can be calculated on the basis of gas composition. Gas composition is usually determined in a laboratory and reported in mole fractions of components in the gas. Let y_i be the mole fraction of component i, and the apparent molecular weight of the gas can be formulated using a mixing rule such as

$$MW_a = \sum_{i=1}^{N_c} y_i MW_i, \quad (2.19)$$

where MW_i is the molecular weight of component i, and N_c is number of components. The molecular weights of compounds (MW_i) can be found in textbooks on organic chemistry or petroleum fluids such as that by Ahmed (1989). Gas-specific gravity varies between 0.55 and 0.9.

2.3.2 Gas Pseudo-Critical Pressure and Temperature

Similar to gas apparent molecular weight, the critical properties of a gas can be determined on the basis of the critical properties of compounds in the gas using the mixing rule. The gas critical properties determined in such a way are called "pseudo-critical properties." Gas pseudo-critical pressure (p_{pc}) and pseudo-critical temperature (T_{pc}) are, respectively, expressed as

$$p_{pc} = \sum_{i=1}^{N_c} y_i p_{ci} \quad (2.20)$$

and

$$T_{pc} = \sum_{i=1}^{N_c} y_i T_{ci}, \quad (2.21)$$

where p_{ci} and T_{ci} are critical pressure and critical temperature of component i, respectively.

Example Problem 2.2 For the gas composition given in the following text, determine apparent molecular weight, specific gravity, pseudo-critical pressure, and pseudo-critical temperature of the gas.

Component	Mole Fraction
C_1	0.775
C_2	0.083
C_3	0.021
i-C_4	0.006
n-C_4	0.002
i-C_5	0.003
n-C_5	0.008
C_6	0.001
C_{7+}	0.001
N_2	0.050
CO_2	0.030
H_2S	0.020

Solution Example Problem 2.2 is solved with the spreadsheet program *MixingRule.xls*. Results are shown in Table 2.2.

If the gas composition is not known but gas-specific gravity is given, the pseudo-critical pressure and temperature can be determined from various charts or correlations developed based on the charts. One set of simple correlations is

$$p_{pc} = 709.604 - 58.718\gamma_g \quad (2.22)$$

$$T_{pc} = 170.491 + 307.344\gamma_g, \quad (2.23)$$

which are valid for $H_2S < 3\%$, $N_2 < 5\%$, and total content of inorganic compounds less than 7%.

Corrections for impurities in sour gases are always necessary. The corrections can be made using either charts or correlations such as the Wichert and Aziz (1972) correction expressed as follows:

$$A = y_{H_2S} + y_{CO_2} \quad (2.24)$$

$$B = y_{H_2S} \quad (2.25)$$

Table 2.2 Results Given by the Spreadsheet Program MixingRule.xls

MixingRule.xls
Description: This spreadsheet calculates gas apparent molecular weight, specific gravity, pseudo-critical pressure, and pseudo-critical temperature.
Instruction: (1) Update gas composition data (y_i); (2) read result.

Compound	y_i	MW_i	$y_i MW_i$	p_{ci} (psia)	$y_i p_{ci}$ (psia)	T_{ci}, (°R)	$y_i T_{ci}$ (°R)
C_1	0.775	16.04	12.43	673	521.58	344	266.60
C_2	0.083	30.07	2.50	709	58.85	550	45.65
C_3	0.021	44.10	0.93	618	12.98	666	13.99
i-C_4	0.006	58.12	0.35	530	3.18	733	4.40
n-C_4	0.002	58.12	0.12	551	1.10	766	1.53
i-C_5	0.003	72.15	0.22	482	1.45	830	2.49
n-C_5	0.008	72.15	0.58	485	3.88	847	6.78
C_6	0.001	86.18	0.09	434	0.43	915	0.92
C_{7+}	0.001	114.23	0.11	361	0.36	1024	1.02
N_2	0.050	28.02	1.40	227	11.35	492	24.60
CO_2	0.030	44.01	1.32	1,073	32.19	548	16.44
H_2S	0.020	34.08	0.68	672	13.45	1306	26.12
	1.000	$MW_a =$	20.71	$p_{pc} =$	661	$T_{pc} =$	411
		$\gamma_g =$	0.71				

$$\varepsilon_3 = 120(A^{0.9} - A^{1.6}) + 15(B^{0.5} - B^{4.0}) \tag{2.26}$$

$$T_{pc'} = T_{pc} - \varepsilon_3 \text{(corrected } T_{pc}\text{)} \tag{2.27}$$

$$P_{pc'} = \frac{P_{pc} T_{pc'}}{T_{pc} + B(1-B)\varepsilon_3} \text{(corrected } p_{pc}\text{)} \tag{2.28}$$

Correlations with impurity corrections for mixture pseudo-criticals are also available (Ahmed, 1989):

$$p_{pc} = 678 - 50(\gamma_g - 0.5) - 206.7 y_{N_2} + 440 y_{CO_2}$$
$$+ 606.7 y_{H_2S} \tag{2.29}$$

$$T_{pc} = 326 + 315.7(\gamma_g - 0.5) - 240 y_{N_2} - 83.3 y_{CO_2}$$
$$+ 133.3 y_{H_2S}. \tag{2.30}$$

Applications of the pseudo-critical pressure and temperature are normally found in petroleum engineering through pseudo-reduced pressure and temperature defined as

$$p_{pr} = \frac{p}{p_{pc}} \tag{2.31}$$

$$T_{pr} = \frac{T}{T_{pc}}. \tag{2.32}$$

2.3.3 Viscosity of Gas

Dynamic viscosity (μ_g) in centipoises (cp) is usually used in petroleum engineering. Kinematic viscosity (ν_g) is related to the dynamic viscosity through density (ρ_g),

$$\nu_g = \frac{\mu_g}{\rho_g}. \tag{2.33}$$

Kinematic viscosity is not typically used in natural gas engineering.

Direct measurements of gas viscosity are preferred for a new gas. If gas composition and viscosities of gas components are known, the mixing rule can be used to determine the viscosity of the gas mixture:

$$\mu_g = \frac{\sum (\mu_{gi} y_i \sqrt{MW_i})}{\sum (y_i \sqrt{MW_i})} \tag{2.34}$$

Viscosity of gas is very often estimated with charts or correlations developed based on the charts. Gas viscosity correlation of Carr et al. 1954 involves a two-step procedure: The gas viscosity at temperature and atmospheric pressure is estimated first from gas-specific gravity and inorganic compound content. The atmospheric value is then adjusted to pressure conditions by means of a correction factor on the basis of reduced temperature and pressure state of the gas. The atmospheric pressure viscosity (μ_1) can be expressed as

$$\mu_1 = \mu_{1HC} + \mu_{1N_2} + \mu_{1CO_2} + \mu_{1H_2S}, \tag{2.35}$$

where

$$\mu_{1HC} = 8.188 \times 10^{-3} - 6.15 \times 10^{-3} \log(\gamma_g)$$
$$+ (1.709 \times 10^{-5} - 2.062 \times 10^{-6} \gamma_g) T, \tag{2.36}$$

$$\mu_{1N_2} = [9.59 \times 10^{-3} + 8.48 \times 10^{-3} \log(\gamma_g)] y_{N_2}, \tag{2.37}$$

$$\mu_{1CO_2} = [6.24 \times 10^{-3} + 9.08 \times 10^{-3} \log(\gamma_g)] y_{CO_2}, \tag{2.38}$$

$$\mu_{1H_2S} = [3.73 \times 10^{-3} + 8.49 \times 10^{-3} \log(\gamma_g)] y_{H_2S}, \tag{2.39}$$

Dempsey (1965) developed the following relation:

$$\mu_r = \ln\left(\frac{\mu_g}{\mu_1} T_{pr}\right)$$
$$= a_0 + a_1 p_{pr} + a_2 p_{pr}^2 + a_3 p_{pr}^3 + T_{pr}(a_4 + a_5 p_{pr}$$
$$+ a_6 p_{pr}^2 + a_7 p_{pr}^3) + T_{pr}^2(a_8 + a_9 p_{pr} + a_{10} p_{pr}^2$$
$$+ a_{11} p_{pr}^3) + T_{pr}^3(a_{12} + a_{13} p_{pr} + a_{14} p_{pr}^2$$
$$+ a_{15} p_{pr}^3), \tag{2.40}$$

where

$a_0 = -2.46211820$
$a_1 = 2.97054714$
$a_2 = -0.28626405$
$a_3 = 0.00805420$
$a_4 = 2.80860949$
$a_5 = -3.49803305$
$a_6 = 0.36037302$
$a_7 = -0.01044324$
$a_8 = -0.79338568$
$a_9 = 1.39643306$
$a_{10} = -0.14914493$
$a_{11} = 0.00441016$
$a_{12} = 0.08393872$
$a_{13} = -0.18640885$
$a_{14} = 0.02033679$
$a_{15} = -0.00060958$

Thus, once the value of μ_r is determined from the right-hand side of this equation, gas viscosity at elevated pressure can be readily calculated using the following relation:

$$\mu_g = \frac{\mu_1}{T_{pr}} e^{\mu_r} \tag{2.41}$$

Other correlations for gas viscosity include that of Dean and Stiel (1958) and Lee et al. (1966).

Example Problem 2.3 A 0.65 specific-gravity natural gas contains 10% nitrogen, 8% carbon dioxide, and 2% hydrogen sulfide. Estimate viscosity of the gas at 10,000 psia and 180°F.

Solution Example Problem 2.3 is solved with the spreadsheet *Carr-Kobayashi-Burrows-GasViscosity.xls*, which is attached to this book. The result is shown in Table 2.3.

2.3.4 Gas Compressibility Factor

Gas compressibility factor is also called "deviation factor" or "z-factor." Its value reflects how much the real gas deviates from the ideal gas at a given pressure and temperature. Definition of the compressibility factor is expressed as

$$z = \frac{V_{actual}}{V_{ideal\ gas}}. \tag{2.42}$$

Introducing the z-factor to the gas law for ideal gas results in the gas law for real gas as

$$pV = nzRT, \tag{2.43}$$

where n is the number of moles of gas. When pressure p is entered in psia, volume V in ft^3, and temperature in °R, the gas constant R is equal to $10.73 \frac{psia - ft^3}{mole - °R}$.

Gas compressibility factor can be determined on the basis of measurements in PVT laboratories. For a given amount of gas, if temperature is kept constant and volume is measured at 14.7 psia and an elevated pressure p_1, z-factor can then be determined with the following formula:

$$z = \frac{p_1}{14.7} \frac{V_1}{V_0}, \tag{2.44}$$

Table 2.3 Results Given by the Spreadsheet Carr-Kobayashi-Burrows-GasViscosity.xls

Carr-Kobayashi-Burrows-GasViscosity.xls
Description: This spreadsheet calculates gas viscosity with correlation of Carr et al.
Instruction: (1) Select a unit system; (2) update data in the Input data section;
(3) review result in the Solution section.

Input data	U.S. Field units	SI units
Pressure:	10,000 psia	
Temperature:	180 °F	
Gas-specific gravity:	0.65 air = 1	
Mole fraction of N_2:	0.1	
Mole fraction of CO_2:	0.08	
Mole fraction of H_2S:	0.02	
Solution		
Pseudo-critical pressure	= 697.164 psia	
Pseudo-critical temperature	= 345.357 °R	
Uncorrected gas viscosity at 14.7 psia	= 0.012174 cp	
N_2 correction for gas viscosity at 14.7 psia	= 0.000800 cp	
CO_2 correction for gas viscosity at 14.7 psia	= 0.000363 cp	
H_2S correction for gas viscosity at 14.7 psia	= 0.000043 cp	
Corrected gas viscosity at 14.7 psia (μ_1)	= 0.013380 cp	
Pseudo-reduced pressure	= 14.34	
Pseudo-reduced temperature	= 1.85	
$ln(\mu_g/\mu_1 * T_{pr})$	= 1.602274	
Gas viscosity	= 0.035843 cp	

where V_0 and V_1 are gas volumes measured at 14.7 psia and p_1, respectively.

Very often the z-factor is estimated with the chart developed by Standing and Katz (1954). This chart has been set up for computer solution by a number of individuals. Brill and Beggs (1974) yield z-factor values accurate enough for many engineering calculations. Brill and Beggs' z-factor correlation is expressed as follows:

$$A = 1.39(T_{pr} - 0.92)^{0.5} - 0.36T_{pr} - 0.10, \quad (2.45)$$

$$B = (0.62 - 0.23T_{pr})p_{pr}$$
$$+ \left(\frac{0.066}{T_{pr} - 0.86} - 0.037\right)p_{pr}^2 + \frac{0.32 p_{pr}^6}{10^E}, \quad (2.46)$$

$$C = 0.132 - 0.32 \log(T_{pr}), \quad (2.47)$$

$$D = 10^F, \quad (2.48)$$

$$E = 9(T_{pr} - 1), \quad (2.49)$$

$$F = 0.3106 - 0.49T_{pr} + 0.1824T_{pr}^2, \quad (2.50)$$

and

$$z = A + \frac{1-A}{e^B} + Cp_{pr}^D. \quad (2.51)$$

Example Problem 2.4 For the natural gas described in Example Problem 2.3, estimate z-factor at 5,000 psia and 180 °F.

Solution Example Problem 2.4 is solved with the spreadsheet program *Brill-Beggs-Z.xls*. The result is shown in Table 2.4.

Hall and Yarborough (1973) presented a more accurate correlation to estimate z-factor of natural gas. This correlation is summarized as follows:

$$t_r = \frac{1}{T_{pr}} \quad (2.52)$$

$$A = 0.06125 t_r e^{-1.2(1-t_r)^2} \quad (2.53)$$

$$B = t_r(14.76 - 9.76 t_r + 4.58 t_r^2) \quad (2.54)$$

$$C = t_r(90.7 - 242.2 t_r + 42.4 t_r^2) \quad (2.55)$$

$$D = 2.18 + 2.82 t_r \quad (2.56)$$

and

$$z = \frac{Ap_{pr}}{Y}, \quad (2.57)$$

where Y is the reduced density to be solved from

$$f(Y) = \frac{Y + Y^2 + Y^3 - Y^4}{(1-Y)^3} - Ap_{pr} - BY^2 + CY^D$$
$$= 0. \quad (2.58)$$

If the Newton and Raphson iteration method is used to solve Eq. (2.58) for Y, the following derivative is needed:

$$\frac{df(Y)}{dY} = \frac{1 + 4Y + 4Y^2 - 4Y^3 + Y^4}{(1-Y)^4} - 2BY$$
$$+ CDY^{D-1} \quad (2.59)$$

2.3.5 Density of Gas

Because gas is compressible, its density depends on pressure and temperature. Gas density can be calculated from gas law for real gas with good accuracy:

$$\rho_g = \frac{m}{V} = \frac{MW_a p}{zRT}, \quad (2.60)$$

where m is mass of gas and ρ_g is gas density. Taking air molecular weight 29 and $R = 10.73 \frac{\text{psia} - \text{ft}^3}{\text{mole} - °R}$, Eq. (2.60) is rearranged to yield

$$\rho_g = \frac{2.7 \gamma_g p}{zT}, \quad (2.61)$$

where the gas density is in lb_m/ft^3.

PROPERTIES OF OIL AND NATURAL GAS

Table 2.4 Results Given by the Spreadsheet Program Brill-Beggs-Z.xls

Brill-Beggs-Z.xls
Description: This spreadsheet calculates gas compressibility factor based on the Brill and Beggs correlation.
Instruction: (1) Select a unit system; (2) update data in the Input data section; (3) review result in the Solution section.

Input data	U.S. Field units	SI units
Pressure:	5,000 psia	
Temperature:	180 °F	
Gas specific gravity:	0.65 air = 1	
Mole fraction of N_2:	0.1	
Mole fraction of CO_2:	0.08	
Mole fraction of H_2S:	0.02	
Solution		
Pseudo-critical pressure	= 697 psia	
Pseudo-critical temperature	= 345 °R	
Pseudo-reduced pressure	= 7.17	
Pseudo-reduced temperature	= 1.95	
A	= 0.6063	
B	= 2.4604	
C	= 0.0395	
D	= 1.1162	
Gas compressibility factor z	= 0.9960	

Example Problem 2.5 A gas from oil has a specific gravity of 0.65, estimate z-factor and gas density at 5,000 psia and 180 °F.

Solution Example Problem 2.5 is solved with the spreadsheet program *Hall-Yarborogh-z.xls*. The result is shown in Table 2.5.

2.3.6 Formation Volume Factor of Gas

Gas formation volume factor is defined as the ratio of gas volume at reservoir condition to the gas volume at standard condition, that is,

$$B_g = \frac{V}{V_{sc}} = \frac{p_{sc}}{p} \frac{T}{T_{sc}} \frac{z}{z_{sc}} = 0.0283 \frac{zT}{p}, \quad (2.62)$$

Table 2.5 Results Given by the Spreadsheet Program Hall-Yarborogh-z.xls

Hall-Yarborogh-z.xls
Description: This spreadsheet computes gas compressibility factor with the Hall–Yarborough method.
Instruction: (1) Select a unit system; (2) update data in the Input data section; (3) click Solution button; (4) view result.

Input data	U.S. Field units	SI units
Temperature:	200 °F	
Pressure:	2,000 psia	
Gas-specific gravity:	0.7 air = 1	
Nitrogen mole fraction:	0.05	
Carbon dioxide fraction:	0.05	
Hydrogen sulfite fraction:	0.02	
Solution		
$T_{pc} = 326 + 315.7(\gamma_g - 0.5) - 240 y_{N_2} - 83.3 y_{CO_2} + 133.3 y_{H_2S}$	= 375.641 °R	
$p_{pc} = 678 - 50(\gamma_g - 0.5) - 206.7 y_{N_2} + 440 y_{CO_2} + 606.7 y_{H_2S}$	= 691.799 psia	
$T_{pr} = \frac{T}{T_{pc}}$	= 1.618967	
$t_r = \frac{1}{T_{pr}}$	= 0.617678	
$p_{pr} = \frac{p}{p_{pc}}$	= 2.891013	
$A = 0.06125 t_r e^{-1.2(1-t_r)^2}$	= 0.031746	
$B = t_r(14.76 - 9.76 t_r + 4.58 t_r^2)$	= 6.472554	
$C = t_r(90.7 - 242.2 t_r + 42.4 t_r^2)$	= −26.3902	
$D = 2.18 + 2.82 t_r$	= 3.921851	
Y = ASSUMED	= 0.109759	
$f(Y) = \frac{Y + Y^2 + Y^3 - Y^4}{(1-Y)^3} - A p_{pr} - BY^2 + CY^D = 0$	= 4.55E-06	
$z = \frac{A p_{pr}}{Y}$	= 0.836184	
$\rho_g = \frac{2.7 \gamma_g p}{zT}$	= 6.849296 lb_m/ft^3	

where the unit of formation volume factor is ft^3/scf. If expressed in rb/scf, it takes the form

$$B_g = 0.00504 \frac{zT}{p}. \tag{2.63}$$

Gas formation volume factor is frequently used in mathematical modeling of gas well inflow performance relationship (IPR).

Another way to express this parameter is to use gas expansion factor defined, in scf/ft^3, as

$$E = \frac{1}{B_g} = 35.3 \frac{P}{ZT} \tag{2.64}$$

or

$$E = 198.32 \frac{p}{zT}, \tag{2.65}$$

in scf/rb. It is normally used for estimating gas reserves.

2.3.7 Gas Compressibility

Gas compressibility is defined as

$$c_g = -\frac{1}{V} \left(\frac{\partial V}{\partial p} \right)_T. \tag{2.66}$$

Because the gas law for real gas gives $V = \frac{nzRT}{p}$,

$$\left(\frac{\partial V}{\partial p} \right) = nRT \left(\frac{1}{p} \frac{\partial z}{\partial p} - \frac{z}{p^2} \right). \tag{2.67}$$

Substituting Eq. (2.67) into Eq. (2.66) yields

$$c_g = \frac{1}{p} - \frac{1}{z} \frac{\partial z}{\partial p}. \tag{2.68}$$

Since the second term in the right-hand side is usually small, gas compressibility is approximately equal to the reciprocal of pressure.

Summary

This chapter presented definitions and properties of crude oil and natural gas. It also provided a few empirical correlations for determining the value of these properties. These correlations are coded in spreadsheet programs that are available with this book. Applications of these fluid properties are found in the later chapters.

References

AHMED, T. *Hydrocarbon Phase Behavior*. Houston: Gulf Publishing Company, 1989.

American Petroleum Institute: "Bulletin on Performance Properties of Casing, Tubing, and Drill Pipe," 20th edition. Washington, DC: American Petroleum Institute. API Bulletin 5C2, May 31, 1987.

BEAL, C. The viscosity of air, water, natural gas, crude oils and its associated gases at oil field temperatures and pressures. *Trans. AIME* 1946;165:94–112.

BEGGS, H.D. and ROBINSON, J.R. Estimating the viscosity of crude oil systems. *J. Petroleum Technol.* Sep. 1975:1140–1141.

BRILL, J.P. and BEGGS, H.D. Two-phase flow in pipes. INTERCOMP Course, The Hague, 1974.

CARR, N.L., KOBAYASHI, R., and BURROWS, D.B. Viscosity of hydrocarbon gases under pressure. *Trans. AIME* 1954;201:264–272.

DEAN, D.E. and STIEL, L.I. The viscosity of non-polar gas mixtures at moderate and high pressures. *AIChE J.* 1958;4:430–436.

DEMPSEY, J.R. Computer routine treats gas viscosity as a variable. *Oil Gas J.* Aug. 16, 1965:141.

GLASO, O. Generalized pressure-volume-temperature correlations. *J. Petroleum Technol.* May 1985:785–795.

HALL, K.R. and YARBOROUGH, L. A new equation of state for Z-factor calculations. *Oil Gas J.* June 18, 1973:82.

KHAN, S.A. Viscosity correlations for Saudi Arabian crude oils. Presented at the 50th Middle East Conference and Exhibition held 7–10 March 1987, in Manama, Bahrain. Paper SPE 15720.

LEE, A.L., GONZALEZ, M.H., and EAKIN, B.E. The viscosity of natural gases. *J. Petroleum Technol.* Aug. 1966:997–1000.

STANDING, M.B. *Volume and Phase Behavior of Oil Field Hydrocarbon Systems*, 9th edition. Dallas: Society of Petroleum Engineers, 1981.

STANDING, M.B. and KATZ, D.L. Density of natural gases. *Trans. AIME* 1954;146:140–149.

WICHERT, E. and AZIZ, K. Calculate Zs for sour gases. *Hydrocarbon Processing* 1972;51(May):119.

Problems

2.1 Estimate the density of a 25-API gravity dead oil at 100 °F.

2.2 The solution gas–oil ratio of a crude oil is 800 scf/stb at 3,000 psia and 120 °F. Given the following PVT data:

   Bubble-point pressure: 2,500 psia
   Oil gravity: 35 °API
   Gas-specific gravity: 0.77 (air = 1),

   estimate densities and viscosities of the crude oil at 120 °F, 2,500 psia, and 3,000 psia.

2.3 For the gas composition given below, determine apparent molecular weight, specific gravity, pseudo-critical pressure, and pseudo-critical temperature of the gas:

Component	Mole Fraction
C_1	0.765
C_2	0.073
C_3	0.021
i-C_4	0.006
n-C_4	0.002
i-C_5	0.003
n-C_5	0.008
C_6	0.001
C_{7+}	0.001
N_2	0.060
CO_2	0.040
H_2S	0.020

2.4 Estimate gas viscosities of a 0.70-specific gravity gas at 200 °F and 100 psia, 1,000 psia, 5,000 psia, and 10,000 psia.

2.5 Calculate gas compressibility factors and densities of a 0.65-specific gravity gas at 150 °F and 50 psia, 500 psia, and 5,000 psia with the Hall–Yarborough method. Compare the results with that given by the Brill and Beggs correlation. What is your conclusion?

2.6 For a 0.65-specific gravity gas at 250 °F, calculate and plot pseudo-pressures in a pressure range from 14.7 and 8,000 psia. Under what condition is the pseudo-pressure linearly proportional to pressure?

2.7 Estimate the density of a 0.8-specific gravity dead oil at 40 °C.

2.8 The solution gas–oil ratio of a crude oil is 4,000 sm^3/m^3 at 20 MPa and 50 °C. Given the following PVT data:

Bubble-point pressure: 15 MPa
Oil-specific gravity: 0.8 water = 1
Gas-specific gravity: 0.77 air = 1,

estimate densities and viscosities of the crude oil at 50 °C, 15 MPa, and 20 MPa.

2.9 For the gas composition given below, determine apparent molecular weight, specific gravity, pseudo-critical pressure, and pseudo-critical temperature of the gas.

Component	Mole Fraction
C_1	0.755
C_2	0.073
C_3	0.011
i-C_4	0.006
n-C_4	0.002
i-C_5	0.003
n-C_5	0.008
C_6	0.001
C_{7+}	0.001
N_2	0.070
CO_2	0.050
H_2S	0.020

2.10 Estimate gas viscosities of a 0.70-specific gravity gas at 90 °C and 1 MPa, 5 MPa, 10 MPa, and 50 MPa.

2.11 Calculate gas compressibility factors and densities of a 0.65-specific gravity gas at 80 °C and 1 MPa, 5 MPa, 10 MPa, and 50 MPa with the Hall–Yarborough method. Compare the results with that given by the Brill and Beggs correlation. What is your conclusion?

2.12 For a 0.65-specific gravity gas at 110 °C, calculate and plot pseudo-pressures in a pressure range from 0.1 to 30 MPa. Under what condition is the pseudo-pressure linearly proportional to pressure?

3 Reservoir Deliverability

Contents
3.1 Introduction 3/30
3.2 Flow Regimes 3/30
3.3 Inflow Performance Relationship 3/32
3.4 Construction of IPR Curves Using Test Points 3/35
3.5 Composite IPR of Stratified Reservoirs 3/37
3.6 Future IPR 3/39
Summary 3/42
References 3/42
Problems 3/43

3.1 Introduction

Reservoir deliverability is defined as the oil or gas production rate achievable from reservoir at a given bottom-hole pressure. It is a major factor affecting well deliverability. Reservoir deliverability determines types of completion and artificial lift methods to be used. A thorough knowledge of reservoir productivity is essential for production engineers.

Reservoir deliverability depends on several factors including the following:

- Reservoir pressure
- Pay zone thickness and permeability
- Reservoir boundary type and distance
- Wellbore radius
- Reservoir fluid properties
- Near-wellbore condition
- Reservoir relative permeabilities

Reservoir deliverability can be mathematically modeled on the basis of flow regimes such as transient flow, steady state flow, and pseudo–steady state flow. An analytical relation between bottom-hole pressure and production rate can be formulated for a given flow regime. The relation is called "inflow performance relationship" (IPR). This chapter addresses the procedures used for establishing IPR of different types of reservoirs and well configurations.

3.2 Flow Regimes

When a vertical well is open to produce oil at production rate q, it creates a pressure funnel of radius r around the wellbore, as illustrated by the dotted line in Fig. 3.1a. In this reservoir model, the h is the reservoir thickness, k is the effective horizontal reservoir permeability to oil, μ_o is viscosity of oil, B_o is oil formation volume factor, r_w is wellbore radius, p_{wf} is the flowing bottom hole pressure, and p is the pressure in the reservoir at the distance r from the wellbore center line. The flow stream lines in the cylindrical region form a horizontal radial flow pattern as depicted in Fig. 3.1b.

3.2.1 Transient Flow

"Transient flow" is defined as a flow regime where/when the radius of pressure wave propagation from wellbore has not reached any boundaries of the reservoir. During transient flow, the developing pressure funnel is small relative to the reservoir size. Therefore, the reservoir acts like an infinitively large reservoir from transient pressure analysis point of view.

Assuming single-phase oil flow in the reservoir, several analytical solutions have been developed for describing the transient flow behavior. They are available from classic textbooks such as that of Dake (1978). A constant-rate solution expressed by Eq. (3.1) is frequently used in production engineering:

$$p_{wf} = p_i - \frac{162.6 q B_o \mu_o}{kh}$$
$$\times \left(\log t + \log \frac{k}{\phi \mu_o c_t r_w^2} - 3.23 + 0.87 S \right), \quad (3.1)$$

where
 p_{wf} = flowing bottom-hole pressure, psia
 p_i = initial reservoir pressure, psia
 q = oil production rate, stb/day
 μ_o = viscosity of oil, cp
 k = effective horizontal permeability to oil, md
 h = reservoir thickness, ft
 t = flow time, hour

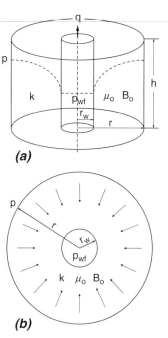

Figure 3.1 A sketch of a radial flow reservoir model: (**a**) lateral view, (**b**) top view.

 ϕ = porosity, fraction
 c_t = total compressibility, psi^{-1}
 r_w = wellbore radius to the sand face, ft
 S = skin factor
 Log = 10-based logarithm $\log_{10}$

Because oil production wells are normally operated at constant bottom-hole pressure because of constant wellhead pressure imposed by constant choke size, a constant bottom-hole pressure solution is more desirable for well-inflow performance analysis. With an appropriate inner boundary condition arrangement, Earlougher (1977) developed a constant bottom-hole pressure solution, which is similar to Eq. (3.1):

$$q = \frac{kh(p_i - p_{wf})}{162.6 B_o \mu_o \left(\log t + \log \frac{k}{\phi \mu_o c_t r_w^2} - 3.23 + 0.87 S \right)}, \quad (3.2)$$

which is used for transient well performance analysis in production engineering.

Equation (3.2) indicates that oil rate decreases with flow time. This is because the radius of the pressure funnel, over which the pressure drawdown $(p_i - p_{wf})$ acts, increases with time, that is, the overall pressure gradient in the reservoir drops with time.

For gas wells, the transient solution is

$$q_g = \frac{kh[m(p_i) - m(p_{wf})]}{1{,}638 T \left(\log t + \log \frac{k}{\phi \mu_o c_t r_w^2} - 3.23 + 0.87 S \right)}, \quad (3.3)$$

where q_g is production rate in Mscf/d, T is temperature in °R, and $m(p)$ is real gas pseudo-pressure defined as

$$m(p) = \int_{pb}^{p} \frac{2p}{\mu z} dp. \quad (3.4)$$

The real gas pseudo-pressure can be readily determined with the spreadsheet program *PseudoPressure.xls*.

3.2.2 Steady-State Flow

"Steady-state flow" is defined as a flow regime where the pressure at any point in the reservoir remains constant over time. This flow condition prevails when the pressure funnel shown in Fig. 3.1 has propagated to a constant-pressure boundary. The constant-pressure boundary can be an aquifer or a water injection well. A sketch of the reservoir model is shown in Fig. 3.2, where p_e represents the pressure at the constant-pressure boundary. Assuming single-phase flow, the following theoretical relation can be derived from Darcy's law for an oil reservoir under the steady-state flow condition due to a circular constant-pressure boundary at distance r_e from wellbore:

$$q = \frac{kh(p_e - p_{wf})}{141.2 B_o \mu_o \left(\ln \frac{r_e}{r_w} + S \right)}, \tag{3.5}$$

where "ln" denotes 2.718-based natural logarithm $\log_e$. Derivation of Eq. (3.5) is left to readers for an exercise.

3.2.3 Pseudo–Steady-State Flow

"Pseudo–steady-state" flow is defined as a flow regime where the pressure at any point in the reservoir declines at the same constant rate over time. This flow condition prevails after the pressure funnel shown in Fig. 3.1 has propagated to all no-flow boundaries. A no-flow boundary can be a sealing fault, pinch-out of pay zone, or boundaries of drainage areas of production wells. A sketch of the reservoir model is shown in Fig. 3.3, where p_e represents the pressure at the no-flow boundary at time t_4. Assuming single-phase flow, the following theoretical relation can be derived from Darcy's law for an oil reservoir under pseudo–steady-state flow condition due to a circular no-flow boundary at distance r_e from wellbore:

$$q = \frac{kh(p_e - p_{wf})}{141.2 B_o \mu_o \left(\ln \frac{r_e}{r_w} - \frac{1}{2} + S \right)}. \tag{3.6}$$

The flow time required for the pressure funnel to reach the circular boundary can be expressed as

$$t_{pss} = 1{,}200 \frac{\phi \mu_o c_t r_e^2}{k}. \tag{3.7}$$

Because the p_e in Eq. (3.6) is not known at any given time, the following expression using the average reservoir pressure is more useful:

$$q = \frac{kh(\bar{p} - p_{wf})}{141.2 B_o \mu_o \left(\ln \frac{r_e}{r_w} - \frac{3}{4} + S \right)}, \tag{3.8}$$

where $\bar{p}$ is the average reservoir pressure in psia. Derivations of Eqs. (3.6) and (3.8) are left to readers for exercises.

If the no-flow boundaries delineate a drainage area of noncircular shape, the following equation should be used for analysis of pseudo–steady-state flow:

$$q = \frac{kh(\bar{p} - p_{wf})}{141.2 B_o \mu_o \left(\frac{1}{2} \ln \frac{4A}{\gamma C_A r_w^2} + S \right)}, \tag{3.9}$$

where

A = drainage area, ft^2
$\gamma = 1.78$ = Euler's constant
C_A = drainage area shape factor, 31.6 for a circular boundary.

The value of the shape factor C_A can be found from Fig. 3.4.

For a gas well located at the center of a circular drainage area, the pseudo–steady-state solution is

$$q_g = \frac{kh[m(\bar{p}) - m(p_{wf})]}{1{,}424 T \left(\ln \frac{r_e}{r_w} - \frac{3}{4} + S + D q_g \right)}, \tag{3.10}$$

where

D = non-Darcy flow coefficient, d/Mscf.

3.2.4 Horizontal Well

The transient flow, steady-state flow, and pseudo–steady-state flow can also exist in reservoirs penetrated by horizontal wells. Different mathematical models are available from

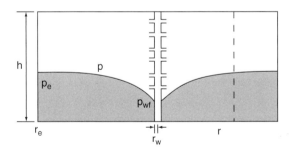

Figure 3.2 A sketch of a reservoir with a constant-pressure boundary.

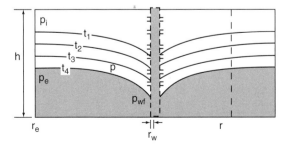

Figure 3.3 A sketch of a reservoir with no-flow boundaries.

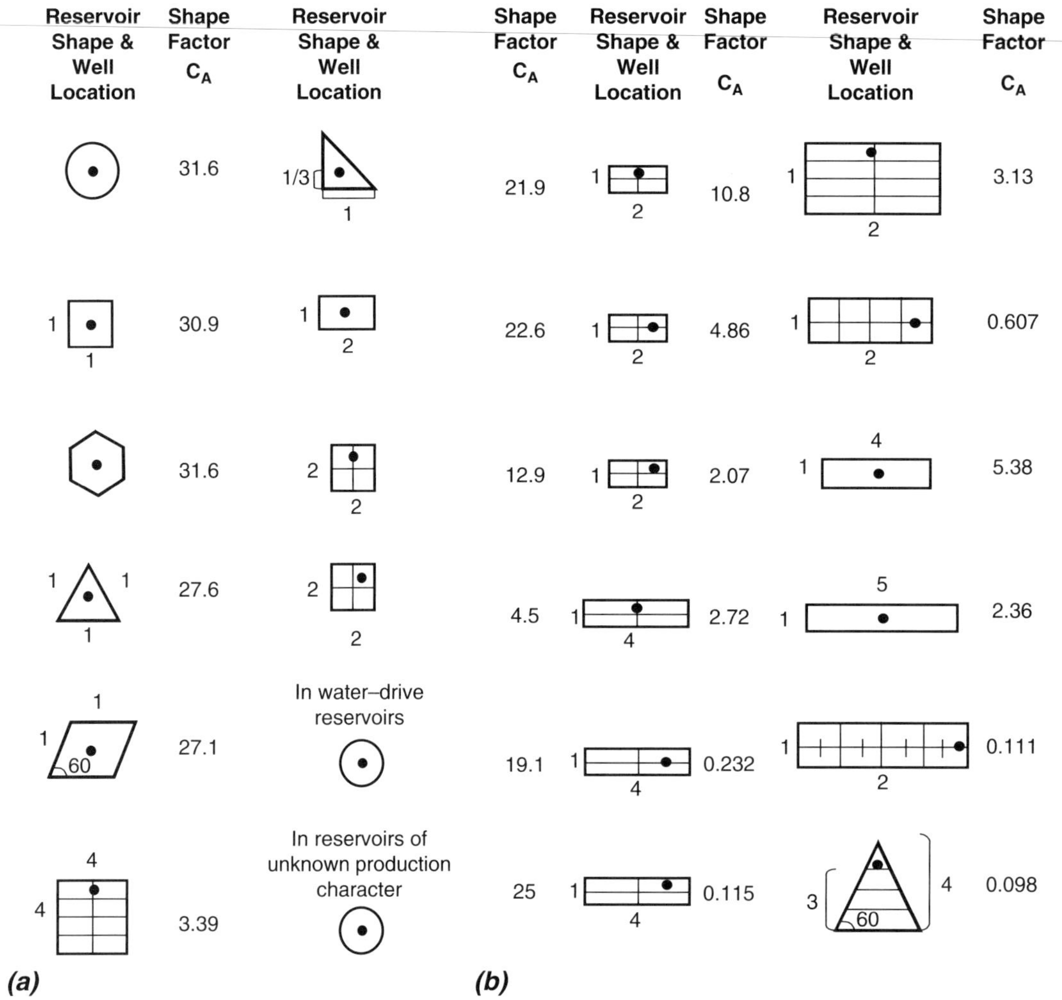

Figure 3.4 (a) Shape factors for closed drainage areas with low-aspect ratios. (b) Shape factors for closed drainage areas with high-aspect ratios (Dietz, 1965).

literature. Joshi (1988) presented the following relationship considering steady-state flow of oil in the horizontal plane and pseudo–steady-state flow in the vertical plane:

$$q = \frac{k_H h(p_e - p_{wf})}{141.2 B\mu \left\{ \ln\left[\frac{a + \sqrt{a^2 - (L/2)^2}}{L/2}\right] + \frac{I_{ani} h}{L} \ln\left[\frac{I_{ani} h}{r_w(I_{ani} + 1)}\right] \right\}}, \quad (3.11)$$

where

$$a = \frac{L}{2} \sqrt{\frac{1}{2} + \sqrt{\frac{1}{4} + \left(\frac{r_{eH}}{L/2}\right)^4}}, \quad (3.12)$$

$$I_{ani} = \sqrt{\frac{k_H}{k_V}}, \quad (3.13)$$

and

k_H = the average horizontal permeability, md
k_V = vertical permeability, md
r_{eH} = radius of drainage area, ft
L = length of horizontal wellbore ($L/2 < 0.9 r_{eH}$), ft.

3.3 Inflow Performance Relationship

IPR is used for evaluating reservoir deliverability in production engineering. The IPR curve is a graphical presentation of the relation between the flowing bottom-hole pressure and liquid production rate. A typical IPR curve is shown in Fig. 3.5. The magnitude of the slope of the IPR curve is called the "productivity index" (PI or J), that is,

$$J = \frac{q}{(p_e - p_{wf})}, \quad (3.14)$$

where J is the productivity index. Apparently J is not a constant in the two-phase flow region.

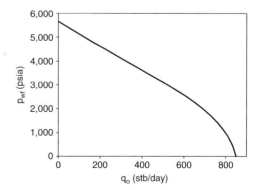

Figure 3.5 A typical IPR curve for an oil well.

Well IPR curves are usually constructed using reservoir inflow models, which can be from either a theoretical basis or an empirical basis. It is essential to validate these models with test points in field applications.

3.3.1 LPR for Single (Liquid)-Phase Reservoirs

All reservoir inflow models represented by Eqs. (3.1), (3.3), (3.7), and (3.8) were derived on the basis of the assumption of single-phase liquid flow. This assumption is valid for undersaturated oil reservoirs, or reservoir portions where the pressure is above the bubble-point pressure. These equations define the productivity index (J^*) for flowing bottom-hole pressures above the bubble-point pressure as follows:

$$J^* = \frac{q}{(p_i - p_{wf})}$$

$$= \frac{kh}{162.6 B_o \mu_o \left(\log t + \log \frac{k}{\phi \mu_o c_t r_w^2} - 3.23 + 0.87S \right)} \quad (3.15)$$

for radial transient flow around a vertical well,

$$J^* = \frac{q}{(p_e - p_{wf})} = \frac{kh}{141.2 B_o \mu_o \left(\ln \frac{r_e}{r_w} + S \right)} \quad (3.16)$$

for radial steady-state flow around a vertical well,

$$J^* = \frac{q}{(\bar{p} - p_{wf})} = \frac{kh}{141.2 B_o \mu_o \left(\frac{1}{2} \ln \frac{4A}{\gamma C_A r_w^2} + S \right)} \quad (3.17)$$

for pseudo–steady-state flow around a vertical well, and

$$J^* = \frac{q}{(p_e - p_{wf})}$$

$$= \frac{k_H h}{141.2 B \mu \left\{ \ln \left[\frac{a + \sqrt{a^2 - (L/2)^2}}{L/2} \right] + \frac{I_{ani} h}{L} \ln \left[\frac{I_{ani} h}{r_w (I_{ani} + 1)} \right] \right\}} \quad (3.18)$$

for steady-state flow around a horizontal well.

Since the productivity index (J^*) above the bubble-point pressure is independent of production rate, the IPR curve for a single (liquid)-phase reservoir is simply a straight line drawn from the reservoir pressure to the bubble-point pressure. If the bubble-point pressure is 0 psig, the absolute open flow (AOF) is the productivity index (J^*) times the reservoir pressure.

Example Problem 3.1 Construct IPR of a vertical well in an oil reservoir. Consider (1) transient flow at 1 month, (2) steady-state flow, and (3) pseudo–steady-state flow. The following data are given:

Porosity:	$\phi = 0.19$
Effective horizontal permeability:	$k = 8.2$ md
Pay zone thickness:	$h = 53$ ft
Reservoir pressure:	p_e or $\bar{p} = 5,651$ psia
Bubble-point pressure:	$p_b = 50$ psia
Fluid formation volume factor:,	$B_o = 1.1$
Fluid viscosity:	$\mu_o = 1.7$ cp
Total compressibility,	$c_t = 0.0000129$ psi^{-1}
Drainage area:	$A = 640$ acres ($r_e = 2,980$ ft)
Wellbore radius:	$r_w = 0.328$ ft
Skin factor:	$S = 0$

Solution

1. For transient flow, calculated points are

$$J^* = \frac{kh}{162.6 B \mu \left(\log t + \log \frac{k}{\phi \mu c_t r_w^2} - 3.23 \right)}$$

$$= \frac{(8.2)(53)}{162.6(1.1)(1.7) \left(\log[(30)(24)] + \log \frac{(8.2)}{(0.19)(1.7)(0.0000129)(0.328)^2} - 3.23 \right)}$$

$$= 0.2075 \text{ STB/d-psi}$$

Transient IPR curve is plotted in Fig. 3.6.

2. For steady state flow:

$$J^* = \frac{kh}{141.2 B \mu \left(\ln \frac{r_e}{r_w} + S \right)}$$

$$= \frac{(8.2)(53)}{141.2(1.1)(1.7) \ln \left(\frac{2,980}{0.328} \right)}$$

$$= 0.1806 \text{ STB/d-psi}$$

Calculated points are:

p_{wf}(psi)	q_o(stb/day)
50	1,011
5,651	0

Steady state IPR curve is plotted in Fig. 3.7.

3. For pseudosteady state flow:

$$J^* = \frac{kh}{141.2 B \mu \left(\ln \frac{r_e}{r_w} - \frac{3}{4} + S \right)}$$

$$= \frac{(8.2)(53)}{141.2(1.1)(1.7) \left(\ln \frac{2,980}{0.328} - 0.75 \right)}$$

$$= 0.1968 \text{ STB/d-psi}$$

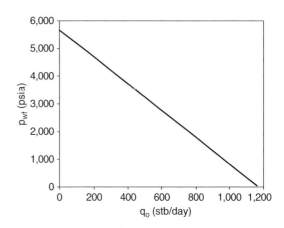

Figure 3.6 Transient IPR curve for Example Problem 3.1.

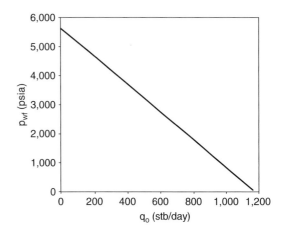

Figure 3.7 Steady-state IPR curve for Example Problem 3.1.

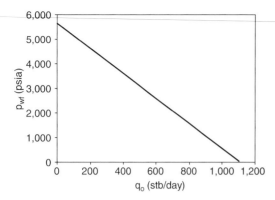

Figure 3.8 Pseudo–steady-state IPR curve for Example Problem 3.1.

Calculated points are:

p_{wf} (psi)	q_o (stb/day)
50	1,102
5,651	0

Pseudo–steady-state IPR curve is plotted in Fig. 3.8.

3.3.2 LPR for Two-Phase Reservoirs

The linear IPR model presented in the previous section is valid for pressure values as low as bubble-point pressure. Below the bubble-point pressure, the solution gas escapes from the oil and become free gas. The free gas occupies some portion of pore space, which reduces flow of oil. This effect is quantified by the reduced relative permeability. Also, oil viscosity increases as its solution gas content drops. The combination of the relative permeability effect and the viscosity effect results in lower oil production rate at a given bottom-hole pressure. This makes the IPR curve deviating from the linear trend below bubble-point pressure, as shown in Fig. 3.5. The lower the pressure, the larger the deviation. If the reservoir pressure is below the initial bubble-point pressure, oil and gas two-phase flow exists in the whole reservoir domain and the reservoir is referred as a "two-phase reservoir."

Only empirical equations are available for modeling IPR of two-phase reservoirs. These empirical equations include Vogel's (1968) equation extended by Standing (1971), the Fetkovich (1973) equation, Bandakhlia and Aziz's (1989) equation, Zhang's (1992) equation, and Retnanto and Economides' (1998) equation. Vogel's equation is still widely used in the industry. It is written as

$$q = q_{max}\left[1 - 0.2\left(\frac{p_{wf}}{\bar{p}}\right) - 0.8\left(\frac{p_{wf}}{\bar{p}}\right)^2\right] \quad (3.19)$$

or

$$p_{wf} = 0.125\bar{p}\left[\sqrt{81 - 80\left(\frac{q}{q_{max}}\right)} - 1\right], \quad (3.20)$$

where q_{max} is an empirical constant and its value represents the maximum possible value of reservoir deliverability, or AOF. The q_{max} can be theoretically estimated based on reservoir pressure and productivity index above the bubble-point pressure. The pseudo–steady-state flow follows that

$$q_{max} = \frac{J^*\bar{p}}{1.8}. \quad (3.21)$$

Derivation of this relation is left to the reader for an exercise.

Fetkovich's equation is written as

$$q = q_{max}\left[1 - \left(\frac{p_{wf}}{\bar{p}}\right)^2\right]^n \quad (3.22)$$

or

$$q = C(\bar{p}^2 - p_{wf}^2)^n, \quad (3.23)$$

where C and n are empirical constants and is related to q_{max} by $C = q_{max}/\bar{p}^{2n}$. As illustrated in Example Problem 3.5, the Fetkovich equation with two constants is more accurate than Vogel's equation IPR modeling.

Again, Eqs. (3.19) and (3.23) are valid for average reservoir pressure $\bar{p}$ being at and below the initial bubble-point pressure. Equation (3.23) is often used for gas reservoirs.

Example Problem 3.2 Construct IPR of a vertical well in a saturated oil reservoir using Vogel's equation. The following data are given:

Porosity:	$\phi = 0.19$
Effective horizontal permeability:	$k = 8.2$ md
Pay zone thickness:	$h = 53$ ft
Reservoir pressure:	$\bar{p} = 5,651$ psia
Bubble point pressure:	$p_b = 5,651$ psia
Fluid formation volume factor:	$B_o = 1.1$
Fluid viscosity:	$\mu_o = 1.7$ cp
Total compressibility:	$c_t = 0.0000129$ psi^{-1}
Drainage area:	$A = 640$ acres
	($r_e = 2,980$ ft)
Wellbore radius:	$r_w = 0.328$ ft
Skin factor:	$S = 0$

Solution

$$J^* = \frac{kh}{141.2B\mu\left(\ln\frac{r_e}{r_w} - \frac{3}{4} + S\right)}$$

$$= \frac{(8.2)(53)}{141.2(1.1)(1.7)\left(\ln\frac{2,980}{0.328} - 0.75\right)}$$

$$= 0.1968 \text{ STB/d-psi}$$

$$q_{max} = \frac{J^*\bar{p}}{1.8} = \frac{(0.1968)(5,651)}{1.8} = 618 \text{ stb/day}$$

p_{wf} (psi)	q_o (stb/day)
5,651	0
5,000	122
4,500	206
4,000	283
3,500	352
3,000	413
2,500	466
2,000	512
1,500	550
1,000	580
500	603
0	618

Calculated points by Eq. (3.19) are
The IPR curve is plotted in Fig. 3.9.

3.3.3 IPR for Partial Two-Phase Oil Reservoirs

If the reservoir pressure is above the bubble-point pressure and the flowing bottom-hole pressure is below the bubble-point pressure, a generalized IPR model can be formulated. This can be done by combining the straight-line IPR model for single-phase flow with Vogel's IPR model for two-phase flow. Figure 3.10 helps to understand the formulation.

According to the linear IPR model, the flow rate at bubble-point pressure is

$$q_b = J^*(\bar{p} - p_b), \quad (3.24)$$

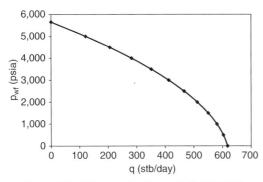

Figure 3.9 IPR curve for Example Problem 3.2.

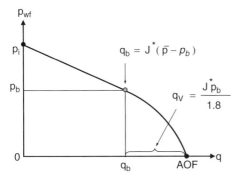

Figure 3.10 Generalized Vogel IPR model for partial two-phase reservoirs.

Based on Vogel's IPR model, the additional flow rate caused by a pressure below the bubble-point pressure is expressed as

$$\Delta q = q_v \left[1 - 0.2 \left(\frac{p_{wf}}{p_b} \right) - 0.8 \left(\frac{p_{wf}}{p_b} \right)^2 \right] \quad (3.25)$$

Thus, the flow rate at a given bottom-hole pressure that is below the bubble-point pressure is expressed as

$$q = q_b + q_v \left[1 - 0.2 \left(\frac{p_{wf}}{p_b} \right) - 0.8 \left(\frac{p_{wf}}{p_b} \right)^2 \right]. \quad (3.26)$$

Because

$$q_v = \frac{J^* p_b}{1.8}, \quad (3.27)$$

Eq. (3.26) becomes

$$q = J^*(\bar{p} - p_b) + \frac{J^* p_b}{1.8}$$

$$\times \left[1 - 0.2 \left(\frac{p_{wf}}{p_b} \right) - 0.8 \left(\frac{p_{wf}}{p_b} \right)^2 \right]. \quad (3.28)$$

Example Problem 3.3 Construct IPR of a vertical well in an undersaturated oil reservoir using the generalized Vogel equation. The following data are given:

Porosity:	$\phi = 0.19$
Effective horizontal permeability:	$k = 8.2$ md
Pay zone thickness:	$h = 53$ ft
Reservoir pressure:	$\bar{p} = 5,651$ psia
Bubble point pressure:	$p_b = 3,000$ psia
Fluid formation volume factor:	$B_o = 1.1$
Fluid viscosity:	$\mu_o = 1.7$ cp
Total compressibility:	$c_t = 0.0000129$ psi^{-1}
Drainage area:	$A = 640$ acres
	($r_e = 2,980$ ft)
Wellbore radius:	$r_w = 0.328$ ft
Skin factor:	$S = 0$

Solution

$$J^* = \frac{kh}{141.2 B \mu \left(\ln \frac{r_e}{r_w} - \frac{3}{4} + S \right)}$$

$$= \frac{(8.2)(53)}{141.2(1.1)(1.7)\left(\ln \frac{2,980}{0.328} - 0.75 \right)}$$

$$= 0.1968 \text{ STB/d-psi}$$

$$q_b = J^*(\bar{p} - p_b)$$
$$= (0.1968)(5,651 - 3,000)$$
$$= 522 \text{ sbt/day}$$

$$q_v = \frac{J^* p_b}{1.8}$$
$$= \frac{(0.1968)(3,000)}{1.8}$$
$$= 328 \text{ stb/day}$$

Calculated points by Eq. (3.28) are

p_{wf} (psi)	q_o (stb/day)
0	850
565	828
1,130	788
1,695	729
2,260	651
2,826	555
3,000	522
5,651	0

The IPR curve is plotted in Fig. 3.11.

3.4 Construction of IPR Curves Using Test Points

It has been shown in the previous section that well IPR curves can be constructed using reservoir parameters including formation permeability, fluid viscosity, drainage area, wellbore radius, and well skin factor. These parameters determine the constants (e.g., productivity index) in the IPR model. However, the values of these parameters are not always available. Thus, test points (measured values of production rate and flowing bottom-hole pressure) are frequently used for constructing IPR curves.

Constructing IPR curves using test points involves backing-calculation of the constants in the IPR models. For a single-phase (unsaturated oil) reservoir, the model constant J^* can be determined by

$$J^* = \frac{q_1}{(\bar{p} - p_{wf1})}, \quad (3.29)$$

where q_1 is the tested production rate at tested flowing bottom-hole pressure p_{wf1}.

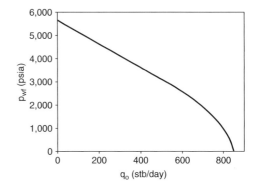

Figure 3.11 IPR curve for Example Problem 3.3.

For a partial two-phase reservoir, model constant J^* in the generalized Vogel equation must be determined based on the range of tested flowing bottom-hole pressure. If the tested flowing bottom-hole pressure is greater than bubble-point pressure, the model constant J^* should be determined by

$$J^* = \frac{q_1}{(\bar{p} - p_{wf1})}. \tag{3.30}$$

If the tested flowing bottom-hole pressure is less than bubble-point pressure, the model constant J^* should be determined using Eq. (3.28), that is,

$$J^* = \frac{q_1}{\left((\bar{p} - p_b) + \frac{p_b}{1.8}\left[1 - 0.2\left(\frac{p_{wf1}}{p_b}\right) - 0.8\left(\frac{p_{wf1}}{p_b}\right)^2\right]\right)}. \tag{3.31}$$

Example Problem 3.4 Construct IPR of two wells in an undersaturated oil reservoir using the generalized Vogel equation. The following data are given:

Reservoir pressure:	$\bar{p} = 5{,}000$ psia
Bubble point pressure:	$p_b = 3{,}000$ psia
Tested flowing bottom-hole pressure in Well A:	$p_{wf1} = 4{,}000$ psia
Tested production rate from Well A:	$q_1 = 300$ stb/day
Tested flowing bottom hole pressure in Well B:	$p_{wf1} = 2{,}000$ psia
Tested production rate from Well B:	$q_1 = 900$ stb/day

Solution

Well A:

$$J^* = \frac{q_1}{(\bar{p} - p_{wf1})}$$
$$= \frac{300}{(5{,}000 - 4{,}000)}$$
$$= 0.3000 \text{ stb/day-psi}$$

Calculated points are

p_{wf} (psia)	q (stb/day)
0	1,100
500	1,072
1,000	1,022
1,500	950
2,000	856
2,500	739
3,000	600
5,000	0

The IPR curve is plotted in Fig. 3.12.

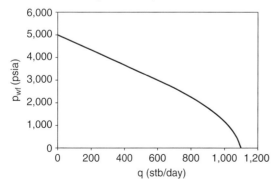

Figure 3.12 IPR curves for Example Problem 3.4, Well A.

Well B:

$$J^* = \frac{q_1}{\left((\bar{p} - p_b) + \frac{p_b}{1.8}\left[1 - 0.2\left(\frac{p_{wf1}}{p_b}\right) - 0.8\left(\frac{p_{wf1}}{p_b}\right)^2\right]\right)}$$

$$= \frac{900}{\left((5{,}000 - 3{,}000) + \frac{3{,}000}{1.8}\left[1 - 0.2\left(\frac{2{,}000}{3{,}000}\right) - 0.8\left(\frac{2{,}000}{3{,}000}\right)^2\right]\right)}$$

$$= 0.3156 \text{ stb/day-psi}$$

Calculated points are

p_{wf} (psia)	q (stb/day)
0	1,157
500	1,128
1,000	1,075
1,500	999
2,000	900
2,500	777
3,000	631
5,000	0

The IPR curve is plotted in Fig. 3.13.

For a two-phase (saturated oil) reservoir, if the Vogel equation, Eq. (3.20), is used for constructing the IPR curve, the model constant q_{max} can be determined by

$$q_{max} = \frac{q_1}{1 - 0.2\left(\frac{p_{wf1}}{\bar{p}}\right) - 0.8\left(\frac{p_{wf1}}{\bar{p}}\right)^2}. \tag{3.32}$$

The productivity index at and above bubble-point pressure, if desired, can then be estimated by

$$J^* = \frac{1.8 q_{max}}{\bar{p}}. \tag{3.33}$$

If Fetkovich's equation, Eq. (3.22), is used, two test points are required for determining the values of the two model constant, that is,

$$n = \frac{\log\left(\frac{q_1}{q_2}\right)}{\log\left(\frac{\bar{p}^2 - p_{wf1}^2}{\bar{p}^2 - p_{wf2}^2}\right)} \tag{3.34}$$

and

$$C = \frac{q_1}{(\bar{p}^2 - p_{wf1}^2)^n}, \tag{3.35}$$

where q_1 and q_2 are the tested production rates at tested flowing bottom-hole pressures p_{wf1} and p_{wf1}, respectively.

Example Problem 3.5 Construct IPR of a well in a saturated oil reservoir using both Vogel's equation and Fetkovich's equation. The following data are given:

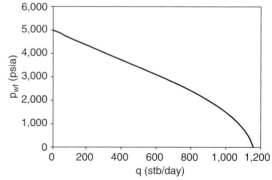

Figure 3.13 IPR curves for Example Problem 3.4, Well B.

Reservoir pressure, $\bar{p} = 3,000$ psia
Tested flowing bottom-hole pressure, $p_{wf1} = 2,000$ psia
Tested production rate at p_{wf1}, $q_1 = 500$ stb/day
Tested flowing bottom-hole pressure, $p_{wf2} = 1,000$ psia
Tested production rate at p_{wf2}, $q_2 = 800$ stb/day

Solution

Vogel's equation:

$$q_{max} = \frac{q_1}{1 - 0.2\left(\frac{p_{wf1}}{\bar{p}}\right) - 0.8\left(\frac{p_{wf1}}{\bar{p}}\right)^2}$$

$$= \frac{500}{1 - 0.2\left(\frac{2,000}{3,000}\right) - 0.8\left(\frac{2,000}{3,000}\right)^2}$$

$$= 978 \text{ stb/day}$$

Calculated data points are

p_{wf} (psia)	q (stb/day)
0	978
500	924
1,000	826
1,500	685
2,000	500
2,500	272
3,000	0

Fetkovich's equation:

$$n = \frac{\log\left(\frac{q_1}{q_2}\right)}{\log\left(\frac{\bar{p}^2 - p_{wf1}^2}{\bar{p}^2 - p_{wf2}^2}\right)} = \frac{\log\left(\frac{500}{800}\right)}{\log\left(\frac{(3,000)^2 - (2,000)^2}{(3,000)^2 - (1,000)^2}\right)} = 1.0$$

$$C = \frac{q_1}{(\bar{p}^2 - p_{wf1}^2)^n}$$

$$= \frac{500}{((3,000)^2 - (2,000)^2)^{1.0}}$$

$$= 0.0001 \text{ stb/day-psi}^{2n}$$

Calculated data points are

p_{wf} (psia)	q (stb/day)
0	900
500	875
1,000	800
1,500	675
2,000	500
2,500	275
3,000	0

The IPR curves are plotted in Fig. 3.14, which indicates that Fetkovich's equation with two constants catches more details than Vogel's equation.

3.5 Composite IPR of Stratified Reservoirs

Nearly all producing formations are stratified to some extent. This means that the vertical borehole in the production zone has different layers having different reservoir pressures, permeabilities, and producing fluids. If it is assumed that there are no other communication between these formations (other than the wellbore), the production will come mainly from the higher permeability layers.

As the well's rate of production is gradually increased, the less consolidated layers will begin to produce one by one (at progressively lower GOR), and so the overall ratio of production will fall as the rate is increased. If, however, the most highly depleted layers themselves produce at high ratios because of high free gas saturations, the overall GOR will eventually start to rise as the rate is increased and this climb will be continued (after the most permeable zone has come onto production). Thus, it is to be expected that a well producing from a stratified formation will exhibit a minimum GOR as the rate of production is increased.

One of the major concerns in a multiplayer system is that interlayer cross-flow may occur if reservoir fluids are produced from commingled layers that have unequal initial pressures. This cross-flow greatly affects the composite IPR of the well, which may result in an optimistic estimate of production rate from the commingled layers.

El-Banbi and Wattenbarger (1996, 1997) investigated productivity of commingled gas reservoirs based on history matching to production data. However, no information was given in the papers regarding generation of IPR curves.

3.5.1 Composite IPR Models

The following assumptions are made in this section:

1. Pseudo–steady-state flow prevails in all the reservoir layers.
2. Fluids from/into all the layers have similar properties.
3. Pressure losses in the wellbore sections between layers are negligible (these pressure losses are considered in Chapter 6 where multilateral wells are addressed).
4. The IPR of individual layers is known.

On the basis of Assumption 1, under steady-flow conditions, the principle of material balance dictates

net mass flow rate from layers to the well
= mass flow rate at well head

or

$$\sum_{i=1}^{n} \rho_i q_i = \rho_{wh} q_{wh}, \qquad (3.36)$$

where

ρ_i = density of fluid from/into layer i,
q_i = flow rate from/into layer i,
ρ_{wh} = density of fluid at wellhead,
q_{wh} = flow rate at wellhead, and
n = number of layers.

Fluid flow from wellbore to reservoir is indicated by negative q_i. Using Assumption 2 and ignoring density change from bottom hole to well head, Eq. (3.36) degenerates to

$$\sum_{i=1}^{n} q_i = q_{wh} \qquad (3.37)$$

or

$$\sum_{i=1}^{n} J_i(\bar{p}_i - p_{wf}) = q_{wh}, \qquad (3.38)$$

where J_i is the productivity index of layer i.

3.5.1.1 Single-Phase Liquid Flow

For reservoir layers containing undersaturated oils, if the flowing bottom-hole pressure is above the bubble-point pressures of oils in all the layers, single-phase flow in all the layers is expected. Then Eq. (3.38) becomes

$$\sum_{i=1}^{n} J_i^*(\bar{p}_i - p_{wf}) = q_{wh}, \qquad (3.39)$$

where J_i^* is the productivity index of layer i at and above the bubble-point pressure. Equations (3.39) represents a linear composite IPR of the well. A straight-line IPR can be

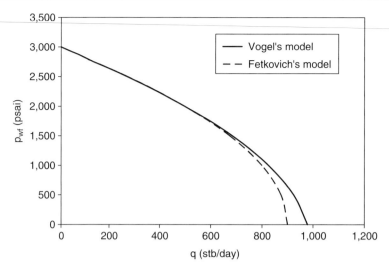

Figure 3.14 IPR curves for Example Problem 3.5.

drawn through two points at AOF and shut-in bottom-hole pressure (p_{wfo}). It is apparent from Eq. (3.39) that

$$AOF = \sum_{i=1}^{n} J_i^* \bar{p}_i = \sum_{i=1}^{n} AOF_i \qquad (3.40)$$

and

$$p_{wfo} = \frac{\sum_{i=1}^{n} J_i^* \bar{p}_i}{\sum_{i=1}^{n} J_i^*}. \qquad (3.41)$$

It should be borne in mind that p_{wfo} is a dynamic bottom-hole pressure because of cross-flow between layers.

3.5.1.2 Two-Phase Flow
For reservoir layers containing saturated oils, two-phase flow is expected. Then Eq. (3.38) takes a form of polynomial of order greater than 1. If Vogel's IPR model is used, Eq. (3.38) becomes

$$\sum_{i=1}^{n} \frac{J_i^* \bar{p}_i}{1.8}\left[1 - 0.2\left(\frac{p_{wf}}{\bar{p}_i}\right) - 0.8\left(\frac{p_{wf}}{\bar{p}_i}\right)^2\right] = q_{wh}, \qquad (3.42)$$

which gives

$$AOF = \sum_{i=1}^{n} \frac{J_i^* \bar{p}_i}{1.8} = \sum_{i=1}^{n} AOF_i \qquad (3.43)$$

and

$$p_{wfo} = \frac{\sqrt{80\sum_{i=1}^{n} J_i^* \bar{p}_i \sum_{i=1}^{n} \frac{J_i^*}{\bar{p}_i} + \left(\sum_{i=1}^{n} J_i^*\right)^2} - \sum_{i=1}^{n} J_i^*}{8\sum_{i=1}^{n} \frac{J_i^*}{\bar{p}_i}}. \qquad (3.44)$$

Again, p_{wfo} is a dynamic bottom-hole pressure because of cross-flow between layers.

3.5.1.3 Partial Two-Phase Flow
The generalized Vogel IPR model can be used to describe well inflow from multilayer reservoirs where reservoir pressures are greater than oil bubble pressures and the wellbore pressure is below these bubble-point pressures. Equation (3.38) takes the form

$$\sum_{i=1}^{n} J_i^*$$
$$\times \left\{(\bar{p}_i - p_{bi}) + \frac{p_{bi}}{1.8}\left[1 - 0.2\left(\frac{p_{wf}}{p_{bi}}\right) - 0.8\left(\frac{p_{wf}}{p_{bi}}\right)^2\right]\right\}$$
$$= q_{wh}, \qquad (3.45)$$

which gives

$$AOF = \sum_{i=1}^{n} J_i^*(\bar{p}_i - 0.44 p_{bi}) = \sum_{i=1}^{n} AOF_i \qquad (3.46)$$

and

$$p_{wfo} = \frac{\sqrt{147\left[0.56\sum_{i=1}^{n} J_i^* p_{bi} + \sum_{i=1}^{n} J_i^*(\bar{p}_i - p_{bi})\right]\sum_{i=1}^{n} \frac{J_i^*}{p_{bi}} + \left(\sum_{i=1}^{n} J_i^*\right)^2} - \sum_{i=1}^{n} J_i^*}{8\sum_{i=1}^{n} \frac{J_i^*}{p_{bi}}}. \qquad (3.47)$$

Again, p_{wfo} is a dynamic bottom-hole pressure because of cross-flow between layers.

3.5.2 Applications
The equations presented in the previous section can be readily used for generation of a composite IPR curve if all J_i^* are known. Although numerous equations have been proposed to estimate J_i^* for different types of wells, it is always recommended to determine J_i^* based on flow tests of individual layers. If the tested rate (q_i) was obtained at a wellbore pressure (p_{wfi}) that is greater than the bubble-point pressure in layer i, the productivity index J_i^* can be determined by

$$J_i^* = \frac{q_i}{\bar{p}_i - p_{wfi}}. \qquad (3.48)$$

If the tested rate (q_i) was obtained at a wellbore pressure (p_{wfi}) that is less than the bubble-point pressure in layer i, the productivity index J_i^* should be determined by

$$J_i^* = \frac{q_i}{(\bar{p}_i - p_{bi}) + \frac{p_{bi}}{1.8}\left[1 - 0.2\left(\frac{p_{wfi}}{p_{bi}}\right) - 0.8\left(\frac{p_{wfi}}{p_{bi}}\right)^2\right]}. \qquad (3.49)$$

With J_i^*, $\bar{p}_i$, and p_{bi} known, the composite IPR can be generated using Eq. (3.45).

Table 3.1 Summary of Test Points for Nine Oil Layers

Layer no.:	D3-D4	C1	B4-C2	B1	A5	A4
Layer pressure (psi)	3,030	2,648	2,606	2,467	2,302	2,254
Bubble point (psi)	26.3	4.1	4.1	56.5	31.2	33.8
Test rate (bopd)	3,200	3,500	3,510	227	173	122
Test pressure (psi)	2,936	2,607	2,571	2,422	2,288	2,216
J^* (bopd/psi)	34	85.4	100.2	5.04	12.4	3.2

Case Study

An exploration well in the south China Sea penetrated eight oil layers with unequal pressures within a short interval. These oil layers were tested in six groups. Layers B4 and C2 were tested together and Layers D3 and D4 were tested together. Test data and calculated productivity index (J_i^*) are summarized in Table 3.1. The IPR curves of the individual layers are shown in Fig. 3.15. It is seen from this figure that productivities of Layers A4, A5, and B1 are significantly lower than those of other layers. It is expected that wellbore cross-flow should occur if the bottom pressure is above the lowest reservoir pressure of 2,254 psi. Layers B4, C1, and C2 should be the major thief zones because of their high injectivities (assuming to be equal to their productivities) and relatively low pressures.

The composite IPR of these layers is shown in Fig. 3.16 where the net production rate from the well is plotted against bottom-hole pressure. It is seen from this figure that net oil production will not be available unless the bottom-hole pressure is reduced to below 2,658 psi.

Figure 3.15 suggests that the eight oil layers be produced separately in three layer groups:

Group 1: Layers D3 and D4
Group 2: Layers B4, C1, and C2
Group 3: Layers B1, A4 and A5

The composite IPR for Group 1 (D3 and D4) is the same as shown in Fig. 3.15 because these two layers were the commingle-tested. Composite IPRs of Group 2 and Group 3 are plotted in Figs. 3.17 and 3.18. Table 3.2 compares production rates read from Figs. 3.16, 3.17, and 3.18 at some pressures. This comparison indicates that significant production from Group 1 can be achieved at bottom-hole pressures higher than 2658 psi, while Group 2 and Group 3 are shut-in. A significant production from Group 1 and Group 2 can be achieved at bottom-hole pressures higher than 2,625 psi while Group 3 is shut-in. The grouped-layer production will remain beneficial until bottom-hole pressure is dropped to below 2,335 psi where Group 3 can be open for production.

3.6 Future IPR

Reservoir deliverability declines with time. During transient flow period in single-phase reservoirs, this decline is because the radius of the pressure funnel, over which the pressure drawdown ($p_i - p_{wf}$) acts, increases with time, i.e., the overall pressure gradient in the reservoir drops with time. In two-phase reservoirs, as reservoir pressure depletes, reservoir deliverability drops due to reduced relative permeability to oil and increased oil viscosity. Future IPR can be predicted by both Vogel's method and Fetkovich's method.

3.6.1 Vogel's Method

Let J_p^* and J_f^* be the present productivity index and future productivity index, respectively. The following relation can be derived:

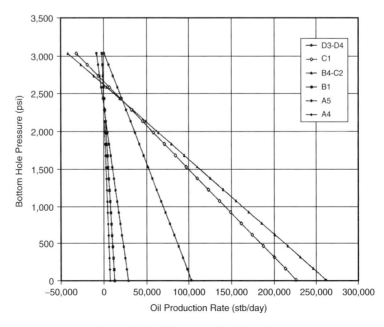

Figure 3.15 IPR curves of individual layers.

Table 3.2 Comparison of Commingled and Layer-Grouped Productions

Bottom-hole pressure (psi)	Production rate (stb/day)				
	All layers commingled	Grouped layers			
		Group 1	Group 2	Group 3	Total
2,658	0	12,663	Shut-in	Shut-in	12,663
2,625	7866	13,787	0	Shut-in	13,787
2,335	77,556	23,660	53,896	0	77,556
2,000	158,056	35,063	116,090	6,903	158,056

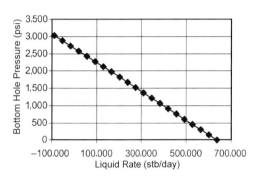

Figure 3.16 Composite IPR curve for all the layers open to flow.

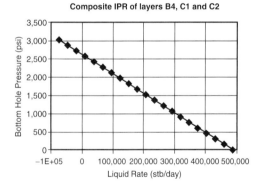

Figure 3.17 Composite IPR curve for Group 2 (Layers B4, C1, and C2).

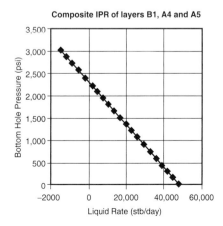

Figure 3.18 Composite IPR curve for Group 3 (Layers B1, A4, and A5).

$$\frac{J_f^*}{J_p^*} = \frac{\left(\frac{k_{ro}}{B_o \mu_o}\right)_f}{\left(\frac{k_{ro}}{B_o \mu_o}\right)_p} \tag{3.50}$$

or

$$J_f^* = J_p^* \frac{\left(\frac{k_{ro}}{B_o \mu_o}\right)_f}{\left(\frac{k_{ro}}{B_o \mu_o}\right)_p}. \tag{3.51}$$

Thus,

$$q = \frac{J_f^* \bar{p}_f}{1.8}\left[1 - 0.2\frac{p_{wf}}{\bar{p}_f} - 0.8\left(\frac{p_{wf}}{\bar{p}_f}\right)^2\right], \tag{3.52}$$

where $\bar{p}_f$ is the reservoir pressure in a future time.

Example Problem 3.6 Determine the IPR for a well at the time when the average reservoir pressure will be 1,800 psig. The following data are obtained from laboratory tests of well fluid samples:

Reservoir properties	Present	Future
Average pressure (psig)	2,250	1,800
Productivity index J^* (stb/day-psi)	1.01	
Oil viscosity (cp)	3.11	3.59
Oil formation volume factor (rb/stb)	1.173	1.150
Relative permeability to oil	0.815	0.685

Solution

$$J_f^* = J_p^* \frac{\left(\frac{k_{ro}}{B_o \mu_o}\right)_f}{\left(\frac{k_{ro}}{B_o \mu_o}\right)_p}$$

$$= 1.01 \frac{\left(\frac{0.685}{3.59(1.150)}\right)}{\left(\frac{0.815}{3.11(1.173)}\right)}$$

$$= 0.75 \, \text{stb/day-psi}$$

Vogel's equation for future IPR:

$$q = \frac{J_f^* \bar{p}_f}{1.8}\left[1 - 0.2\frac{p_{wf}}{\bar{p}_f} - 0.8\left(\frac{p_{wf}}{\bar{p}_f}\right)^2\right]$$

$$= \frac{(0.75)(1,800)}{1.8}\left[1 - 0.2\frac{p_{wf}}{1,800} - 0.8\left(\frac{p_{wf}}{1,800}\right)^2\right]$$

Calculated data points are as follows:
Present and future IPR curves are plotted in Fig. 3.19.

Reservoir pressure = 2,250 psig		Reservoir pressure = 1,800 psig	
p_{wf} (psig)	q (stb/day)	p_{wf} (psig)	q (stb/day)
2,250	0	1,800	0
2,025	217	1,620	129
1,800	414	1,440	246
1,575	591	1,260	351
1,350	747	1,080	444
1,125	884	900	525
900	1000	720	594
675	1096	540	651
450	1172	360	696
225	1227	180	729
0	1263	0	750

3.6.2 Fetkovich's Method

The integral form of reservoir inflow relationship for multiphase flow is expressed as

$$q = \frac{0.007082kh}{\ln\left(\frac{r_e}{r_w}\right)} \int_{p_{wf}}^{p_e} f(p) dp, \quad (3.53)$$

where $f(p)$ is a pressure function. The simplest two-phase flow case is that of constant pressure p_e at the outer boundary (r_e), with p_e less than the bubble-point pressure so that there is two-phase flow throughout the reservoir. Under these circumstances, $f(p)$ takes on the value $\frac{k_{ro}}{\mu_o B_o}$, where k_{ro} is the relative permeability to oil at the saturation conditions in the formation corresponding to the pressure p. In this method, Fetkovich makes the key assumption that to a good degree of approximation, the expression $\frac{k_{ro}}{\mu_o B_o}$ is a linear function of p, and is a straight line passing through the origin. If p_i is the initial formation pressure (i.e., $\sim p_e$), then the straight-line assumption is

$$\frac{k_{ro}}{\mu_o B_o} = \left(\frac{k_{ro}}{\mu_o B_o}\right)_i \frac{p}{p_i}. \quad (3.54)$$

Substituting Eq. (3.54) into Eq. (3.53) and integrating the latter gives

$$q_o = \frac{0.007082kh}{\ln\left(\frac{r_e}{r_w}\right)} \left(\frac{k_{ro}}{\mu_o B_o}\right)_i \frac{1}{2p_i} (p_i^2 - p_{wf}^2) \quad (3.55)$$

or

$$q_o = J_i'(p_i^2 - p_{wf}^2), \quad (3.56)$$

where

$$J_i' = \frac{0.007082kh}{\ln\left(\frac{r_e}{r_w}\right)} \left(\frac{k_{ro}}{\mu_o B_o}\right)_i \frac{1}{2p_i}. \quad (3.57)$$

The derivative of Eq. (3.45) with respect to the flowing bottom-hole pressure is

$$\frac{dq_o}{dp_{wf}} = -2J_i'p_{wf}. \quad (3.58)$$

This implies that the rate of change of q with respect to p_{wf} is lower at the lower values of the inflow pressure.

Next, we can modify Eq. (3.58) to take into account that in practice p_e is not constant but decreases as cumulative production increases. The assumption made is that J_i' will decrease in proportion to the decrease in average reservoir (drainage area) pressure. Thus, when the static pressure is $p_e (< p_i)$, the IPR equation is

$$q_o = J_i'\frac{p_e}{p_i}(p_e^2 - p_{wf}^2) \quad (3.59)$$

or, alternatively,

$$q_o = J'(p_e^2 - p_{wf}^2), \quad (3.60)$$

where

$$J' = J_i'\frac{p_e}{p_i}. \quad (3.61)$$

These equations may be used to extrapolate into the future.

Example Problem 3.7 Using Fetkovich's method plot the IPR curves for a well in which p_i is 2,000 psia and $J_i' = 5 \times 10^{-4}$ stb/day-psia^2. Predict the IPRs of the well at well shut-in static pressures of 1,500 and 1,000 psia.

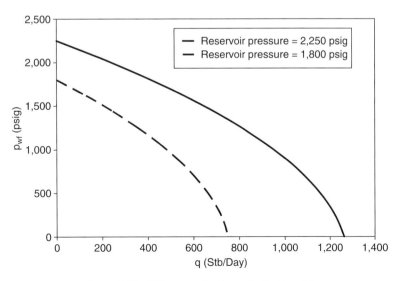

Figure 3.19 IPR curves for Example Problem 3.6.

Solution The value of J'_o at 1,500 psia is

$$J'_o = 5 \times 10^{-4}\left(\frac{1,500}{2,000}\right)$$
$$= 3.75 \times 10^{-4} \text{ stb/day (psia)}^2,$$

and the value of J'_o at 1,000 psia is

$$J'_o = 5 \times 10^{-4}\left(\frac{1,000}{2,000}\right) = 2.5 \times 10^{-4} \text{ stb/day(psia)}^2.$$

Using the above values for J'_o and the accompanying p_e in Eq. (3.46), the following data points are calculated:

$p_e = 2,000$ psig		$p_e = 1,500$ psig		$p_e = 1,000$ psig	
p_{wf} (psig)	q (stb/day)	p_{wf} (psig)	q (stb/day)	p_{wf} (psig)	q (stb/day)
2,000	0	1,500	0	1,000	0
1,800	380	1,350	160	900	48
1,600	720	1,200	304	800	90
1,400	1,020	1,050	430	700	128
1,200	1,280	900	540	600	160
1,000	1,500	750	633	500	188
800	1,680	600	709	400	210
600	1,820	450	768	300	228
400	1,920	300	810	200	240
200	1,980	150	835	100	248
0	2,000	0	844	0	250

IPR curves are plotted in Fig. 3.20.

Summary

This chapter presented and illustrated various mathematical models for estimating deliverability of oil and gas reservoirs. Production engineers should make selections of the models based on the best estimate of his/her reservoir conditions, that is, flow regime and pressure level. The selected models should be validated with actual well production rate and bottom-hole pressure. At least one test point is required to validate a straight-line (single-liquid flow) IPR model. At least two test points are required to validate a curvic (single-gas flow or two-phase flow) IPR model.

References

BANDAKHLIA, H. and AZIZ, K. Inflow performance relationship for solution-gas drive horizontal wells. Presented at the 64th SPE Annual Technical Conference and Exhibition held 8–11 October 1989, in San Antonio, Texas. Paper SPE 19823.

CHANG, M. Analysis of inflow performance simulation of solution-gas drive for horizontal/slant vertical wells. Presented at the SPE Rocky Mountain Regional Meeting held 18–21 May 1992, in Casper, Wyoming. Paper SPE 24352.

DIETZ, D.N. Determination of average reservoir pressure from build-up surveys. *J. Pet. Tech.* 1965; August.

DAKE, L.P. *Fundamentals of Reservoir Engineering*. New York: Elsevier, 1978.

EARLOUGHER, R.C. *Advances in Well Test Analysis*. Dallad: Society of Petroleum Engineers, 1977.

EL-BANBI, A.H. and WATTENBARGER, R.A. Analysis of commingled tight gas reservoirs. Presented at the SPE Annual Technical Conference and Exhibition held 6–9 October 1996, in Denver, Colorado. Paper SPE 36736.

EL-BANBI, A.H. and WATTENBARGER, R.A. Analysis of commingled gas reservoirs with variable bottom-hole flowing pressure and non-Darcy flow. Presented at the SPE Annual Technical Conference and Exhibition held 5–8 October 1997, in San Antonio, Texas. Paper SPE 38866.

FETKOVICH, M.J. The isochronal testing of oil wells. Presented at the SPE Annual Technical Conference and Exhibition held 30 September–3 October 1973, Las Vegas, Nevada. Paper SPE 4529.

JOSHI, S.D. Augmentation of well productivity with slant and horizontal wells. *J. Petroleum Technol.* 1988; June:729–739.

RETNANTO, A. and ECONOMIDES, M. Inflow performance relationships of horizontal and multibranched wells in a solution gas drive reservoir. Presented at the 1998 SPE Annual Technical Conference and Exhibition held 27–30 September 1998, in New Orleans, Louisiana. Paper SPE 49054.

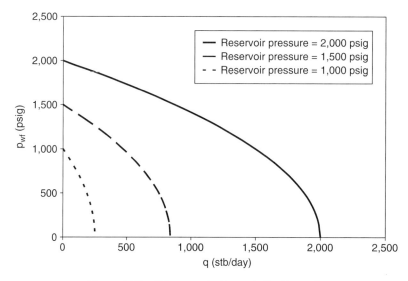

Figure 3.20 *IPR curves for Example Problem 3.7.*

STANDING, M.B. Concerning the calculation of inflow performance of wells producing from solution gas drive reservoirs. *J. Petroleum Technol.* 1971; Sep.:1141–1142.

VOGEL, J.V. Inflow performance relationships for solution-gas drive wells. *J. Petroleum Technol.* 1968; Jan.:83–92.

Problems

3.1 Construct IPR of a vertical well in an oil reservoir. Consider (1) transient flow at 1 month, (2) steady-state flow, and (3) pseudo–steady-state flow. The following data are given:

Porosity, $\phi = 0.25$
Effective horizontal permeability, $k = 10$ md
Pay zone thickness, $h = 50$ ft
Reservoir pressure, p_e or $\bar{p} = 5,000$ psia
Bubble point pressure, $p_b = 100$ psia
Fluid formation volume factor, $B_o = 1.2$
Fluid viscosity, $\mu_o = 1.5$ cp
Total compressibility, $c_t = 0.0000125\,\text{psi}^{-1}$
Drainage area, $A = 640$ acres ($r_e = 2,980$ ft)
Wellbore radius, $r_w = 0.328$ ft
Skin factor, $S = 5$

3.2 Construct IPR of a vertical well in a saturated oil reservoir using Vogel's equation. The following data are given:

Porosity, $\phi = 0.2$
Effective horizontal permeability, $k = 80$ md
Pay zone thickness, $h = 55$ ft
Reservoir pressure, $\bar{p} = 4,500$ psia
Bubble point pressure, $p_b = 4,500$ psia
Fluid formation volume factor, $B_o = 1.1$
Fluid viscosity, $\mu_o = 1.8$ cp
Total compressibility, $c_t = 0.000013\,\text{psi}^{-1}$
Drainage area, $A = 640$ acres ($r_e = 2,980$ ft)
Wellbore radius, $r_w = 0.328$ ft
Skin factor, $S = 2$

3.3 Construct IPR of a vertical well in an unsaturated oil reservoir using generalized Vogel's equation. The following data are given:

Porosity, $\phi = 0.25$
Effective horizontal permeability, $k = 100$ md
Pay zone thickness, $h = 55$ ft
Reservoir pressure, $\bar{p} = 5,000$ psia
Bubble point pressure, $p_b = 3,000$ psia
Fluid formation volume factor, $B_o = 1.2$
Fluid viscosity, $\mu_o = 1.8$ cp
Total compressibility, $c_t = 0.000013\,\text{psi}^{-1}$
Drainage area, $A = 640$ acres ($r_e = 2,980$ ft)
Wellbore radius, $r_w = 0.328$ ft
Skin factor, $S = 5.5$

3.4 Construct IPR of two wells in an unsaturated oil reservoir using generalized Vogel's equation. The following data are given:

Reservoir pressure, $\bar{p} = 5,500$ psia
Bubble point pressure, $p_b = 3,500$ psia
Tested flowing bottom-hole pressure in Well A, $p_{wf1} = 4,000$ psia
Tested production rate from Well A, $q_1 = 400$ stb/day
Tested flowing bottom-hole pressure in Well B, $p_{wf1} = 2,000$ psia
Tested production rate from Well B, $q_1 = 1,000$ stb/day

3.5 Construct IPR of a well in a saturated oil reservoir using both Vogel's equation and Fetkovich's equation. The following data are given:

Reservoir pressure, $\bar{p} = 3,500$ psia
Tested flowing bottom-hole pressure, $p_{wf1} = 2,500$ psia
Tested production rate at p_{wf1}, $q_1 = 600$ stb/day
Tested flowing bottom-hole pressure, $p_{wf2} = 1,500$ psia
Tested production rate at p_{wf2}, $q_2 = 900$ stb/day

3.6 Determine the IPR for a well at the time when the average reservoir pressure will be 1,500 psig. The following data are obtained from laboratory tests of well fluid samples:

Reservoir properties	Present	Future
Average pressure (psig)	2,200	1,500
Productivity index J^* (stb/day-psi)	1.25	
Oil viscosity (cp)	3.55	3.85
Oil formation volume factor (rb/stb)	1.20	1.15
Relative permeability to oil	0.82	0.65

3.7 Using Fetkovich's method, plot the IPR curve for a well in which p_i is 3,000 psia and $J'_o = 4 \times 10^{-4}$ stb/day-psia^2. Predict the IPRs of the well at well shut-in static pressures of 2,500 psia, 2,000 psia, 1,500 psia, and 1,000 psia.

4 Wellbore Performance

Contents
4.1 Introduction 4/46
4.2 Single-Phase Liquid Flow 4/46
4.3 Multiphase Flow in Oil Wells 4/48
4.4 Single-Phase Gas Flow 4/53
4.5 Mist Flow in Gas Wells 4/56
Summary 4/56
References 4/57
Problems 4/57

4.1 Introduction

Chapter 3 described reservoir deliverability. However, the achievable oil production rate from a well is determined by wellhead pressure and the flow performance of production string, that is, tubing, casing, or both. The flow performance of production string depends on geometries of the production string and properties of fluids being produced. The fluids in oil wells include oil, water, gas, and sand. Wellbore performance analysis involves establishing a relationship between tubular size, wellhead and bottom-hole pressure, fluid properties, and fluid production rate. Understanding wellbore flow performance is vitally important to production engineers for designing oil well equipment and optimizing well production conditions.

Oil can be produced through tubing, casing, or both in an oil well depending on which flow path has better performance. Producing oil through tubing is a better option in most cases to take the advantage of gas-lift effect. The traditional term *tubing performance relationship* (TPR) is used in this book (other terms such as *vertical lift performance* have been used in the literature). However, the mathematical models are also valid for casing flow and casing-tubing annular flow as long as hydraulic diameter is used. This chapter focuses on determination of TPR and pressure traverse along the well string. Both single-phase and multiphase fluids are considered. Calculation examples are illustrated with hand calculations and computer spreadsheets that are provided with this book.

4.2 Single-Phase Liquid Flow

Single-phase liquid flow exists in an oil well only when the wellhead pressure is above the bubble-point pressure of the oil, which is usually not a reality. However, it is convenient to start from single-phase liquid for establishing the concept of fluid flow in oil wells where multiphase flow usually dominates.

Consider a fluid flowing from point 1 to point 2 in a tubing string of length L and height Δz (Fig. 4.1). The first law of thermodynamics yields the following equation for pressure drop:

$$\Delta P = P_1 - P_2 = \frac{g}{g_c}\rho\Delta z + \frac{\rho}{2g_c}\Delta u^2 + \frac{2f_F \rho u^2 L}{g_c D} \quad (4.1)$$

where

ΔP = pressure drop, lb_f/ft^2
P_1 = pressure at point 1, lb_f/ft^2
P_2 = pressure at point 2, lb_f/ft^2

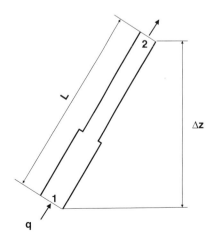

Figure 4.1 *Flow along a tubing string.*

g = gravitational acceleration, $32.17\,ft/s^2$
g_c = unit conversion factor, $32.17\,lb_m\text{-}ft/lb_f\text{-}s^2$
ρ = fluid density lb_m/ft^3
Δz = elevation increase, ft
u = fluid velocity, ft/s
f_F = Fanning friction factor
L = tubing length, ft
D = tubing inner diameter, ft

The first, second, and third term in the right-hand side of the equation represent pressure drops due to changes in elevation, kinetic energy, and friction, respectively.

The Fanning friction factor (f_F) can be evaluated based on Reynolds number and relative roughness. Reynolds number is defined as the ratio of inertial force to viscous force. The Reynolds number is expressed in consistent units as

$$N_{Re} = \frac{Du\rho}{\mu} \quad (4.2)$$

or in U.S. field units as

$$N_{Re} = \frac{1.48q\rho}{d\mu} \quad (4.3)$$

where

N_{Re} = Reynolds number
q = fluid flow rate, bbl/day
ρ = fluid density lb_m/ft^3
d = tubing inner diameter, in.
μ = fluid viscosity, cp

For laminar flow where $N_{Re} < 2,000$, the Fanning friction factor is inversely proportional to the Reynolds number, or

$$f_F = \frac{16}{N_{Re}} \quad (4.4)$$

For turbulent flow where $N_{Re} > 2,100$, the Fanning friction factor can be estimated using empirical correlations. Among numerous correlations developed by different investigators, Chen's (1979) correlation has an explicit form and gives similar accuracy to the Colebrook–White equation (Gregory and Fogarasi, 1985) that was used for generating the friction factor chart used in the petroleum industry. Chen's correlation takes the following form:

$$\frac{1}{\sqrt{f_F}} = -4 \times \log\left\{\frac{\varepsilon}{3.7065} - \frac{5.0452}{N_{Re}}\log\left[\frac{\varepsilon^{1.1098}}{2.8257} + \left(\frac{7.149}{N_{Re}}\right)^{0.8981}\right]\right\}$$
(4.5)

where the relative roughness is defined as $\varepsilon = \frac{\delta}{d}$, and δ is the absolute roughness of pipe wall.

The Fanning friction factor can also be obtained based on Darcy–Wiesbach friction factor shown in Fig. 4.2. The Darcy–Wiesbach friction factor is also referred to as the *Moody friction factor* (f_M) in some literatures. The relation between the Moody and the Fanning friction factor is expressed as

$$f_F = \frac{f_M}{4}. \quad (4.6)$$

Example Problem 4.1 Suppose that 1,000 bbl/day of 40°API, 1.2 cp oil is being produced through $2\frac{7}{8}$-in., 8.6-lb_m/ft tubing in a well that is 15 degrees from vertical. If the tubing wall relative roughness is 0.001, calculate the pressure drop over 1,000 ft of tubing.

WELLBORE PERFORMANCE 4/47

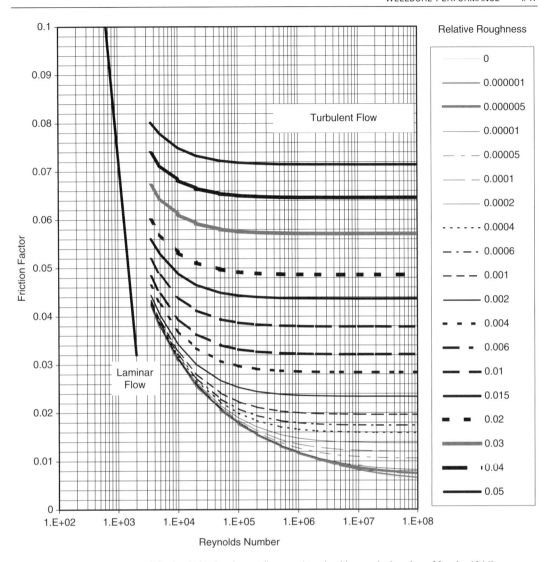

Figure 4.2 Darcy–Wiesbach friction factor diagram (used, with permission, from Moody, 1944).

Solution Oil-specific gravity:

$$\gamma_o = \frac{141.5}{{}^\circ API + 131.5}$$
$$= \frac{141.5}{40 + 131.5}$$
$$= 0.825$$

Oil density:

$$\rho = 62.4\gamma_o$$
$$= (62.5)(0.825)$$
$$= 51.57 \, \text{lb}_m/\text{ft}^3$$

Elevation increase:

$$\Delta Z = \cos(\alpha)L$$
$$= \cos(15)(1,000)$$
$$= 966 \, \text{ft}$$

The $2\frac{7}{8}$-in., 8.6-lb$_m$/ft tubing has an inner diameter of 2.259 in. Therefore,

$$D = \frac{2.259}{12}$$
$$= 0.188 \, \text{ft}.$$

Fluid velocity can be calculated accordingly:

$$u = \frac{4q}{\pi D^2}$$
$$= \frac{4(5.615)(1,000)}{\pi (0.188)^2 (86,400)}$$
$$= 2.34 \, \text{ft/s}.$$

Reynolds number:

$$N_{\text{Re}} = \frac{1.48 q \rho}{d \mu}$$
$$= \frac{1.48(1,000)(51.57)}{(2.259)(1.2)}$$
$$= 28,115 > 2,100, \text{ turbulent flow}$$

Chen's correlation gives

$$\frac{1}{\sqrt{f_F}} = -4\log\left\{\frac{\varepsilon}{3.7065} - \frac{5.0452}{N_{Re}}\log\left[\frac{\varepsilon^{1.1098}}{2.8257} + \left(\frac{7.149}{N_{Re}}\right)^{0.8981}\right]\right\}$$
$$= 12.3255$$
$$f_F = 0.006583$$

If Fig. 4.2 is used, the chart gives a Moody friction factor of 0.0265. Thus, the Fanning friction factor is estimated as

$$f_F = \frac{0.0265}{4}$$
$$= 0.006625$$

Finally, the pressure drop is calculated:

$$\Delta P = \frac{g}{g_c}\rho\Delta z + \frac{\rho}{2g_c}\Delta u^2 + \frac{2f_F \rho u^2 L}{g_c D}$$
$$= \frac{32.17}{32.17}(51.57)(966) + \frac{51.57}{2(32.17)}(0)^2 + \frac{2(0.006625)(51.57)(2.34)^2(1000)}{(32.17)(0.188)}$$
$$= 50{,}435\ \text{lb}_f/\text{ft}^2$$
$$= 350\ \text{psi}$$

4.3 Multiphase Flow in Oil Wells

In addition to oil, almost all oil wells produce a certain amount of water, gas, and sometimes sand. These wells are called *multiphase-oil wells*. The TPR equation for single-phase flow is not valid for multiphase oil wells. To analyze TPR of multiphase oil wells rigorously, a multiphase flow model is required.

Multiphase flow is much more complicated than single-phase flow because of the variation of flow regime (or flow pattern). Fluid distribution changes greatly in different flow regimes, which significantly affects pressure gradient in the tubing.

4.3.1 Flow Regimes

As shown in Fig. 4.3, at least four flow regimes have been identified in gas-liquid two-phase flow. They are bubble, slug, churn, and annular flow. These flow regimes occur as a progression with increasing gas flow rate for a given liquid flow rate. In bubble flow, gas phase is dispersed in the form of small bubbles in a continuous liquid phase. In slug flow, gas bubbles coalesce into larger bubbles that eventually fill the entire pipe cross-section. Between the large bubbles are slugs of liquid that contain smaller bubbles of entrained gas. In churn flow, the larger gas bubbles become unstable and collapse, resulting in a highly turbulent flow pattern with both phases dispersed. In annular flow, gas becomes the continuous phase, with liquid flowing in an annulus, coating the surface of the pipe and with droplets entrained in the gas phase.

4.3.2 Liquid Holdup

In multiphase flow, the amount of the pipe occupied by a phase is often different from its proportion of the total volumetric flow rate. This is due to density difference between phases. The density difference causes dense phase to slip down in an upward flow (i.e., the lighter phase moves faster than the denser phase). Because of this, the *in situ* volume fraction of the denser phase will be greater than the input volume fraction of the denser phase (i.e., the denser phase is "held up" in the pipe relative to the lighter phase). Thus, liquid "holdup" is defined as

$$y_L = \frac{V_L}{V}, \tag{4.7}$$

where

y_L = liquid holdup, fraction
V_L = volume of liquid phase in the pipe segment, ft^3
V = volume of the pipe segment, ft^3

Liquid holdup depends on flow regime, fluid properties, and pipe size and configuration. Its value can be quantitatively determined only through experimental measurements.

4.3.3 TPR Models

Numerous TPR models have been developed for analyzing multiphase flow in vertical pipes. Brown (1977) presents a thorough review of these models. TPR models for multiphase flow wells fall into two categories: (1) homogeneous-flow models and (2) separated-flow models. Homogeneous models treat multiphase as a homogeneous mixture and do not consider the effects of liquid holdup (no-slip assumption). Therefore, these models are less accurate and are usually calibrated with local operating conditions in field applications. The major advantage of these models comes from their mechanistic nature. They can handle gas-oil-water three-phase and gas-oil-water-sand four-phase systems. It is easy to code these mechanistic models in computer programs.

Separated-flow models are more realistic than the homogeneous-flow models. They are usually given in the form of empirical correlations. The effects of liquid holdup (slip) and flow regime are considered. The major disadvantage of the separated flow models is that it is difficult to code them in computer programs because most correlations are presented in graphic form.

4.3.3.1 Homogeneous-Flow Models

Numerous homogeneous-flow models have been developed for analyzing the TPR of multiphase wells since the pioneering works of Poettmann and Carpenter (1952). Poettmann–Carpenter's model uses empirical two-phase friction factor for friction pressure loss calculations without considering the effect of liquid viscosity. The effect of liquid viscosity was considered by later researchers including Cicchitti (1960) and Dukler et al. (1964). A comprehensive review of these models was given by Hasan and Kabir (2002). Guo and Ghalambor (2005) presented work addressing gas-oil-water-sand four-phase flow.

Assuming no slip of liquid phase, Poettmann and Carpenter (1952) presented a simplified gas-oil-water three-phase flow model to compute pressure losses in wellbores by estimating mixture density and friction factor. According to Poettmann and Carpenter, the following equation can be used to calculate pressure traverse in a vertical tubing when the acceleration term is neglected:

$$\Delta p = \left(\bar{\rho} + \frac{\bar{k}}{\bar{\rho}}\right)\frac{\Delta h}{144} \tag{4.8}$$

where

Δp = pressure increment, psi
$\bar{\rho}$ = average mixture density (specific weight), lb/ft^3
Δh = depth increment, ft

and

$$\bar{k} = \frac{f_{2F} q_o^2 M^2}{7.4137 \times 10^{10} D^5} \tag{4.9}$$

where

f_{2F} = Fanning friction factor for two-phase flow
q_o = oil production rate, stb/day
M = total mass associated with 1 stb of oil
D = tubing inner diameter, ft

The average mixture density $\bar{\rho}$ can be calculated by

$$\bar{\rho} = \frac{\rho_1 + \rho_2}{2} \tag{4.10}$$

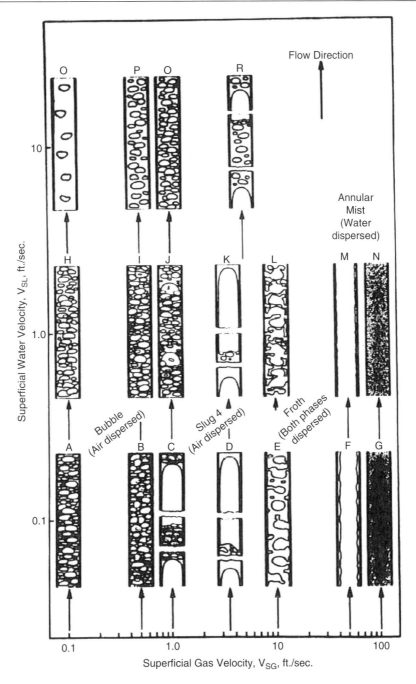

Figure 4.3 Flow regimes in gas-liquid flow (used, with permission, from Govier and Aziz, 1977).

where

ρ_1 = mixture density at top of tubing segment, lb/ft^3
ρ_2 = mixture density at bottom of segment, lb/ft^3

The mixture density at a given point can be calculated based on mass flow rate and volume flow rate:

$$\rho = \frac{M}{V_m} \qquad (4.11)$$

where

$$M = 350.17(\gamma_o + WOR\,\gamma_w) + GOR\rho_{air}\gamma_g \qquad (4.12)$$

$$V_m = 5.615(B_o + WOR\,B_w) + (GOR$$
$$- R_s)\left(\frac{14.7}{p}\right)\left(\frac{T}{520}\right)\left(\frac{z}{1.0}\right) \qquad (4.13)$$

and where

γ_o = oil specific gravity, 1 for freshwater
WOR = producing water–oil ratio, bbl/stb
γ_w = water-specific gravity, 1 for freshwater
GOR = producing gas–oil ratio, scf/stb

ρ_{air} = density of air, lb$_m$/ft^3
γ_g = gas-specific gravity, 1 for air
V_m = volume of mixture associated with 1 stb of oil, ft^3
B_o = formation volume factor of oil, rb/stb
B_w = formation volume factor of water, rb/bbl
R_s = solution gas–oil ratio, scf/stb
p = in situ pressure, psia
T = in situ temperature, °R
z = gas compressibility factor at p and T.

If data from direct measurements are not available, solution gas–oil ratio and formation volume factor of oil can be estimated using the following correlations:

$$R_s = \gamma_g \left[\frac{p}{18} \frac{10^{0.0125 API}}{10^{0.00091 t}} \right]^{1.2048} \quad (4.14)$$

$$B_o = 0.9759 + 0.00012 \left[R_s \left(\frac{\gamma_g}{\gamma_o} \right)^{0.5} + 1.25 t \right]^{1.2} \quad (4.15)$$

where t is in situ temperature in °F. The two-phase friction factor f_{2F} can be estimated from a chart recommended by Poettmann and Carpenter (1952). For easy coding in computer programs, Guo and Ghalambor (2002) developed the following correlation to represent the chart:

$$f_{2F} = 10^{1.444 - 2.5 \log(D\rho v)}, \quad (4.16)$$

where $(D\rho v)$ is the numerator of Reynolds number representing inertial force and can be formulated as

$$(D\rho v) = \frac{1.4737 \times 10^{-5} M q_o}{D}. \quad (4.17)$$

Because the Poettmann–Carpenter model takes a finite-difference form, this model is accurate for only short-depth incremental Δh. For deep wells, this model should be used in a piecewise manner to get accurate results (i.e., the tubing string should be "broken" into small segments and the model is applied to each segment).

Because iterations are required to solve Eq. (4.8) for pressure, a computer spreadsheet program *Poettmann-CarpenterBHP.xls* has been developed. The program is available from the attached CD.

Example Problem 4.2 For the following given data, calculate bottom-hole pressure:

Tubing head pressure:	500 psia
Tubing head temperature:	100 °F
Tubing inner diameter:	1.66 in.
Tubing shoe depth (near bottom hole):	5,000 ft
Bottom hole temperature:	150 °F
Liquid production rate:	2,000 stb/day
Water cut:	25%
Producing GLR:	1,000 scf/stb
Oil gravity:	30 °API
Water specific gravity:	1.05 1 for freshwater
Gas specific gravity:	0.65 1 for air

Solution This problem can be solved using the computer program *Poettmann-CarpenterBHP.xls*. The result is shown in Table 4.1.

The gas-oil-water-sand four-phase flow model proposed by Guo and Ghalambor (2005) is similar to the gas-oil-water three-phase flow model presented by Poettmann and Carpenter (1952) in the sense that no slip of liquid phase was assumed. But the Guo–Ghalambor model takes a closed (integrated) form, which makes it easy to use. The Guo–Ghalambor model can be expressed as follows:

$$144 b (p - p_{hf}) + \frac{1 - 2bM}{2} \ln \left| \frac{(144 p + M)^2 + N}{(144 p_{hf} + M)^2 + N} \right|$$
$$- \frac{M + \frac{b}{c} N - bM^2}{\sqrt{N}}$$
$$\times \left[\tan^{-1} \left(\frac{144 p + M}{\sqrt{N}} \right) - \tan^{-1} \left(\frac{144 p_{hf} + M}{\sqrt{N}} \right) \right]$$
$$= a (\cos \theta + d^2 e) L, \quad (4.18)$$

where the group parameters are defined as

$$a = \frac{0.0765 \gamma_g q_g + 350 \gamma_o q_o + 350 \gamma_w q_w + 62.4 \gamma_s q_s}{4.07 T_{av} q_g}, \quad (4.19)$$

$$b = \frac{5.615 q_o + 5.615 q_w + q_s}{4.07 T_{av} Q_g}, \quad (4.20)$$

$$c = 0.00678 \frac{T_{av} q_g}{A}, \quad (4.21)$$

$$d = \frac{0.00166}{A} (5.615 q_o + 5.615 q_w + q_s), \quad (4.22)$$

$$e = \frac{f_M}{2 g D_H}, \quad (4.23)$$

$$M = \frac{cde}{\cos \theta + d^2 e}, \quad (4.24)$$

$$N = \frac{c^2 e \cos \theta}{(\cos \theta + d^2 e)^2}, \quad (4.25)$$

where

A = cross-sectional area of conduit, ft^2
D_H = hydraulic diameter, ft
f_M = Darcy–Wiesbach friction factor (Moody factor)
g = gravitational acceleration, 32.17 ft/s^2
L = conduit length, ft
p = pressure, psia
p_{hf} = wellhead flowing pressure, psia
q_g = gas production rate, scf/d
q_o = oil production rate, bbl/d
q_s = sand production rate, ft^3/day
q_w = water production rate, bbl/d
T_{av} = average temperature, °R
γ_g = specific gravity of gas, air = 1
γ_o = specific gravity of produced oil, freshwater = 1
γ_s = specific gravity of produced solid, fresh water = 1
γ_w = specific gravity of produced water, fresh water = 1

The Darcy–Wiesbach friction factor (f_M) can be obtained from diagram (Fig. 4.2) or based on Fanning friction factor (f_F) obtained from Eq. (4.16). The required relation is $f_M = 4 f_F$.

Because iterations are required to solve Eq. (4.18) for pressure, a computer spreadsheet program *Guo-GhalamborBHP.xls* has been developed.

Example Problem 4.3 For the following data, estimate bottom-hole pressure with the Guo–Ghalambor method:

Total measured depth:	7,000 ft
The average inclination angle:	20 deg
Tubing inner diameter:	1.995 in.
Gas production rate:	1 MMscfd
Gas-specific gravity:	0.7 air = 1
Oil production rate:	1,000 stb/d
Oil-specific gravity:	0.85 H_2O = 1
Water production rate:	300 bbl/d
Water-specific gravity:	1.05 H_2O = 1
Solid production rate:	1 ft^3/d
Solid specific gravity:	2.65 H_2O = 1
Tubing head temperature:	100 °F
Bottom hole temperature:	224 °F
Tubing head pressure:	300 psia

Solution This example problem is solved with the spreadsheet program *Guo-GhalamborBHP.xls*. The result is shown in Table 4.2.

4.3.3.2 Separated-Flow Models

A number of separated-flow models are available for TPR calculations. Among many others are the Lockhart and Martinelli correlation (1949), the Duns and Ros correlation (1963), and the Hagedorn and Brown method (1965).

Based on comprehensive comparisons of these models, Ansari et al. (1994) and Hasan and Kabir (2002) recommended the Hagedorn–Brown method with modifications for near-vertical flow.

The modified Hagedorn–Brown (mH-B) method is an empirical correlation developed on the basis of the original work of Hagedorn and Brown (1965). The modifications include using the no-slip liquid holdup when the original correlation predicts a liquid holdup value less than the no-slip holdup and using the Griffith correlation (Griffith and Wallis, 1961) for the bubble flow regime.

The original Hagedorn–Brown correlation takes the following form:

$$\frac{dP}{dz} = \frac{g}{g_c}\bar{\rho} + \frac{2f_F \bar{\rho} u_m^2}{g_c D} + \bar{\rho}\frac{\Delta(u_m^2)}{2g_c \Delta z}, \quad (4.26)$$

which can be expressed in U.S. field units as

$$144\frac{dp}{dz} = \bar{\rho} + \frac{f_F M_t^2}{7.413 \times 10^{10} D^5 \bar{\rho}} + \bar{\rho}\frac{\Delta(u_m^2)}{2g_c \Delta z}, \quad (4.27)$$

where

M_t = total mass flow rate, lb_m/d
$\bar{\rho}$ = *in situ* average density, lb_m/ft^3

Table 4.1 Result Given by Poettmann-CarpenterBHP.xls for Example Problem 4.2

Poettmann–CarpenterBHP.xls
Description: This spreadsheet calculates flowing bottom-hole pressure based on tubing head pressure and tubing flow performance using the Poettmann–Carpenter method.
Instruction: (1) Select a unit system; (2) update parameter values in the Input data section;
(3) Click "Solution" button; and (4) view result in the Solution section.

Input data	U.S. Field units	
Tubing ID:	1.66	in
Wellhead pressure:	500	psia
Liquid production rate:	2,000	stb/d
Producing gas–liquid ratio (GLR):	1,000	scf/stb
Water cut (WC):	25	%
Oil gravity:	30	°API
Water-specific gravity:	1.05	freshwater =1
Gas-specific gravity:	0.65	1 for air
N_2 content in gas:	0	mole fraction
CO_2 content in gas:	0	mole fraction
H_2S content in gas:	0	mole fraction
Formation volume factor for water:	1.2	rb/stb
Wellhead temperature:	100	°F
Tubing shoe depth:	5,000	ft
Bottom-hole temperature:	150	°F
Solution		
Oil-specific gravity	= 0.88	freshwater = 1
Mass associated with 1 stb of oil	= 495.66	lb
Solution gas ratio at wellhead	= 78.42	scf/stb
Oil formation volume factor at wellhead	= 1.04	rb/stb
Volume associated with 1 stb oil @ wellhead	= 45.12	cf
Fluid density at wellhead	= 10.99	lb/cf
Solution gas–oil ratio at bottom hole	= 301.79	scf/stb
Oil formation volume factor at bottom hole	= 1.16	rb/stb
Volume associated with 1 stb oil @ bottom hole	= 17.66	cf
Fluid density at bottom hole	= 28.07	lb/cf
The average fluid density	= 19.53	lb/cf
Inertial force ($D\rho v$)	= 79.21	lb/day-ft
Friction factor	= 0.002	
Friction term	= 293.12	$(lb/cf)^2$
Error in depth	= 0.00	ft
Bottom hole pressure	= 1,699	psia

Table 4.2 Result Given by Guo-GhalamborBHP.xls for Example Problem 4.3

Guo-GhalamborBHP.xls
Description: This spreadsheet calculates flowing bottom-hole pressure based on tubing head pressure and tubing flow performance using the Guo–Ghalambor Method.
Instruction: (1) Select a unit system; (2) update parameter values in the Input data section; (3) click "Solution" button; and (4) view result in the Solution section.

Input data	U.S. Field units	SI units
Total measured depth:	7,000 ft	
Average inclination angle:	20 degrees	
Tubing inside diameter:	1.995 in.	
Gas production rate:	1,000,000 scfd	
Gas-specific gravity:	0.7 air = 1	
Oil production rate:	1000 stb/d	
Oil-specific gravity:	0.85 $H_2O = 1$	
Water production rate:	300 bbl/d	
Water-specific gravity:	1.05 $H_2O = 1$	
Solid production rate:	1 ft^3/d	
Solid specific gravity:	2.65 $H_2O = 1$	
Tubing head temperature:	100 °F	
Bottom-hole temperature:	224 °F	
Tubing head pressure:	300 psia	
Solution		
$A =$	3.1243196 in.^{2}	
$D =$	0.16625 ft	
$T_{av} =$	622 °R	
$\cos(\theta) =$	0.9397014	
$(D\rho v) =$	40.908853	
$f_M =$	0.0415505	
$a =$	0.0001713	
$b =$	2.884E-06	
$c =$	1349785.1	
$d =$	3.8942921	
$e =$	0.0041337	
$M =$	20447.044	
$N =$	6.669E+09	
Bottom-hole pressure, $p_{wf} =$	1,682 psia	

u_m = mixture velocity, ft/s

and

$$\bar{\rho} = y_L \rho_L + (1 - y_L)\rho_G, \tag{4.28}$$

$$u_m = u_{SL} + u_{SG}, \tag{4.29}$$

where

ρ_L = liquid density, lb_m/ft^3
ρ_G = in situ gas density, lb_m/ft^3
u_{SL} = superficial velocity of liquid phase, ft/s
u_{SG} = superficial velocity of gas phase, ft/s

The superficial velocity of a given phase is defined as the volumetric flow rate of the phase divided by the pipe cross-sectional area for flow. The third term in the right-hand side of Eq. (4.27) represents pressure change due to kinetic energy change, which is in most instances negligible for oil wells.

Obviously, determination of the value of liquid holdup y_L is essential for pressure calculations. The mH-B correlation uses liquid holdup from three charts using the following dimensionless numbers:

Liquid velocity number, N_{vL}:

$$N_{vL} = 1.938 \, u_{SL} \sqrt[4]{\frac{\rho_L}{\sigma}} \tag{4.30}$$

Gas velocity number, N_{vG}:

$$N_{vG} = 1.938 \, u_{SG} \sqrt[4]{\frac{\rho_L}{\sigma}} \tag{4.31}$$

Pipe diameter number, N_D:

$$N_D = 120.872 \, D \sqrt{\frac{\rho_L}{\sigma}} \tag{4.32}$$

Liquid viscosity number, N_L:

$$N_L = 0.15726 \, \mu_L \sqrt[4]{\frac{1}{\rho_L \sigma^3}}, \tag{4.33}$$

where

D = conduit inner diameter, ft
σ = liquid–gas interfacial tension, dyne/cm
μ_L = liquid viscosity, cp
μ_G = gas viscosity, cp

The first chart is used for determining parameter (CN_L) based on N_L. We have found that this chart can be replaced by the following correlation with acceptable accuracy:

$$(CN_L) = 10^Y, \tag{4.34}$$

where

$$Y = -2.69851 + 0.15841 X_1 - 0.55100 X_1^2 + 0.54785 X_1^3 - 0.12195 X_1^4 \tag{4.35}$$

and

$$X_1 = \log[(N_L) + 3]. \quad (4.36)$$

Once the value of parameter (CN_L) is determined, it is used for calculating the value of the group $\dfrac{N_{vL} p^{0.1}(CN_L)}{N_{vG}^{0.575} p_a^{0.1} N_D}$, where p is the absolute pressure at the location where pressure gradient is to be calculated, and p_a is atmospheric pressure. The value of this group is then used as an entry in the second chart to determine parameter (y_L/ψ). We have found that the second chart can be represented by the following correlation with good accuracy:

$$\dfrac{y_L}{\psi} = -0.10307 + 0.61777[\log(X_2) + 6]$$
$$- 0.63295[\log(X_2) + 6]^2 + 0.29598[\log(X_2) + 6]^3 - 0.0401[\log(X_2) + 6]^4, \quad (4.37)$$

where

$$X_2 = \dfrac{N_{vL} p^{0.1}(CN_L)}{N_{vG}^{0.575} p_a^{0.1} N_D}. \quad (4.38)$$

According to Hagedorn and Brown (1965), the value of parameter ψ can be determined from the third chart using a value of group $\dfrac{N_{vG} N_L^{0.38}}{N_D^{2.14}}$.

We have found that for $\dfrac{N_{vG} N_L^{0.38}}{N_D^{2.14}} > 0.01$ the third chart can be replaced by the following correlation with acceptable accuracy:

$$\psi = 0.91163 - 4.82176 X_3 + 1{,}232.25 X_3^2 - 22{,}253.6 X_3^3 + 116174.3 X_3^4, \quad (4.39)$$

where

$$X_3 = \dfrac{N_{vG} N_L^{0.38}}{N_D^{2.14}}. \quad (4.40)$$

However, $\psi = 1.0$ should be used for $\dfrac{N_{vG} N_L^{0.38}}{N_D^{2.14}} \leq 0.01$.

Finally, the liquid holdup can be calculated by

$$y_L = \psi \left(\dfrac{y_L}{\psi} \right). \quad (4.41)$$

The Fanning friction factor in Eq. (4.27) can be determined using either Chen's correlation Eq. (4.5) or (4.16). The Reynolds number for multiphase flow can be calculated by

$$N_{Re} = \dfrac{2.2 \times 10^{-2} m_t}{D \mu_L^{y_L} \mu_G^{(1-y_L)}}, \quad (4.42)$$

where m_t is mass flow rate. The modified mH-B method uses the Griffith correlation for the bubble-flow regime. The bubble-flow regime has been observed to exist when

$$\lambda_G < L_B, \quad (4.43)$$

where

$$\lambda_G = \dfrac{u_{sG}}{u_m} \quad (4.44)$$

and

$$L_B = 1.071 - 0.2218 \left(\dfrac{u_m^2}{D} \right), \quad (4.45)$$

which is valid for $L_B \geq 0.13$. When the L_B value given by Eq. (4.45) is less than 0.13, $L_B = 0.13$ should be used.

Neglecting the kinetic energy pressure drop term, the Griffith correlation in U.S. field units can be expressed as

$$144 \dfrac{dp}{dz} = \bar{\rho} + \dfrac{f_F m_L^2}{7.413 \times 10^{10} D^5 \rho_L y_L^2}, \quad (4.46)$$

where m_L is mass flow rate of liquid only. The liquid holdup in Griffith correlation is given by the following expression:

$$y_L = 1 - \dfrac{1}{2}\left[1 + \dfrac{u_m}{u_s} - \sqrt{\left(1 + \dfrac{u_m}{u_s}\right)^2 - 4\dfrac{u_{sG}}{u_s}}\right], \quad (4.47)$$

where $\mu_s = 0.8\,\text{ft/s}$. The Reynolds number used to obtain the friction factor is based on the *in situ* average liquid velocity, that is,

$$N_{Re} = \dfrac{2.2 \times 10^{-2} m_L}{D \mu_L}. \quad (4.48)$$

To speed up calculations, the Hagedorn–Brown correlation has been coded in the spreadsheet program *HagedornBrownCorrelation.xls*.

Example Problem 4.4 For the data given below, calculate and plot pressure traverse in the tubing string:

Tubing shoe depth:	9,700 ft
Tubing inner diameter:	1.995 in.
Oil gravity:	40 °API
Oil viscosity:	5 cp
Production GLR:	75 scf/bbl
Gas-specific gravity:	0.7 air = 1
Flowing tubing head pressure:	100 psia
Flowing tubing head temperature:	80 °F
Flowing temperature at tubing shoe:	180 °F
Liquid production rate:	758 stb/day
Water cut:	10%
Interfacial tension:	30 dynes/cm
Specific gravity of water:	1.05 H$_2$O = 1

Solution This example problem is solved with the spreadsheet program *HagedornBrownCorrelation.xls*. The result is shown in Table 4.3 and Fig. 4.4.

4.4 Single-Phase Gas Flow

The first law of thermodynamics (conservation of energy) governs gas flow in tubing. The effect of kinetic energy change is negligible because the variation in tubing diameter is insignificant in most gas wells. With no shaft work device installed along the tubing string, the first law of thermodynamics yields the following mechanical balance equation:

$$\dfrac{dP}{\rho} + \dfrac{g}{g_c} dZ + \dfrac{f_M v^2 dL}{2 g_c D_i} = 0 \quad (4.49)$$

Because $dZ = \cos\theta\, dL$, $\rho = \dfrac{29\gamma_g P}{ZRT}$, and $v = \dfrac{4 q_{sc} P_{sc} T}{\pi D_i^2 T_{sc} P}$, Eq. (4.49) can be rewritten as

$$\dfrac{zRT}{29\gamma_g} \dfrac{dP}{P} + \left\{ \dfrac{g}{g_c} \cos\theta + \dfrac{8 f_M Q_{sc}^2 P_{sc}^2}{\pi^2 g_c D_i^5 T_{sc}^2} \left[\dfrac{zT}{P}\right]^2 \right\} dL = 0, \quad (4.50)$$

which is an ordinary differential equation governing gas flow in tubing. Although the temperature T can be approximately expressed as a linear function of length L through geothermal gradient, the compressibility factor z is a function of pressure P and temperature T. This makes it difficult to solve the equation analytically. Fortunately, the pressure P at length L is not a strong function of temperature and compressibility factor. Approximate solutions to Eq. (4.50) have been sought and used in the natural gas industry.

Table 4.3 Result Given by HagedornBrownCorrelation.xls for Example Problem 4.4

HagedornBrownCorrelation.xls
Description: This spreadsheet calculates flowing pressures in tubing string based on tubing head pressure using the Hagedorn–Brown correlation.
Instruction: (1) Select a unit system; (2) update parameter values in the Input data section; (3) click "Solution" button; and (4) view result in the Solution section and charts.

Input data	U.S. Field units	SI units
Depth (D):	9,700 ft	
Tubing inner diameter (d_{ti}):	1.995 in.	
Oil gravity (API):	40 °API	
Oil viscosity (μ_o):	5 cp	
Production GLR (GLR):	75 scf/bbl	
Gas-specific gravity (γ_g):	0.7 air =1	
Flowing tubing head pressure (p_{hf}):	100 psia	
Flowing tubing head temperature (t_{hf}):	80 °F	
Flowing temperature at tubing shoe (t_{wf}):	180 °F	
Liquid production rate (q_L):	758 stb/day	
Water cut (WC):	10%	
Interfacial tension (σ):	30 dynes/cm	
Specific gravity of water (γ_w):	1.05 H$_2$O = 1	

Solution

Depth		Pressure	
(ft)	(m)	(psia)	(MPa)
0	0	100	0.68
334	102	183	1.24
669	204	269	1.83
1,003	306	358	2.43
1,338	408	449	3.06
1,672	510	543	3.69
2,007	612	638	4.34
2,341	714	736	5.01
2,676	816	835	5.68
3,010	918	936	6.37
3,345	1,020	1,038	7.06
3,679	1,122	1,141	7.76
4,014	1,224	1,246	8.48
4,348	1,326	1,352	9.20
4,683	1,428	1,459	9.93
5,017	1,530	1,567	10.66
5,352	1,632	1,676	11.40
5,686	1,734	1,786	12.15
6,021	1,836	1,897	12.90
6,355	1,938	2,008	13.66
6,690	2,040	2,121	14.43
7,024	2,142	2,234	15.19
7,359	2,243	2,347	15.97
7,693	2,345	2,461	16.74
8,028	2,447	2,576	17.52
8,362	2,549	2,691	18.31
8,697	2,651	2,807	19.10
9,031	2,753	2,923	19.89
9,366	2,855	3,040	20.68
9,700	2,957	3,157	21.48

4.4.1 Average Temperature and Compressibility Factor Method

If single average values of temperature and compressibility factor over the entire tubing length can be assumed, Eq. (4.50) becomes

$$\frac{\bar{z}R\bar{T}}{29\gamma_g}\frac{dP}{P} + \left\{\frac{g}{g_c}\cos\theta + \frac{8f_M Q_{cs}^2 P_{sc}^2 \bar{z}^2 \bar{T}^2}{\pi^2 g_c D_i^5 T_{sc}^2 P^2}\right\}dL = 0. \quad (4.51)$$

By separation of variables, Eq. (4.51) can be integrated over the full length of tubing to yield

$$P_{wf}^2 = Exp(s)P_{hf}^2 + \frac{8f_M[Exp(s) - 1]Q_{sc}^2 P_{sc}^2 \bar{z}^2 \bar{T}^2}{\pi^2 g_c D_i^5 T_{sc}^2 \cos\theta}, \quad (4.52)$$

where

$$s = \frac{58\gamma_g g L \cos\theta}{g_c R \bar{z} \bar{T}}. \quad (4.53)$$

Equations (4.52) and (4.53) take the following forms when U.S. field units (q_{sc} in Mscf/d) are used (Katz et al., 1959):

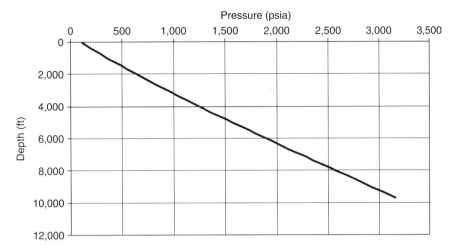

Figure 4.4 Pressure traverse given by HagedornBrownCorrelation.xls for Example Problem 4.4.

$$p_{wf}^2 = Exp(s)p_{hf}^2$$
$$+ \frac{6.67 \times 10^{-4}[Exp(s) - 1]f_M q_{sc}^2 \bar{z}^2 \bar{T}^2}{d_i^5 \cos\theta} \quad (4.54)$$

and

$$s = \frac{0.0375\gamma_g L \cos\theta}{\bar{z}\bar{T}} \quad (4.55)$$

The Darcy–Wiesbach (Moody) friction factor f_M can be found in the conventional manner for a given tubing diameter, wall roughness, and Reynolds number. However, if one assumes fully turbulent flow, which is the case for most gas wells, then a simple empirical relation may be used for typical tubing strings (Katz and Lee 1990):

$$f_M = \frac{0.01750}{d_i^{0.224}} \quad \text{for } d_i \le 4.277 \text{ in.} \quad (4.56)$$

$$f_M = \frac{0.01603}{d_i^{0.164}} \quad \text{for } d_i > 4.277 \text{ in.} \quad (4.57)$$

Guo (2001) used the following Nikuradse friction factor correlation for fully turbulent flow in rough pipes:

$$f_M = \left[\frac{1}{1.74 - 2\log\left(\frac{2\varepsilon}{d_i}\right)}\right]^2 \quad (4.58)$$

Because the average compressibility factor is a function of pressure itself, a numerical technique such as Newton–Raphson iteration is required to solve Eq. (4.54) for bottom-hole pressure. This computation can be performed automatically with the spreadsheet program *AverageTZ.xls*. Users need to input parameter values in the Input data section and run Macro Solution to get results.

Example Problem 4.5 Suppose that a vertical well produces 2 MMscf/d of 0.71 gas-specific gravity gas through a $2\frac{7}{8}$ in. tubing set to the top of a gas reservoir at a depth of 10,000 ft. At tubing head, the pressure is 800 psia and the temperature is 150 °F; the bottom-hole temperature is 200 °F. The relative roughness of tubing is about 0.0006. Calculate the pressure profile along the tubing length and plot the results.

Solution Example Problem 4.5 is solved with the spreadsheet program *AverageTZ.xls*. Table 4.4 shows the appearance of the spreadsheet for the Input data and Result sections. The calculated pressure profile is plotted in Fig. 4.5.

4.4.2 Cullender and Smith Method

Equation (4.50) can be solved for bottom-hole pressure using a fast numerical algorithm originally developed by Cullender and Smith (Katz et al., 1959). Equation (4.50) can be rearranged as

Table 4.4 Spreadsheet AverageTZ.xls: the Input Data and Result Sections

AverageTZ.xls
Description: This spreadsheet calculates tubing pressure traverse for gas wells.
Instructions:
Step 1: Input your data in the Input data section.
Step 2: Click "Solution" button to get results.
Step 3: View results in table and in graph sheet "Profile".

Input data

$\gamma_g =$	0.71
$d =$	2.259 in.
$\varepsilon/d =$	0.0006
$L =$	10.000 ft
$\theta =$	0 degrees
$p_{hf} =$	800 psia
$T_{hf} =$	150 °F
$T_{wf} =$	200 °F
$q_{sc} =$	2,000 Mscf/d

Solution

$f_M =$ 0.017396984

Depth (ft)	T (°R)	p (psia)	Z_{av}
0	610	800	0.9028
1,000	615	827	0.9028
2,000	620	854	0.9027
3,000	625	881	0.9027
4,000	630	909	0.9026
5,000	635	937	0.9026
6,000	640	965	0.9026
7,000	645	994	0.9026
8,000	650	1023	0.9027
9,000	655	1053	0.9027
10,000	660	1082	0.9028

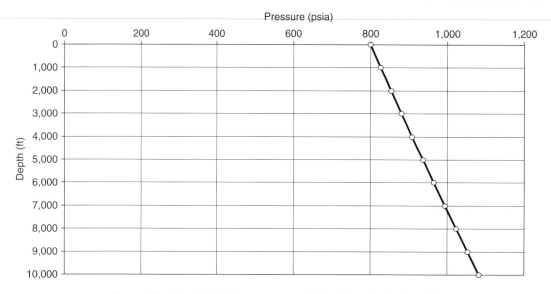

Figure 4.5 Calculated tubing pressure profile for Example Problem 4.5.

$$\frac{\frac{P}{zT}dp}{\frac{g}{g_c}\cos\theta\left(\frac{P}{zT}\right)^2 + \frac{8f_M Q_{sc}^2 P_{sc}^2}{\pi^2 g_c D_i^5 T_{sc}^2}} = -\frac{29\gamma_g}{R}dL \quad (4.59)$$

that takes an integration form of

$$\int_{P_{hf}}^{P_{wf}} \left[\frac{\frac{P}{zT}}{\frac{g}{g_c}\cos\theta\left(\frac{P}{zT}\right)^2 + \frac{8f_M Q_{sc}^2 P_{sc}^2}{\pi^2 g_c D_i^5 T_{sc}^2}} \right] dp = \frac{29\gamma_g L}{R}. \quad (4.60)$$

In U.S. field units (q_{msc} in MMscf/d), Eq. (4.60) has the following form:

$$\int_{P_{hf}}^{P_{wf}} \left[\frac{\frac{p}{zT}}{0.001\cos\theta\left(\frac{P}{zT}\right)^2 + 0.6666\frac{f_M q_{msc}^2}{d_i^5}} \right] dp = 18.75\gamma_g L \quad (4.61)$$

If the integrant is denoted with symbol I, that is,

$$I = \frac{\frac{p}{zT}}{0.001\cos\theta\left(\frac{P}{zT}\right)^2 + 0.6666\frac{f_M q_{sc}^2}{d_i^5}}, \quad (4.62)$$

Eq. (4.61) becomes

$$\int_{P_{hf}}^{P_{wf}} I\,dp = 18.75\gamma_g L. \quad (4.63)$$

In the form of numerical integration, Eq. (4.63) can be expressed as

$$\frac{(p_{mf} - p_{hf})(I_{mf} + I_{hf})}{2} + \frac{(p_{wf} - p_{mf})(I_{wf} + I_{mf})}{2}$$
$$= 18.75\gamma_g L, \quad (4.64)$$

where p_{mf} is the pressure at the mid-depth. The I_{hf}, I_{mf}, and I_{wf} are integrant Is evaluated at p_{hf}, p_{mf}, and p_{wf}, respectively. Assuming the first and second terms in the right-hand side of Eq. (4.64) each represents half of the integration, that is,

$$\frac{(p_{mf} - p_{hf})(I_{mf} + I_{hf})}{2} = \frac{18.75\gamma_g L}{2} \quad (4.65)$$

$$\frac{(p_{wf} - p_{mf})(I_{wf} + I_{mf})}{2} = \frac{18.75\gamma_g L}{2}, \quad (4.66)$$

the following expressions are obtained:

$$p_{mf} = p_{hf} + \frac{18.75\gamma_g L}{I_{mf} + I_{hf}} \quad (4.67)$$

$$p_{wf} = p_{mf} + \frac{18.75\gamma_g L}{I_{wf} + I_{mf}} \quad (4.68)$$

Because I_{mf} is a function of pressure p_{mf} itself, a numerical technique such as Newton–Raphson iteration is required to solve Eq. (4.67) for p_{mf}. Once p_{mf} is computed, p_{wf} can be solved numerically from Eq. (4.68). These computations can be performed automatically with the spreadsheet program *Cullender-Smith.xls*. Users need to input parameter values in the *Input Data* section and run Macro Solution to get results.

Example Problem 4.6 Solve the problem in Example Problem 4.5 with the Cullender and Smith Method.

Solution Example Problem 4.6 is solved with the spreadsheet program *Cullender-Smith.xls*. Table 4.5 shows the appearance of the spreadsheet for the Input data and Result sections. The pressures at depths of 5,000 ft and 10,000 ft are 937 psia and 1,082 psia, respectively. These results are exactly the same as that given by the Average Temperature and Compressibility Factor Method.

4.5 Mist Flow in Gas Wells

In addition to gas, almost all gas wells produce certain amount of liquids. These liquids are formation water and/or gas condensate (light oil). Depending on pressure and temperature, in some wells, gas condensate is not seen at surface, but it exists in the wellbore. Some gas wells produce sand and coal particles. These wells are called *multiphase-gas wells*. The four-phase flow model in Section 4.3.3.1 can be applied to mist flow in gas wells.

Summary

This chapter presented and illustrated different mathematical models for describing wellbore/tubing performance. Among many models, the mH-B model has been found to give results with good accuracy. The industry practice is to conduct a flow gradient (FG) survey to measure the

Table 4.5. Spreadsheet Cullender-Smith.xls: the Input Data and Result Sections

Cullender-SmithBHP.xls
Description: This spreadsheet calculates bottom-hole pressure with the Cullender–Smith method.
Instructions:
Step 1: Input your data in the Input data section.
Step 2: Click Solution button to get results.

Input data

γ_g	=0.71
d	=2.259 in.
ε/d	=0.0006
L	=10,000 ft
θ	=0 degrees
p_{hf}	=800 psia
T_{hf}	=150 °F
T_{wf}	=200 °F
q_{msc}	=2 MMscf/d

Solution

f_M =0.017397

Depth (ft)	T (°R)	p (psia)	Z	p/ZT	I
0	610	800	0.9028	1.45263	501.137
5,000	635	937	0.9032	1.63324	472.581
10,000	660	1,082	0.9057	1.80971	445.349

flowing pressures along the tubing string. The FG data are then employed to validate one of the models and tune the model if necessary before the model is used on a large scale.

References

ANSARI, A.M., SYLVESTER, N.D., SARICA, C., SHOHAM, O., AND BRILL, J.P. A comprehensive mechanistic model for upward two-phase flow in wellbores. *SPE Production and Facilities* (May 1994) 143, *Trans. AIME* 1994; May:297.

BROWN, K.E. *The Technology of Artificial Lift Methods*, Vol. 1. Tulsa, OK: PennWell Books, 1977, pp. 104–158.

CHEN, N.H. An explicit equation for friction factor in pipe. *Ind. Eng. Chem. Fund.* 1979;18:296.

CICCHITTI, A. Two-phase cooling experiments—pressure drop, heat transfer and burnout measurements. *Energia Nucleare* 1960;7(6):407.

DUKLER, A.E., WICKS, M., and CLEVELAND, R.G. Frictional pressure drop in two-phase flow: a comparison of existing correlations for pressure loss and hold-up. *AIChE J.* 1964:38–42.

DUNS, H. and ROS, N.C.J. Vertical flow of gas and liquid mixtures in wells. Proceedings of the 6th World Petroleum Congress, Tokyo, 1963.

GOIER, G.W. and AZIZ, K. *The Flow of Complex Mixtures in Pipes*. Huntington, NY: Robert E. Drieger Publishing Co., 1977.

GREGORY, G.A. and FOGARASI, M. Alternate to standard friction factor equation. *Oil Gas J.* 1985;April 1:120–127.

GRIFFITH, P. and WALLIS, G.B. Two-phase slug flow. *Trans. ASME* 1961;83(Ser. C):307–320.

GUO, B. and GHALAMBOR, A. *Gas Volume Requirements for Underbalanced Drilling Deviated Holes*. Tulsa, OK: PennWell Corporation, 2002, pp. 132–133.

GUO, B. and GHALAMBOR, A. *Natural Gas Engineering Handbook*. Houston: Gulf Publishing Company, 2005, pp. 59–61.

HAGEDORN, A.R. and BROWN, K.E. Experimental study of pressure gradients occurring during continuous two-phase flow in small-diameter conduits. *J. Petroleum Technol.* 1965;475.

HASAN, A.R. and KABIR, C.S. *Fluid Flow and Heat Transfer in Wellbores*. Richardson, TX: Society of Petroleum Engineers, 2002, pp. 10–15.

KATZ, D.L., CORNELL, D., KOBAYASHI, R., POETTMANN, F.H., VARY, J.A., ELENBAAS, J.R., and WEINAUG, C.F. *Handbook of Natural Gas Engineering*. New York: McGraw-Hill Publishing Company, 1959.

KATZ, D.L. and LEE, R.L. *Natural Gas Engineering—Production and Storage*. New York: McGraw-Hill Publishing Company, 1990.

LOCKHART, R.W. and MARTINELLI, R.C. Proposed correlation of data for isothermal two-phase, two-component flow in pipes. *Chem. Eng. Prog.* 1949;39.

POETTMANN, F.H. and CARPENTER, P.G. The multiphase flow of gas, oil, and water through vertical strings. *API Dril. Prod. Prac.* 1952:257–263.

Problems

4.1 Suppose that 1,000 bbl/day of 16 °API, 5-cp oil is being produced through $2\frac{7}{8}$-in., 8.6-lb_m/ft tubing in a well that is 3 degrees from vertical. If the tubing wall relative roughness is 0.001, assuming no free gas in tubing string, calculate the pressure drop over 1,000 ft of tubing.

4.2 For the following given data, calculate bottom-hole pressure using the Poettmann–Carpenter method:

Tubing head pressure:	300 psia
Tubing head temperature:	100 °F
Tubing inner diameter:	1.66 in.
Tubing shoe depth (near bottom hole):	8,000 ft
Bottom-hole temperature:	170 °F
Liquid production rate:	2,000 stb/day
Water cut:	30%
Producing GLR:	800 scf/stb
Oil gravity:	40 °API
Water-specific gravity:	1.05 1 for freshwater
Gas-specific gravity:	0.70 1 for air

4.3 For the data given below, estimate bottom-hole pressure with the Guo–Ghalambor method.

Total measured depth:	8,000 ft
The average inclination angle:	5 degrees
Tubing inner diameter:	1.995 in.
Gas production rate:	0.5 MMscfd
Gas specific gravity:	0.75 air = 1
Oil production rate:	2,000 stb/d
Oil-specific gravity:	0.85 H_2O = 1
Water production rate:	500 bbl/d
Water-specific gravity:	1.05 H_2O = 1
Solid production rate:	4 ft^3/d
Solid-specific gravity:	2.65 H_2O = 1
Tubing head temperature:	100 °F
Bottom-hole temperature:	170 °F
Tubing head pressure:	500 psia

(continued)

Tubing shoe depth:	6,000 ft
Tubing inner diameter:	1.995 in.
Oil gravity:	30 °API
Oil viscosity:	2 cp
Production GLR:	500 scf/bbl
Gas-specific gravity:	0.65 air = 1
Flowing tubing head pressure:	100 psia
Flowing tubing head temperature:	80 °F
Flowing temperature at tubing shoe:	140 °F
Liquid production rate:	1,500 stb/day
Water cut:	20%
Interfacial tension:	30 dynes/cm
Specific gravity of water:	1.05 H_2O = 1

4.4 For the data given below, calculate and plot pressure traverse in the tubing string using the Hagedorn–Brown correlation:

4.5 Suppose 3 MMscf/d of 0.75 specific gravity gas is produced through a $3\frac{1}{2}$-in. tubing string set to the top of a gas reservoir at a depth of 8,000 ft. At the tubing head, the pressure is 1,000 psia and the temperature is 120 °F; the bottom-hole temperature is 180 °F. The relative roughness of tubing is about 0.0006. Calculate the flowing bottom-hole pressure with three methods: (a) the average temperature and compressibility factor method; (b) the Cullender–Smith method; and (c) the four-phase flow method. Make comments on your results.

4.6 Solve Problem 4.5 for gas production through a K-55, 17-lb/ft, $5\frac{1}{2}$-in casing.

4.7 Suppose 2 MMscf/d of 0.65 specific gravity gas is produced through a $2\frac{7}{8}$-in. (2.259-in. inside diameter) tubing string set to the top of a gas reservoir at a depth of 5,000 ft. Tubing head pressure is 300 psia and the temperature is 100 °F; the bottom-hole temperature is 150 °F. The relative roughness of tubing is about 0.0006. Calculate the flowing bottom pressure with the average temperature and compressibility factor method.

5 Choke Performance

Contents
5.1 Introduction 5/60
5.2 Sonic and Subsonic Flow 5/60
5.3 Single-Phase Liquid Flow 5/60
5.4 Single-Phase Gas Flow 5/60
5.5 Multiphase Flow 5/63
Summary 5/66
References 5/66
Problems 5/66

5.1 Introduction

Wellhead chokes are used to limit production rates for regulations, protect surface equipment from slugging, avoid sand problems due to high drawdown, and control flow rate to avoid water or gas coning. Two types of wellhead chokes are used. They are (1) positive (fixed) chokes and (2) adjustable chokes.

Placing a choke at the wellhead means fixing the wellhead pressure and, thus, the flowing bottom-hole pressure and production rate. For a given wellhead pressure, by calculating pressure loss in the tubing the flowing bottom-hole pressure can be determined. If the reservoir pressure and productivity index is known, the flow rate can then be determined on the basis of inflow performance relationship (IPR).

5.2 Sonic and Subsonic Flow

Pressure drop across well chokes is usually very significant. There is no universal equation for predicting pressure drop across the chokes for all types of production fluids. Different choke flow models are available from the literature, and they have to be chosen based on the gas fraction in the fluid and flow regimes, that is, subsonic or sonic flow.

Both sound wave and pressure wave are mechanical waves. When the fluid flow velocity in a choke reaches the traveling velocity of sound in the fluid under the *in situ* condition, the flow is called "sonic flow." Under sonic flow conditions, the pressure wave downstream of the choke cannot go upstream through the choke because the medium (fluid) is traveling in the opposite direction at the same velocity. Therefore, a pressure discontinuity exists at the choke, that is, the downstream pressure does not affect the upstream pressure. Because of the pressure discontinuity at the choke, any change in the downstream pressure cannot be detected from the upstream pressure gauge. Of course, any change in the upstream pressure cannot be detected from the downstream pressure gauge either. This sonic flow provides a unique choke feature that stabilizes well production rate and separation operation conditions.

Whether a sonic flow exists at a choke depends on a downstream-to-upstream pressure ratio. If this pressure ratio is less than a critical pressure ratio, sonic (critical) flow exists. If this pressure ratio is greater than or equal to the critical pressure ratio, subsonic (subcritical) flow exists. The critical pressure ratio through chokes is expressed as

$$\left(\frac{p_{outlet}}{p_{up}}\right)_c = \left(\frac{2}{k+1}\right)^{\frac{k}{k-1}}, \qquad (5.1)$$

where p_{outlet} is the pressure at choke outlet, p_{up} is the upstream pressure, and $k = C_p/C_v$ is the specific heat ratio. The value of the k is about 1.28 for natural gas. Thus, the critical pressure ratio is about 0.55 for natural gas. A similar constant is used for oil flow. A typical choke performance curve is shown in Fig. 5.1.

5.3 Single-Phase Liquid Flow

When the pressure drop across a choke is due to kinetic energy change, for single-phase liquid flow, the second term in the right-hand side of Eq. (4.1) can be rearranged as

$$q = C_D A \sqrt{\frac{2g_c \Delta P}{\rho}}, \qquad (5.2)$$

where

q = flow rate, ft^3/s
C_D = choke discharge coefficient
A = choke area, ft^2
g_c = unit conversion factor, 32.17 lb$_m$-ft/lb$_f$-s^2

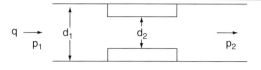

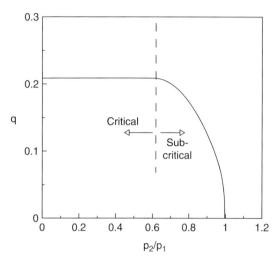

Figure 5.1 *A typical choke performance curve.*

ΔP = pressure drop, lb$_f$/ft^2
ρ = fluid density, lb$_m$/ft^3

If U.S. field units are used, Eq. (5.2) is expressed as

$$q = 8074 C_D d_2^2 \sqrt{\frac{\Delta p}{\rho}}, \qquad (5.3)$$

where

q = flow rate, bbl/d
d_2 = choke diameter, in.
Δp = pressure drop, psi

The choke discharge coefficient C_D can be determined based on Reynolds number and choke/pipe diameter ratio (Figs. 5.2 and 5.3). The following correlation has been found to give reasonable accuracy for Reynolds numbers between 10^4 and 10^6 for nozzle-type chokes (Guo and Ghalambor, 2005):

$$C_D = \frac{d_2}{d_1} + \frac{0.3167}{\left(\frac{d_2}{d_1}\right)^{0.6}} + 0.025[\log(N_{Re}) - 4], \qquad (5.4)$$

where

d_1 = upstream pipe diameter, in.
d_2 = choke diameter, in.
N_{Re} = Reynolds number based on d_2

5.4 Single-Phase Gas Flow

Pressure equations for gas flow through a choke are derived based on an isentropic process. This is because there is no time for heat to transfer (adiabatic) and the friction loss is negligible (assuming reversible) at chokes. In addition to the concern of pressure drop across the chokes, temperature drop associated with choke flow is also an important issue for gas wells, because hydrates may form that may plug flow lines.

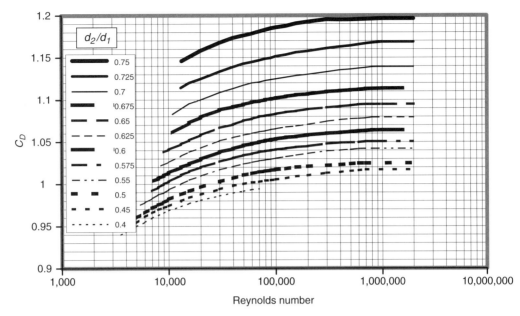

Figure 5.2 Choke flow coefficient for nozzle-type chokes (data used, with permission, from Crane, 1957).

5.4.1 Subsonic Flow

Under subsonic flow conditions, gas passage through a choke can be expressed as

$$q_{sc} = 1,248 C_D A_2 p_{up}$$
$$\times \sqrt{\frac{k}{(k-1)\gamma_g T_{up}} \left[\left(\frac{p_{dn}}{p_{up}}\right)^{\frac{2}{k}} - \left(\frac{p_{dn}}{p_{up}}\right)^{\frac{k+1}{k}} \right]}, \quad (5.5)$$

where

q_{sc} = gas flow rate, Mscf/d
p_{up} = upstream pressure at choke, psia
A_2 = cross-sectional area of choke, in.^{2}
T_{up} = upstream temperature, °R
g = acceleration of gravity, 32.2 ft/s^2
γ_g = gas-specific gravity related to air

The Reynolds number for determining C_D is expressed as

$$N_{Re} = \frac{20 q_{sc} \gamma_g}{\mu d_2}, \quad (5.6)$$

where μ is gas viscosity in cp.

Gas velocity under subsonic flow conditions is less than the sound velocity in the gas at the *in situ* conditions:

$$\nu = \sqrt{\nu_{up}^2 + 2g_c C_p T_{up} \left[1 - \frac{z_{up}}{z_{dn}} \left(\frac{p_{down}}{p_{up}}\right)^{\frac{k-1}{k}} \right]}, \quad (5.7)$$

where C_p = specific heat of gas at constant pressure (187.7 lbf-ft/lbm-R for air).

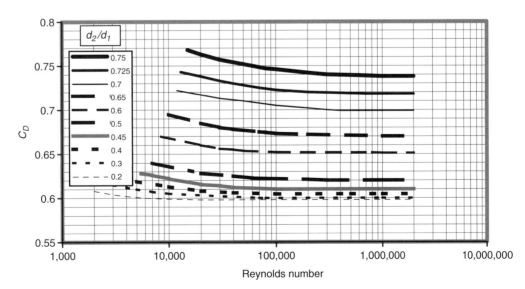

Figure 5.3 Choke flow coefficient for orifice-type chokes (data used, with permission, from Crane, 1957).

5.4.2 Sonic Flow

Under sonic flow conditions, the gas passage rate reaches its maximum value. The gas passage rate is expressed in the following equation for ideal gases:

$$Q_{sc} = 879 C_D A p_{up} \sqrt{\left(\frac{k}{\gamma_g T_{up}}\right)\left(\frac{2}{k+1}\right)^{\frac{k+1}{k-1}}} \quad (5.8)$$

The choke flow coefficient C_D is not sensitive to the Reynolds number for Reynolds number values greater than 10^6. Thus, the C_D value at the Reynolds number of 10^6 can be assumed for C_D values at higher Reynolds numbers.

Gas velocity under sonic flow conditions is equal to sound velocity in the gas under the *in situ* conditions:

$$v = \sqrt{v_{up}^2 + 2 g_c C_p T_{up}\left[1 - \frac{z_{up}}{z_{outlet}}\left(\frac{2}{k+1}\right)\right]} \quad (5.9)$$

or

$$v \approx 44.76 \sqrt{T_{up}} \quad (5.10)$$

5.4.3 Temperature at Choke

Depending on the upstream-to-downstream pressure ratio, the temperature at choke can be much lower than expected. This low temperature is due to the Joule–Thomson cooling effect, that is, a sudden gas expansion below the nozzle causes a significant temperature drop. The temperature can easily drop to below ice point, resulting in ice-plugging if water exists. Even though the temperature still can be above ice point, hydrates can form and cause plugging problems. Assuming an isentropic process for an ideal gas flowing through chokes, the temperature at the choke downstream can be predicted using the following equation:

$$T_{dn} = T_{up} \frac{z_{up}}{z_{outlet}} \left(\frac{p_{outlet}}{p_{up}}\right)^{\frac{k-1}{k}} \quad (5.11)$$

The outlet pressure is equal to the downstream pressure in subsonic flow conditions.

5.4.4 Applications

Equations (5.5) through (5.11) can be used for estimating

- Downstream temperature
- Gas passage rate at given upstream and downstream pressures
- Upstream pressure at given downstream pressure and gas passage
- Downstream pressure at given upstream pressure and gas passage

To estimate the gas passage rate at given upstream and downstream pressures, the following procedure can be taken:

Step 1: Calculate the critical pressure ratio with Eq. (5.1).
Step 2: Calculate the downstream-to-upstream pressure ratio.
Step 3: If the downstream-to-upstream pressure ratio is greater than the critical pressure ratio, use Eq. (5.5) to calculate gas passage. Otherwise, use Eq. (5.8) to calculate gas passage.

Example Problem 5.1 A 0.6 specific gravity gas flows from a 2-in. pipe through a 1-in. orifice-type choke. The upstream pressure and temperature are 800 psia and 75 °F, respectively. The downstream pressure is 200 psia (measured 2 ft from the orifice). The gas-specific heat ratio is 1.3. (a) What is the expected daily flow rate? (b) Does heating need to be applied to ensure that the frost does not clog the orifice? (c) What is the expected pressure at the orifice outlet?

Solution (a)

$$\left(\frac{P_{outlet}}{P_{up}}\right)_c = \left(\frac{2}{k+1}\right)^{\frac{k}{k-1}} = \left(\frac{2}{1.3+1}\right)^{\frac{1.3}{1.3-1}} = 0.5459$$

$$\frac{P_{dn}}{P_{up}} = \frac{200}{800} = 0.25 < 0.5459 \quad \text{Sonic flow exists.}$$

$$\frac{d_2}{d_1} = \frac{1''}{2''} = 0.5$$

Assuming $N_{Re} > 10^6$, Fig. 5.2 gives $C_D = 0.62$.

$$q_{sc} = 879 C_D A P_{up} \sqrt{\left(\frac{k}{\gamma_g T_{up}}\right)\left(\frac{2}{k+1}\right)^{\frac{k+1}{k-1}}}$$

$$q_{sc} = (879)(0.62)[\pi(1)^2/4](800)\sqrt{\left(\frac{1.3}{(0.6)(75+460)}\right)\left(\frac{2}{1.3+1}\right)^{\frac{1.3+1}{1.3-1}}}$$

$$q_{sc} = 12{,}743 \text{ Mscf/d}$$

Check N_{Re}:
$\mu = 0.01245$ cp by the Carr–Kobayashi–Burrows correlation.

$$N_{Re} = \frac{20 q_{sc} \gamma_g}{\mu d_2} = \frac{(20)(12{,}743)(0.6)}{(0.01245)(1)} = 1.23 \times 10^7 > 10^6$$

(b)

$$T_{dn} = T_{up} \frac{z_{up}}{z_{outlet}} \left(\frac{P_{outlet}}{P_{up}}\right)^{\frac{k-1}{k}} = (75+460)(1)(0.5459)^{\frac{1.3-1}{1.3}}$$

$$= 465\,°R = 5\,°F < 32\,°F$$

Therefore, heating is needed to prevent icing.

(c)

$$P_{outlet} = P_{up}\left(\frac{P_{outlet}}{P_{up}}\right) = (800)(0.5459) = 437 \text{ psia}$$

Example Problem 5.2 A 0.65 specific gravity natural gas flows from a 2-in. pipe through a 1.5-in. nozzle-type choke. The upstream pressure and temperature are 100 psia and 70 °F, respectively. The downstream pressure is 80 psia (measured 2 ft from the nozzle). The gas-specific heat ratio is 1.25. (a) What is the expected daily flow rate? (b) Is icing a potential problem? (c) What is the expected pressure at the nozzle outlet?

Solution (a)

$$\left(\frac{P_{outlet}}{P_{up}}\right)_c = \left(\frac{2}{k+1}\right)^{\frac{k}{k-1}} = \left(\frac{2}{1.25+1}\right)^{\frac{1.25}{1.25-1}} = 0.5549$$

$$\frac{P_{dn}}{P_{up}} = \frac{80}{100} = 0.8 > 0.5549 \quad \text{Subsonic flow exists.}$$

$$\frac{d_2}{d_1} = \frac{1.5''}{2''} = 0.75$$

Assuming $N_{Re} > 10^6$, Fig. 5.1 gives $C_D = 1.2$.

$$q_{sc} = 1{,}248 C_D A P_{up} \sqrt{\frac{k}{(k-1)\gamma_g T_{up}}\left[\left(\frac{P_{dn}}{P_{up}}\right)^{\frac{2}{k}} - \left(\frac{P_{dn}}{P_{up}}\right)^{\frac{k+1}{k}}\right]}$$

$$q_{sc} = (1{,}248)(1.2)[\pi(1.5)^2/4](100)$$

$$\times \sqrt{\frac{1.25}{(1.25-1)(0.65)(530)}\left[\left(\frac{80}{100}\right)^{\frac{2}{1.25}} - \left(\frac{80}{100}\right)^{\frac{1.25+1}{1.25}}\right]}$$

$$q_{sc} = 5{,}572 \text{ Mscf/d}$$

Check N_{Re}:
$\mu = 0.0108$ cp by the Carr–Kobayashi–Burrows correlation.

$$N_{Re} = \frac{20q_{sc}\gamma_g}{\mu d} = \frac{(20)(5,572)(0.65)}{(0.0108)(1.5)} = 4.5 \times 10^6 > 10^6$$

(b)

$$T_{dn} = T_{up}\frac{z_{up}}{z_{outlet}}\left(\frac{P_{outlet}}{P_{up}}\right)^{\frac{k-1}{k}} = (70+460)(1)(0.8)^{\frac{1.25-1}{1.25}}$$
$$= 507\,°R = 47\,°F > 32\,°F$$

Heating may not be needed, but the hydrate curve may need to be checked.
(c)

$$P_{outlet} = P_{dn} = 80\,\text{psia for subcritical flow.}$$

To estimate upstream pressure at a given downstream pressure and gas passage, the following procedure can be taken:

Step 1: Calculate the critical pressure ratio with Eq. (5.1).
Step 2: Calculate the minimum upstream pressure required for sonic flow by dividing the downstream pressure by the critical pressure ratio.
Step 3: Calculate gas flow rate at the minimum sonic flow condition with Eq. (5.8).
Step 4: If the given gas passage is less than the calculated gas flow rate at the minimum sonic flow condition, use Eq. (5.5) to solve upstream pressure numerically. Otherwise, Eq. (5.8) to calculate upstream pressure.

Example Problem 5.3 For the following given data, estimate upstream pressure at choke:

Downstream pressure:	300 psia
Choke size:	32 1/64 in.
Flowline ID:	2 in.
Gas production rate:	5,000 Mscf/d
Gas-specific gravity:	0.75 1 for air
Gas-specific heat ratio:	1.3
Upstream temperature:	110 °F
Choke discharge coefficient:	0.99

Solution Example Problem 5.3 is solved with the spreadsheet program *GasUpChokePressure.xls*. The result is shown in Table 5.1.

Downstream pressure cannot be calculated on the basis of given upstream pressure and gas passage under sonic flow conditions, but it can be calculated under subcritical flow conditions. The following procedure can be followed:

Step 1: Calculate the critical pressure ratio with Eq. (5.1).
Step 2: Calculate the maximum downstream pressure for minimum sonic flow by multiplying the upstream pressure by the critical pressure ratio.
Step 3: Calculate gas flow rate at the minimum sonic flow condition with Eq. (5.8).
Step 4: If the given gas passage is less than the calculated gas flow rate at the minimum sonic flow condition, use Eq. (5.5) to solve downstream pressure numerically. Otherwise, the downstream pressure cannot be calculated. The maximum possible downstream pressure for sonic flow can be estimated by multiplying the upstream pressure by the critical pressure ratio.

Table 5.1 Solution Given by the Spreadsheet Program GasUpChokePressure.xls

GasUpChokePressure.xls
Description: This spreadsheet calculates upstream pressure at choke for dry gases.
Instructions: (1) Update parameter values in blue; (2) click Solution button; (3) view results.

Input data

Downstream pressure:	300 psia
Choke size:	32 1/64 in.
Flowline ID:	2 in.
Gas production rate:	5,000 Mscf/d
Gas-specific gravity:	0.75 1 for air
Gas-specific heat ratio (k):	1.3
Upstream temperature:	110 °F
Choke discharge coefficient:	0.99

Solution

Choke area:	0.19625 in.^{2}
Critical pressure ratio:	0.5457
Minimum upstream pressure required for sonic flow:	549.72 psia
Flow rate at the minimum sonic flow condition:	3,029.76 Mscf/d
Flow regime (1 = sonic flow; −1 = subsonic flow):	1
Upstream pressure given by sonic flow equation:	907.21 psia
Upstream pressure given by subsonic flow equation:	1,088.04 psia
Estimated upstream pressure:	907.21 psia

Example Problem 5.4 For the following given data, estimate downstream pressure at choke:

Upstream pressure:	600 psia
Choke size:	32 1/64 in.
Flowline ID:	2 in.
Gas production rate:	2,500 Mscf/d
Gas-specific gravity:	0.75 1 for air
Gas-specific heat ratio:	1.3
Upstream temperature:	110 °F
Choke discharge coefficient:	0.99

Solution Example Problem 5.4 is solved with the spreadsheet program *GasDownChokePressure.xls*. The result is shown in Table 5.2.

5.5 Multiphase Flow

When the produced oil reaches the wellhead choke, the wellhead pressure is usually below the bubble-point pressure of the oil. This means that free gas exists in the fluid stream flowing through choke. Choke behaves differently depending on gas content and flow regime (sonic or subsonic flow).

5.5.1 Critical (Sonic) Flow
Tangren et al. (1949) performed the first investigation on gas-liquid two-phase flowthrough restrictions. They presented an analysis of the behavior of an expanding gas-liquid system. They showed that when gas bubbles are added to an incompressible fluid, above a critical flow velocity, the medium becomes incapable of transmitting pressure change upstream against the flow. Several

Table 5.2 Solution Given by the Spreadsheet Program GasDownChokePressure.xls

GasDownChokePressure.xls

Description: This spreadsheet calculates upstream pressure at choke for dry gases.

Instructions: (1) Update values in the Input data section; (2) click Solution button; (3) view results.

Input data

Upstream pressure:	700 psia
Choke size:	32 1/64 in.
Flowline ID:	2 in.
Gas production rate:	2,500 Mscf/d
Gas-specific gravity:	0.75 1 for air
Gas-specific heat ratio (k):	1.3
Upstream temperature:	110 °F
Choke discharge coefficient:	0.99

Solution

Choke area:	0.19625 in.^{2}
Critical pressure ratio:	0.5457
Minimum downstream pressure for minimum sonic flow:	382 psia
Flow rate at the minimum sonic flow condition:	3,857 Mscf/d
Flow regime (1 = sonic flow; −1 = subsonic flow):	−1
The maximum possible downstream pressure in sonic flow:	382 psia
Downstream pressure given by subsonic flow equation:	626 psia
Estimated downstream pressure:	626 psia

empirical choke flow models have been developed in the past half century. They generally take the following form for sonic flow:

$$p_{wh} = \frac{CR^m q}{S^n},\quad (5.12)$$

where

p_{wh} = upstream (wellhead) pressure, psia
q = gross liquid rate, bbl/day
R = producing gas-liquid ratio, Scf/bbl
S = choke size, 1/64 in.

and C, m, and n are empirical constants related to fluid properties. On the basis of the production data from Ten Section Field in California, Gilbert (1954) found the values for C, m, and n to be 10, 0.546, and 1.89, respectively. Other values for the constants were proposed by different researchers including Baxendell (1957), Ros (1960), Achong (1961), and Pilehvari (1980). A summary of these values is presented in Table 5.3. Poettmann and Beck (1963) extended the work of Ros (1960) to develop charts for different API crude oils. Omana (1969) derived dimensionless choke correlations for water-gas systems.

Table 5.3 A Summary of C, m, and n Values Given by Different Researchers

Correlation	C	m	n
Gilbert	10	0.546	1.89
Ros	17.4	0.5	2
Baxendell	9.56	0.546	1.93
Achong	3.82	0.65	1.88
Pilehvari	46.67	0.313	2.11

5.5.2 Subcritical (Subsonic) Flow

Mathematical modeling of subsonic flow of multiphase fluid through choke has been controversial over decades. Fortunati (1972) was the first investigator who presented a model that can be used to calculate critical and subcritical two-phase flow through chokes. Ashford (1974) also developed a relation for two-phase critical flow based on the work of Ros (1960). Gould (1974) plotted the critical–subcritical boundary defined by Ashford, showing that different values of the polytropic exponents yield different boundaries. Ashford and Pierce (1975) derived an equation to predict the critical pressure ratio. Their model assumes that the derivative of flow rate with respect to the downstream pressure is zero at critical conditions. One set of equations was recommended for both critical and subcritical flow conditions. Pilehvari (1980, 1981) also studied choke flow under subcritical conditions. Sachdeva (1986) extended the work of Ashford and Pierce (1975) and proposed a relationship to predict critical pressure ratio. He also derived an expression to find the boundary between critical and subcritical flow. Surbey et al. (1988, 1989) discussed the application of multiple orifice valve chokes for both critical and subcritical flow conditions. Empirical relations were developed for gas and water systems. Al-Attar and Abdul-Majeed (1988) made a comparison of existing choke flow models. The comparison was based on data from 155 well tests. They indicated that the best overall comparison was obtained with the Gilbert correlation, which predicted measured production rate within an average error of 6.19%. On the basis of energy equation, Perkins (1990) derived equations that describe isentropic flow of multiphase mixtures through chokes. Osman and Dokla (1990) applied the least-square method to field data to develop empirical correlations for gas condensate choke flow. Gilbert-type relationships were generated. Applications of these choke flow models can be found elsewhere (Wallis, 1969; Perry, 1973; Brown and Beggs, 1977; Brill and Beggs, 1978; Ikoku, 1980; Nind, 1981; Bradley, 1987; Beggs, 1991; Rastion et al., 1992; Saberi, 1996).

Sachdeva's multiphase choke flow mode is representative of most of these works and has been coded in some commercial network modeling software. This model uses the following equation to calculate the critical–subcritical boundary:

$$y_c = \left\{ \frac{\frac{k}{k-1} + \frac{(1-x_1)V_L(1-y_c)}{x_1 V_{G1}}}{\frac{k}{k-1} + \frac{n}{2} + \frac{n(1-x_1)V_L}{x_1 V_{G2}} + \frac{n}{2}\left[\frac{(1-x_1)V_L}{x_1 V_{G2}}\right]^2} \right\}^{\frac{k}{k-1}},\quad (5.13)$$

where

y_c = critical pressure ratio
$k = C_p/C_v$, specific heat ratio
n = polytropic exponent for gas
x_1 = free gas quality at upstream, mass fraction
V_L = liquid specific volume at upstream, ft^3/lbm
V_{G1} = gas specific volume at upstream, ft^3/lbm
V_{G2} = gas specific volume at downstream, ft^3/lbm.

The polytropic exponent for gas is calculated using

$$n = 1 + \frac{x_1(C_p - C_v)}{x_1 C_v + (1-x_1)C_L}.\quad (5.14)$$

The gas-specific volume at upstream (V_{G1}) can be determined using the gas law based on upstream pressure and temperature. The gas-specific volume at downstream (V_{G2}) is expressed as

$$V_{G2} = V_{G1} y_c^{-\frac{1}{k}}.\quad (5.15)$$

The critical pressure ratio y_c can be solved from Eq. (5.13) numerically.

The actual pressure ratio can be calculated by

$$y_a = \frac{p_2}{p_1}, \quad (5.16)$$

where

y_a = actual pressure ratio
p_1 = upstream pressure, psia
p_2 = downstream pressure, psia

If $y_a < y_c$, critical flow exists, and the y_c should be used ($y = y_c$). Otherwise, subcritical flow exists, and y_a should be used ($y = y_a$).

Table 5.4 An Example Calculation with Sachdeva's Choke Model

Input data

Choke diameter (d_2):	$24\frac{1}{64}$ in.
Discharge coefficient (C_D):	0.75
Downstream pressure (p_2):	50 psia
Upstream pressure (p_1):	80 psia
Upstream temperature (T_1):	100 °F
Downstream temperature (T_2):	20 °F
Free gas quality (x_1):	0.001 mass fraction
Liquid-specific gravity:	0.9 water = 1
Gas-specific gravity:	0.7 air = 1
Specific heat of gas at constant pressure (C_p):	0.24
Specific heat of gas at constant volume (C_v):	0.171429
Specific heat of liquid (C_L):	0.8

Precalculations

Gas-specific heat ratio ($k = C_p/C_v$):	1.4
Liquid-specific volume (V_L):	0.017806 ft³/lbm
Liquid density (ρ_L):	56.16 lb/ft³
Upstream gas density (ρ_{G1}):	0.27 lb/ft³
Downstream gas density (ρ_{G2}):	0.01 lb/ft⁴
Upstream gas-specific volume (V_{G1}):	3.70 ft³/lbm
Polytropic exponent of gas (n):	1.000086

Critical pressure ratio computation

$k/(k-1) =$	3.5
$(1 - x_1)/x_1 =$	999
$n/2 =$	0.500043
$V_L/V_{G1} =$	0.004811
Critical pressure ratio (y_c):	0.353134
$V_{G2} =$	7.785109 ft³/lbm
$V_L/V_{G2} =$	0.002287
Equation residue (goal seek 0 by changing y_c):	0.000263

Flow rate calculations

Pressure ratio (y_{actual}):	0.625
Critical flow index:	−1
Subcritical flow index:	1
Pressure ratio to use (y):	0.625
Downstream mixture density (ρ_{m2}):	43.54 lb/ft³
Downstream gas-specific volume (V_{G2}):	5.178032
Choke area (A_2) =	0.000767 ft²
Mass flux (G_2) =	1432.362 lbm/ft²/s
Mass flow rate (M) =	1.098051 lbm/s
Liquid mass flow rate (M_L) =	1.096953 lbm/s
Liquid flow rate =	300.5557 bbl/d
Gas mass flow rate (M_G) =	0.001098 lbm/s
Gas flow rate =	0.001772 MMscfd

The total mass flux can be calculated using the following equation:

$$G_2 = C_D \left\{ 288 g_c p_1 \rho_{m2}^2 \left[\frac{(1-x_1)(1-y)}{\rho_L} + \frac{x_1 k}{k-1}(V_{G1} - y V_{G2}) \right] \right\}^{0.5}, \quad (5.17)$$

where

G_2 = mass flux at downstream, lbm/ft²/s
C_D = discharge coefficient, 0.62–0.90
ρ_{m2} = mixture density at downstream, lbm/ft³
ρ_L = liquid density, lbm/ft³

The mixture density at downstream (ρ_{m2}) can be calculated using the following equation:

$$\frac{1}{\rho_{m2}} = x_1 V_{G1} y^{-\frac{1}{k}} + (1-x_1)V_L \quad (5.18)$$

Once the mass flux is determined from Eq. (5.17), mass flow rate can be calculated using the following equation:

$$M_2 = G_2 A_2, \quad (5.19)$$

where

A_2 = choke cross-sectional area, ft²
M_2 = mass flow rate at down stream, lbm/s

Liquid mass flow rate is determined by

$$M_{L2} = (1-x_2)M_2. \quad (5.20)$$

At typical velocities of mixtures of 50–150 ft/s flowing through chokes, there is virtually no time for mass transfer between phases at the throat. Thus, $x_2 = x_1$ can be assumed. Liquid volumetric flow rate can then be determined based on liquid density.

Gas mass flow rate is determined by

$$M_{G2} = x_2 M_2. \quad (5.21)$$

Gas volumetric flow rate at choke downstream can then be determined using gas law based on downstream pressure and temperature.

The major drawback of Sachdeva's multiphase choke flow model is that it requires free gas quality as an input parameter to determine flow regime and flow rates, and this parameter is usually not known before flow rates are known. A trial-and-error approach is, therefore, needed in flow rate computations. Table 5.4 shows an example calculation with Sachdeva's choke model. Guo et al. (2002) investigated the applicability of Sachdeva's choke flow model in southwest Louisiana gas condensate wells. A total of 512 data sets from wells in southwest Louisiana were gathered for this study. Out of these data sets, 239 sets were collected from oil wells and 273 from condensate wells. Each of the data sets includes choke size, gas rate, oil rate, condensate rate, water rate, gas–liquid ratio, upstream and downstream pressures, oil API gravity, and gas deviation factor (z-factor). Liquid and gas flow rates from these wells were also calculated using Sachdeva's choke model. The overall performance of the model was studied in predicting the gas flow rate from both oil and gas condensate wells. Out of the 512 data sets, 48 sets failed to comply with the model. Mathematical errors occurred in finding square roots of negative numbers. These data sets were from the condensate wells where liquid densities ranged from 46.7 to 55.1 lb/ft³ and recorded pressure differential across the choke less than 1,100 psi. Therefore, only 239 data sets from oil wells and 235 sets from condensate wells were used. The total number of data sets is 474. Different values of discharge coefficient C_D were used to improve the model performance. Based on the cases studied, Guo et al. (2002) draw the following conclusions:

1. The accuracy of Sachdeva's choke model can be improved by using different discharge coefficients for different fluid types and well types.
2. For predicting liquid rates of oil wells and gas rates of gas condensate wells, a discharge coefficient of $C_D = 1.08$ should be used.
3. A discharge coefficient $C_D = 0.78$ should be used for predicting gas rates of oil wells.
4. A discharge coefficient $C_D = 1.53$ should be used for predicting liquid rates of gas condensate wells.

Summary

This chapter presented and illustrated different mathematical models for describing choke performance. While the choke models for gas flow have been well established with fairly good accuracy in general, the models for two-phase flow are subject to tuning to local oil properties. It is essential to validate two-phase flow choke models before they are used on a large scale.

References

ACHONG, I.B. "Revised Bean and Performance Formula for Lake Maracaibo Wells," Shell Internal Report, October 1961.

AL-ATTAR, H.H. and ABDUL-MAJEED, G. Revised bean performance equation for east Baghdad oil wells. *SPE Production Eng.* 1988;February:127–131.

ASHFORD, F.E. An evaluation of critical multiphase flow performance through wellhead chokes. *J. Petroleum Technol.* 1974;26(August):843–848.

ASHFORD, F.E. and PIERCE, P.E. Determining multiphase pressure drop and flow capabilities in down hole safety valves. *J. Petroleum Technol.* 1975;27(September):1145–1152.

BAXENDELL, P.B. Bean performance-lake wells. *Shell Internal Report*, October 1957.

BEGGS, H.D. *Production Optimization Using Nodal Analysis.* Tulsa, OK: OGTC Publications, 1991.

BRADLEY, H.B. *Petroleum Engineering Handbook.* Richardson, TX: Society of Petroleum Engineers, 1987.

BRILL, J.P. and BEGGS, H.D. *Two-Phase Flow in Pipes.* Tulsa, OK: The University of Tulsa Press, 1978.

BROWN, K.E. and BEGGS, H.D. *The Technology of Artificial Lift Methods.* Tulsa, OK: PennWell Books, 1977.

Crane, Co. "Flow of Fluids through Valves, Fittings, and Pipe. Technical paper No. 410. Chicago, 1957.

FORTUNATI, F. Two-phase flow through wellhead chokes. Presented at the SPE European Spring Meeting held 16–18 May 1972 in Amsterdam, the Netherlands. SPE paper 3742.

GILBERT, W.E. Flowing and gas-lift well performance. *API Drilling Production Practice* 1954;20:126–157.

GOULD, T.L. Discussion of an evaluation of critical multiphase flow performance through wellhead chokes. *J. Petroleum Technol.* 1974;26(August):849–850.

GUO, B. and GHALAMBOR, A. *Natural Gas Engineering Handbook.* Houston, TX: Gulf Publishing Company, 2005.

GUO, B., AL-BEMANI, A., and GHALAMBOR, A. Applicability of Sachdeva's choke flow model in southwest Louisiana gas condensate wells. Presented at the SPE Gas technology Symposium held 30 April–2 May 2002 in Calgary, Canada. Paper SPE 75507.

IKOKU, C.U. *Natural Gas Engineering.* Tulsa, OK: PennWell Books, 1980.

NIND, T.E.W. *Principles of Oil Well Production*, 2nd edition. New York: McGraw-Hill Book Co., 1981.

OMANA, R., HOUSSIERE, C., JR., BROWN, K.E., BRILL, J.P., and THOMPSON, R.E. Multiphase flow through chokes. Presented at the SPE 44th Annual Meeting held 28–31 September 1969 in Denver, Colorado. SPE paper 2682.

OSMAN, M.E. and DOKLA, M.E. Has condensate flow through chokes. Presented at 23 April 1990. SPE paper 20988.

PERKINS, T.K. Critical and subcritical flow of multiphase mixtures through chokes. Presented at the SPE 65th Annual Technical Conference and Exhibition held 23–26 September 1990 in New Orleans, Louisiana. SPE paper 20633.

PERRY, R.H. *Chemical Engineers' Handbook*, 5th edition. New York: McGraw-Hill Book Co., 1973.

PILEHVARI, A.A. Experimental study of subcritical two-phase flow through wellhead chokes. Tulsa, OK: University of Tulsa Fluid Flow Projects Report, September 1980.

PILEHVARI, A.A. Experimental study of critical two-phase flow through wellhead chokes. Tulsa, OK: University of Tulsa Fluid Flow Projects Report, June 1981.

POETTMANN, F.H. and BECK, R.L. New charts developed to predict gas-liquid flow through chokes. *World Oil* 1963;March:95–101.

POETTMANN, F.H., BECK, R.L., and BECK, R.L. A review of multiphase flow through chokes. Paper presented at the ASME Winter Annual Meeting held 8–13 November 1992, Anaheim, California, pp. 51–62.

ROS, N.C.J. An analysis of critical simultaneous gas/liquid flow through a restriction and its application to flow metering. *Applied Sci. Res.* 1960; Section A(9):374–389.

SABERI, M. A study on flow through wellhead chokes and choke size selection, MS thesis, University of Southwestern Louisiana, Lafayette, 1996, pp. 78–89.

SACHDEVA, R., SCHMIDT, Z., BRILL, J.P., and BLAIS, R.M. Two-phase flow through chokes. Paper presented at the SPE 61st Annual Technical Conference and Exhibition held 5–8 October 1986 in New Orleans, Louisiana. SPE paper 15657.

SECEN, J.A. Surface-choke measurement equation improved by field testing and analysis. *Oil Gas J.* 1976;30(August):65–68.

SURBEY, D.W., KELKAR, B.G., and BRILL, J.P. Study of subcritical flow through multiple orifice valves. *SPE Production Eng.* 1988;February:103–108.

SURBEY, D.W., KELKAR, B.G., and BRILL, J.P. Study of multiphase critical flow through wellhead chokes. *SPE Production Eng.* 1989;May:142–146.

TANGREN, R.F., DODGE, C.H., and SEIFERT, H.S. Compressibility effects in two-phase flow. *J. Applied Physics* 1947;20:637–645.

WALLIS, G.B. *One Dimensional Two-Phase Flow.* New York: McGraw-Hill Book Co., 1969.

Problems

5.1 A well is producing 40 °API oil at 200 stb/d and no gas. If the beam size is 1 in., pipe size is 2 in., temperature is 100 °F, estimate pressure drop across a nozzle-type choke.

5.2 A well is producing at 200 stb/d of liquid along with a 900 scf/stb of gas. If the beam size is ½ in., assuming sonic flow, calculate the flowing wellhead pressure using Gilbert's formula.

5.3 A 0.65 specific gravity gas flows from a 2-in. pipe through a 1-in. orifice-type choke. The upstream pressure and temperature are 850 psia and 85 °F, respectively. The downstream pressure is 210 psia (measured 2 ft from the orifice). The gas-specific heat ratio is 1.3. (a) What is the expected daily flow rate? (b) Does heating need to be applied to ensure that the frost does not clog the orifice? (c) What is the expected pressure at the orifice outlet?

5.4 A 0.70 specific gravity natural gas flows from a 2-in. pipe through a 1.5-in. nozzle-type choke. The upstream pressure and temperature are 120 psia and 75 °F, respectively. The downstream pressure is 90 psia (measured 2 ft from the nozzle). The gas-specific heat ratio is 1.25. (a) What is the expected daily flow rate? (b) Is icing a potential problem? (c) What is the expected pressure at the nozzle outlet?

5.5 For the following given data, estimate upstream gas pressure at choke:

Downstream pressure:	350 psia
Choke size:	32 $\frac{1}{64}$ in.
Flowline ID:	2 in.
Gas production rate:	4,000 Mscf/d
Gas-specific gravity:	0.70 1 for air
Gas-specific heat ratio:	1.25
Upstream temperature:	100 °F
Choke discharge coefficient:	0.95

5.6 For the following given data, estimate downstream gas pressure at choke:

Upstream pressure:	620 psia
Choke size:	32 $\frac{1}{64}$ in.
Flowline ID:	2 in.
Gas production rate:	2,200 Mscf/d
Gas-specific gravity:	0.65 1 for air
Gas-specific heat ratio:	1.3
Upstream temperature:	120 °F
Choke discharge coefficient:	0.96

5.7 For the following given data, assuming subsonic flow, estimate liquid and gas production rate:

Choke diameter:	32 $\frac{1}{64}$ in.
Discharge coefficient:	0.85
Downstream pressure:	60 psia
Upstream pressure:	90 psia
Upstream temperature:	120 °F
Downstream temperature:	30 °F
Free gas quality:	0.001 mass fraction
Liquid-specific gravity:	0.85 water = 1
Gas-specific gravity:	0.75 air = 1
Specific heat of gas at constant pressure:	0.24
Specific heat of gas at constant volume:	0.171429
Specific heat of liquid:	0.8

6 Well Deliverability

Contents
6.1 Introduction 6/70
6.2 Nodal Analysis 6/70
6.3 Deliverability of Multilateral Well 6/79
Summary 6/84
References 6/85
Problems 6/85

6.1 Introduction

Well deliverability is determined by the combination of well inflow performance (see Chapter 3) and wellbore flow performance (see Chapter 4). Whereas the former describes the deliverability of the reservoir, the latter presents the resistance to flow of production string. This chapter focuses on prediction of achievable fluid production rates from reservoirs with specified production string characteristics. The technique of analysis is called "Nodal analysis" (a Schlumberger patent). Calculation examples are illustrated with computer spreadsheets that are provided with this book.

6.2 Nodal Analysis

Fluid properties change with the location-dependent pressure and temperature in the oil and gas production system. To simulate the fluid flow in the system, it is necessary to "break" the system into discrete nodes that separate system elements (equipment sections). Fluid properties at the elements are evaluated locally. The system analysis for determination of fluid production rate and pressure at a specified node is called "Nodal analysis" in petroleum engineering. Nodal analysis is performed on the principle of pressure continuity, that is, there is only one unique pressure value at a given node regardless of whether the pressure is evaluated from the performance of upstream equipment or downstream equipment. The performance curve (pressure–rate relation) of upstream equipment is called "inflow performance curve"; the performance curve of downstream equipment is called "outflow performance curve." The intersection of the two performance curves defines the operating point, that is, operating flow rate and pressure, at the specified node. For the convenience of using pressure data measured normally at either the bottom-hole or the wellhead, Nodal analysis is usually conducted using the bottom-hole or wellhead as the solution node. This chapter illustrates the principle of Nodal analysis with simplified tubing string geometries (i.e., single-diameter tubing strings).

6.2.1 Analysis with the Bottom-Hole Node

When the bottom-hole is used as a solution node in Nodal analysis, the inflow performance is the well inflow performance relationship (IPR) and the outflow performance is the tubing performance relationship (TPR), if the tubing shoe is set to the top of the pay zone. Well IPR can be established with different methods presented in Chapter 3. TPR can be modeled with various approaches as discussed in Chapter 4.

Traditionally, Nodal analysis at the bottom-hole is carried out by plotting the IPR and TPR curves and graphically finding the solution at the intersection point of the two curves. With modern computer technologies, the solution can be computed quickly without plotting the curves, although the curves are still plotted for visual verification.

6.2.1.1 Gas Well
Consider the bottom-hole node of a gas well. If the IPR of the well is defined by

$$q_{sc} = C(\bar{p}^2 - p_{wf}^2)^n, \quad (6.1)$$

and if the outflow performance relationship of the node (i.e., the TPR) is defined by

$$p_{wf}^2 = Exp(s)p_{hf}^2 + \frac{6.67 \times 10^{-4}[Exp(s) - 1]f_M q_{sc}^2 \bar{z}^2 \bar{T}^2}{d_i^5 \cos\theta}, \quad (6.2)$$

then the operating flow rate q_{sc} and pressure p_{wf} at the bottom-hole node can be determined graphically by plotting Eqs. (6.1) and (6.2) and finding the intersection point.

The operating point can also be solved analytically by combining Eqs. (6.1) and (6.2). In fact, Eq. (6.1) can be rearranged as

$$p_{wf}^2 = \bar{p}^2 - \left(\frac{q_{sc}}{C}\right)^{\frac{1}{n}}. \quad (6.3)$$

Substituting Eq. (6.3) into Eq. (6.2) yields

$$\bar{p}^2 - \left(\frac{q_{sc}}{C}\right)^{\frac{1}{n}} - Exp(s)p_{hf}^2$$
$$- \frac{6.67 \times 10^{-4}[Exp(s) - 1]f_M q_{sc}^2 \bar{z}^2 \bar{T}^2}{D_i^5 \cos\theta} = 0, \quad (6.4)$$

which can be solved with a numerical technique such as the Newton–Raphson iteration for gas flow rate q_{sc}. This computation can be performed automatically with the spreadsheet program *BottomHoleNodalGas.xls*.

Example Problem 6.1 Suppose that a vertical well produces 0.71 specific gravity gas through a $2\frac{7}{8}$-in. tubing set to the top of a gas reservoir at a depth of 10,000 ft. At tubing head, the pressure is 800 psia and the temperature is 150°F, whereas the bottom-hole temperature is 200°F. The relative roughness of tubing is about 0.0006. Calculate the expected gas production rate of the well using the following data for IPR:

Reservoir pressure: 2,000 psia
IPR model parameter C: 0.1 Mscf/d-psi^{2n}
IPR model parameter n: 0.8

Solution Example Problem 6.1 is solved with the spreadsheet program *BottomHoleNodalGas.xls*. Table 6.1 shows the appearance of the spreadsheet for the Input data and Result sections. It indicates that the expected gas flow rate is 1478 Mscf/d at a bottom-hole pressure of 1059 psia. The inflow and outflow performance curves plotted in Fig. 6.1 confirm this operating point.

6.2.1.2 Oil Well
Consider the bottom-hole node of an oil well. As discussed in Chapter 3, depending on reservoir pressure range, different IPR models can be used. For instance, if the reservoir pressure is above the bubble-point pressure, a straight-line IPR can be used:

$$q = J^*(\bar{p} - p_{wf}) \quad (6.5)$$

The outflow performance relationship of the node (i.e., the TPR) can be described by a different model. The simplest model would be Poettmann–Carpenter model defined by Eq. (4.8), that is,

$$p_{wf} = p_{wh} + \left(\bar{\rho} + \frac{\bar{k}}{\bar{\rho}}\right)\frac{L}{144} \quad (6.6)$$

where p_{wh} and L are tubing head pressure and well depth, respectively, then the operating flow rate q and pressure p_{wf} at the bottom-hole node can be determined graphically by plotting Eqs. (6.5) and (6.6) and finding the intersection point.

The operating point can also be solved analytically by combining Eqs. (6.5) and (6.6). In fact, substituting Eq. (6.6) into Eq. (6.5) yields

$$q = J^*\left[\bar{p} - p_{wh} + \left(\bar{\rho} + \frac{\bar{k}}{\bar{\rho}}\right)\frac{L}{144}\right], \quad (6.7)$$

which can be solved with a numerical technique such as the Newton–Raphson iteration for liquid flow rate q. This computation can be performed automatically with the spreadsheet program *BottomHoleNodalOil-PC.xls*.

Table 6.1 Result Given by BottomHoleNodalGas.xls for Example Problem 6.1

BottomHoleNodalGas.xls
Description: This spreadsheet calculates gas well deliverability with bottom-hole node.
Instructions: (1) Input your data in the Input data section; (2) click Solution button; (3) view results in table and in graph sheet "Plot."

Input data

Gas-specific gravity (γ_g):	0.71
Tubing inside diameter (D):	2.259 in.
Tubing relative roughness (e/D):	0.0006
Measured depth at tubing shoe (L):	10,000 ft
Inclination angle (Θ):	0 degrees
Wellhead pressure (p_{hf}):	800 psia
Wellhead temperature (T_{hf}):	150 °F
Bottom-hole temperature (T_{wf}):	200 °F
Reservoir pressure ($p \sim$):	2000 psia
C-constant in back-pressure IPR model:	0.01 Mscf/d-psi^{2n}
n-exponent in back-pressure IPR model:	0.8

Solution

$T_{av} =$	635 °R
$Z_{av} =$	0.8626
$s =$	0.486062358
$e^s =$	1.62590138
$f_M =$	0.017396984
$AOF =$	1912.705 Mscf/d

q_{sc} (Mscf/d)	IPR	TPR
0	2,000	1,020
191	1,943	1,021
383	1,861	1,023
574	1,764	1,026
765	1,652	1,031
956	1,523	1,037
1,148	1,374	1,044
1,339	1,200	1,052
1,530	987	1,062
1,721	703	1,073
1,817	498	1,078
1,865	353	1,081
1,889	250	1,083
1,913	0	1,084
Operating flow rate	= 1,470 Mscf/d	
Residual of objective function	= −0.000940747	
Operating pressure	= 1,059 psia	

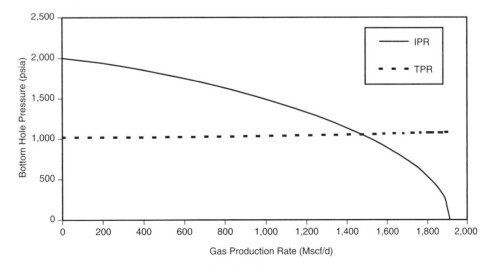

Figure 6.1 Nodal analysis for Example Problem 6.1.

Example Problem 6.2 For the data given in the following table, predict the operating point:

Reservoir pressure:	3,000 psia
Tubing ID:	1.66 in.
Wellhead pressure:	500 psia
Productivity index above bubble point:	1 stb/d-psi
Producing gas–liquid ratio (GLR):	1,000 scf/stb
Water cut (WC):	25 %
Oil gravity:	30 °API
Water-specific gravity:	1.05 1 for fresh-water
Gas-specific gravity:	0.65 1 for air
N_2 content in gas:	0 mole fraction
CO_2 content in gas:	0 mole fraction
H_2S content in gas:	0 mole fraction
Formation volume factor of oil:	1.2 rb/stb
Wellhead temperature:	100 °F
Tubing shoe depth:	5,000 ft
Bottom-hole temperature:	150 °F

Solution Example Problem 6.2 is solved with the spreadsheet program *BottomHoleNodalOil-PC.xls*. Table 6.2 shows the appearance of the spreadsheet for the Input data and Result sections. It indicates that the expected oil flow rate is 1127 stb/d at a bottom-hole pressure of 1,873 psia.

If the reservoir pressure is below the bubble-point pressure, Vogel's IPR can be used

$$q = q_{max}\left[1 - 0.2\left(\frac{p_{wf}}{\bar{p}}\right) - 0.8\left(\frac{p_{wf}}{\bar{p}}\right)^2\right] \quad (6.8)$$

or

$$p_{wf} = 0.125p_b\left[\sqrt{81 - 80\left(\frac{q}{q_{max}}\right)} - 1\right] \quad (6.9)$$

If the outflow performance relationship of the node (i.e., the TPR) is described by the Guo–Ghalambor model defined by Eq. (4.18), that is,

Table 6.2 Result Given by BottomHoleNodalOil-PC.xls for Example Problem 6.2

BottomHoleNodalOil-PC.xls
Description: This spreadsheet calculates the operating point using the Poettmann–Carpenter method with bottom-hole node.
Instruction: (1) Select a unit system; (2) update parameter values in the Input data section; (3) click Solution button; and (4) view result in the Solution section.

Input data	U.S. Field units	SI units
Reservoir pressure:	3,000 psia	
Tubing ID:	1.66 in.	
Wellhead pressure:	500 psia	
Productivity index above bubble point:	1 stb/d-psi	
Producing gas–liquid ratio (GLR):	1,000 scf/stb	
Water cut:	25 %	
Oil gravity:	30 °API	
Water-specific gravity:	1.05, 1 for water	
Gas-specific gravity:	0.65, 1 for air	
N_2 content in gas:	0 mole fraction	
CO_2 content in gas:	0 mole fraction	
H_2S content in gas:	0 mole fraction	
Formation volume factor of oil:	1.2 rb/stb	
Wellhead temperature:	100 °F	
Tubing shoe depth:	5,000 ft	
Bottom-hole temperature:	150 °F	
Solution		
Oil-specific gravity	= 0.88, 1 for water	
Mass associated with 1 stb of oil	= 495.66 lb	
Solution–gas ratio at wellhead	= 78.42 scf/stb	
Oil formation volume factor at wellhead	= 1.04 rb/stb	
Volume associated with 1 stb of oil at wellhead	= 45.12 cf	
Fluid density at wellhead	= 10.99 lb/cf	
Solution gas–oil ratio at bottom-hole	= 339.39 scf/stb	
Oil formation volume factor at bottom-hole	= 1.18 rb/stb	
Volume associated with 1 stb of oil at bottom-hole	= 16.56 cf	
Fluid density at bottom-hole	= 29.94 lb/cf	
The average fluid density	= 20.46 lb/cf	
Inertial force ($D\rho v$)	= 44.63 lb/day-ft	
Friction factor	= 0.0084	
Friction term	= 390.50 $(lb/cf)^2$	
Error in liquid rate	= 0.00 stb/d	
Bottom-hole pressure	= 1,873 psia	
Liquid production rate:	1,127 stb/d	

$$144b(p_{wf} - p_{hf}) + \frac{1 - 2bM}{2}$$

$$\ln \left| \frac{(144p_{wf} + M)^2 + N}{(144p_{hf} + M)^2 + N} \right| - \frac{M + \frac{b}{c}N - bM^2}{\sqrt{N}}$$

$$\left[\tan^{-1} \left(\frac{144p_{wf} + M}{\sqrt{N}} \right) - \tan^{-1} \left(\frac{144p_{hf} + M}{\sqrt{N}} \right) \right]$$

$$= a(\cos\theta + d^2 e)L, \qquad (6.10)$$

substituting Eq. (6.9) into Eq. (6.10) will give an equation to solve for liquid production rate q. The equation can be solved with a numerical technique such as the Newton–Raphson iteration. This computation is performed automatically with the spreadsheet program *BottomHoleNodalOil-GG.xls*.

Example Problem 6.3 For the data given in the following table, predict the operating point:

Reservoir pressure:	3,000 psia
Total measured depth:	7,000 ft
Average inclination angle:	20 degree
Tubing ID:	1.995 in.
Gas production rate:	1,000,000 scfd
Gas-specific gravity:	0.7 air = 1
Oil-specific gravity:	0.85 H$_2$O = 1
Water cut:	30 %
Water-specific gravity:	1.05 H$_2$O = 1
Solid production rate:	1 ft^3/d
Solid-specific gravity:	2.65 H$_2$O = 1
Tubing head temperature:	100 °F
Bottom-hole temperature:	160 °F
Tubing head pressure:	300 psia
Absolute open flow (AOF):	2,000 bbl/d

Solution Example Problem 6.3 is solved with the spreadsheet program *BottomHoleNodalOil-GG.xls*. Table 6.3 shows the appearance of the spreadsheet for the Input data and Result sections. It indicates that the expected oil flow rate is 1,268 stb/d at a bottom-hole pressure of 1,688 psia.

If the reservoir pressure is above the bubble-point pressure, but the flowing bottom-hole pressure is in the range of below bubble-point pressure, the generalized Vogel's IPR can be used:

$$q = q_b + q_v \left[1 - 0.2 \left(\frac{p_{wf}}{p_b} \right) - 0.8 \left(\frac{p_{wf}}{p_b} \right)^2 \right] \qquad (6.11)$$

If the outflow performance relationship of the node (i.e., TPR) is described by Hagedorn-Brown correlation, Eq. (4.27) can be used for generating the TPR curve. Combining Eqs. (6.11) and (4.27) can be solved with a numerical technique such as the Newton–Raphson iteration for liquid flow rate

Table 6.3 Result Given by BottomHoleNodalOil-GG.xls for Example Problem 6.2

BottomHoleNodalOil-GG.xls
Description: This spreadsheet calculates flowing bottom-hole pressure based on tubing head pressure and tubing flow performance using the Guo–Ghalambor method.
Instruction: (1) Select a unit system; (2) update parameter values in the Input data section; (3) click Result button; and (4) view result in the Result section.

Input data	U.S. Field units	SI units
Reservoir pressure:	3,000 psia	
Total measured depth:	7,000 ft	
Average inclination angle:	20 degrees	
Tubing ID:	1.995 in.	
Gas production rate:	1,000,000 scfd	
Gas-specific gravity:	0.7 air = 1	
Oil-specific gravity:	0.85 H$_2$O = 1	
Water cut:	30%	
Water-specific gravity:	1.05 H$_2$O = 1	
Solid production rate:	1 ft^3/d	
Solid-specific gravity:	2.65 H$_2$O = 1	
Tubing head temperature:	100 °F	
Bottom-hole temperature:	160 °F	
Tubing head pressure:	300 psia	
Absolute open flow (AOF):	2000 bbl/d	
Solution		
A	= 3.1243196 in.^{2}	
D	= 0.16625 ft	
T_{av}	= 622 °R	
$\cos(\theta)$	= 0.9397014	
$(D\rho v)$	= 40.576594	
f_M	= 0.0424064	
a	= 0.0001699	
b	= 2.814E-06	
c	= 1,349,785.1	
d	= 3.7998147	
e	= 0.0042189	
M	= 20,395.996	
N	= 6.829E+09	
Liquid production rate, q	= 1,268 bbl/d	
Bottom hole pressure, p_{wf}	= 1,688 psia	

q. This computation can be performed automatically with the spreadsheet program *BottomHoleNodalOil-HB.xls*.

Example Problem 6.4 For the data given in the following table, predict the operating point:

Depth:	9,850 ft
Tubing inner diameter:	1.995 in.
Oil gravity:	45 °API
Oil viscosity:	2 cp
Production GLR:	500 scf/bbl
Gas-specific gravity:	0.7 air = 1
Flowing tubing head pressure:	450 psia
Flowing tubing head temperature:	80 °F
Flowing temperature at tubing shoe:	180 °F
Water cut:	10%
Reservoir pressure:	5,000 psia
Bubble-point pressure:	4,000 psia
Productivity index above bubble point:	1.5 stb/d-psi

Solution Example Problem 6.4 is solved with the spreadsheet program *BottomHoleNodalOil-HB.xls*. Table 6.4 shows the appearance of the spreadsheet for the Input data and Result sections. Figure 6.2 indicates that the expected gas flow rate is 2200 stb/d at a bottom-hole pressure of 3500 psia.

6.2.2 Analysis with Wellhead Node

When the wellhead is used as a solution node in Nodal analysis, the inflow performance curve is the "wellhead performance relationship" (WPR), which is obtained by transforming the IPR to wellhead through the TPR. The outflow performance curve is the wellhead choke performance relationship (CPR). Some TPR models are presented in Chapter 4. CPR models are discussed in Chapter 5.

Nodal analysis with wellhead being a solution node is carried out by plotting the WPR and CPR curves and finding the solution at the intersection point of the two curves. Again, with modern computer technologies, the solution can be computed quickly without plotting the curves, although the curves are still plotted for verification.

6.2.2.1 Gas Well

If the IPR of a well is defined by Eq. (6.1) and the TPR is represented by Eq. (6.2), substituting Eq. (6.2) into Eq. (6.1) gives

$$q_{sc} = C \left[\bar{p}^2 - \left(Exp(s) p_{hf}^2 \right. \right.$$
$$\left. \left. + \frac{6.67 \times 10^{-4} [Exp(s) - 1] f_M q_{sc}^2 \bar{z}^2 \bar{T}^2}{d_i^5 \cos\theta} \right) \right]^n, \quad (6.12)$$

which defines a relationship between wellhead pressure p_{hf} and gas production rate q_{sc}, that is, WPR. If the CPR is defined by Eq. (5.8), that is,

$$q_{sc} = 879 C A p_{hf} \sqrt{\left(\frac{k}{\gamma_g T_{up}}\right)\left(\frac{2}{k+1}\right)^{\frac{k+1}{k-1}}}, \quad (6.13)$$

Table 6.4 Solution Given by BottomHoleNodalOil-HB.xls

BottomHoleNodalOil-HB.xls
Description: This spreadsheet calculates operating point using the Hagedorn–Brown correlation.
Instruction: (1) Select a unit system; (2) update parameter values in the Input data section; (3) click Solution button; and (4) view result in the Result section and charts.

Input data	U.S. Field units	SI units
Depth (D):	9,850 ft	
Tubing inner diameter (d_{ti}):	1.995 in.	
Oil gravity (API):	45 °API	
Oil viscosity (μ_o):	2 cp	
Production GLR (GLR):	500 scf/bbl	
Gas-specific gravity (γ_g):	0.7 air = 1	
Flowing tubing head pressure (p_{hf}):	450 psia	
Flowing tubing head temperature (t_{hf}):	80 °F	
Flowing temperature at tubing shoe (t_{wf}):	180 °F	
Water cut:	10%	
Reservoir pressure (p_e):	5,000 psia	
Bubble-point pressure (p_b):	4,000 psia	
Productivity index above bubble point (J^*):	1.5 stb/d-psi	

Solution

US Field units:

q_b	= 1,500		
q_{max}	= 4,833		
	q (stb/d)	p_{wf} (psia)	
		IPR	TPR
	0	4,908	
	537	4,602	2,265
	1,074	4,276	2,675
	1,611	3,925	3,061
	2,148	3,545	3,464
	2,685	3,125	3,896
	3,222	2,649	4,361
	3,759	2,087	4,861
	4,296	1,363	5,397
	4,833	0	5,969

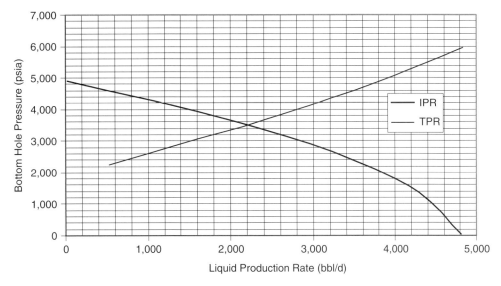

Figure 6.2 Nodal analysis for Example Problem 6.4.

then the operating flow rate q_{sc} and pressure p_{hf} at the wellhead node can be determined graphically by plotting Eqs. (6.12) and (6.13) and finding the intersection point.

The operating point can also be solved numerically by combining Eqs. (6.12) and (6.13). In fact, Eq. (6.13) can be rearranged as

$$p_{hf} = \frac{q_{sc}}{879CA\sqrt{\left(\frac{k}{\gamma_g T_{up}}\right)\left(\frac{2}{k+1}\right)^{\frac{k+1}{k-1}}}}. \quad (6.14)$$

Substituting Eq. (6.14) into Eq. (6.12) gives

$$q_{sc} = C\left[\bar{p}^2 - \left(Exp(s)\left(\frac{q_{sc}}{879CA\sqrt{\left(\frac{k}{\gamma_g T_{up}}\right)\left(\frac{2}{k+1}\right)^{\frac{k+1}{k-1}}}}\right)^2 \right.\right.$$
$$\left.\left. + \frac{6.67 \times 10^{-4}[Exp(s)-1]f_M q_{sc}^2 \bar{z}^2 \bar{T}^2}{d_i^5 \cos\theta}\right)\right]^n, \quad (6.15)$$

which can be solved numerically for gas flow rate q_{sc}. This computation can be performed automatically with the spreadsheet program *WellheadNodalGas-SonicFlow.xls*.

Example Problem 6.5 Use the data given in the following table to estimate gas production rate of a gas well:

Gas-specific gravity:	0.71
Tubing inside diameter:	2.259 in.
Tubing wall relative roughness:	0.0006
Measured depth at tubing shoe:	10,000 ft
Inclination angle:	0 degrees
Wellhead choke size:	16 $1/64$ 4 in.
Flowline diameter:	2 in.
Gas-specific heat ratio:	1.3
Gas viscosity at wellhead:	0.01 cp
Wellhead temperature:	150 °F
Bottom-hole temperature:	200 °F
Reservoir pressure:	2,000 psia
C-constant in IPR model:	0.01 Mscf/ d-psi^{2n}
n-exponent in IPR model:	0.8

Solution Example Problem 6.5 is solved with the spreadsheet program *WellheadNodalGas-SonicFlow.xls*. Table 6.5 shows the appearance of the spreadsheet for the Input data and Result sections. It indicates that the expected gas flow rate is 1,478 Mscf/d at a bottom-hole pressure of 1,050 psia. The inflow and outflow performance curves plotted in Fig. 6.3 confirm this operating point.

6.2.2.2 Oil Well

As discussed in Chapter 3, depending on reservoir pressure range, different IPR models can be used. For instance, if the reservoir pressure is above the bubble-point pressure, a straight-line IPR can be used:

$$q = J^*(\bar{p} - p_{wf}) \quad (6.16)$$

If the TPR is described by the Poettmann–Carpenter model defined by Eq. (4.8), that is,

$$p_{wf} = p_{wh} + \left(\bar{\rho} + \frac{\bar{k}}{\bar{\rho}}\right)\frac{L}{144} \quad (6.17)$$

substituting Eq. (6.17) into Eq. (6.16) gives

$$q = J^*\left[\bar{p} - \left(p_{wh} + \left(\bar{\rho} + \frac{\bar{k}}{\bar{\rho}}\right)\frac{L}{144}\right)\right], \quad (6.18)$$

which describes inflow for the wellhead node and is called the WPR. If the CPR is given by Eq. (5.12), that is,

$$p_{wh} = \frac{CR^m q}{S^n}, \quad (6.19)$$

the operating point can be solved analytically by combining Eqs. (6.18) and (6.19). In fact, substituting Eq. (6.19) into Eq. (6.18) yields

$$q = J^*\left[\bar{p} - \left(\frac{CR^m q}{S^n} + \left(\bar{\rho} + \frac{\bar{k}}{\bar{\rho}}\right)\frac{L}{144}\right)\right], \quad (6.20)$$

which can be solved with a numerical technique. Because the solution procedure involves loop-in-loop iterations, it cannot be solved in MS Excel in an easy manner. A special computer program is required. Therefore, a computer-assisted graphical solution method is used in this text.

The operating flow rate q and pressure p_{wh} at the wellhead node can be determined graphically by plotting Eqs. (6.18) and (6.19) and finding the intersection point. This computation can be performed automatically with the spreadsheet program *WellheadNodalOil-PC.xls*.

Table 6.5 Solution Given by WellheadNodalGas-SonicFlow.xls

WellheadNodalGas-SonicFlow.xls
Description: This spreadsheet calculates well deliverability with wellhead node.
Instructions:
Step 1: Input your data in the Input data section.
Step 2: Click Solution button to get results.
Step 3: View results in table and in the plot graph sheet.

Input data

Gas-specific gravity (γ_g):	0.71
Tubing inside diameter (D):	2.259 in.
Tubing relative roughness (ε/D):	0.0006
Measured depth at tubing shoe (L):	10,000 ft
Inclination angle (θ):	0 degrees
Wellhead choke size (D_{ck}):	16 1/64 in.
Flowline diameter (D_{fl}):	2 in.
Gas-specific heat ratio (k):	1.3
Gas viscosity at wellhead (μ_g):	0.01 cp
Wellhead temperature (T_{hf}):	120 °F
Bottom-hole temperature (T_{wf}):	180 °F
Reservoir pressure ($p \sim$):	2,000 psia
C-constant in back-pressure IPR model:	0.01 Mscf/d-psi^{2n}
n-exponent in back-pressure IPR model:	0.8

Solution

T_{av}	= 610 °R
Z_{av}	= 0.8786
s	= 0.4968
e^s	= 1.6434
f_m	= 0.0174
AOF	= 1,913 Mscf/d
D_{ck}/D_{fl}	= 0.125
Re	= 8,348,517
C_{ck}	= 1.3009 in.^{2}
A_{ck}	= 0.0490625

q_{sc} (Mscf/d)	WPR	CPR
0	1,600	0
191	1,554	104
383	1,489	207
574	1,411	311
765	1,321	415
956	1,218	518
1,148	1,099	622
1,339	960	726
1,530	789	830
1,721	562	933
1,817	399	985
1,865	282	1,011
1,889	200	1,024
1,913	1	1,037
Operating flow rate =	1,470 Mscf/d	
Operating pressure =	797 psia	

Example Problem 6.6 Use the following data to estimate the liquid production rate of an oil well:

Reservoir pressure:	6,000 psia
Tubing ID:	3.5 in.
Choke size:	64 1/64 in.
Productivity index above bubble point:	1 stb/d-psi
Producing gas–liquid ratio (GLR):	1000 scf/stb
Water cut:	25%
Oil gravity:	30 °API
Water-specific gravity:	1.05 1 for freshwater
Gas-specific gravity:	0.65 1 for air
Choke constant:	10
Choke GLR exponent:	0.546
Choke-size exponent:	1.89
Formation volume factor of oil:	1 rb/stb
Wellhead temperature:	100 °F
Tubing shoe depth:	12,000 ft
Bottom-hole temperature:	150 °F

Solution Example Problem 6.6 is solved with the spreadsheet program *WellheadNodalOil-PC.xls*. Table 6.6 shows the appearance of the spreadsheet for the Input data and Result sections. The inflow and outflow performance curves are plotted in Fig. 6.4, which indicates that the expected oil flow rate is 3280 stb/d at a wellhead pressure of 550 psia.

If the reservoir pressure is below the bubble-point pressure, Vogel's IPR can be rearranged to be

$$p_{wf} = 0.125\bar{p}\left[\sqrt{81 - 80\left(\frac{q}{q_{max}}\right)} - 1\right] \quad (6.21)$$

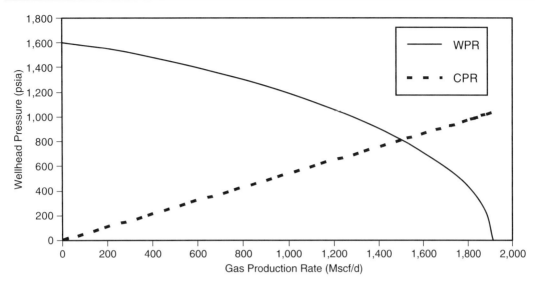

Figure 6.3 Nodal analysis for Example Problem 6.5.

If the TPR is described by the Guo–Ghalambor model defined by Eq. (4.18), that is,

$$144b(p_{wf} - p_{hf}) + \frac{1-2bM}{2}$$

$$\ln\left|\frac{(144p_{wf}+M)^2+N}{(144p_{hf}+M)^2+N}\right| - \frac{M+\frac{b}{c}N-bM^2}{\sqrt{N}}$$

$$\left[\tan^{-1}\left(\frac{144p_{wf}+M}{\sqrt{N}}\right) - \tan^{-1}\left(\frac{144p_{hf}+M}{\sqrt{N}}\right)\right]$$

$$= a(\cos\theta + d^2 e)L, \quad (6.22)$$

and the CPR is given by Eq. (5.12), that is,

$$p_{hf} = \frac{CR^m q}{S^n}, \quad (6.23)$$

solving Eqs. (6.21), (6.22), and (6.23) simultaneously will give production rate q and wellhead pressure p_{hf}. The solution procedure has been coded in the spreadsheet program *WellheadNodalOil-GG.xls*.

Example Problem 6.7 Use the following data to estimate the liquid production rate of an oil well:

Choke size:	64 1/64 in.
Reservoir pressure:	3,000 psia
Total measured depth:	7,000 ft
Average inclination angle:	20 degrees
Tubing ID:	1.995 in.
Gas production rate:	1,000,000 scfd
Gas-specific gravity:	0.7 air = 1
Oil-specific gravity:	0.85 H_2O = 1
Water cut:	30%
Water specific gravity:	1.05 H_2O = 1
Solid production rate:	1 ft^3/d
Solid-specific gravity:	2.65 H_2O = 1
Tubing head temperature:	100 °F
Bottom-hole temperature:	160 °F
Absolute open flow (AOF):	2,000 bbl/d
Choke flow constant:	10
Choke GLR exponent:	0.546
Choke-size exponent:	1.89

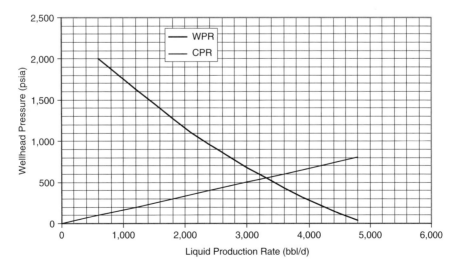

Figure 6.4 Nodal analysis for Example Problem 6.6.

Table 6.6 Solution Given by WellheadNodalOil-PC.xls

WellheadNodalOil-PC.xls
Description: This spreadsheet calculates operating point using the Poettmann–Carpenter method with wellhead node.
Instruction: (1) Select a unit system; (2) update parameter values in the Input data section; (3) click Solution button; and (4) view result in the Solution section and charts.

Input data	U.S. Field Units	SI Units
Reservoir pressure:	6,000 psia	
Tubing ID:	3.5 in.	
Choke size:	64 1/64 in.	
Productivity index above bubble point:	1 stb/d-psi	
Producing gas–liquid ratio:	1,000 scf/stb	
Water cut:	25%	
Oil gravity:	30 °API	
Water-specific gravity:	1.05 1 for freshwater	
Gas-specific gravity:	0.65 1 for air	
Choke constant:	10	
Choke gas–liquid ratio exponent:	0.546	
Choke-size exponent:	1.89	
Formation volume factor for water:	1 rb/stb	
Wellhead temperature:	100 °F	
Tubing shoe depth:	12,000 ft	
Bottom-hole temperature:	150 °F	

Solution:

q (stb/d)	p_{wf} (psia)	p_{wh} (psia) WPR	p_{wh} (psia) CPR
0	6,000		0
600	5,400	2,003	101
1,200	4,800	1,630	201
1,800	4,200	1,277	302
2,400	3,600	957	402
3,000	3,000	674	503
3,600	2,400	429	603
4,200	1,800	220	704
4,800	1,200	39	805

Solution Example Problem 6.7 is solved with the spreadsheet program *WellheadNodalOil-GG.xls*. Table 6.7 shows the appearance of the spreadsheet for the *Data Input* and *Result* sections. It indicates that the expected oil flow rate is 1,289 stb/d at a wellhead pressure of 188 psia.

If the reservoir pressure is above the bubble-point pressure, but the flowing bottom-hole pressure is in the range of below bubble-point pressure, the generalized Vogel's IPR can be used:

$$q = q_b + q_v \left[1 - 0.2\left(\frac{p_{wf}}{p_b}\right) - 0.8\left(\frac{p_{wf}}{p_b}\right)^2\right] \quad (6.24)$$

Hagedorn–Brown correlation, Eq. (4.27), can be used for translating the IPR to the WPR. Again, if the CPR is given by Eq. (5.12), that is,

$$p_{hf} = \frac{CR^m q}{S^n}, \quad (6.25)$$

solving Eqs. (6.24), (4.27), and (6.25) simultaneously will give production rate q and wellhead pressure p_{hf}. Because the solution procedure involves loop-in-loop iterations, it cannot be solved in MS Excel in an easy manner. A special computer program is required. Therefore, a computer-assisted graphical solution method is used in this text.

The operating flow rate q and pressure p_{hf} at the wellhead node can be determined graphically. This computation can be performed automatically with the spreadsheet program *WellheadNodalOil-HB.xls*.

Example Problem 6.8 For the following data, predict the operating point:

Depth:	7,000 ft
Tubing inner diameter:	3.5 in.
Oil gravity:	45 °API
Oil viscosity:	0.5 cp
Production gas–liquid ratio (GLR):	500 scf/bbl
Gas-specific gravity:	0.7 air = 1
Choke size:	32 1/64 in.
Flowing tubing head temperature:	80 °F
Flowing temperature at tubing shoe:	150 °F
Water cut:	10 %
Reservoir pressure:	4,000 psia
Bubble-point pressure:	3,800 psia
Productivity index above bubble point:	5 stb/d-psi
Choke flow constant:	10.00
Choke GLR exponent:	0.546
Choke-size exponent:	1.89

Solution Example Problem 6.8 is solved with the spreadsheet program *WellheadNodalOil-HB.xls*. Table 6.8 shows the appearance of the spreadsheet for the Input data and Result sections. Figure 6.5 indicates that the expected oil flow rate is 4,200 stb/d at a wellhead pressure of 1,800 psia.

Table 6.7 Solution Given by WellheadNodalOil-GG.xls

WellheadNodalOil-GG.xls
Description: This spreadsheet calculates operating point based on CPR and Guo–Ghalambor TPR.
Instruction: (1) Select a unit system; (2) update parameter values in the Input data section; (3) click Solution button; and (4) view result in the Solution section.

Input data	U.S. Field units	SI units
Choke size.	64 1/64 in.	
Reservoir pressure:	3,000 psia	
Total measured depth:	7,000 ft	
Average inclination angle:	20 degrees	
Tubing ID:	1.995 in.	
Gas production rate:	1,000,000 scfd	
Gas-specific gravity:	0.7 air = 1	
Oil-specific gravity:	0.85 $H_2O = 1$	
Water cut:	30%	
Water-specific gravity:	1.05 $H_2O = 1$	
Solid production rate:	1 ft^3/d	
Solid-specific gravity:	2.65 $H_2O = 1$	
Tubing head temperature:	100 °F	
Bottom-hole temperature:	160 °F	
Absolute open flow (AOF):	2,000 bbl/d	
Choke flow constant:	10	
Choke GLR exponent:	0.546	
Choke-size exponent:	1.89	
Solution		
A	= 3.1243196 in.^{2}	
D	= 0.16625 ft	
T_{av}	= 622 °R	
$cos(\theta)$	= 0.9397014	
$(D\rho v)$	= 41.163012	
f_M	= 0.0409121	
a	= 0.0001724	
b	= 2.86E−06	
c	− 1349785.1	
d	= 3.8619968	
e	= 0.0040702	
M	= 20003.24	
N	= 6.591E+09	
Liquid production rate, q	= 1,289 bbl/d	205 m^3/d
Bottom hole pressure, p_{wf}	= 1,659 psia	11.29 MPa
Wellhead pressure, p_{hf}	= 188 psia	1.28 MPa

6.3 Deliverability of Multilateral Well

Following the work of Pernadi et al. (1996), several mathematical models have been proposed to predict the deliverability of multilateral wells. Some of these models are found from Salas et al. (1996), Larsen (1996), and Chen et al. (2000). Some of these models are oversimplified and some others are too complex to use.

Consider a multilateral well trajectory depicted in Fig. 6.6. Nomenclatures are illustrated in Fig. 6.7. Suppose the well has n laterals and each lateral consists of three sections: horizontal, curvic, and vertical. Let L_i, R_i, and H_i denote the length of the horizontal section, radius of curvature of the curvic section, and length of the vertical section of lateral i, respectively. Assuming the pressure losses in the horizontal sections are negligible, pseudo–steady IPR of the laterals can be expressed as follows:

$$q_i = f_{L_i}(p_{wf_i}) \quad i = 1, 2, \ldots, n, \quad (6.26)$$

where

q_i = production rate from lateral i
f_{Li} = inflow performance function of the horizontal section of lateral i
p_{wf_i} = the average flowing bottom-lateral pressure in lateral i.

The fluid flow in the curvic sections can be described by

$$p_{wf_i} = f_{Ri}(p_{kf_i}, q_i) \quad i = 1, 2, \ldots, n, \quad (6.27)$$

where

f_{Ri} = flow performance function of the curvic section of lateral i
p_{kf_i} = flowing pressure at the kick-out-point of lateral i.

The fluid flow in the vertical sections may be described by

$$p_{kf_i} = f_{hi}\left(p_{hf_i}, \sum_{j=1}^{i} q_j\right) \quad i = 1, 2, \ldots, n, \quad (6.28)$$

where

f_{hi} = flow performance function of the vertical section of lateral i
p_{hf_i} = flowing pressure at the top of lateral i.

The following relation holds true at the junction points:

$$p_{kf_i} = p_{hf_{i-1}} \quad i = 1, 2, \ldots, n \quad (6.29)$$

Table 6.8 Solution Given by WellheadNodalOil-HB.xls

WellheadNodalOil-HB.xls
Description: This spreadsheet calculates operating point using Hagedorn–Brown correlation.
Instruction: (1) Select a unit system; (2) update parameter values in the Input data section; (3) click Solution button; and (4) view result in the Solution section and charts.

Input data	U.S. Field units	SI units
Depth (D):	7,000 ft	
Tubing inner diameter (d_{ti}):	3.5 in.	
Oil gravity (API):	45 °API	
Oil viscosity (μ_o):	0.5 cp	
Production gas–liquid ratio:	500 scf/bbl	
Gas-specific gravity (γ_g):	0.7 air = 1	
Choke size (S):	32 1/64 in.	
Flowing tubing head temperature (t_{hf}):	80 °F	
Flowing temperature at tubing shoe (t_{wf}):	150 °F	
Water cut:	10%	
Reservoir pressure (p_e):	4,000 psia	
Bubble-point pressure (p_b):	3,800 psia	
Productivity above bubble point (J^*):	5 stb/d-psi	
Choke flow constant (C):	10.00	
Choke gas–liquid ratio exponent (m):	0.546	
Choke-size exponent (n):	1.89	

Solution

q (stb/d)	p_{wf} (psia)	p_{hf} (psia) WPR	p_{hf} (psia) CPR
0	3,996		0
1,284	3,743	2,726	546
2,568	3,474	2,314	1,093
3,852	3,185	1,908	1,639
5,136	2,872	1,482	2,185
6,420	2,526	1,023	2,732
7,704	2,135	514	3,278
8,988	1,674	0	3,824

Equations (6.26) through (6.29) contain $(4n - 1)$ equations. For a given flowing pressure p_{hf_n} at the top of lateral n, the following $(4n - 1)$ unknowns can be solved from the $(4n - 1)$ equations:

$$q_1, q_2, \ldots q_n$$

$$p_{wf_1}, p_{wf_2}, \ldots p_{wf_n}$$

$$p_{kf_1}, p_{kf_2}, \ldots p_{kf_n}$$

$$p_{hf_1}, p_{hf_2}, \ldots p_{hf_{n-1}}$$

Then the production rate of the multilateral well can be determined by

$$q = \sum_{i=1}^{n} q_i. \qquad (6.30)$$

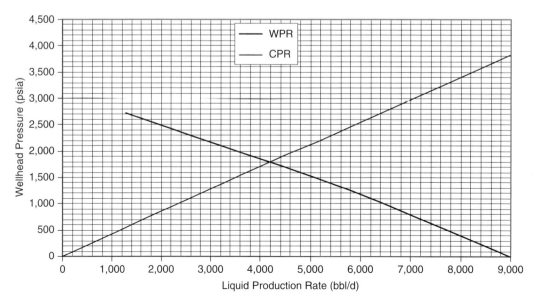

Figure 6.5 Nodal analysis for Example Problem 6.8.

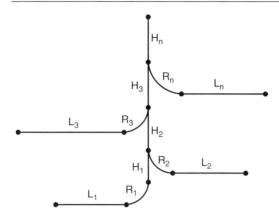

Figure 6.6 Schematic of a multilateral well trajectory.

Thus, the composite IPR,

$$q = f(p_{hf_n}), \qquad (6.31)$$

can be established implicitly.

It should be noted that the composite IPR model described here is general. If the vertical section of the top lateral is the production string (production through tubing or/and casing), then p_{hf_n} will be the flowing wellhead pressure. In this case, the relation expression (Eq. [6.31]) represents the WPR.

6.3.1 Gas well
For gas wells, Eq. (6.26) becomes

$$q_{g_i} = C_i(\bar{p}_i^2 - p_{wf_i}^2)^{n_i}, \qquad (6.32)$$

where

C_i = productivity coefficient of lateral i
n_i = productivity exponent of lateral i

As described in Chapter 4, Eq. (6.27), in U.S. field units (q_{gi} in Mscf/d), can be approximated as (Katz et al., 1959)

$$p_{wf_i}^2 = e^{S_i} p_{kf_i}^2 + \frac{6.67 \times 10^{-4}(e^{S_i} - 1)f_{Mi}q_{gi}^2 \bar{z}_i^2 \overline{T}_i^2}{d_i^5 \cos(45°)}, \qquad (6.33)$$

where

$$S_i = \frac{0.0375\pi \gamma_g R_i \cos(45°)}{2\bar{z}_i \overline{T}}. \qquad (6.34)$$

The friction factor f_{Mi} can be found in the conventional manner for a given tubing diameter, wall roughness, and Reynolds number. However, if one assumes fully turbulent flow, which is the case for most gas wells, then a simple empirical relation may be used for typical tubing strings (Katz and Lee, 1990):

$$f_{Mi} = \frac{0.01750}{d_i^{0.224}} \quad \text{for } d_i \leq 4.277\,\text{in}. \qquad (6.35)$$

$$f_{Mi} = \frac{0.01603}{d_i^{0.164}} \quad \text{for } d_i > 4.277\,\text{in}. \qquad (6.36)$$

Guo (2001) used the following Nikuradse friction factor correlation for fully turbulent flow in rough pipes:

$$f_{Mi} = \left[\frac{1}{1.74 - 2\log\left(\frac{2\varepsilon_i}{d_i}\right)}\right]^2 \qquad (6.37)$$

For gas wells, Eq. (6.28) can be expressed as (Katz et al., 1959)

$$p_{hf_i}^2 = e^{S_i} p_{hf_i}^2$$

$$+ \frac{6.67 \times 10^{-4}(e^{S_i} - 1)f_{Mi}\left(\sum_{j=1}^{i} q_{gi}\right)^2 \bar{z}_i^2 \overline{T}_i^2}{d_i^5}, \qquad (6.38)$$

where

$$S_i = \frac{0.0375\gamma_g H_i}{\bar{z}_i \overline{T}_i}. \qquad (6.39)$$

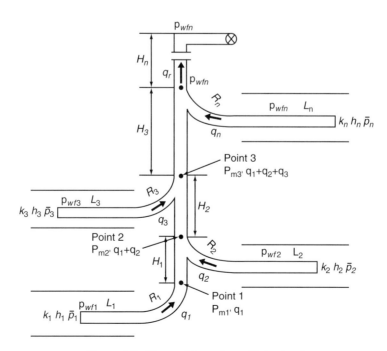

Figure 6.7 Nomenclature of a multilateral well.

At the junction points,

$$p_{kf_i} = p_{hf_{i-1}}. \quad (6.40)$$

Equations (6.32), (6.33), (6.38), and (6.40) contain $(4n-1)$ equations. For a given flowing pressure p_{hf_n} at the top of lateral n, the following $(4n-1)$ unknowns can be solved from the $(4n-1)$ equations:

$$q_{g1}, q_{g2}, \ldots q_{gn}$$

$$p_{wf_1}, p_{wf_2}, \ldots p_{wf_n}$$

$$p_{kf_1}, p_{kf_2}, \ldots p_{kf_n}$$

$$p_{hf_1}, p_{hf_2}, \ldots p_{hf_{n-1}}$$

Then the gas production rate of the multilateral well can be determined by

$$q_g = \sum_{i=1}^{n} q_{gi}. \quad (6.41)$$

Thus, the composite IPR,

$$q_g = f(p_{hf_n}), \quad (6.42)$$

can be established implicitly. The solution procedure has been coded in the spreadsheet program *MultilateralGasWellDeliverability(C-nIPR).xls*. It has been found that the program does not allow cross-flow to be computed because of difficulty of computing roof of negative number with Eq. (6.32). Therefore, another spreadsheet was developed to solve the problem. The second spreadsheet is *MultilateralGasWellDeliverability(Radial-FlowIPR).xls* and it employs the following IPR model for individual laterals:

$$q_g = \frac{k_h h(\bar{p}^2 - p_{wf}^2)}{1424 \mu Z T} \left[\frac{1}{\ln\left(\frac{0.472 r_{eh}}{L/4}\right)} \right] \quad (6.43)$$

Example Problem 6.9 For the data given in the following table, predict gas production rate against 1,000 psia wellhead pressure and 100 °F wellhead temperature:

Solution Example Problem 6.9 is solved with the spreadsheet program *MultilateralGasWellDeliverability(Radial-FlowIPR).xls*. Table 6.9 shows the appearance of the spreadsheet for the Input data and Result sections. It indicates that the expected total gas flow rate is 4,280 Mscf/d from the four laterals. Lateral 3 will steals 6,305 Mscf/d.

6.3.2 Oil well
The inflow performance function for oil wells can be expressed as

$$q_{o_i} = J_i(\bar{p}_i - p_{wf_i}), \quad (6.44)$$

where J_i = productivity index of lateral i.

The fluid flow in the curvic sections can be approximated as

$$p_{wf_i} = p_{kf_i} + \nabla_{R_i} R_i, \quad (6.45)$$

where ∇_{R_i} = vertical pressure gradient in the curvic section of lateral i.

The pressure gradient ∇_{R_i} may be estimated by the Poettmann–Carpenter method:

$$\nabla_{R_i} = \frac{\bar{\rho}_i^2 + \bar{k}_i}{144 \bar{\rho}_i}, \quad (6.46)$$

where

$$\bar{\rho}_i = \frac{M_F}{V_{m_i}}, \quad (6.47)$$

$$M_F = 350.17(\gamma_o + WOR\, \gamma_w) + 0.0765 GOR\, \gamma_g, \quad (6.48)$$

$$V_{m_i} = 5.615(B_o + WOR\, B_w) + \bar{z}(GOR - R_s)$$

$$\times \left(\frac{29.4}{p_{wf_i} + p_{xf_i}}\right)\left(\frac{T_i}{520}\right), \quad (6.49)$$

$$\bar{k}_i = \frac{f_{2F} q_{oi}^2 M_F}{7.4137 \times 10^{10} d_i^5}. \quad (6.50)$$

The fluid flow in the vertical sections may be expressed as

$$p_{kf_i} = p_{hf_i} + \nabla_{h_i} H_i \quad i = 1, 2, \ldots, n, \quad (6.51)$$

where ∇_{h_i} = pressure gradient in the vertical section of lateral i.

Horizontal sections

Lateral no.:	1	2	3	4	
Length of horizontal section (L)	500	600	700	400	ft
Horizontal permeability (k)	1	2	3	4	md
Net pay thickness (h)	20	20	20	20	ft
Reservoir pressure (p-bar)	3700	3500	1,800	2,800	psia
Radius of drainage (r_{eh})	2,000	2,500	1,700	2,100	ft
Gas viscosity (μ_g)	0.02	0.02	0.02	0.02	cp
Wellbore diameter (Di)	8.00	8.00	8.00	8.00	in.
Bottom-hole temperature (T)	270	260	250	230	°F
Gas compressibility factor (z)	0.85	0.90	0.95	0.98	
Gas-specific gravity (γ_g)	0.85	0.83	0.80	0.75	air $= 1$

Curvic sections

Lateral no.:	1	2	3	4	
Radius of curve (R)	250	300	200	270	ft
Average inclination angle (θ)	45	45	45	45	°F
Tubing diameter (di)	3	3	3	3	in.
Pipe roughness (e)	0.0018	0.0018	0.0018	0.0018	in.

Vertical sections

Lateral no.:	1	2	3	4	
Interval length (H)	250	300	200	8,000	ft
Tubing diameter (d_i)	3	3	3	3	in.
Pipe roughness (e)	0.0018	0.0018	0.0018	0.0018	in.

Table 6.9 Solution Given by MultilateralGasWellDeliverability(Radial-FlowIPR).xls

Horizontal sections

Lateral no.:	1	2	3	4	
Length of horizontal section (L)	500	600	700	400	ft
Bottom-hole pressure (p_{wf})	2,701	2,686	2,645	2,625	psia
Horizontal permeability (k)	1	2	3	4	md
Net pay thickness (h)	20	20	20	20	ft
Reservoir pressure (p-bar)	3,700	3,500	1,800	2,800	psia
Radius of drainage (r_{eh})	2,000	2,500	1,700	2,100	ft
Gas viscosity (μ_g)	0.02	0.02	0.02	0.02	cp
Wellbore diameter (D_i)	8.00	8.00	8.00	8.00	in.
Bottom-hole temperature (T)	270	260	250	230	°F
Gas compressibility factor (z)	0.85	0.90	0.95	0.98	
Gas-specific gravity (γ_g)	0.85	0.83	0.80	0.75	air = 1

Curvic sections

Lateral no.:	1	2	3	4	
Radius of curve (R)	250	300	200	270	ft
Average inclination angle (θ)	45	45	45	45	degrees
Tubing diameter (d_i)	3	3	3	3	in.
Pipe roughness (e)	0.0018	0.0018	0.0018	0.0018	in.

Vertical sections

Lateral no.:	1	2	3	4	
Interval length ()	250	300	200	8,000	ft
Tubing diameter (d_i)	3	3	3	3	in.
Pipe roughness (e)	0.0018	0.0018	0.0018	0.0018	in.

Kick-off points

	1	2	3	4	
Flow rate (q)	3,579	8,870	2,564	4,280	Mscf/d
Pressure (p)	2,682	2,665	2,631	2,609	psia
Temperature (T)	265	250	240	230	°F

Total

Production rate (q) =	3,579	5,290	(6,305)	1,716	4,280 Mscf/d

Based on the Poettmann–Carpenter method, the pressure gradient ∇_{h_i} may be estimated by the follow equation:

$$\nabla_{h_i} = \frac{\bar{\rho}_i^2 + \bar{k}_i}{144\bar{\rho}_i}, \quad (6.52)$$

where

$$\bar{\rho}_i = \frac{M_F}{V_{m_i}}, \quad (6.53)$$

$$M_F = 350.17(\gamma_o + WOR\ \gamma_w) + 0.0765 GOR\ \gamma_g, \quad (6.54)$$

$$V_{m_i} = 5.615(B_o + WOR\ B_w) + \bar{z}(GOR\ \ R_s)$$
$$\times \left(\frac{29.4}{p_{xf_i} + p_{hf_i}}\right)\left(\frac{T_i}{520}\right), \quad (6.55)$$

$$\bar{k}_i = \frac{f_{2F}\left(\sum_{j=1}^{i} q_{o_i}\right)^2 M_F}{7.4137 \times 10^{10} d_i^5}. \quad (6.56)$$

Horizontal sections

Lateral no.:	1	2	3	4	
Reservoir pressure (p-bar)	3,700	3,500	3,300	2,800	psia
Oil formation factor (B_o)	1.20	1.15	1.10	1.1	stb/rb
Water formation factor (B_w)	1.00	1.00	1.00	1.00	stb/rb
Bottom-hole temperature (T)	270	260	250	230	°F
Gas compressibility factor (z)	0.85	0.90	0.95	0.98	
Gas-specific gravity (γ_g)	0.85	0.83	0.80	0.75	air = 1
Oil-specific gravity (γ_o)	0.80	0.78	0.87	0.85	water = 1
Water-specific gravity (γ_w)	1.07	1.06	1.05	1.04	water = 1
Water–oil ratio (WOR)	0.10	0.40	0.20	0.30	stb/stb
Gas–oil ratio (GOR)	1000	1,500	2,000	2,500	scf/stb
Solution–gas–oil ratio (R_s)	800	1,200	1,500	2,000	scf/stb
Productivity index (J)	1	0.8	0.7	0.6	stb/d/psi

Curvic sections

Lateral no.:	1	2	3	4	
Radius of curve (R)	200	200	200	200	ft
Average inclination angle (θ)	45	45	45	45	degrees
Tubing diameter (d_i)	5	5	5	5	in.
Pipe roughness (e)	0.0018	0.0018	0.0018	0.0018	in.

Vertical sections

Lateral no.:	1	2	3	4	
Interval length (H)	500	400	300	3,000	ft
Tubing diameter (d_i)	5	5	5	5	in.
Pipe roughness (e)	0.0018	0.0018	0.0018	0.0018	in.

Table 6.10 Data Input and Result Sections of the Spreadsheet MultilateralOilWellDeliverability.xls

MultilateralOilWellDeliverability.xls
Instruction: (1) Update parameter values in the Input data section; (2) click Calculate button; and (3) view result.

Input data

Top node

Pressure (p_{wh})	1,800 psia	
Temperature (T_{wh})	100 °F	Calculate

Horizontal sections

Lateral no.:	1	2	3	4	
Initial guess for p_{wf}	3,249	3,095	2,961	2,865	psia
Reservoir pressure (p-bar)	3,700	3,500	3,300	2,800	psia
Oil formation factor (B_o)	1.20	1.15	1.10	1.1	stb/rb
Water formation factor (B_w)	1.00	1.00	1.00	1.00	stb/rb
Bottom-hole temperature (T)	270	260	250	230	°F
Gas compressibility factor (z)	0.85	0.90	0.95	0.98	
Gas-specific gravity (γ_g)	0.85	0.83	0.80	0.75	air = 1
Oil-specific gravity (γ_o)	0.80	0.78	0.87	0.85	water = 1
Water-specific gravity (γ_w)	1.07	1.06	1.05	1.04	water = 1
Water–oil ratio (WOR)	0.10	0.40	0.20	0.30	stb/stb
Gas-oil ratio (GOR)	1,000	1,500	2,000	2,500	scf/stb
Solution–gas–oil ratio (R_s)	800	1,200	1,500	2,000	scf/stb
Productivity index (J)	1	0.8	0.7	0.6	stb/d/psi

Curvic sections

Lateral no.:	1	2	3	4	
Radius of curve (R)	200	200	200	200	ft
Average inclination angle (θ)	45	45	45	45	°F
Tubing diameter (d_i)	3	3	3	3	in.
Pipe roughness (e)	0.0018	0.0018	0.0018	0.0018	in.

Vertical sections

Lateral no.:	1	2	3	4	
Interval length (H)	500	400	300	3,000	ft
Tubing diameter (d_i)	3	3	3	3	in.
Pipe roughness (e)	0.0018	0.0018	0.0018	0.0018	in.

Kick off points

		1	2	3	4	
Flow rate (q)		451	775	1,012	973	stb/d
Pressure (p)		3,185	3,027	2,895	2,797	psia
Temperature (T)		265	250	240	230	°F
Total:	973	451	451	237	(39)	stb/d

At the junction points,

$$p_{kf_i} = p_{hf_{i-1}}. \quad (6.57)$$

Equations (6.44), (4.45), (6.51), and (6.57) contain $(4n-1)$ equations. For a given flowing pressure p_{hf_n} at the top of lateral n, the following $(4n-1)$ unknowns can be solved from the $(4n-1)$ equations:

$$q_{o_1}, q_{o_2}, \ldots q_{o_n}$$
$$p_{wf_1}, p_{wf_2}, \ldots p_{wf_n}$$
$$p_{kf_1}, p_{kf_2}, \ldots p_{kf_n}$$
$$p_{hf_1}, p_{hf_2}, \ldots p_{hf_{n-1}}$$

Then the oil production rate of the multilateral well can be determined by

$$q_o = \sum_{i=1}^{n} q_{oi}. \quad (6.58)$$

Thus, the composite IPR,

$$q_o = f(p_{hf_n}), \quad (6.59)$$

can be established implicitly. The solution procedure has been coded in spreadsheet program *MultilateralOilWellDeliverability.xls*.

Example Problem 6.10 For the data given in the last page, predict the oil production rate against 1,800 psia wellhead pressure and 100 °F wellhead temperature.

Solution Example Problem 6.10 is solved with the spreadsheet program *MultilateralOilWellDeliverability.xls*. Table 6.10 shows the appearance of the spreadsheet for the data Input and Result sections. It indicates that the expected total oil production rate is 973 stb/d. Lateral 4 would steal 39 stb/d.

Summary

This chapter illustrated the principle of system analysis (Nodal analysis) with simplified well configurations. In the industry, the principle is applied with a piecewise approach to handle local flow path dimension, fluid properties, and heat transfer to improve accuracy. It is vitally important to validate IPR and TPR models before performing Nodal analysis on a large scale. A Nodal analysis model is not considered to be reliable before it can match well production rates at two bottom-hole pressures.

References

CHEN, W., ZHU, D., and HILL, A.D. A comprehensive model of multilateral well deliverability. Presented at the SPE International Oil and Gas Conference and Exhibition held 7–10 November 2000 in Beijing, China. Paper SPE 64751.

GREENE, W.R. Analyzing the performance of gas wells. *J. Petroleum Technol.* 1983:31–39.

LARSEN, L. Productivity computations for multilateral, branched and other generalized and extended well concepts. Presented at the SPE Annual Technical Conference and Exhibition held 6–9 October 1996 in Denver, Colorado. Paper SPE 36753.

NIND, T.E.W. *Principles of Oil Well Production*, 2nd edition. New York: McGraw-Hill, 1981.

PERNADI, P., WIBOWO, W., and PERMADI, A.K. Inflow performance of stacked multilateral well. Presented at the SPE Asia Pacific Conference on Integrated Modeling for Asset Management held 23–24 March 1996 in Kuala Lumpur, Malaysia. Paper SPE 39750.

RUSSELL, D.G., GOODRICH, J.H., PERRY, G.E., and BRUSKOTTER, J.F. Methods for predicting gas well performance. *J. Petroleum Technol.* January 1966:50–57.

SALAS, J.R., CLIFFORD, P.J., and HILL, A.D. Multilateral well performance prediction. Presented at the SPE Western Regional Meeting held 22–24 May 1996 in Anchorage, Alaska. Paper SPE 35711.

Problems

6.1 Suppose that a vertical well produces 0.65 specific gravity gas through a $2\frac{7}{8}$-in. tubing set to the top of a gas reservoir at a depth of 8,000 ft. At tubing head, the pressure is 600 psia and the temperature is 120 °F, and the bottom-hole temperature is 180 °F. The relative roughness of tubing is about 0.0006. Calculate the expected gas production rate of the well using the following data for IPR:

Reservoir pressure:	1,800 psia
IPR model parameter C:	0.15 Mscf/d-psi^{2n}
IPR model parameter n:	0.82

6.2 For the data given in the following table, predict the operating point using the bottom-hole as a solution node:

Reservoir pressure:	3,200 psia
Tubing ID:	1.66 in.
Wellhead pressure:	600 psia
Productivity index above bubble point:	1.5 stb/d-psi
Producing gas–liquid ratio (GLR):	800 scf/stb
Water cut (WC):	30%
Oil gravity:	40°API
Water-specific gravity:	1.05 1 for freshwater
Gas-specific gravity:	0.75 1 for air
N_2 content in gas:	0.05 mole fraction
CO_2 content in gas:	0.03 mole fraction
H_2S content in gas:	0.02 mole fraction
Formation volume factor for water:	1.25 rb/stb
Wellhead temperature:	110 °F
Tubing shoe depth:	6,000 ft
Bottom-hole temperature:	140 °F

6.3 For the data given in the following table, predict the operating point using the bottom-hole as the solution node:

Reservoir pressure:	3,500 psia
Total measured depth:	8,000 ft
Average inclination angle:	10 degrees
Tubing ID:	1.995 in.
Gas production rate:	500,000 scfd
Gas-specific gravity:	0.7 air = 1
Oil-specific gravity:	0.82 H_2O = 1
Water cut:	20%
Water-specific gravity:	1.07 H_2O = 1
Solid production rate:	2 ft^3/d
Solid-specific gravity:	2.65 H_2O = 1
Tubing head temperature:	120 °F
Bottom-hole temperature:	160 °F
Tubing head pressure:	400 psia
Absolute open flow (AOF):	2,200 bbl/d

6.4 For the data given in the following table, predict the operating point using the bottom-hole as the solution node:

Depth:	9,500 ft
Tubing inner diameter:	1.995 in.
Oil gravity:	40 °API
Oil viscosity:	3 cp
Production gas–liquid ratio:	600 scf/bbl
Gas-specific gravity:	0.75 air = 1
Flowing tubing head pressure:	500 psia
Flowing tubing head temperature:	90 °F
Flowing temperature at tubing shoe:	190 °F
Water cut:	20%
Reservoir pressure:	5,250 psia
Bubble-point pressure:	4,200 psia
Productivity above bubble point:	1.2 stb/d-psi

6.5 Use the following data to estimate the gas production rate of a gas well:

Gas-specific gravity:	0.75
Tubing inside diameter:	2.259 in.
Tubing wall relative roughness:	0.0006
Measured depth at tubing shoe:	8,000 ft
Inclination angle:	0 degrees
Wellhead choke size:	24 $\frac{1}{64}$ in.
Flowline diameter:	2 in.
Gas-specific heat ratio:	1.3
Gas viscosity at wellhead:	0.01 cp
Wellhead temperature:	140 °F
Bottom-hole temperature:	180 °F
Reservoir pressure:	2,200 psia
C-constant in backpressure IPR model:	0.01 Mscf d-psi^{2n}
n-exponent in backpressure IPR model:	0.84

6.6 Use the following data to estimate liquid production rate of an oil well:

Reservoir pressure:	6,500 psia
Tubing ID:	3.5 in
Choke size:	64 1/64 in.
Productivity index above bubble point:	1.2 stb/d-psi
Producing gas–liquid ratio:	800 scf/stb
Water cut:	35 %
Oil gravity:	40 °API
Water-specific gravity:	1.05 1 for freshwater
Gas-specific gravity:	0.75 1 for air
Choke constant:	10
Choke gas–liquid ratio exponent:	0.546
Choke-size exponent:	1.89
Formation volume factor for water:	1 rb/stb
Wellhead temperature:	110 °F
Tubing shoe depth:	10,000 ft
Bottom-hole temperature:	200 °F

6.7 Use the following data to estimate the liquid production rate of an oil well:

Choke size:	48 1/64 in.
Reservoir pressure:	3,200 psia
Total measured depth:	7,000 ft
Average inclination angle:	10 degrees
Tubing ID:	1.995 in.
Gas production rate:	600,000 scfd
Gas-specific gravity:	0.7 air = 1
Oil-specific gravity:	0.85 H_2O = 1
Water cut:	20%
Water-specific gravity:	1.05 H_2O = 1
Solid production rate:	0.5 ft^3/d
Solid-specific gravity:	2.65 H_2O = 1
Tubing head temperature:	120 °F
Bottom-hole temperature:	180 °F
Absolute open flow (AOF):	2,200 bbl/d
Choke flow constant:	10
Choke gas–liquid ratio exponent:	0.546
Choke size exponent:	1.89

6.8 For the following data, predict the oil production rate:

Depth:	7,500 ft
Tubing inner diameter:	3.5 in.
Oil gravity:	40 °API
Oil viscosity:	0.8 cp
Production GLR:	700 scf/bbl
Gas-specific gravity:	0.7 air = 1
Choke size:	48 1/64 in.
Flowing tubing head temperature:	90 °F
Flowing temperature at tubing shoe:	160 °F
Water cut:	20%
Reservoir pressure:	4,200 psia
Bubble-point pressure:	4,000 psia
Productivity above bubble point:	4 stb/d-psi
Choke flow constant:	10
Choke gas–liquid ratio exponent:	0.546
Choke-size exponent:	1.89

6.9 For the following data, predict the gas production rate against 1,200 psia wellhead pressure and 90 °F wellhead temperature:

Horizontal sections

Lateral no.:	1	2	3	
Length of horizontal section (L)	1,000	1,100	1,200	ft
Horizontal permeability (k)	8	5	4	md
Net pay thickness (h)	40	50	30	ft
Reservoir pressure (p-bar)	3,500	3,450	3,400	psia
Radius of drainage area (r_{eh})	2,000	2,200	2,400	ft
Gas viscosity (μ_g)	0.02	0.02	0.02	cp
Wellbore diameter (D_i)	6.00	6.00	6.00	in.
Bottom-hole temperature (T)	150	140	130	°F
Gas compressibility factor (z)	0.95	0.95	0.95	
Gas specific gravity (γ_g)	0.80	0.80	0.80	air = 1

Curvic sections

Lateral no.:	1	2	3	
Radius of curve (R)	333	400	500	ft
Average inclination angle (θ)	45	45	45	degrees
Tubing diameter (d_i)	1.995	1.995	1.995	in.
Pipe roughness (e)	0.0018	0.0018	0.0018	in.

Vertical sections

Lateral no.:	1	2	3	
Interval length (H)	500	500	6,000	ft
Tubing diameter (d_i)	1.995	1.995	1.995	in.
Pipe roughness (e)	0.0018	0.0018	0.0018	in.

6.10 For the following data, predict the gas production rate against 2,000 psia wellhead pressure and 80 °F wellhead temperature:

Horizontal sections

Lateral no.:	1	2	3	4	
Reservoir pressure (p-bar)	3,500	3,300	3,100	2,900	psia
Oil formation factor (B_o)	1.25	1.18	1.19	1.16	stb/rb
Water formation factor (B_w)	1.00	1.00	1.00	1.00	stb/rb
Bottom-hole temperature (T)	170	160	150	130	°F
Gas compressibility factor (z)	0.9	0.90	0.90	0.90	
Gas-specific gravity (γ_g)	0.75	0.73	0.70	0.75	air = 1
Oil-specific gravity (γ_o)	0.85	0.88	0.87	0.8	6 water = 1
Water-specific gravity (γ_w)	1.07	1.06	1.05	1.0	4 water = 1
Water–oil ratio (WOR)	0.30	0.20	0.10	0.1	0 stb/stb
Gas-oil ratio (GOR)	1,000	1,200	1,500	2,000	scf/stb
Solution–gas–oil ratio (Rs)	600	1,000	1,200	1,800	scf/stb
Productivity index (J)	2	1.8	1.7	1.6	stb/d/psi

Curvic sections

Lateral no.:	1	2	3	4	
Radius of curve (R)	400	400	400	400	ft
Average inclination angle (θ)	45	45	45	45	degrees
Tubing diameter (d_i)	2.441	2.441	2.441	2.441	in.
Pipe roughness (e)	0.0018	0.0018	0.0018	0.0018	in.

Vertical sections

Lateral no.:	1	2	3	4	
Interval length (H)	100	100	100	5,000	ft
Tubing diameter (d_i)	2.441	2.441	2.441	2.441	in.
Pipe roughness (e)	0.0018	0.0018	0.0018	0.0018	in.

7 Forecast of Well Production

Contents
7.1 Introduction 7/88
7.2 Oil Production during Transient Flow Period 7/88
7.3 Oil Production during Pseudo–Steady Flow Period 7/88
7.4 Gas Production during Transient Flow Period 7/92
7.5 Gas Production during Pseudo–Steady-State Flow Period 7/92
Summary 7/94
References 7/94
Problems 7/95

7.1 Introduction

With the knowledge of Nodal analysis, it is possible to forecast well production, that is, future production rate and cumulative production of oil and gas. Combined with information of oil and gas prices, the results of a production forecast can be used for field economics analyses.

A production forecast is performed on the basis of principle of material balance. The remaining oil and gas in the reservoir determine future inflow performance relationship (IPR) and, therefore, production rates of wells. Production rates are predicted using IPR (see Chapter 3) and tubing performance relationship (TPR) (see Chapter 4) in the future times. Cumulative productions are predicted by integrations of future production rates.

A complete production forecast should be carried out in different flow periods identified on the basis of flow regimes and drive mechanisms. For a volumetric oil reservoir, these periods include the following:

- Transient flow period
- Pseudo–steady one-phase flow period
- Pseudo–steady two-phase flow period

7.2 Oil Production during Transient Flow Period

The production rate during the transient flow period can be predicted by Nodal analysis using transient IPR and steady flow TPR. IPR model for oil wells is given by Eq. (3.2), that is,

$$q = \frac{kh(p_i - p_{wf})}{162.6 B_o \mu_o \left(\log t + \log \frac{k}{\phi \mu_o c_t r_w^2} - 3.23 + 0.87 S\right)}. \quad (7.1)$$

Equation 7.1 can be used for generating IPR curves for future time t before any reservoir boundary is reached by the pressure wave from the wellbore. After all reservoir boundaries are reached, either pseudo–steady-state flow or steady-state flow should prevail depending on the types of reservoir boundaries. The time required for the pressure wave to reach a circular reservoir boundary can be with $t_{pss} \approx 1{,}200 \frac{\phi \mu c_t r_e^2}{k}$.

The same TPR is usually used in the transient flow period assuming fluid properties remain the same in the well over the period. Depending on the producing gas–liquid ratio (GLR), the TPR model can be chosen from simple ones such as Poettmann–Carpenter and sophisticated ones such as the modified Hagedorn–Brown. It is essential to validate the selected TPR model based on measured data such as flow gradient survey from local wells.

Example Problem 7.1 Suppose a reservoir can produce oil under transient flow for the next 6 months. Predict oil production rate and cumulative oil production over the 6 months using the following data:

Reservoir porosity (ϕ):	0.2
Effective horizontal permeability (k):	10 md
Pay zone thickness (h):	50 ft
Reservoir pressure (p_i):	5,500 psia
Oil formation volume factor (B_o):	1.2 rb/stb
Total reservoir compressibility (c_t):	0.000013 psi^{-1}
Wellbore radius (r_w):	0.328 ft
Skin factor (S):	0
Well depth (H):	10,000 ft
Tubing inner diameter (d):	2.441
Oil gravity (API):	30 API
Oil viscosity (μ_o):	1.5 cp
Producing GLR (GLR):	300 scf/bbl
Gas-specific gravity (γ_g):	0.7 air = 1
Flowing tubing head pressure (p_{hf}):	800 psia
Flowing tubing head temperature (T_{hf}):	150 °F
Flowing temperature at tubing shoe (T_{wf}):	180 °F
Water cut:	10%
Interfacial tension (σ):	30 dynes/cm
Specific gravity of water (γ_w):	1.05

Solution To solve Example Problem 7.1, the spreadsheet program *TransientProductionForecast.xls* was used to perform Nodal analysis for each month. Operating points are shown in Fig. 7.1. The production forecast result is shown in Table 7.1, which also includes calculated cumulative production at the end of each month. The data in Table 7.1 are plotted in Fig. 7.2.

7.3 Oil Production during Pseudo–Steady Flow Period

It is generally believed that oil production during a pseudo–steady-state flow period is due to fluid expansion in undersaturated oil reservoirs and solution-gas drive in saturated oil reservoirs. An undersaturated oil reservoir becomes a saturated oil reservoir when the reservoir pressure drops to below the oil bubble-point pressure. Single-phase flow dominates in undersaturated oil reservoirs and two-phase flow prevails in saturated oil reservoirs. Different mathematical models have been used for time projection in production forecast for these two types of reservoirs, or the same reservoir at different stages of development based on reservoir pressure. IPR changes over time due to the changes in gas saturation and fluid properties.

7.3.1 Oil Production During Single-Phase Flow Period

Following a transient flow period and a transition time, oil reservoirs continue to deliver oil through single-phase flow under a pseudo–steady-state flow condition. The IPR changes with time because of the decline in reservoir pressure, while the TPR may be considered constant because fluid properties do not significantly vary above the bubble-point pressure. The TPR model can be chosen from simple ones such as Poettmann–Carpenter and sophisticated ones such as the modified Hagedorn–Brown. The IPR model is given by Eq. (3.7), in Chapter 3, that is,

$$q = \frac{kh(\bar{p} - p_{wf})}{141.2 B_o \mu_o \left(\frac{1}{2} \ln \frac{4A}{\gamma C_A r_w^2} + S\right)}. \quad (7.2)$$

The driving mechanism above the bubble-point pressure is essentially the oil expansion because oil is slightly compressible. The isothermal compressibility is defined as

$$c = -\frac{1}{V} \frac{\partial V}{\partial p}, \quad (7.3)$$

where V is the volume of reservoir fluid and p is pressure. The isothermal compressibility c is small and essentially constant for a given oil reservoir. The value of c can be measured experimentally. By separating variables, integration of Eq. (7.3) from the initial reservoir pressure p_i to the current average-reservoir pressure $\bar{p}$ results in

$$\frac{V}{V_i} = e^{c(p_i - \bar{p})}, \quad (7.4)$$

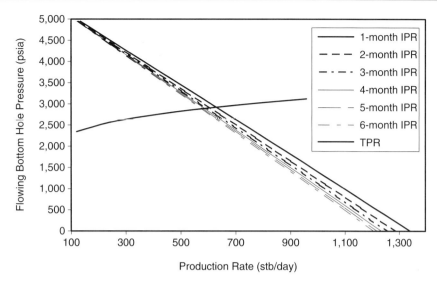

Figure 7.1 Nodal analysis plot for Example Problem 7.1.

Table 7.1 Production Forecast Given by TransientProductionForecast.xls

Time (mo)	Production rate (stb/d)	Cumulative production (stb)
1	639	19,170
2	618	37,710
3	604	55,830
4	595	73,680
5	588	91,320
6	583	108,795

where V_i is the reservoir volume occupied by the reservoir fluid. The fluid volume V at lower pressure $\bar{p}$ includes the volume of fluid that remains in the reservoir (still V_i) and the volume of fluid that has been produced, that is,

$$V = V_i + V_p. \tag{7.5}$$

Substituting Eq. (7.5) into Eq. (7.4) and rearranging the latter give

$$r = \frac{V_p}{V_i} = e^{c(p_i - \bar{p})} - 1, \tag{7.6}$$

where r is the recovery ratio. If the original oil in place N is known, the cumulative recovery (cumulative production) is simply expressed as $N_p = rN$.

For the case of an undersaturated oil reservoir, formation water and rock also expand as reservoir pressure drops. Therefore, the compressibility c should be the total compressibility c_t, that is,

$$c_t = c_o S_o + c_w S_w + c_f, \tag{7.7}$$

where c_o, c_w, and c_f are the compressibilities of oil, water, and rock, respectively, and S_o and S_w are oil and water saturations, respectively.

The following procedure is taken to perform the production forecast during the single-phase flow period:

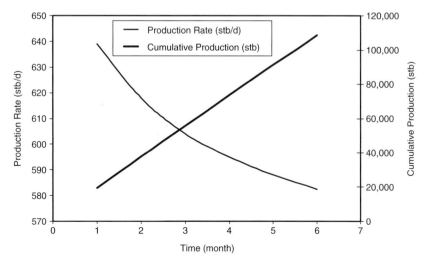

Figure 7.2 Production forecast for Example Problem 7.1.

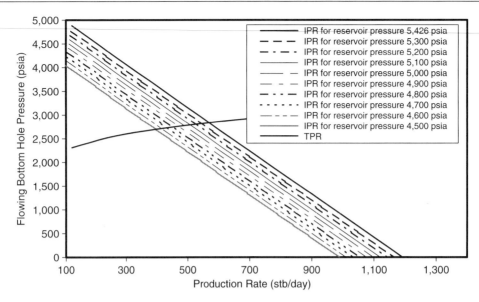

Figure 7.3 Nodal analysis plot for Example Problem 7.2.

1. Assume a series of average-reservoir pressure $\bar{p}$ values between the initial reservoir pressure p_i and oil bubble-point pressure p_b. Perform Nodal analyses to estimate production rate q at each average-reservoir pressure and obtain the average production rate $\bar{q}$ over the pressure interval.
2. Calculate recovery ratio r, cumulative production N_p at each average-reservoir pressure, and the incremental cumulative production ΔN_p within each average-reservoir pressure interval.
3. Calculate production time Δt for each average-reservoir pressure interval by $\Delta t = \Delta N_p / \bar{q}$ and the cumulative production time by $t = \sum \Delta t$.

Example Problem 7.2 Suppose the reservoir described in Example Problem 7.1 begins to produce oil under pseudo–steady-state flow conditions immediately after the 6-month transient flow. If the bubble-point pressure is 4,500 psia, predict the oil production rate and cumulative oil production over the time interval before the reservoir pressure declines to bubble-point pressure.

Solution Based on the transient flow IPR, Eq. (7.1), the productivity index will drop to 0.2195 stb/d-psi and production rate will drop to 583 stb/d at the end of the 6 months. If a pseudo–steady-state flow condition assumes immediately after the 6-month transient flow, the same production rate should be given by the pseudo–steady-state flow IPR, Eq. (7.2). These conditions require that the average-reservoir pressure be 5,426 psia by $\bar{p} = p_e^{-35.3} \frac{q\mu}{kh}$ and drainage be 1458 acres by Eq. (3.9). Assuming an initial water saturation of 0.35, the original oil in place (OOIP) in the drainage area is estimated to be 87,656,581 stb.

Using these additional data, Nodal analyses were performed with spreadsheet program *Pseudo-Steady-1Phase ProductionForecast.xls* at 10 average-reservoir pressures from 5,426 to bubble-point pressure of 4,500 psia. Operating points are shown in Fig. 7.3. The production forecast result is shown in Table 7.2. The production rate and cumulative production data in Table 7.2 are plotted in Fig. 7.4.

7.3.2 Oil Production during Two-Phase Flow Period

Upon the average-reservoir pressure drops to bubble-point pressure, a significant amount of solution gas becomes free gas in the reservoir, and solution-gas drive becomes a dominating mechanism of fluid production. The gas–oil two-phase pseudo–steady-state flow begins to prevail the reservoir. Both IPR and TPR change with time because of the significant variations of fluid properties, relative permeabilities, and gas–liquid ratio (GLR). The Hagedorn–Brown correlation should be used to model the TPR. The IPR can be described with Vogel's model by Eq. (3.19), in Chapter 3, that is,

Table 7.2 Production Forecast for Example Problem 7.2

Reservoir pressure (psia)	Production rate (stb/d)	Recovery ratio	Cumulative production (stb)	Incremental production (stb)	Incremental production time (days)	Pseudo–steady-state production time (days)
5,426	583	0.0010	84,366			0
5,300	563	0.0026	228,204	143,837	251	251
5,200	543	0.0039	342,528	114,325	207	458
5,100	523	0.0052	457,001	114,473	215	673
5,000	503	0.0065	571,624	114,622	223	896
4,900	483	0.0078	686,395	114,771	233	1,129
4,800	463	0.0091	801,315	114,921	243	1,372
4,700	443	0.0105	916,385	115,070	254	1,626
4,600	423	0.0118	1,031,605	115,220	266	1,892
4,500	403	0.0131	1,146,975	115,370	279	2,171

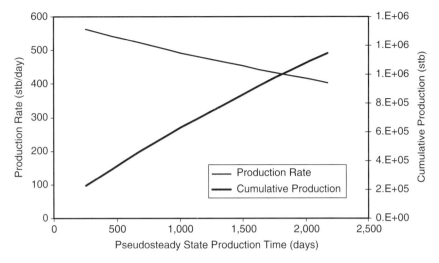

Figure 7.4 Production forecast for Example Problem 7.2.

$$q = \frac{J^*\bar{p}}{1.8}\left[1 - 0.2\left(\frac{p_{wf}}{\bar{p}}\right) - 0.8\left(\frac{p_{wf}}{\bar{p}}\right)^2\right]. \quad (7.8)$$

To perform production forecast for solution-gas drive reservoirs, material balance models are used for establishing the relation of the cumulative production to time. The commonly used material balance model is found in Craft and Hawkins (1991), which was based on the original work of Tarner (1944).

The following procedure is taken to carry out a production forecast during the two-phase flow period:

Step 1: Assume a series of average-reservoir pressure $\bar{p}$ values between the bubble-point pressure p_b and abandonment reservoir pressure p_a.
Step 2: Estimate fluid properties at each average-reservoir pressure, and calculate incremental cumulative production ΔN_p and cumulative production N_p within each average-reservoir pressure interval.
Step 3: Perform Nodal analyses to estimate production rate q at each average-reservoir pressure.
Step 4: Calculate production time Δt for each average-reservoir pressure interval by $\Delta t = \Delta N_p/q$ and the cumulative production time by $t = \sum \Delta t$.

Step 2 is further described in the following procedure:

1. Calculate coefficients Φ_n and Φ_g for the two pressure values that define the pressure interval, and obtain average values $\bar{\Phi}_n$ and $\bar{\Phi}_g$ in the interval. The Φ_n and Φ_g are calculated using

$$\Phi_n = \frac{B_o - R_s B_g}{(B_o - B_{oi}) + (R_{si} - R_s)B_g}, \quad (7.9)$$

$$\Phi_g = \frac{B_g}{(B_o - B_{oi}) + (R_{si} - R_s)B_g}, \quad (7.10)$$

where B_g should be in rb/scf if R_s is in scf/stb.

2. Assume an average gas–oil ratio $\bar{R}$ in the interval, and calculate incremental oil and gas production per stb of oil in place by

$$\Delta N_p^1 = \frac{1 - \bar{\Phi}_n N_p^1 - \bar{\Phi}_g G_p^1}{\bar{\Phi}_n + \bar{R}\bar{\Phi}_g}, \quad (7.11)$$

$$\Delta G_p^1 = \Delta N_p^1 \bar{R}, \quad (7.12)$$

where N_p^1 and G_p^1 are the cumulative oil and gas production per stb of oil in place at the beginning of the interval.

3. Calculate cumulative oil and gas production at the end of the interval by adding ΔN_p^1 and ΔG_p^1 to N_p^1 and G_p^1, respectively.
4. Calculate oil saturation by

$$S_o = \frac{B_o}{B_{oi}}(1 - S_w)(1 - N_p^1). \quad (7.13)$$

5. Obtain the relative permeabilities k_{rg} and k_{ro} based on S_o.
6. Calculate the average gas–oil ratio by

$$\bar{R} = R_s + \frac{k_{rg}\mu_o B_o}{k_{ro}\mu_g B_g}, \quad (7.14)$$

where again B_g should be in rb/scf if R_s is in scf/stb.

7. Compare the calculated $\bar{R}$ with the value assumed in Step 2. Repeat Steps 2 through 6 until $\bar{R}$ converges.

Example Problem 7.3 For the oil reservoir described in Example Problem 7.2, predict the oil production rate and cumulative oil production over the time interval during which reservoir pressure declines from bubble-point pressure to abandonment reservoir pressure of 2,500. The following additional data are given:

Reservoir pressure (psia)	B_o (rb /stb)	B_g (rb /scf)	R_s (rb /scf)	μ_g (cp)
4,500	1.200	6.90E−04	840	0.01
4,300	1.195	7.10E−04	820	0.01
4,100	1.190	7.40E−04	770	0.01
3,900	1.185	7.80E−04	730	0.01
3,700	1.180	8.10E−04	680	0.01
3,500	1.175	8.50E−04	640	0.01
3,300	1.170	8.90E−04	600	0.01
3,100	1.165	9.30E−04	560	0.01
2,900	1.160	9.80E−04	520	0.01
2,700	1.155	1.00E−03	480	0.01
2,500	1.150	1.10E−03	440	0.01

$$k_{ro} = 10^{-(4.8455S_g + 0.301)}$$
$$k_{rg} = 0.730678 S_g^{1.892}$$

Solution Example Problem 7.3 is solved using spreadsheets *Pseudo-Steady-2PhaseProductionForecast.xls* and *Pseudo-steady2PhaseForecastPlot.xls*. The former computes operating points and the latter performs material balance calculations. The results are shown in Tables 7.3, 7.4, and 7.5. Production forecast curves are given in Fig. 7.5.

7.4 Gas Production during Transient Flow Period

Similar to oil production, the gas production rate during a transient flow period can be predicted by Nodal analysis using transient IPR and steady-state flow TPR. The IPR model for gas wells is described in Chapter 3, that is,

$$q = \frac{kh[m(p_i) - m(p_{wf})]}{1638T\left(\log t + \log \frac{k}{\phi \mu_o c_t r_w^2} - 3.23 + 0.87S\right)}. \quad (7.15)$$

Equation (7.15) can be used for generating IPR curves for future time t before any reservoir boundary is "felt." After all reservoir boundaries are reached, a pseudo–steady-state flow should prevail for a volumetric gas reservoir. For a circular reservoir, the time required for the pressure wave to reach the reservoir boundary can be estimated with $t_{pss} \approx 1200 \frac{\phi \mu c_t r_e^2}{k}$.

The same TPR is usually used in the transient flow period assuming fluid properties remain the same in the well over the period. The average temperature–average z-factor method can be used for constructing TPR.

7.5 Gas Production during Pseudo–Steady-State Flow Period

Gas production during pseudo–steady-state flow period is due to gas expansion. The IPR changes over time due to the change in reservoir pressure. An IPR model is described in Chapter 3, that is,

$$q = \frac{kh[m(\bar{p}) - m(p_{wf})]}{1424T\left(\ln \frac{r_e}{r_w} - \frac{3}{4} + S + Dq\right)}. \quad (7.16)$$

Table 7.3 Oil Production Forecast for N = 1

p-bar (psia)	B_o (rb/stb)	B_g (rb/scf)	R_s (rb/scf)	Φ_n	Φ_g	R_{av} (rb/scf)	ΔN_p^1 (stb)	N_p^1 (stb)
4,500	1.200	6.9E−04	840					
	1.195	7.1E−04	820	66.61	0.077	859	7.52E−03	7.52E−03
4,300								7.52E−03
	1.190	7.4E−04	770	14.84	0.018	1,176	2.17E−02	2.92E−02
4,100								2.92E−02
	1.185	7.8E−04	730	8.69	0.011	1,666	1.45E−02	4.38E−02
3,900								4.38E−02
	1.180	8.1E−04	680	5.74	0.007	2,411	1.41E−02	5.79E−02
3,700								5.79E−02
	1.175	8.5E−04	640	4.35	0.006	3,122	9.65E−03	6.76E−02
3,500								6.76E−02
	1.170	8.9E−04	600	3.46	0.005	3,877	8.18E−03	7.57E−02
3,300								7.57E−02
	1.165	9.3E−04	560	2.86	0.004	4,658	7.05E−03	8.28E−02
3,100								8.28E−02
	1.160	9.8E−04	520	2.38	0.004	5,436	6.43E−03	8.92E−02
2,900								8.92E−02
	1.155	1.0E−03	480	2.07	0.003	6,246	5.47E−03	9.47E−02
2,700								9.47E−02
	1.150	1.1E−03	440	1.83	0.003	7,066	4.88E−03	9.96E−02
2,500								9.96E−02

Table 7.4 Gas Production Forecast for N = 1

p-bar (psia)	ΔG_p^1 (scf)	G_p^1 (scf)	S_o	S_g	k_{ro}	k_{rg}	R_{av} (rb/scf)
4,500							
	6.46E+00	6.46E+00	0.642421	0.007579	0.459492	7.11066E−05	859
4,300		6.46E+00					
	2.55E+01	3.20E+01	0.625744	0.024256	0.381476	0.000642398	1,176
4,100		3.20E+01					
	2.42E+01	5.62E+01	0.61378	0.03622	0.333809	0.001371669	1,666
3,900		5.62E+01					
	3.41E+01	9.03E+01	0.602152	0.047848	0.293192	0.002322907	2,411
3,700		9.03E+01					
	3.01E+01	1.20E+02	0.593462	0.056538	0.266099	0.003185377	3,122
3,500		1.20E+02					
	3.17E+01	1.52E+02	0.585749	0.064251	0.244159	0.004057252	3,877
3,300		1.52E+02					
	3.28E+01	1.85E+02	0.578796	0.071204	0.225934	0.004927904	4,658
3,100		1.85E+02					
	3.50E+01	2.20E+02	0.572272	0.077728	0.210073	0.005816961	5,436
2,900		2.20E+02					
	3.41E+01	2.54E+02	0.566386	0.083614	0.19672	0.006678504	6,246
2,700		2.54E+02					
	3.45E+01	2.89E+02	0.560892	0.089108	0.185024	0.007532998	7,066
2,500		2.89E+02					

Table 7.5 Production Schedule Forecast

p-bar (psia)	q_o (stb/d)	ΔN_p (stb)	N_p (stb)	ΔG_p (scf)	G_p (scf)	Δt (d)	t (d)
4,500							
	393	2.8E+04		2.37E+07		70	
4,300			27,601		2.37E+07		7.02E+01
	363	8.0E+04		9.36E+07		219	
4,100			107,217		1.17E+08		2.90E+02
	336	5.3E+04		8.89E+07		159	
3,900			160,565		2.06E+08		4.48E+02
	305	5.2E+04		1.25E+08		170	
3,700			212,442		3.31E+08		6.18E+02
	276	3.5E+04		1.10E+08		128	
3,500			247,824		4.42E+08		7.47E+02
	248	3.0E+04		1.16E+08		121	
3,300			277,848		5.58E+08		8.68E+02
	217	2.6E+04		1.21E+08		119	
3,100			303,716		6.79E+08		9.87E+02
	187	2.4E+04		1.28E+08		126	
2,900			327,302		8.07E+08		1.11E+03
	155	2.0E+04		1.25E+08		129	
2,700			347,354		9.32E+08		1.24E+03
	120	1.8E+04		1.27E+08		149	
2,500			365,268		1.06E+09		1.39E+03

Constant TPR is usually assumed if liquid loading is not a problem and the wellhead pressure is kept constant over time.

The gas production schedule can be established through the material balance equation,

$$G_p = G_i \left(1 - \frac{\frac{\bar{p}}{z}}{\frac{p_i}{z_i}}\right), \tag{7.17}$$

where G_p and G_i are the cumulative gas production and initial "gas in place," respectively.

If the gas production rate is predicted by Nodal analysis at a given reservoir pressure level and the cumulative gas production is estimated with Eq. (7.17) at the same reservoir pressure level, the corresponding production time can be calculated and, thus, production forecast can be carried out.

Example Problem 7.4 Use the following data and develop a forecast of a well production after transient flow until the average reservoir pressure declines to 2,000 psia:

Reservoir depth:	10,000 ft
Initial reservoir pressure:	4,613 psia
Reservoir temperature:	180 °F
Pay zone thickness:	78 ft
Formation permeability:	0.17 md
Formation porosity:	0.14
Water saturation:	0.27
Gas-specific gravity:	0.7 air = 1
Total compressibility:	1.5×10^{-4} psi^{-1}
Darcy skin factor:	0
Non-Darcy flow coefficient:	0
Drainage area:	40 acres
Wellbore radius:	0.328 ft
Tubing inner diameter:	2.441 in.
Desired flowing bottom-hole pressure:	1,500 psia

Solution The spreadsheet program *Carr-Kobayashi-Burrows-GasViscosity.xls* gives a gas viscosity value of 0.0251 cp at the initial reservoir pressure of 4,613 psia and temperature of 180 °F for the 0.7 specific gravity gas. The spreadsheet program *Hall-Yarborogh-z.xls* gives a z-factor value of 1.079 at the same conditions. Formation volume factor at the initial reservoir pressure is calculated with Eq. (2.62):

$$B_{gi} = 0.0283 \frac{(1.079)(180 + 460)}{4,613} = 0.004236 \, \text{ft}^3/\text{scf}$$

The initial "gas in place" within the 40 acres is

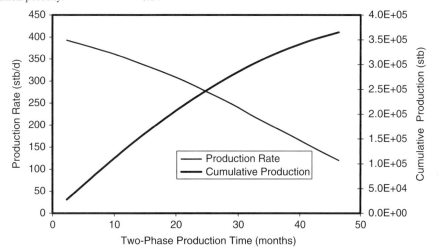

Figure 7.5 Production forecast for Example Problem 7.3.

Table 7.6 Result of Production Forecast for Example Problem 7.4

Reservoir pressure (psia)	z	Pseudo-pressure (10^8 psi^2/cp)	G_p (MMscf)	ΔG_p (MMscf)	q (Mscf/d)	Δt (day)	t (day)
4,409	1.074	11.90	130				
4,200	1.067	11.14	260	130	1,942	67	67
4,000	1.060	10.28	385	125	1,762	71	138
3,800	1.054	9.50	514	129	1,598	81	218
3,600	1.048	8.73	645	131	1,437	91	309
3,400	1.042	7.96	777	132	1,277	103	413
3,200	1.037	7.20	913	136	1,118	122	534
3,000	1.032	6.47	1,050	137	966	142	676
2,800	1.027	5.75	1,188	139	815	170	846
2,600	1.022	5.06	1,328	140	671	209	1,055
2,400	1.018	4.39	1,471	143	531	269	1,324
2,200	1.014	3.76	1,615	144	399	361	1,686
2,000	1.011	3.16	1,762	147	274	536	2,222

$$G_i = \frac{(43{,}560)(40)(78)(0.14)(1 - 0.27)}{0.004236} = 3.28 \times 10^9 \text{ scf}.$$

Assuming a circular drainage area, the equivalent radius of the 40 acres is 745 ft. The time required for the pressure wave to reach the reservoir boundary is estimated as

$$t_{pss} \approx 1200 \frac{(0.14)(0.0251)(1.5 \times 10^{-4})(745)^2}{0.17}$$

$$= 2{,}065 \text{ hours} = 86 \text{ days}.$$

The spreadsheet program *PseudoPressure.xls* gives

$$m(p_i) = m(4613) = 1.27 \times 10^9 \text{ psi}^2/\text{cp}$$
$$m(p_{wf}) = m(1500) = 1.85 \times 10^8 \text{ psi}^2/\text{cp}.$$

Substituting these and other given parameter values into Eq. (7.15) yields

$$q = \frac{(0.17)(78)[1.27 \times 10^9 - 1.85 \times 10^8]}{1638(180 + 460)\left(\log(2065) + \log\frac{0.17}{(0.14)(0.0251)(1.5 \times 10^{-4})(0.328)^2} - 3.23\right)}$$

$$= 2{,}092 \text{ Mscf/day}.$$

Substituting $q = 2{,}092$ Mscf/day into Eq. (7.16) gives

$$2{,}092 = \frac{(0.17)(78)[m(\bar{p}) - 1.85 \times 10^8]}{1424(180 + 460)\left(\ln\frac{745}{0.328} - \frac{3}{4} + 0\right)},$$

which results in $m(\bar{p}) = 1.19 \times 10^9$ psi^2/cp. The spreadsheet program *PseudoPressure.xls* gives $\bar{p} = 4{,}409$ psia at the beginning of the pseudo–steady-state flow period.

If the flowing bottom-hole pressure is maintained at a level of 1,500 psia during the pseudo–steady-state flow period (after 86 days of transient production), Eq. (7.16) is simplified as

$$q = \frac{(0.17)(78)[m(\bar{p}) - 1.85 \times 10^8]}{1424(180 + 460)\left(\ln\frac{745}{0.328} - \frac{3}{4} + 0\right)}$$

or

$$q = 2.09 \times 10^{-6}[m(\bar{p}) - 1.85 \times 10^8],$$

which, combined with Eq. (7.17), gives the production forecast shown in Table 7.6, where z-factors and real gas pseudo-pressures were obtained using spreadsheet programs *Hall-Yarborogh-z.xls* and *PseudoPressure.xls*, respectively. The production forecast result is also plotted in Fig. 7.6.

Summary

This chapter illustrated how to perform production forecast using the principle of Nodal analysis and material balance. Accuracy of the forecast strongly depends on the quality of fluid property data, especially for the two-phase flow period. It is always recommended to use fluid properties derived from PVT lab measurements in production forecast calculations.

References

CRAFT, B.C. and HAWKINS, M. *Applied Petroleum Reservoir Engineering*, 2nd edition. Englewood Cliffs, NJ: Prentice Hall, 1991.

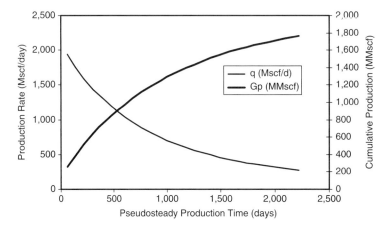

Figure 7.6 Result of production forecast for Example Problem 7.4.

TARNER, J. How different size gas caps and pressure maintenance programs affect amount of recoverable oil. *Oil Weekly* June 12, 1944;144:32–34.

Problems

7.1 Suppose an oil reservoir can produce under transient flow for the next 1 month. Predict oil production rate and cumulative oil production over the 1 month using the following data:

Reservoir porosity (ϕ):	0.25
Effective horizontal permeability (k):	50 md
Pay zone thickness (h):	75 ft
Reservoir pressure (p_i):	5000 psia
Oil formation volume factor (B_o):	1.3 rb/stb
Total reservoir compressibility (c_t):	0.000012 psi^{-1}
Wellbore radius (r_w):	0.328 ft
Skin factor (S):	0
Well depth (H):	8,000 ft
Tubing inner diameter (d):	2.041
Oil gravity (API):	35 API
Oil viscosity (μ_o):	1.3 cp
Producing gas–liquid ratio:	400 scf/bbl
Gas specific gravity (γ_g):	0.7 air = 1
Flowing tubing head pressure (p_{hf}):	500 psia
Flowing tubing head temperature (T_{hf}):	120 °F
Flowing temperature at tubing shoe (T_{wf}):	160 °F
Water cut:	10%
Interfacial tension (σ):	30 dynes/cm
Specific gravity of water (γ_w):	1.05

7.2 Suppose the reservoir described in Problem 7.1 begins to produce oil under a pseudo–steady-state flow condition immediately after the 1-month transient flow. If the bubble-point pressure is 4,000 psia, predict oil production rate and cumulative oil production over the time interval before reservoir pressure declines to bubble-point pressure.

Reservoir pressure (psia)	B_o(rb/stb)	B_g (rb/scf)	R_s (rb/scf)	μ_g (cp)
4,000	1.300	6.80E−04	940	0.015
3,800	1.275	7.00E−04	920	0.015
3,600	1.250	7.20E−04	870	0.015
3,400	1.225	7.40E−04	830	0.015
3,200	1.200	8.00E−04	780	0.015
3,000	1.175	8.20E−04	740	0.015
2,800	1.150	8.50E−04	700	0.015
2,600	1.125	9.00E−04	660	0.015
2,400	1.120	9.50E−04	620	0.015
2,200	1.115	1.00E−03	580	0.015
2,000	1.110	1.10E−03	540	0.015

7.3 For the oil reservoir described in Problem 7.2, predict oil production rate and cumulative oil production over the time interval during which reservoir pressure declines from bubble-point pressure to abandonment reservoir pressure of 2,000. The following additional data are given:

$$k_{ro} = 10^{-(4.55S_g + 0.3)}$$

$$k_{rg} = 0.75 S_g^{1.8}$$

7.4 Assume that a 0.328-ft radius well in a gas reservoir drains gas from an area of 40 acres at depth 8,000 ft through a 2.441 inside diameter (ID) tubing against a wellhead pressure 500 psia. The reservoir has a net pay of 78 ft, porosity of 0.14, permeability of 0.17 md, and water saturation of 0.27. The initial reservoir pressure is 4,613 psia. Reservoir temperature is 180 °F. Gas-specific gravity is 0.65. The total system compressibility is 0.00015 psi^{-1}. Both Darcy and non-Darcy skin are negligible. Considering both transient and pseudo–steady-state flow periods, generate a gas production forecast until the reservoir pressure drops to 3,600 psia.

7.5 Use the following data and develop a forecast of a gas well production during the transient flow period:

Reservoir depth:	9,000 ft
Initial reservoir pressure:	4,400 psia
Reservoir temperature:	170°F
Pay zone thickness:	60 ft
Formation permeability:	0.25 md
Formation porosity:	0.15
Water saturation:	0.30
Gas-specific gravity:	0.7 air = 1
Total compressibility:	1.6×10^{-4} psi^{-1}
Darcy skin factor:	0
Non-Darcy flow coefficient:	0
Drainage area:	40 acres
Wellbore radius:	0.328 ft
Tubing inner diameter:	2.441 in.
Desired flowing bottom-hole pressure:	1,100 psia

7.6 Use the following data and develop a forecast of a gas well production after transient flow until the average reservoir pressure declines to 2,000 psia:

Reservoir depth:	8,000 ft
Initial reservoir pressure:	4,300 psia
Reservoir temperature:	160°F
Pay zone thickness:	50 ft
Formation permeability:	0.20 md
Formation porosity:	0.15
Water saturation:	0.30
Gas-specific gravity:	0.7 air = 1
Total compressibility:	1.6×10^{-4} psi^{-1}
Darcy skin factor:	0
Non-Darcy flow coefficient:	0
Drainage area:	160 acres
Wellbore radius:	0.328 ft
Tubing inner diameter:	1.995 in.
Desired flowing bottom-hole pressure:	1,200 psia

7.7 Use the following data and develop a forecast of a gas well production after transient flow until the average reservoir pressure declines to 2,000 psia:

Reservoir depth:	8,000 ft
Initial reservoir pressure:	4,300 psia
Reservoir temperature:	160°F
Pay zone thickness:	50 ft
Formation permeability:	0.20 md

Formation porosity:	0.15	Non-Darcy flow coefficient:	0
Water saturation:	0.30	Drainage area:	160 acres
Gas-specific gravity:	0.7 air = 1	Wellbore radius:	0.328 ft
Total compressibility:	$1.6 \times 10^{-4} \text{psi}^{-1}$	Tubing inner diameter:	1.995 in.
Darcy skin factor:	0	Desired flowing wellhead pressure:	800 psia

8

Production Decline Analysis

Contents
8.1 Introduction 8/98
8.2 Exponential Decline 8/98
8.3 Harmonic Decline 8/100
8.4 Hyperbolic Decline 8/100
8.5 Model Identification 8/100
8.6 Determination of Model Parameters 8/101
8.7 Illustrative Examples 8/101
Summary 8/104
References 8/104
Problems 8/104

8.1 Introduction

Production decline analysis is a traditional means of identifying well production problems and predicting well performance and life based on real production data. It uses empirical decline models that have little fundamental justifications. These models include the following:

- Exponential decline (constant fractional decline)
- Harmonic decline
- Hyperbolic decline

Although the hyperbolic decline model is more general, the other two models are degenerations of the hyperbolic decline model. These three models are related through the following relative decline rate equation (Arps, 1945):

$$\frac{1}{q}\frac{dq}{dt} = -bq^d, \qquad (8.1)$$

where b and d are empirical constants to be determined based on production data. When $d = 0$, Eq. (8.1) degenerates to an exponential decline model, and when $d = 1$, Eq. (8.1) yields a harmonic decline model. When $0 < d < 1$, Eq. (8.1) derives a hyperbolic decline model. The decline models are applicable to both oil and gas wells.

8.2 Exponential Decline

The relative decline rate and production rate decline equations for the exponential decline model can be derived from volumetric reservoir model. Cumulative production expression is obtained by integrating the production rate decline equation.

8.2.1 Relative Decline Rate

Consider an oil well drilled in a volumetric oil reservoir. Suppose the well's production rate starts to decline when a critical (lowest permissible) bottom-hole pressure is reached. Under the pseudo–steady-state flow condition, the production rate at a given decline time t can be expressed as

$$q = \frac{kh(\bar{p}_t - p_{wf}^c)}{141.2 B_o \mu \left[\ln\left(\frac{0.472 r_e}{r_w}\right) + s\right]}, \qquad (8.2)$$

where

$\bar{p}_t =$ average reservoir pressure at decline time t,
$p_{wf}^c =$ the critical bottom-hole pressure maintained during the production decline.

The cumulative oil production of the well after the production decline time t can be expressed as

$$N_p = \int_0^t \frac{kh(\bar{p}_t - p_{wf}^c)}{141.2 B_o \mu \left[\ln\left(\frac{0.472 r_e}{r_w}\right) + s\right]} dt. \qquad (8.3)$$

The cumulative oil production after the production decline upon decline time t can also be evaluated based on the total reservoir compressibility:

$$N_p = \frac{c_t N_i}{B_o}(\bar{p}_0 - \bar{p}_t), \qquad (8.4)$$

where

$c_t =$ total reservoir compressibility,
$N_i =$ initial oil in place in the well drainage area,
$\bar{p}_0 =$ average reservoir pressure at decline time zero.

Substituting Eq. (8.3) into Eq. (8.4) yields

$$\int_0^t \frac{kh(\bar{p}_t - p_{wf}^c)}{141.2 B_o \mu \left[\ln\left(\frac{0.472 r_e}{r_w}\right) + s\right]} dt = \frac{c_t N_i}{B_o}(\bar{p}_0 - \bar{p}_t). \qquad (8.5)$$

Taking derivative on both sides of this equation with respect to time t gives the differential equation for reservoir pressure:

$$\frac{kh(\bar{p}_t - p_{wf}^c)}{141.2 \mu \left[\ln\left(\frac{0.472 r_e}{r_w}\right) + s\right]} = -c_t N_i \frac{d\bar{p}_t}{dt} \qquad (8.6)$$

Because the left-hand side of this equation is q and Eq. (8.2) gives

$$\frac{dq}{dt} = \frac{kh}{141.2 B_o \mu \left[\ln\left(\frac{0.472 r_e}{r_w}\right) + s\right]} \frac{d\bar{p}_t}{dt}, \qquad (8.7)$$

Eq. (8.6) becomes

$$q = \frac{-141.2 c_t N_i \mu \left[\ln\left(\frac{0.472 r_e}{r_w}\right) + s\right]}{kh} \frac{dq}{dt} \qquad (8.8)$$

or the relative decline rate equation of

$$\frac{1}{q}\frac{dq}{dt} = -b, \qquad (8.9)$$

where

$$b = \frac{kh}{141.2 \mu c_t N_i \left[\ln\left(\frac{0.472 r_e}{r_w}\right) + s\right]}. \qquad (8.10)$$

8.2.2 Production rate decline

Equation (8.6) can be expressed as

$$-b(\bar{p}_t - p_{wf}^c) = \frac{d\bar{p}_t}{dt}. \qquad (8.11)$$

By separation of variables, Eq. (8.11) can be integrated,

$$-\int_0^t b\, dt = \int_{\bar{p}_0}^{\bar{p}_t} \frac{d\bar{p}_t}{(\bar{p}_t - p_{wf}^c)}, \qquad (8.12)$$

to yield an equation for reservoir pressure decline:

$$\bar{p}_t = p_{wf}^c + (\bar{p}_0 - p_{wf}^c) e^{-bt} \qquad (8.13)$$

Substituting Eq. (8.13) into Eq. (8.2) gives the well production rate decline equation:

$$q = \frac{kh(\bar{p}_0 - p_{wf}^c)}{141.2 B_o \mu \left[\ln\left(\frac{0.472 r_e}{r_w}\right) + s\right]} e^{-bt} \qquad (8.14)$$

or

$$q = \frac{b c_t N_i}{B_o}(\bar{p}_0 - p_{wf}^c) e^{-bt}, \qquad (8.15)$$

which is the exponential decline model commonly used for production decline analysis of solution-gas-drive reservoirs. In practice, the following form of Eq. (8.15) is used:

$$q = q_i e^{-bt}, \qquad (8.16)$$

where q_i is the production rate at $t = 0$.

It can be shown that $\frac{q_2}{q_1} = \frac{q_3}{q_2} = \ldots = \frac{q_n}{q_{n-1}} = e^{-b}$. That is, the fractional decline is constant for exponential decline. As an exercise, this is left to the reader to prove.

8.2.3 Cumulative production

Integration of Eq. (8.16) over time gives an expression for the cumulative oil production since decline of

$$N_p = \int_0^t q\, dt = \int_0^t q_i e^{-bt} dt, \qquad (8.17)$$

that is,
$$N_p = \frac{q_i}{b}\left(1 - e^{-bt}\right). \quad (8.18)$$

Since $q = q_i e^{-bt}$, Eq. (8.18) becomes
$$N_p = \frac{1}{b}(q_i - q). \quad (8.19)$$

8.2.4 Determination of decline rate

The constant b is called the *continuous decline rate*. Its value can be determined from production history data. If production rate and time data are available, the b value can be obtained based on the slope of the straight line on a semi-log plot. In fact, taking logarithm of Eq. (8.16) gives
$$\ln(q) = \ln(q_i) - bt, \quad (8.20)$$
which implies that the data should form a straight line with a slope of $-b$ on the $\log(q)$ versus t plot, if exponential decline is the right model. Picking up any two points, (t_1, q_1) and (t_2, q_2), on the straight line will allow analytical determination of b value because
$$\ln(q_1) = \ln(q_i) - bt_1 \quad (8.21)$$
and
$$\ln(q_2) = \ln(q_i) - bt_2 \quad (8.22)$$
give
$$b = \frac{1}{(t_2 - t_1)}\ln\left(\frac{q_1}{q_2}\right). \quad (8.23)$$

If production rate and cumulative production data are available, the b value can be obtained based on the slope of the straight line on an N_p versus q plot. In fact, rearranging Eq. (8.19) yields
$$q = q_i - bN_p. \quad (8.24)$$
Picking up any two points, (N_{p1}, q_1) and (N_{p2}, q_2), on the straight line will allow analytical determination of the b value because
$$q_1 = q_i - bN_{p1} \quad (8.25)$$
and
$$q_2 = q_i - bN_{p2} \quad (8.26)$$
give
$$b = \frac{q_1 - q_2}{N_{p2} - N_{p1}}. \quad (8.27)$$

Depending on the unit of time t, the b can have different units such as month^{-1} and year^{-1}. The following relation can be derived:
$$b_a = 12b_m = 365b_d, \quad (8.28)$$
where b_a, b_m, and b_d are annual, monthly, and daily decline rates, respectively.

8.2.5 Effective decline rate

Because the exponential function is not easy to use in hand calculations, traditionally the effective decline rate has been used. Since $e^{-x} \approx 1 - x$ for small x-values based on Taylor's expansion, $e^{-b} \approx 1 - b$ holds true for small values of b. The b is substituted by b', the effective decline rate, in field applications. Thus, Eq. (8.16) becomes
$$q = q_i(1 - b')^t. \quad (8.29)$$

Again, it can be shown that $\frac{q_2}{q_1} = \frac{q_3}{q_2} = \ldots\ldots = \frac{q_n}{q_{n-1}} = 1 - b'$.

Depending on the unit of time t, the b' can have different units such as month^{-1} and year^{-1}. The following relation can be derived:
$$(1 - b'_a) = (1 - b'_m)^{12} = (1 - b'_d)^{365}, \quad (8.30)$$

where b'_a, b'_m, and b'_d are annual, monthly, and daily effective decline rates, respectively.

Example Problem 8.1 Given that a well has declined from 100 stb/day to 96 stb/day during a 1-month period, use the exponential decline model to perform the following tasks:

1. Predict the production rate after 11 more months
2. Calculate the amount of oil produced during the first year
3. Project the yearly production for the well for the next 5 years

Solution

1. Production rate after 11 more months:
$$b_m = \frac{1}{(t_{1m} - t_{0m})}\ln\left(\frac{q_{0m}}{q_{1m}}\right)$$
$$= \left(\frac{1}{1}\right)\ln\left(\frac{100}{96}\right) = 0.04082/\text{month}$$

Rate at end of 1 year:
$$q_{1m} = q_{0m}e^{-b_mt} = 100e^{-0.04082(12)} = 61.27\,\text{stb/day}$$

If the effective decline rate b' is used,
$$b'_m = \frac{q_{0m} - q_{1m}}{q_{0m}} = \frac{100 - 96}{100} = 0.04/\text{month}.$$

From
$$1 - b'_y = (1 - b'_m)^{12} = (1 - 0.04)^{12},$$
one gets
$$b'_y = 0.3875/\text{yr}$$

Rate at end of 1 year:
$$q_1 = q_0(1 - b'_y) = 100(1 - 0.3875) = 61.27\,\text{stb/day}$$

2. The amount of oil produced during the first year:
$$b_y = 0.04082(12) = 0.48986/\text{year}$$
$$N_{p,1} = \frac{q_0 - q_1}{b_y} = \left(\frac{100 - 61.27}{0.48986}\right)365 = 28{,}858\,\text{stb}$$
or
$$b_d = \left[\ln\left(\frac{100}{96}\right)\right]\left(\frac{1}{30.42}\right) = 0.001342\frac{1}{\text{day}}$$
$$N_{p,1} = \frac{100}{0.001342}(1 - e^{-0.001342(365)}) = 28{,}858\,\text{stb}$$

3. Yearly production for the next 5 years:
$$N_{p,2} = \frac{61.27}{0.001342}(1 - e^{-0.001342(365)}) = 17{,}681\,\text{stb}$$
$$q_2 = q_i e^{-bt} = 100e^{-0.04082(12)(2)} = 37.54\,\text{stb/day}$$
$$N_{p,3} = \frac{37.54}{0.001342}(1 - e^{-0.001342(365)}) = 10{,}834\,\text{stb}$$
$$q_3 = q_i e^{-bt} = 100e^{-0.04082(12)(3)} = 23.00\,\text{stb/day}$$
$$N_{p,4} = \frac{23.00}{0.001342}(1 - e^{-0.001342(365)}) = 6639\,\text{stb}$$
$$q_4 = q_i e^{-bt} = 100e^{-0.04082(12)(4)} = 14.09\,\text{stb/day}$$
$$N_{p,5} = \frac{14.09}{0.001342}(1 - e^{-0.001342(365)}) = 4061\,\text{stb}$$

In summary,

Year	Rate at End of Year (stb/day)	Yearly Production (stb)
0	100.00	—
1	61.27	28,858
2	37.54	17,681
3	23.00	10,834
4	14.09	6,639
5	8.64	4,061
		68,073

8.3 Harmonic Decline

When $d = 1$, Eq. (8.1) yields differential equation for a harmonic decline model:

$$\frac{1}{q}\frac{dq}{dt} = -bq, \qquad (8.31)$$

which can be integrated as

$$q = \frac{q_0}{1 + bt}, \qquad (8.32)$$

where q_0 is the production rate at $t = 0$.

Expression for the cumulative production is obtained by integration:

$$N_p = \int_0^t q\,dt,$$

which gives

$$N_p = \frac{q_0}{b}\ln(1 + bt). \qquad (8.33)$$

Combining Eqs. (8.32) and (8.33) gives

$$N_p = \frac{q_0}{b}[\ln(q_0) - \ln(q)]. \qquad (8.34)$$

8.4 Hyperbolic Decline

When $0 < d < 1$, integration of Eq. (8.1) gives

$$\int_{q_0}^{q}\frac{dq}{q^{1+d}} = -\int_0^t b\,dt, \qquad (8.35)$$

which results in

$$q = \frac{q_0}{(1 + dbt)^{1/d}} \qquad (8.36)$$

or

$$q = \frac{q_0}{\left(1 + \frac{b}{a}t\right)^a}, \qquad (8.37)$$

where $a = 1/d$.

Expression for the cumulative production is obtained by integration:

$$N_p = \int_0^t q\,dt,$$

which gives

$$N_p = \frac{aq_0}{b(a-1)}\left[1 - \left(1 + \frac{b}{a}t\right)^{1-a}\right]. \qquad (8.38)$$

Combining Eqs. (8.37) and (8.38) gives

$$N_p = \frac{a}{b(a-1)}\left[q_0 - q\left(1 + \frac{b}{a}t\right)\right]. \qquad (8.39)$$

8.5 Model Identification

Production data can be plotted in different ways to identify a representative decline model. If the plot of $\log(q)$ versus t shows a straight line (Fig. 8.1), according to Eq. (8.20), the decline data follow an exponential decline model. If the plot of q versus N_p shows a straight line (Fig. 8.2), according to Eq. (8.24), an exponential decline model should be adopted. If the plot of $\log(q)$ versus $\log(t)$ shows a straight line (Fig. 8.3), according to Eq. (8.32), the

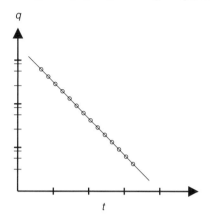

Figure 8.1 A semilog plot of q versus t indicating an exponential decline.

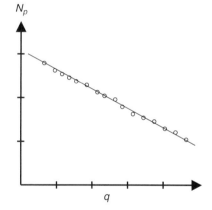

Figure 8.2 A plot of N_p versus q indicating an exponential decline.

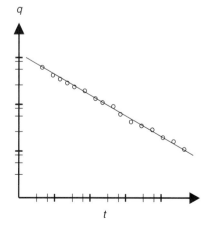

Figure 8.3 A plot of log(q) versus log(t) indicating a harmonic decline.

decline data follow a harmonic decline model. If the plot of N_p versus $\log(q)$ shows a straight line (Fig. 8.4), according to Eq. (8.34), the harmonic decline model should be used. If no straight line is seen in these plots, the hyperbolic decline model may be verified by plotting the relative decline rate defined by Eq. (8.1). Figure 8.5 shows such a plot. This work can be easily performed with computer program *UcomS.exe*.

8.6 Determination of Model Parameters

Once a decline model is identified, the model parameters a and b can be determined by fitting the data to the selected model. For the exponential decline model, the b value can be estimated on the basis of the slope of the straight line in the plot of $\log(q)$ versus t (Eq. [8.23]). The b value can also be determined based on the slope of the straight line in the plot of q versus N_p (Eq. [8.27]).

For the harmonic decline model, the b value can be estimated on the basis of the slope of the straight line in the plot of $\log(q)$ versus $\log(t)$ or Eq. (8.32):

$$b = \frac{\frac{q_0}{q_1} - 1}{t_1}. \tag{8.40}$$

The b value can also be estimated based on the slope of the straight line in the plot of N_p versus $\log(q)$ (Eq. [8.34]).

For the hyperbolic decline model, determination of a and b values is somewhat tedious. The procedure is shown in Fig. 8.6. Computer program *UcomS.exe* can be used for both model identification and model parameter determination, as well as production rate prediction.

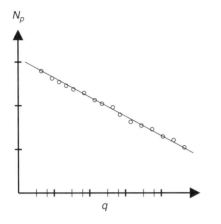

Figure 8.4 A plot of N_p versus $\log(q)$ indicating a harmonic decline.

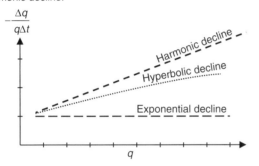

Figure 8.5 A plot of relative decline rate versus production rate.

8.7 Illustrative Examples

Example Problem 8.2 For the data given in Table 8.1, identify a suitable decline model, determine model parameters, and project production rate until a marginal rate of 25 stb/day is reached.

Solution A plot of $\log(q)$ versus t is presented in Fig. 8.7, which shows a straight line. According to Eq. (8.20), the exponential decline model is applicable. This is further evidenced by the relative decline rate shown in Fig. 8.8.

Select points on the trend line:

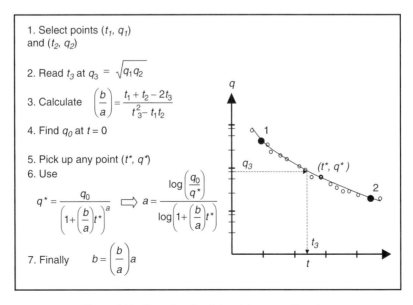

Figure 8.6 Procedure for determining a- and b-values.

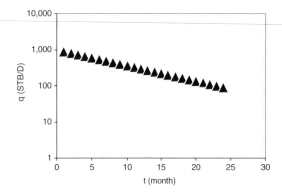

Figure 8.7 A plot of log(q) versus t showing an exponential decline.

Table 8.1 Production Data for Example Problem 8.2

t (mo)	q (stb/day)	t (mo)	q (stb/day)
1.00	904.84	13.00	272.53
2.00	818.73	14.00	246.60
3.00	740.82	15.00	223.13
4.00	670.32	16.00	201.90
5.00	606.53	17.00	182.68
6.00	548.81	18.00	165.30
7.00	496.59	19.00	149.57
8.00	449.33	20.00	135.34
9.00	406.57	21.00	122.46
10.00	367.88	22.00	110.80
11.00	332.87	23.00	100.26
12.00	301.19	24.00	90.720

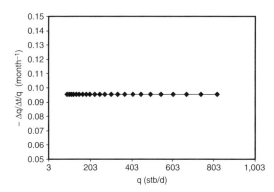

Figure 8.8 Relative decline rate plot showing exponential decline.

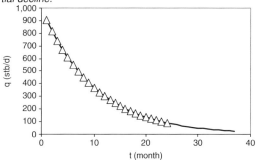

Figure 8.9 Projected production rate by a exponential decline model.

$t_1 = 5$ months, $q_1 = 607$ stb/day
$t_2 = 20$ months, $q_2 = 135$ stb/day

Decline rate is calculated with Eq. (8.23):

$$b = \frac{1}{(5-20)} \ln\left(\frac{135}{607}\right) = 0.11/\text{month}$$

Projected production rate profile is shown in Fig. 8.9.

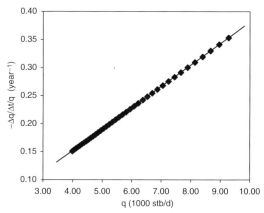

Figure 8.10 Relative decline rate plot showing harmonic decline.

Example Problem 8.3 For the data given in Table 8.2, identify a suitable decline model, determine model parameters, and project production rate until the end of the fifth year.

Solution A plot of relative decline rate is shown in Fig. 8.10, which clearly indicates a harmonic decline model.

On the trend line, select

$$q_0 = 10,000 \text{ stb/day at } t = 0$$

$$q_1 = 5,680 \text{ stb/day at } t = 2 \text{ years}$$

Therefore, Eq. (8.40) gives

Table 8.2 Production Data for Example Problem 8.3

t (yr)	q (1,000 stb/day)	t (yr)	q (1,000 stb/day)
0.20	9.29	2.10	5.56
0.30	8.98	2.20	5.45
0.40	8.68	2.30	5.34
0.50	8.40	2.40	5.23
0.60	8.14	2.50	5.13
0.70	7.90	2.60	5.03
0.80	7.67	2.70	4.94
0.90	7.45	2.80	4.84
1.00	7.25	2.90	4.76
1.10	7.05	3.00	4.67
1.20	6.87	3.10	4.59
1.30	6.69	3.20	4.51
1.40	6.53	3.30	4.44
1.50	6.37	3.40	4.36
1.60	6.22	3.50	4.29
1.70	6.08	3.60	4.22
1.80	5.94	3.70	4.16
1.90	5.81	3.80	4.09
2.00	5.68	3.90	4.03

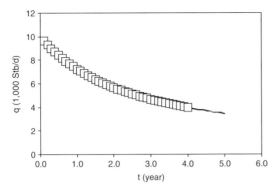

Figure 8.11 Projected production rate by a harmonic decline model.

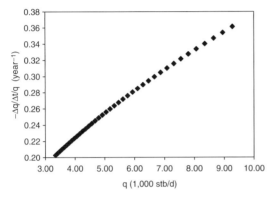

Figure 8.12 Relative decline rate plot showing hyperbolic decline.

$$b = \frac{\frac{10,000}{5,680} - 1}{2} = 0.38 \ 1/\text{yr}.$$

Projected production rate profile is shown in Fig. 8.11.

Example Problem 8.4 For the data given in Table 8.3, identify a suitable decline model, determine model parameters, and project production rate until the end of the fifth year.

Solution A plot of relative decline rate is shown in Fig. 8.12, which clearly indicates a hyperbolic decline model.

Select points

$$t_1 = 0.2 \text{ year}, \ q_1 = 9,280 \text{ stb/day}$$

$$t_2 = 3.8 \text{ years}, \ q_2 = 3,490 \text{ stb/day}$$

$$q_3 = \sqrt{(9,280)(3,490)} = 5,670 \text{ stb/day}$$

$$\left(\frac{b}{a}\right) = \frac{0.2 + 3.8 - 2(1.75)}{(1.75)^2 - (0.2)(3.8)} = 0.217$$

Read from decline curve (Fig. 8.13) $t_3 = 1.75$ years at $q_3 = 5,670$ stb/day.

Read from decline curve (Fig. 8.13) $q_0 = 10,000$ stb/day at $t_0 = 0$.

Pick up point ($t^* = 1.4$ years, $q^* = 6,280$ stb/day).

Table 8.3 Production Data for Example Problem 8.4

t (yr)	q (1,000 stb/day)	t (yr)	q (1,000 stb/day)
0.10	9.63	2.10	5.18
0.20	9.28	2.20	5.05
0.30	8.95	2.30	4.92
0.40	8.64	2.40	4.80
0.50	8.35	2.50	4.68
0.60	8.07	2.60	4.57
0.70	7.81	2.70	4.46
0.80	7.55	2.80	4.35
0.90	7.32	2.90	4.25
1.00	7.09	3.00	4.15
1.10	6.87	3.10	4.06
1.20	6.67	3.20	3.97
1.30	6.47	3.30	3.88
1.40	6.28	3.40	3.80
1.50	6.10	3.50	3.71
1.60	5.93	3.60	3.64
1.70	5.77	3.70	3.56
1.80	5.61	3.80	3.49
1.90	5.46	3.90	3.41
2.00	5.32	4.00	3.34

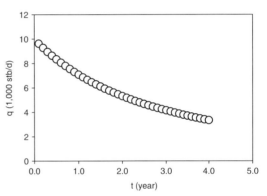

Figure 8.13 Relative decline rate shot showing hyperbolic decline.

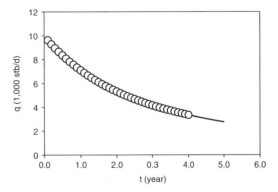

Figure 8.14 Projected production rate by a hyperbolic decline model.

$$a = \frac{\log\left(\frac{10,000}{6,280}\right)}{\log\left(1 + (0.217)(1.4)\right)} = 1.75$$

$$b = (0.217)(1.758) = 0.38$$

Projected production rate profile is shown in Fig. 8.14.

Summary

This chapter presents empirical models and procedure for using the models to perform production decline data analyses. Computer program *UcomS.exe* can be used for model identification, model parameter determination, and production rate prediction.

References

ARPS, J.J. Analysis of decline curves. *Trans. AIME* 1945;160:228–247.

GOLAN, M. AND WHITSON, C.M. *Well Performance*, pp. 122–125. Upper Saddle River, NJ: International Human Resource Development Corp., 1986.

ECONOMIDES, M.J., HILL, A.D., and EHLIG-ECONOMIDES, C. *Petroleum Production Systems*, pp. 516–519. Upper Saddle River, NJ: Prentice Hall PTR, 1994.

Problems

8.1 For the data given in the following table, identify a suitable decline model, determine model parameters, and project production rate until the end of the tenth year. Predict yearly oil productions:

8.2 For the data given in the following table, identify a suitable decline model, determine model parameters, predict the time when the production rate will decline to a marginal value of 500 stb/day, and the reverses to be recovered before the marginal production rate is reached:

Time (yr)	Production Rate (1,000 stb/day)
0.1	9.63
0.2	9.29
0.3	8.98
0.4	8.68
0.5	8.4
0.6	8.14
0.7	7.9
0.8	7.67
0.9	7.45
1	7.25
1.1	7.05
1.2	6.87
1.3	6.69
1.4	6.53
1.5	6.37
1.6	6.22
1.7	6.08
1.8	5.94
1.9	5.81
2	5.68
2.1	5.56
2.2	5.45
2.3	5.34
2.4	5.23
2.5	5.13
2.6	5.03
2.7	4.94
2.8	4.84
2.9	4.76
3	4.67
3.1	4.59
3.2	4.51
3.3	4.44
3.4	4.36

Time (yr)	Production Rate (stb/day)
0.1	9.63
0.2	9.28
0.3	8.95
0.4	8.64
0.5	8.35
0.6	8.07
0.7	7.81
0.8	7.55
0.9	7.32
1	7.09
1.1	6.87
1.2	6.67
1.3	6.47
1.4	6.28
1.5	6.1
1.6	5.93
1.7	5.77
1.8	5.61
1.9	5.46
2	5.32
2.1	5.18
2.2	5.05
2.3	4.92
2.4	4.8
2.5	4.68
2.6	4.57
2.7	4.46
2.8	4.35
2.9	4.25
3	4.15
3.1	4.06
3.2	3.97
3.3	3.88
3.4	3.8

8.3 For the data given in the following table, identify a suitable decline model, determine model parameters, predict the time when the production rate will decline to a marginal value of 50 Mscf/day, and the reverses to be recovered before the marginal production rate is reached:

Time (mo)	Production Rate (Mscf/day)
1	904.84
2	818.73
3	740.82
4	670.32
5	606.53
6	548.81
7	496.59
8	449.33
9	406.57
10	367.88
11	332.87
12	301.19
13	272.53
14	246.6
15	223.13
16	201.9
17	182.68
18	165.3
19	149.57
20	135.34
21	122.46
22	110.8
23	100.26
24	90.72

8.4 For the data given in the following table, identify a suitable decline model, determine model parameters, predict the time when the production rate will decline to a marginal value of 50 stb/day, and yearly oil productions:

Time (mo)	Production Rate (stb/day)
1	1,810
2	1,637
3	1,482
4	1,341
5	1,213
6	1,098
7	993
8	899
9	813
10	736
11	666
12	602
13	545
14	493
15	446
16	404
17	365
18	331
19	299
20	271
21	245
22	222
23	201
24	181

Part II Equipment Design and Selection

The role of a petroleum production engineer is to maximize oil and gas production in a cost-effective manner. Design and selection of the right equipment for production systems is essential for a production engineer to achieve his or her job objective. To perform their design work correctly, production engineers should have thorough knowledge of the principles and rules used in the industry for equipment design and selection. This part of the book provides graduating production engineers with principles and rules used in the petroleum production engineering practice. Materials are presented in the following three chapters:

Chapter 9   Well Tubing 9/109
Chapter 10   Separation Systems 10/117
Chapter 11   Transportation Systems 11/133

9 Well Tubing

Contents
9.1 Introduction 9/110
9.2 Strength of Tubing 9/110
9.3 Tubing Design 9/111
Summary 9/114
References 9/114
Problems 9/114

9.1 Introduction

Most oil wells produce reservoir fluids through tubing strings. This is mainly because tubing strings provide good sealing performance and allow the use of gas expansion to lift oil. Gas wells produce gas through tubing strings to reduce liquid loading problems.

Tubing strings are designed considering tension, collapse, and burst loads under various well operating conditions to prevent loss of tubing string integrity including mechanical failure and deformation due to excessive stresses and buckling. This chapter presents properties of the American Petroleum Institute (API) tubing and special considerations in designing tubing strings.

9.2 Strength of Tubing

The API defines "tubing size" using nominal diameter and weight (per foot). The nominal diameter is based on the internal diameter of tubing body. The weight of tubing determines the tubing outer diameter. Steel grades of tubing are designated to H-40, J-55, C-75, L-80, N-80, C-90, and P-105, where the digits represent the minimum yield strength in 1,000 psi. Table 9.1 gives the tensile requirements of API tubing. The minimum performance properties of API tubing are listed in Appendix B of this book.

The tubing collapse strength data listed in Appendix B do not reflect the effect of biaxial stress. The effect of tension of the collapse resistance is analyzed as follows.

Consider a simple uniaxial test of a metal specimen as shown in Fig. 9.1, Hooke's Law applies to the elastic portion before yield point:

$$\sigma = E\varepsilon, \quad (9.1)$$

where σ, ε, and E are stress, strain, and Young's modulus, respectively. The energy in the elastic portion of the test is

$$U_u = \frac{1}{2}\sigma\varepsilon = \frac{1}{2}\frac{P}{A}\frac{\Delta l}{L} = \frac{1}{2}\frac{(P \cdot \Delta l)}{V}$$

$$U_u = \frac{1}{2}\frac{W}{V}, \quad (9.2)$$

where P, A, L, V, and Δl are force, area, length, volume, and length change, respectively. However, using Hooke's Law, we have

$$U_u = \frac{1}{2}\sigma\varepsilon = \frac{1}{2}\sigma\left(\frac{\sigma}{E}\right) = \frac{1}{2}\frac{\sigma^2}{E}. \quad (9.3)$$

To assess whether a material is going to fail, we use various material failure criteria. One of the most important is the Distortion Energy Criteria. This is for 3D and is

$$U = \frac{1}{2}\left(\frac{1+v}{3E}\right)[(\sigma_1 - \sigma_2)^2 + (\sigma_2 - \sigma_3)^2 + (\sigma_3 - \sigma_1)^2], \quad (9.4)$$

Table 9.1 API Tubing Tensile Requirements

Tubing grade	Yield strength (psi)		Minimum tensile strength (psi)
	Minimum	Maximum	
H-40	40,000	80,000	60,000
J-55	55,000	80,000	75,000
C-75	75,000	90,000	95,000
L-80	80,000	95,000	95,000
N-80	80,000	110,000	100,000
C-90	90,000	105,000	100,000
P-105	105,000	135,000	120,000

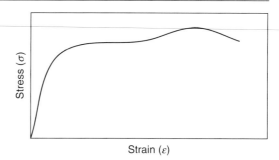

Figure 9.1 A simple uniaxial test of a metal specimen.

where

v = Poison's ratio
σ_1 = axial principal stress, psi
σ_2 = tangential principal stress, psi
σ_3 = radial principal stress, psi.

For our case of the uniaxial test, we would have

$$\sigma_1 = \sigma$$
$$\sigma_2 = 0 \ . \quad (9.5)$$
$$\sigma_3 = 0$$

Then from Eq. (9.4), we would get

$$U = \frac{1}{2}\left(\frac{1+v}{3E}\right)[\sigma^2 + \sigma^2]$$

$$U = \left(\frac{1+v}{3E}\right)\sigma^2. \quad (9.6)$$

If the failure of a material is taken to be when the material is at the yield point, then Eq. (9.6) is written

$$U_f = \left(\frac{1+v}{3E}\right)\sigma_y^2, \quad (9.7)$$

where σ_y is yield stress. The definition of an "equivalent stress" is the energy level in 3D, which is equivalent to the criteria energy level. Thus,

$$\left(\frac{1+v}{3E}\right)\sigma_e^2 = \left(\frac{1+v}{3E}\right)\sigma_y^2$$

and

$$\sigma_e = \sigma_y, \quad (9.8)$$

where σ_e is the equivalent stress. The collapse pressure is expressed as

$$p_c = 2\sigma_y\left[\frac{\left(\frac{D}{t}\right) - 1}{\left(\frac{D}{t}\right)^2}\right], \quad (9.9)$$

where D is the tubing outer diameter (OD) and t is wall thickness.

For the 3D case, we can consider

$$U = \left(\frac{1+v}{3E}\right)\sigma_e^2, \quad (9.10)$$

where σ_e is the equivalent stress for the 3D case of

$$\left(\frac{1+v}{3E}\right)\sigma_e^2 = \frac{1}{2}\left(\frac{1+v}{3E}\right)[(\sigma_1 - \sigma_2)^2 + (\sigma_2 - \sigma_3)^2 + (\sigma_3 - \sigma_1)^2]; \quad (9.11)$$

thus,

$$\sigma_e^2 = \frac{1}{2}\{(\sigma_1 - \sigma_2)^2 + (\sigma_2 - \sigma_3)^2 + (\sigma_3 - \sigma_1)^2\}. \quad (9.12)$$

Consider the case in which we have only tensile axial loads, and compressive pressure on the outside of the tubing, then Eq. (9.12) reduces to

$$\sigma_e^2 = \frac{1}{2}\left\{(\sigma_1 - \sigma_2)^2 + (\sigma_2)^2 + (-\sigma_1)^2\right\} \quad (9.13)$$

or

$$\sigma_e^2 = \sigma_1^2 - \sigma_1\sigma_2 + \sigma_2^2 \quad (9.14)$$

Further, we can define

$$\sigma_1 = \frac{W}{A}$$
$$\frac{\sigma_2}{Y_m} = -\frac{p_{cc}}{p_c}, \quad (9.15)$$

where

Y_m = minimum yield stress
p_{cc} = the collapse pressure corrected for axial load
p_c = the collapse pressure with no axial load.

$$\sigma_e = -Y_m$$

Thus, Eq. (9.14) becomes

$$Y_m^2 = \left(\frac{W}{A}\right)^2 + \left(\frac{W}{A}\right)\frac{p_{cc}}{p_c} \cdot Y_m + \left(\frac{p_{cc}}{p_c}\right)^2 Y_m^2 \quad (9.16)$$

$$\left(\frac{p_{cc}}{p_c}\right)^2 + \frac{W}{AY_m} \cdot \left(\frac{p_{cc}}{p_c}\right) + \left(\frac{W}{AY_m}\right)^2 - 1 = 0. \quad (9.17)$$

We can solve Eq. (9.17) for the term $\frac{p_{cc}}{p_c}$. This yields

$$\frac{p_{cc}}{p_c} = \frac{-\frac{W}{AY_m} \pm \sqrt{\left(\frac{W}{AY_m}\right)^2 - 4\left(\frac{W}{AY_m}\right)^2 + 4}}{2} \quad (9.18)$$

$$p_{cc} = p_c\left\{\sqrt{1 - 0.75\left(\frac{S_A}{Y_m}\right)^2} - 0.5\left(\frac{S_A}{Y_m}\right)\right\}, \quad (9.19)$$

where $S_A = \frac{W}{A}$ is axial stress at any point in the tubing string.

In Eq. (9.19), it can be seen that as W (or S_A) increases, the corrected collapse pressure resistance decreases (from the nonaxial load case).

In general, there are four cases, as shown in Fig. 9.2:

Case 1: Axial tension stress ($\sigma_1 > 0$) and collapse pressure ($\sigma_2 < 0$)

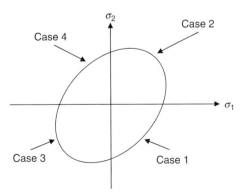

Figure 9.2 Effect of tension stress on tangential stress.

Case 2: Axial tension stress ($\sigma_1 > 0$) and burst pressure ($\sigma_2 > 0$)

Case 3: Axial compression stress ($\sigma_1 < 0$) and collapse pressure ($\sigma_2 < 0$)

Case 4: Axial compression stress ($\sigma_1 < 0$) and burst pressure ($\sigma_2 > 0$)

Example Problem 9.1 Calculate the collapse resistance for a section of $2\frac{7}{8}$ in. API 6.40 lb/ft, Grade J-55, non-upset tubing near the surface of a 10,000-ft string suspended from the surface in a well that is producing gas.

Solution Appendix B shows an inner diameter of tubing of 2.441 in., therefore,

$$t = (2.875 - 2.441)/2 = 0.217 \text{ in.}$$

$$\frac{D}{t} = \frac{2.875}{0.217} = 13.25$$

$$p_c = 2(55,000)\left[\frac{13.25 - 1}{(13.25)^2}\right] = 7,675.3 \text{ psi,}$$

which is consistent with the rounded value of 7,680 psi listed in Appendix B.

$$A = \pi t(D - t) = \pi(0.217)(2.875 - 0.217) = 1.812 \text{ in.}^2$$

$$S_A = \frac{6.40(10,000)}{1.812} = 35,320 \text{ psi.}$$

Using Eq. (9.19), we get

$$p_{cc} = 7675.3\left\{\sqrt{1 - 0.75\left(\frac{35,320}{55,000}\right)^2} - 0.5\left(\frac{35,320}{55,000}\right)\right\}$$

$$= 3,914.5 \text{ psi.}$$

9.3 Tubing Design

Tubing design should consider tubing failure due to tension, collapse, and burst loads under various well operating conditions. Forces affecting tubing strings include the following:

1. Axial tension due to weight of tubing and compression due to buoyancy
2. External pressure (completion fluids, oil, gas, formation water)
3. Internal pressure (oil, gas, formation water)
4. Bending forces in deviated portion of well
5. Forces due to lateral rock pressure
6. Other forces due to thermal gradient or dynamics

9.3.1 Tension, Collapse, and Burst Design

The last three columns of the tables in Appendix B present tubing collapse resistance, internal yield pressure, and joint yield strength. These are the limiting strengths for a given tubing joint without considering the biaxial effect shown in Fig. 9.2. At any point should the net external pressure, net internal pressure, and buoyant tensile load not be allowed to exceed tubing's axial load-corrected collapse resistance, internal yield pressure, and joint yield strength, respectively. Tubing strings should be designed to have strengths higher than the maximum expected loads with safety factors greater than unity. In addition, bending stress should be considered in tension design for deviated and horizontal wells. The tensile stress due to bending is expressed as

$$\sigma_b = \frac{ED_o}{2R_c}, \qquad (9.20)$$

where

σ_b = bending stress, psi
E = Young's modulus, psi
R_c = radius of hole curvature, in.
D_o = OD of tubing, in.

Because of the great variations in well operating conditions, it is difficult to adopt a universal tubing design criterion for all well situations. Probably the best design practice is to consider the worst loading cases for collapse, burst, and tension loads that are possible for the well to experience during the life of the well. It is vitally important to check the remaining strengths of tubing in a subject well before any unexpected well treatment is carried out. Some special considerations in well operations that affect tubing string integrity are addressed in the sections that follow.

9.3.2 Buckling Prevention during Production
A completion fluid is in place in the annular space between the tubing and the casing before a well is put into production. The temperature at depth is $T = T_{sf} + G_T D$, where G_T is geothermal gradient. When the oil is produced, the temperature in the tubing will rise. This will expand (thermal) the tubing length, and if there is not sufficient landing tension, the tubing will buckle. The temperature distribution in the tubing can be predicted on the basis of the work of Ramey (1962), Hasan and Kabir (2002), and Guo et al. (2005). The latter is described in Chapter 11. A conservative approach to temperature calculations is to assume the maximum possible temperature in the tubing string with no heat loss to formation through annulus.

Example Problem 9.2 Consider a $2\frac{7}{8}$ in. API, 6.40 lb/ft Grade P-105 non-upset tubing anchored with a packer set at 10,000 ft. The crude oil production through the tubing from the bottom of the hole is 1,000 stb/day (no gas or water production). A completion fluid is in place in the annular space between the tubing and the casing (9.8 lb/gal KCl water). Assuming surface temperature is 60 °F and geothermal gradient of 0.01 °F/ft, determine the landing tension to avoid buckling.

Solution The temperature of the fluid at the bottom of the hole is estimated to be

$$T_{10,000} = 60 + 0.01(10,000) = 160°F.$$

The average temperature of the tubing before oil production is

$$T_{av1} = \frac{60 + 160}{2} = 110°F.$$

The maximum possible average temperature of the tubing after oil production has started is

$$T_{av2} = \frac{160 + 160}{2} = 160°F.$$

This means that the approximate thermal expansion of the tubing in length will be

$$\Delta L_T \approx \beta(\Delta T_{avg})L,$$

where β is the coefficient of thermal expansion (for steel, this is $\beta_s = 0.0000065$ per °F). Thus,

$$\Delta L_T \approx 0.0000065[160 - 110]10,000 = 3.25\,\text{ft}.$$

To counter the above thermal expansion, a landing tension must be placed on the tubing string that is equivalent to the above. Assuming the tubing is a simple uniaxial element, then

$$A \approx \pi t(D - t) = \pi(0.217)(2.875 - 0.217) = 1.812\,\text{in.}^2$$
$$\sigma = E\varepsilon$$
$$\frac{F}{A} = E \cdot \frac{\Delta L}{L}$$
$$F = \frac{AE\,\Delta L}{L} = \frac{(1.812)(30 \times 10^6)(3.25)}{10,000} = 17,667\,\text{lb}_f.$$

Thus, an additional tension of 17,667 lb_f at the surface must be placed on the tubing string to counter the thermal expansion.

It can be shown that turbulent flow will transfer heat efficiently to the steel wall and then to the completion fluid and then to the casing and out to the formation. While laminar flow will not transfer heat very efficiently to the steel then out to the formation. Thus, the laminar flow situations are the most likely to have higher temperature oil at the exit. Therefore, it is most likely the tubing will be hotter via simple conduction. This effect has been considered in the work of Hasan and Kabir (2002). Obviously, in the case of laminar flow, landing tension beyond the buoyancy weight of the tubing may not be required, but in the case of turbulent flow, the landing tension beyond the buoyancy weight of the tubing is usually required to prevent buckling of tubing string. In general, it is good practice to calculate the buoyant force of the tubing and add approximately 4,000–5,000 lb_f of additional tension when landing.

9.3.3 Considerations for Well Treatment and Stimulation
Tubing strings are designed to withstand the harsh conditions during wellbore treatment and stimulation operations such as hole cleaning, cement squeezing, gravel packing, frac-packing, acidizing, and hydraulic fracturing. Precautionary measures to take depend on tubing–packer relation. If the tubing string is set through a non-restraining packer, the tubing is free to move. Then string buckling and tubing–packer integrity will be major concerns. If the tubing string is set on a restraining packer, the string is not free to move and it will apply force to the packer.

The factors to be considered in tubing design include the following:

- Tubing size, weight, and grade
- Well conditions

  - Pressure effect
  - Temperature effect

- Completion method

  - Cased hole
  - Open hole
  - Multitubing
  - Packer type (restraining, non-restraining)

9.3.3.1 Temperature Effect
As discussed in Example Problem 9.2, if the tubing string is free to move, its thermal expansion is expressed as

$$\Delta L_T = \beta L \Delta T_{avg}. \qquad (9.21)$$

If the tubing string is not free to move, its thermal expansion will generate force. Since Hook's Law gives

$$\Delta L_T = \frac{L\Delta F}{AE}, \qquad (9.22)$$

substitution of Eq. (9.22) into Eq. (9.21) yields

$$\Delta F = AE\beta\Delta T_{avg} \approx 207A\Delta T_{avg} \qquad (9.23)$$

for steel tubing.

9.3.3.2 Pressure Effect
Pressures affect tubing string in different ways including piston effect, ballooning effect, and buckling effect.

Consider the tubing–pack relation shown in Fig. 9.3. The total upward force acting on the tubing string from internal and external pressures is expressed as

$$F_{up} = p_i(A_p - A_i') + p_o(A_o - A_o'), \qquad (9.24)$$

where

p_i = pressure in the tubing, psi
p_o = pressure in the annulus, psi
A_p = inner area of packer, in.^{2}
A_i' = inner area of tubing sleeve, in.^{2}
A_o' = outer area of tubing sleeve, in.^{2}

The total downward force acting on the tubing string is expressed as

$$F_{down} = p_i(A_i - A_i') + p_o(A_p - A_o'), \qquad (9.25)$$

where A_i is the inner area of tubing. The net upward force is then

$$F = F_{up} - F_{down} = p_i(A_p - A_i) - p_o(A_p - A_o). \qquad (9.26)$$

During a well treatment operation, the change (increase) in the net upward force is expressed as

$$\Delta F = [\Delta p_i(A_p - A_i) - \Delta p_o(A_p - A_o)]. \qquad (9.27)$$

If the tubing string is anchored to a restraining packer, this force will be transmitted to the packer, which may cause packer failure. If the tubing string is free to move, this force will cause the tubing string to shorten by

$$\Delta L_P = \frac{L \Delta F}{AE}, \qquad (9.28)$$

which represents tubing string shrinkage due to piston effect.

As shown in Fig. 9.4a, the ballooning effect is due to the internal pressure being higher than the external pressure during a well treatment. The change in tensile force can be expressed as

$$\Delta F_B = 0.6[\Delta p_{i\ avg} A_i - \Delta p_{o\ avg} A_o]. \qquad (9.29)$$

If the tubing string is set through a restraining packer, this force will be transmitted to the packer, which may cause packer failure. If the tubing string is free to move, this force will cause the tubing string to shorten by

$$\Delta L_B = \frac{2L}{10^8}\left[\frac{\Delta p_{i\ avg} - R^2 \Delta p_{o\ avg}}{R^2 - 1}\right], \qquad (9.30)$$

where

$\Delta p_{i\ avg}$ = the average pressure change in the tubing, psi
$\Delta p_{o\ avg}$ = the average pressure change in the annulus, psi

$R^2 = A_o/A_i$.

As illustrated in Fig. 9.4b, the buckling effect is caused by the internal pressure being higher than the external pressure during a well treatment. The tubing string buckles when $F_{BK} = A_p(p_i - p_o) > 0$. If the tubing end is set through a restraining packer, this force will be transmitted to the packer, which may cause packer failure. If the tubing string is not restrained at bottom, this force will cause the tubing string to shorten by

$$\Delta L_{BK} = \frac{r^2 F_{BK}^2}{8EIW}, \qquad (9.31)$$

which holds true only if F_{BK} is greater than 0, and

$$r = \frac{D_{ci} - D_i}{2}$$
$$I = \frac{\pi}{64}(D_o^4 - D_i^4)$$
$$W = W_{air} + W_{fi} - W_{fo},$$

where

D_{ci} = inner diameter of casing, in.
D_i = inner diameter of tubing, in.
D_o = outer diameter of tubing, in.
W_{air} = weight of tubing in air, lb/ft
W_{fi} = weight of fluid inside tubing, lb/ft
W_{fo} = weight of fluid displaced by tubing, lb/ft.

9.3.3.3 Total Effect of Temperature and Pressure

The combination of Eqs. (9.22), (9.28), (9.30), and (9.31) gives

$$\Delta L = \Delta L_T + \Delta L_P + \Delta L_B + \Delta L_{BK}, \qquad (9.32)$$

which represents the tubing shortening with a non-restraining packer. If a restraining packer is used, the total tubing force acting on the packer is expressed as

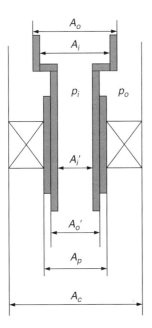

Figure 9.3 Tubing–packer relation.

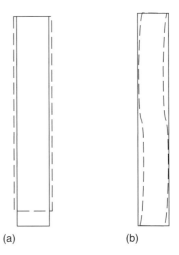

Figure 9.4 Ballooning and buckling effects.

$$\Delta F = \frac{AE\Delta L}{L}. \tag{9.33}$$

Example Problem 9.3 The following data are given for a cement squeezing job:

Tubing:	$2\frac{7}{8}$ in., 6.5 lb/ft (2.441-in. ID)
Casing:	7 in., 32 lb/ft (6.094-in. ID)
Packer:	Bore size $D_p = 3.25$ in., set at 10,000 ft
Initial condition:	Tubing and casing are full of 30 API oil (S.G. = 0.88)
Operation:	Tubing is displaced with 15 ppg cement with an injection pressure 5,000 psi and casing pressure 1,000 psi. The average temperature drop is 20° F.

1. Calculate tubing movement if the tubing is not restrained by the packer, and discuss solutions to the possible operation problems.
2. Calculate the tubing force acting on a restraining packer.

Solution

Temperature Effect:

$$\Delta l_T = \beta L \Delta T_{avg} = (6.9 \times 10^{-6})(10,000)(20) = 1.38 \, \text{ft}$$

Piston Effect:

$$\Delta P_i = (0.052)(10,000)[15 - (0.88)(8.33)] + 5,000 = 8,988 \, \text{psi}$$

$$\Delta p_o = 0 \, \text{psi}$$

$$A_p = 3.14(3.25)^2/4 = 8.30 \, \text{in.}^2$$

$$A_i = 3.14(2.441)^2/4 = 4.68 \, \text{in.}^2$$

$$A_o = 3.14(2.875)^2/4 = 6.49 \, \text{in.}^2$$

$$\Delta F = [\Delta p_i(A_p - A_i) - \Delta p_o(A_p - A_o)]$$
$$= [(8,988)(8.30 - 4.68) - (1,000)(8.30 - 6.49)]$$
$$= 30,727 \, \text{lb}_f$$

$$\Delta L_P = \frac{L \Delta F}{AE} = \frac{(10,000)(30,727)}{(6.49 - 4.68)(30,000,000)} = 5.65 \, \text{ft}$$

Ballooning Effect:

$$\Delta P_{i,avg} = (10,000/2)(0.052)[15 - (0.88)(8.33)] + 5,000 = 6,994 \, \text{psi}$$

$$\Delta P_{o,avg} = 1,000 \, \text{psi}$$

$$R^2 = 6.49/4.68 = 1.387$$

$$\Delta L_B = \frac{2L}{10^8}\left[\frac{\Delta p_{i\,avg} - R^2 \Delta p_{o\,avg}}{R^2 - 1}\right]$$

$$= \frac{2(10,000)}{10^8}\left[\frac{6,994 - 1.387(1,000)}{1.387 - 1}\right] = 2.898 \, \text{ft}$$

Since the tubing internal pressure is higher than the external pressure during the cement squeezing, tubing string buckling should occur.

$$p_i = 5,000 + (0.052)(15)(10,000) = 12,800 \, \text{psi}$$

$$p_o = 1,000 + (0.88)(0.433)(10,000) = 4,810 \, \text{psi}$$

$$r = (6.094 - 2.875)/2 = 1.6095 \, \text{in.}$$

$$F_{BK} = A_p(p_i - p_o) = (8.30)(12,800 - 4,800) = 66,317 \, \text{lb}_f$$

$$I = \frac{\pi}{64}\left((2.875)^4 - (2.441)^4\right) = 1.61 \, \text{in.}^4$$

$$W_{air} = 6.5 \, \text{lb}_f/\text{ft}$$

$$W_{fi} = (15)(7.48)(4.68/144) = 3.65 \, \text{lb}_f/\text{ft}$$

$$W_{fo} = (0.88)(62.4)(6.49/144) = 2.48 \, \text{lb}_f/\text{ft}$$

$$W = 6.5 + 3.65 - 2.48 = 7.67 \, \text{lb}_f/\text{ft}$$

$$\Delta L_{BK} = \frac{r^2 F_{BK}^2}{8EIW}$$

$$= \frac{(1.6095)^2(66,317)^2}{(8)(30,000,000)(1.61)(7.67)} = 3.884 \, \text{ft}$$

1. Tubing is not restrained by the packer. The tubing shortening is

$$\Delta L = \Delta L_T + \Delta L_P + \Delta L_B + \Delta L_{BK}$$
$$= 1.38 + 5.65 + 2.898 + 3.844 = 13.77 \, \text{ft}.$$

Buckling point from bottom:

$$L_{BK} = \frac{F_{BK}}{W}$$
$$= \frac{66,317}{7.67}$$
$$= 8,646 \, \text{ft}$$

To keep the tubing in the packer, one of the following measures needs to be taken:

a. Use a sleeve longer than 13.77 ft
b. Use a restraining packer
c. Put some weight on the packer (slack-hook) before treatment. Buckling due to slacking off needs to be checked.

2. Tubing is restrained by the packer. The force acting on the packer is

$$\Delta F = \frac{AE \Delta L}{L} = \frac{(6.49 - 4.68)(30,000,000)(13.77)}{(10,000)}$$

$$= 74,783 \, \text{lb}_f.$$

Summary

This chapter presents strength of API tubing that can be used for designing tubing strings for oil and gas wells. Tubing design should consider operating conditions in individual wells. Special care should be taken for tubing strings before a well undergoes a treatment or stimulation.

References

GUO, B., SONG, S., CHACKO, J., and GHALAMBOR, A. *Offshore Pipelines*. Burlington: Gulf Professional Publishing, 2005.

HASAN, R. and KABIR, C.S. *Fluid Flow and Heat Transfer in Wellbores*, pp. 79–89. Richardson, TX: SPE, 2002.

RAMEY, H.J., JR. Wellbore heat transmission. *Trans. AIME* April 1962;14:427.

Problems

9.1 Calculate the collapse resistance for a section of 3-in. API 9.20 lb/ft, Grade J-55, non-upset tubing near the surface of a 12,000-ft string suspended from the surface in a well that is producing gas.

9.2 Consider a 2⅞-in. API, 6.40 lb/ft Grade J-55 non-upset tubing anchored with a packer set at 8,000 ft. The crude oil production through the tubing from the bottom of the hole is 1,500 stb/day (no gas or water production). A completion fluid is in place in the annular space between the tubing and the casing (9.6 lb/gal KCl water). Assuming surface temperature is 80 °F and geothermal gradient of 0.01 °F/ft, determine the landing tension to avoid buckling.

9.3 The following data are given for a frac-packing job:
   Tubing: 2⅞ in., 6.5 lb/ft (2.441 in. ID)
   Casing: 7 in., 32 lb/ft (6.094 in. ID)
   Packer: Bore size $D_p = 3.25$ in., set at 8,000 ft
   Initial condition: Tubing and casing are full of 30 API oil (S.G. = 0.88)
   Operation: Tubing is displaced with 12 ppg cement with an injection pressure 4,500 psi and casing pressure 1,200 psi. The average temperature drop is 30 °F.

a. Calculate tubing movement if the tubing is not restrained by the packer, and discuss solutions to the possible operation problems.
b. Calculate the tubing force acting on a restraining packer.

10 Separation Systems

Contents
10.1 Introduction 10/118
10.2 Separation System 10/118
10.3 Dehydration System 10/125
Summary 10/132
References 10/132
Problems 10/132

10.1 Introduction

Oil and gas produced from wells are normally complex mixtures of hundreds of different compounds. A typical well stream is a turbulent mixture of oil, gas, water, and sometimes solid particles. The well stream should be processed as soon as possible after bringing it to the surface. Field separation processes fall into two categories: (1) separation of oil, water, and gas; and (2) dehydration that removes condensable water vapor and other undesirable compounds, such as hydrogen sulfide or carbon dioxide. This chapter focuses on the principles of separation and dehydration and selection of required separators and dehydrators.

10.2 Separation System

Separation of well stream gas from free liquids is the first and most critical stage of field-processing operations. Composition of the fluid mixture and pressure determine what type and size of separator are required. Separators are also used in other locations such as upstream and downstream of compressors, dehydration units, and gas sweetening units. At these locations, separators are referred to as scrubbers, knockouts, and free liquid knockouts. All these vessels are used for the same purpose: to separate free liquids from the gas stream.

10.2.1 Principles of Separation

Separators work on the basis of gravity segregation and/or centrifugal segregation. A separator is normally constructed in such a way that it has the following features:

1. It has a centrifugal inlet device where the primary separation of the liquid and gas is made.
2. It provides a large settling section of sufficient height or length to allow liquid droplets to settle out of the gas stream with adequate surge room for slugs of liquid.
3. It is equipped with a mist extractor or eliminator near the gas outlet to coalesce small particles of liquid that do not settle out by gravity.
4. It allows adequate controls consisting of level control, liquid dump valve, gas backpressure valve, safety relief valve, pressure gauge, gauge glass, instrument gas regulator, and piping.

The centrifugal inlet device makes the incoming stream spin around. Depending on the mixture flow rate, the reaction force from the separator wall can generate a centripetal acceleration of up to 500 times the gravitational acceleration. This action forces the liquid droplets together where they fall to the bottom of the separator into the settling section. The settling section in a separator allows the turbulence of the fluid stream to subside and the liquid droplets to fall to the bottom of the vessel due to gravity segregation. A large open space in the vessel is required for this purpose. Use of internal baffling or plates may produce more liquid to be discharged from the separator. However, the product may not be stable because of the light ends entrained in it. Sufficient surge room is essential in the settling section to handle slugs of liquid without carryover to the gas outlet. This can be achieved by placing the liquid level control in the separator, which in turn determines the liquid level. The amount of surge room required depends on the surge level of the production steam and the separator size used for a particular application.

Small liquid droplets that do not settle out of the gas stream due to little gravity difference between them and the gas phase tend to be entrained and pass out of the separator with the gas. A mist eliminator or extractor near the gas outlet allows this to be almost eliminated. The small liquid droplets will hit the eliminator or extractor surfaces, coalesce, and collect to form larger droplets that will then drain back to the liquid section in the bottom of the separator. A stainless steel woven-wire mesh mist eliminator can remove up to 99.9% of the entrained liquids from the gas stream. Cane mist eliminators can be used in areas where there is entrained solid material in the gas phase that may collect and plug a wire mesh mist eliminator.

10.2.2 Types of Separators

Three types of separators are generally available from manufacturers: vertical, horizontal, and spherical separators. Horizontal separators are further classified into two categories: single tube and double tube. Each type of separator has specific advantages and limitations. Selection of separator type is based on several factors including characteristics of production steam to be treated, floor space availability at the facility site, transportation, and cost.

10.2.2.1 Vertical Separators

Figure 10.1 shows a vertical separator. The inlet diverter baffle is a centrifugal inlet device making the incoming stream spin around. This action forces the liquid droplets to stay together and fall to the bottom of the separator along the separator wall due to gravity. Sufficient surge room is available in the settling section of the vertical separator to handle slugs of liquid without carryover to the gas outlet. A mist eliminator or extractor near the gas outlet allows the entrained liquid in the gas to be almost eliminated.

Vertical separators are often used to treat low to intermediate gas–oil ratio well streams and streams with relatively large slugs of liquid. They handle greater slugs of liquid without carryover to the gas outlet, and the action of the liquid level control is not as critical. Vertical separators occupy less floor space, which is important for facility sites such as those on offshore platforms where space is limited. Because of the large vertical distance between the liquid level and the gas outlet, the chance for liquid to revaporize into the gas phase is limited. However, because of the natural upward flow of gas in a vertical separator against the falling droplets of liquid, adequate separator diameter is required. Vertical separators are more costly to fabricate and ship in skid-mounted assemblies.

10.2.2.2 Horizontal Separators

Figure 10.2 presents a sketch of a horizontal separator. In horizontal separators, gas flows horizontally while liquid droplets fall toward the liquid surface. The moisture gas flows in the baffle surface and forms a liquid film that is drained away to the liquid section of the separator. The baffles need to be longer that the distance of liquid trajectory travel. The liquid-level control placement is more critical in a horizontal separator than in a vertical separator because of limited surge space.

Horizontal separators are usually the first choice because of their low costs. They are almost widely used for high gas–oil ratio well streams, foaming well streams, or liquid-from-liquid separation. They have much greater gas–liquid interface because of a large, long, baffled gas-separation section. Horizontal separators are easier to skid-mount and service and require less piping for field connections. Individual separators can be stacked easily into stage-separation assemblies to minimize space requirements.

Figure 10.3 demonstrates a horizontal double-tube separator consisting of two tube sections. The upper tube section is filled with baffles, gas flows straight through and at higher velocities, and the incoming free liquid is immediately drained away from the upper tube

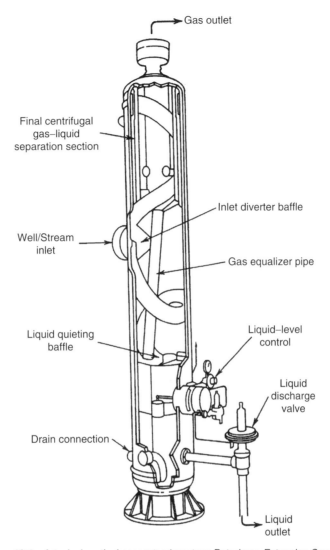

Figure 10.1 A typical vertical separator (courtesy Petroleum Extension Services).

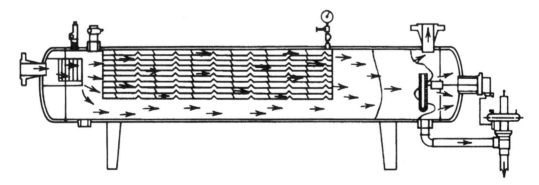

Figure 10.2 A typical horizontal separator (courtesy Petroleum Extension Services).

section into the lower tube section. Horizontal double-tube separators have all the advantages of normal horizontal single-tube separators, plus much higher liquid capacities.

Figure 10.4 illustrates a horizontal oil–gas–water three-phase separator. This type of separator is commonly used for well testing and in instances where free water readily separates from the oil or condensate. Three-phase separation can be accomplished in any type of separator. This can be achieved by installing either special internal baffling to construct a water leg or water siphon arrangement. It can also be achieved by using an interface liquid-level control. In three-phase operations, two liquid dump valves are required.

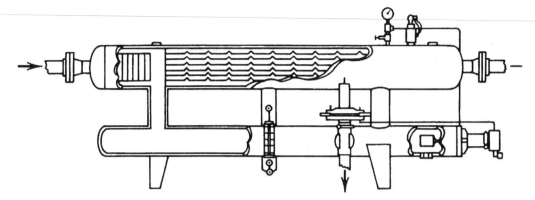

Figure 10.3 A typical horizontal double-tube separator (courtesy Petroleum Extension Services).

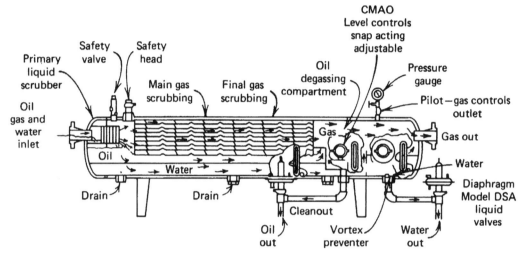

Figure 10.4 A typical horizontal three-phase separator (courtesy Petroleum Extension Services).

10.2.2.3 Spherical Separators
A spherical separator is shown in Fig. 10.5. Spherical separators offer an inexpensive and compact means of separation arrangement. Because of their compact configurations, this type of separator has a very limited surge space and liquid settling section. Also, the placement and action of the liquid-level control in this type of separator is very critical.

10.2.3 Factors Affecting Separation
Separation efficiency is dominated by separator size. For a given separator, factors that affect separation of liquid and gas phases include separator operating pressure, separator operating temperature, and fluid stream composition. Changes in any of these factors will change the amount of gas and liquid leaving the separator. An increase in operating pressure or a decrease in operating temperature generally increases the liquid covered in a separator. However, this is often not true for gas condensate systems in which an optimum pressure may exist that yields the maximum volume of liquid phase. Computer simulation (flash vaporization calculation) of phase behavior of the well stream allows the designer to find the optimum pressure and temperature at which a separator should operate to give maximum liquid recovery (see Chapter 18). However, it is often not practical to operate at the optimum point.

This is because storage system vapor losses may become too great under these optimum conditions.

In field separation facilities, operators tend to determine the optimum conditions for them to maximize revenue. As the liquid hydrocarbon product is generally worth more than the gas, high liquid recovery is often desirable, provided that it can be handled in the available storage system. The operator can control operating pressure to some extent by use of backpressure valves. However, pipeline requirements for Btu content of the gas should also be considered as a factor affecting separator operation.

It is usually unfeasible to try to lower the operating temperature of a separator without adding expensive mechanical refrigeration equipment. However, an indirect heater can be used to heat the gas before pressure reduction to pipeline pressure in a choke. This is mostly applied to high-pressure wells. By carefully operating this indirect heater, the operator can prevent overheating the gas stream ahead of the choke. This adversely affects the temperature of the downstream separator.

10.2.4 Selection of Separators
Petroleum engineers normally do not perform detailed designing of separators but carry out selection of separators suitable for their operations from manufacturers' product catalogs. This section addresses how to determine

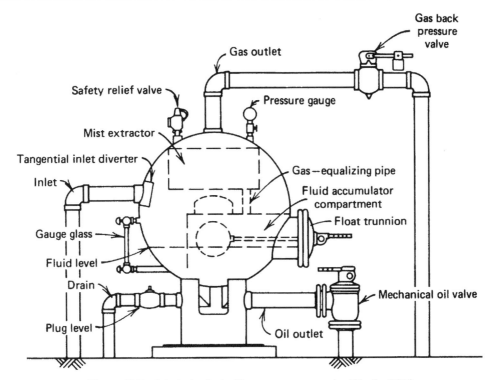

Figure 10.5 A typical spherical low-pressure separator (Sivalls, 1977).

separator specifications based on well stream conditions. The specifications are used for separator selections.

10.2.4.1 Gas Capacity
The following empirical equations proposed by Souders–Brown are widely used for calculating gas capacity of oil/gas separators:

$$v = K\sqrt{\frac{\rho_L - \rho_g}{\rho_g}} \qquad (10.1)$$

and

$$q = Av, \qquad (10.2)$$

where

A = total cross-sectional area of separator, ft^2
v = superficial gas velocity based on total cross-sectional area A, ft/sec
q = gas flow rate at operating conditions, ft^3/sec
ρ_L = density of liquid at operating conditions, lb$_m$/ft^3
ρ_g = density of gas at operating conditions, lb$_m$/ft^3
K = empirical factor

Table 10.1 presents K values for various types of separators. Also listed in the table are K values used for other designs such as mist eliminators and trayed towers in dehydration or gas sweetening units.

Substituting Eq. (10.1) into Eq. (10.2) and applying real gas law gives

$$q_{st} = \frac{2.4D^2 Kp}{z(T+460)}\sqrt{\frac{\rho_L - \rho_g}{\rho_g}}, \qquad (10.3)$$

where

q_{st} = gas capacity at standard conditions, MMscfd
D = internal diameter of vessel, ft
p = operation pressure, psia
T = operating temperature, °F
z = gas compressibility factor

It should be noted that Eq. (10.3) is empirical. Height differences in vertical separators and length differences in horizontal separators are not considered. Field experience has indicated that additional gas capacity can be obtained by increasing height of vertical separators and length of horizontal separators. The separator charts (Sivalls, 1977; Ikoku, 1984) give more realistic values for the gas capacity of separators. In addition, for single-tube horizontal vessels, corrections must be made for the amount of liquid in the bottom of the separator. Although one-half full of liquid is more or less standard for most single-tube horizontal separators, lowering the liquid level to increase the available gas space within the vessel can increase the gas capacity.

Table 10.1 K Values Used for Selecting Separators

Separator type	K	Remarks
Vertical separators	0.06–0.35	
Horizontal separators	0.40–0.50	
Wire mesh mist eliminators	0.35	
Bubble cap trayed columns	0.16	24-in. spacing

10.2.4.2 Liquid Capacity

Retention time of the liquid within the vessel determines liquid capacity of a separator. Adequate separation requires sufficient time to obtain an equilibrium condition between the liquid and gas phase at the temperature and pressure of separation. The liquid capacity of a separator relates to the retention time through the settling volume:

$$q_L = \frac{1,440 V_L}{t} \quad (10.4)$$

where

q_L = liquid capacity, bbl/day
V_L = liquid settling volume, bbl
t = retention time, min

Table 10.2 presents t values for various types of separators tested in fields. It is shown that temperature has a strong impact on three-phase separations at low pressures.

Tables 10.3 through 10.8 present liquid-settling volumes with the conventional placement of liquid-level controls for typical oil/gas separators.

Proper sizing of a separator requires the use of both Eq. (10.3) for gas capacity and Eq. (10.4) for liquid capacity.

Experience shows that for high-pressure separators used for treating high gas/oil ratio well streams, the gas capacity is usually the controlling factor for separator selection. However, the reverse may be true for low-pressure separators used on well streams with low gas/oil ratios.

Example Problem 10.1 Calculate the minimum required size of a standard oil/gas separator for the following conditions. Consider both vertical and horizontal separators.

Gas flow rate: 5.0 MMscfd
Gas-specific gravity: 0.7
Condensate flow rate: 20 bbl/MMscf
Condensate gravity: 60°API
Operating pressure: 800 psia
Operating temperature: 80°F

Solution The total required liquid flow capacity is $(5)(20) = 100$ bbl/day. Assuming a 20-in. × $7\frac{1}{2}$-ft vertical separator, Table 10.1 suggests an average K value of 0.205. The spreadsheet program *Hall-Yarborogh-z.xls* gives $z = 0.8427$ and $\rho_g = 3.38$ lb$_m$/ft^3 at 800 psig and 80°F. Liquid density is calculated as

Table 10.2 Retention Time Required Under Various Separation Conditions

Separation condition	T (°F)	t (min)
Oil/gas separation		1
High-pressure oil/gas/water separation		2–5
Low-pressure oil/gas/water separation	>100	5–10
	90	10–15
	80	15–20
	70	20–25
	60	25–30

Table 10.3 Settling Volumes of Standard Vertical High-Pressure Separators (230–2,000 psi working pressure)

	V_L (bbl)	
Size (D × H)	Oil/Gas separators	Oil/Gas/Water separators
16″ × 5′	0.27	0.44
16″ × 7½′	0.41	0.72
16″ × 10′	0.51	0.94
20″ × 5′	0.44	0.71
20″ × 7½′	0.65	1.15
20″ × 10′	0.82	1.48
24″ × 5′	0.66	1.05
24″ × 7½′	0.97	1.68
24″ × 10′	1.21	2.15
30″ × 5′	1.13	1.76
30″ × 7½′	1.64	2.78
30″ × 10′	2.02	3.54
36″ × 7½′	2.47	4.13
36″ × 10′	3.02	5.24
36″ × 15′	4.13	7.45
42″ × 7½′	3.53	5.80
42″ × 10′	4.29	7.32
42″ × 15′	5.80	10.36
48″ × 7½′	4.81	7.79
48″ × 10′	5.80	9.78
48″ × 15′	7.79	13.76
54″ × 7½	6.33	10.12
54″ × 10′	7.60	12.65
54″ × 15′	10.12	17.70
60″ × 7½′	8.08	12.73
60″ × 10′	9.63	15.83
60″ × 15′	12.73	22.03
60″ × 20′	15.31	27.20

Table 10.4 Settling Volumes of Standard Vertical Low-Pressure Separators (125 psi working pressure)

Size (D × H)	V_L (bbl)	
	Oil/Gas separators	Oil/Gas/Water separators
24″ × 5′	0.65	1.10
24″ × 7½′	1.01	1.82
30″ × 10′	2.06	3.75
36″ × 5′	1.61	2.63
36″ × 7½′	2.43	4.26
36″ × 10′	3.04	5.48
48″ × 10′	5.67	10.06
48″ × 15′	7.86	14.44
60″ × 10′	9.23	16.08
60″ × 15′	12.65	12.93
60″ × 20′	15.51	18.64

Table 10.5 Settling Volumes of Standard Horizontal High-Pressure Separators (230–2,000 psi working pressure)

Size (D × L)	V_L(bbl)		
	½ Full	⅓ Full	¼ Full
12¾″ × 5′	0.38	0.22	0.15
12¾″ × 7½′	0.55	0.32	0.21
12¾″ × 10′	0.72	0.42	0.28
16″ × 5′	0.61	0.35	0.24
16″ × 7½′	0.88	0.50	0.34
16″ × 10′	1.14	0.66	0.44
20″ × 5′	0.98	0.55	0.38
20″ × 7½′	1.39	0.79	0.54
20″ × 10′	1.80	1.03	0.70
24″ × 5′	1.45	0.83	0.55
24″ × 7½′	2.04	1.18	0.78
24″ × 10′	2.63	1.52	1.01
24″ × 15′	3.81	2.21	1.47
30″ × 5′	2.43	1.39	0.91
30″ × 7½′	3.40	1.96	1.29
30″ × 10′	4.37	2.52	1.67
30″ × 15′	6.30	3.65	2.42
36″ × 7½	4.99	2.87	1.90
36″ × 10′	6.38	3.68	2.45
36″ × 15′	9.17	5.30	3.54
36″ × 20′	11.96	6.92	4.63
42″ × 7½′	6.93	3.98	2.61
42″ × 10′	8.83	5.09	3.35
42″ × 15′	12.62	7.30	4.83
42″ × 20′	16.41	9.51	6.32
48″ × 7½′	9.28	5.32	3.51
48″ × 10′	11.77	6.77	4.49
48″ × 15′	16.74	9.67	6.43
48″ × 20′	21.71	12.57	8.38
54″ × 7½′	12.02	6.87	4.49
54″ × 10′	15.17	8.71	5.73
54″ × 15′	12.49	12.40	8.20
54″ × 20′	27.81	16.08	10.68
60″ × 7½′	15.05	8.60	5.66
60″ × 10′	18.93	10.86	7.17
60″ × 15′	26.68	15.38	10.21
60″ × 20′	34.44	19.90	13.24

Table 10.6 Settling Volumes of Standard Horizontal Low-Pressure Separators (125 psi working pressure)

Size (D × L)	V_L (bbl)		
	½ Full	⅓ Full	¼ Full
24″ × 5′	1.55	0.89	0.59
24″ × 7½′	2.22	1.28	0.86
24″ × 10′	2.89	1.67	1.12

(Continued)

Table 10.6 Settling Volumes of Standard Horizontal Low-Pressure Separators (125 psi working pressure)(Continued)

Size (D × L)	V_L (bbl)		
	½ Full	⅓ Full	¼ Full
30" × 5'	2.48	1.43	0.94
30" × 7 1/2'	3.54	2.04	1.36
30" × 10'	4.59	2.66	1.77
36" × 10'	6.71	3.88	2.59
36" × 15'	9.76	5.66	3.79
48" × 10'	12.24	7.07	4.71
48" × 15'	17.72	10.26	6.85
60" × 10'	19.50	11.24	7.47
60" × 15'	28.06	16.23	10.82
60" × 20'	36.63	21.21	14.16

Table 10.7 Settling Volumes of Standard Spherical High-Pressure Separators (230–3,000 psi working pressure)

Size (OD)	V_L (bbl)
24"	0.15
30"	0.30
36"	0.54
42"	0.88
48"	1.33
60"	2.20

Table 10.8 Settling Volumes of Standard Spherical Low-Pressure Separators (125 psi)

Size (OD)	V_L (bbl)
41"	0.77
46"	1.02
54"	1.60

$$\rho_L = 62.4 \frac{141.5}{131.5 + 60} = 46.11 \, \text{lb}_m/\text{ft}^3.$$

Equation (10.3) gives

$$q_{st} = \frac{(2.4)(20/12)^2(0.205)(800)}{(0.8427)(80+460)} \sqrt{\frac{46.11-3.38}{3.38}}$$

$$= 8.70 \, \text{MMscfd}.$$

Sivalls's chart gives 5.4 MMscfd.

From Table 10.3, a 20-in. × 7½-ft separator will handle the following liquid capacity:

$$q_L = \frac{1440(0.65)}{1.0} = 936 \, \text{bbl/day},$$

which is much higher than the liquid load of 100 bbl/day.

Consider a 16-in. × 5-ft horizontal separator and Eq. (10.3) gives

$$q_{st} = \frac{(2.4)(16/12)^2(0.45)(800)}{(0.8427)(80+460)} \sqrt{\frac{46.11-3.38}{3.38}}$$

$$= 12.22 \, \text{MMscfd}.$$

If the separator is one-half full of liquid, it can still treat 6.11 MMscfd of gas. Sivalls's chart indicates that a 16-in. × 5-ft horizontal separator will handle 5.1 MMscfd.

From Table 10.5, a half-full, 16-in. × 5-ft horizontal separator will handle

$$q_L = \frac{1440(0.61)}{1.0} = 878 \, \text{bbl/day},$$

which again is much higher than the liquid load of 100 bbl/day.

This example illustrates a case of high gas/oil ratio well streams where the gas capacity is the controlling factor for separator selection. It suggests that a smaller horizontal separator would be required and would be more economical. The selected separator should have at least a 1,000 psig working pressure.

10.2.5 Stage separation

Stage separation is a process in which hydrocarbon mixtures are separated into vapor and liquid phases by multiple equilibrium flashes at consecutively lower pressures. A two-stage separation requires one separator and a storage tank, and a three-stage separation requires two separators and a storage tank. The storage tank is always counted as the final stage of vapor/liquid separation. Stage separation reduces the pressure a little at a time, in steps or stages, resulting in a more stable stock-tank liquid. Usually a stable stock-tank liquid can be obtained by a stage separation of not more than four stages.

In high-pressure gas-condensate separation systems, a stepwise reduction of the pressure on the liquid condensate can significantly increase the recovery of stock-tank liquids. Prediction of the performance of the various separators in a multistage separation system can be carried out with compositional computer models using the initial well stream composition and the operating temperatures and pressures of the various stages.

Although three to four stages of separation theoretically increase the liquid recovery over a two-stage separation, the incremental liquid recovery rarely pays out the cost of the additional separators. It has been generally recognized that two stages of separation plus the stock tank are practically optimum. The increase in liquid recovery for two-stage separation over single-stage separation usually varies from 2 to 12%, although 20 to 25% increases in liquid recoveries have been reported.

The first-stage separator operating pressure is generally determined by the flowline pressure and operating characteristics of the well. The pressure usually ranges from 600 to 1,200 psi. In situations in which the flowline pressure is greater than 600 psi, it is practical to let the first-stage separator ride the line or operate at the flowline pressure. Pressures at low-stage separations can be determined based on equal pressure ratios between the stages (Campbell, 1976):

$$R_p = \left(\frac{p_1}{p_s}\right)^{\frac{1}{N_{st}}}, \tag{10.5}$$

where

R_p = pressure ratio
N_{st} = number of stages -1
p_1 = first-stage or high-pressure separator pressure, psia
p_s = stock-tank pressure, psia

Pressures at the intermediate stages can then be designed with the following formula:

$$p_i = \frac{p_{i-1}}{R_p}, \qquad (10.6)$$

where p_i = pressure at stage i, psia.

10.3 Dehydration System

All natural gas downstream from the separators still contain water vapor to some degree. Water vapor is probably the most common undesirable impurity found in the untreated natural gas. The main reason for removing water vapor from natural gas is that water vapor becomes liquid water under low-temperature and/or high-pressure conditions. Specifically, water content can affect long-distance transmission of natural gas because of the following facts:

1. Liquid water and natural gas can form hydrates that may plug the pipeline and other equipment.
2. Natural gas containing CO_2 and/or H_2S is corrosive when liquid water is present.
3. Liquid water in a natural gas pipeline potentially causes slugging flow conditions resulting in lower flow efficiency of the pipeline.
4. Water content decreases the heating value of natural gas being transported.

Dehydration systems are designed for further separating water vapor from natural gas before the gas is transported by pipeline.

10.3.1 Water Content of Natural Gas Streams

Solubility of water in natural gas increases with temperature and decreases with pressure. The presence of salt in the liquid water reduces the water content of the gas. Water content of untreated natural gases is normally in the magnitude of a few hundred pounds of water per million standard cubic foot of gas (lb_m/MMscf); while gas pipelines normally require water content to be in the range of $6-8\,lb_m$/MMscf and even lower for offshore pipelines.

The water content of natural gas is indirectly indicated by the "dew point," defined as the temperature at which the natural gas is saturated with water vapor at a given pressure. At the dew point, natural gas is in equilibrium with liquid water; any decrease in temperature or increase in pressure will cause the water vapor to begin condensing. The difference between the dew point temperature of a water-saturated gas stream and the same stream after it has been dehydrated is called "dew-point depression."

It is essential to accurately estimate the saturated water vapor content of natural gas in the design and operation of dehydration equipment. Several methods are available for this purpose including the correlations of McCarthy et al. (1950) and McKetta and Wehe (1958). Dalton's law of partial pressures is valid for estimating water vapor content of gas at near-atmospheric pressures. Readings from the chart by McKetta and Wehe (1958) were re-plotted in Fig. 10.6 by Guo and Ghalambor (2005).

Example Problem 10.2 Estimate water content of a natural gas at a pressure of 3,000 psia and temperature of 150 °F.

Solution The chart in Fig. 10.6 gives water contents of

$$C_{w140F} = 84\,lb_m/MMcf$$
$$C_{w160F} = 130\,lb_m/MMcf$$

Linear interpolation yields:

$$C_{w150F} = 107\,lb_m/MMcf$$

10.3.2 Methods for Dehydration

Dehydration techniques used in the petroleum industry fall into four categories in principle: (a) direct cooling, (b) compression followed by cooling, (c) absorption, and (d) adsorption. Dehydration in the first two methods does

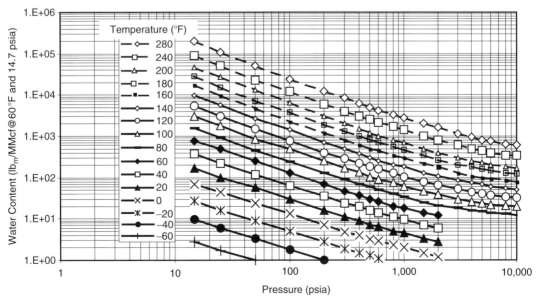

Figure 10.6 Water content of natural gases (Guo and Ghalambor, 2005).

not result in sufficiently low water contents to permit injection into a pipeline. Further dehydration by absorption or adsorption is often required.

10.3.2.1 Dehydration by Cooling

The ability of natural gas to contain water vapor decreases as the temperature is lowered at constant pressure. During the cooling process, the excess water in the vapor state becomes liquid and is removed from the system. Natural gas containing less water vapor at low temperature is output from the cooling unit. The gas dehydrated by cooling is still at its water dew point unless the temperature is raised again or the pressure is decreased. Cooling for the purpose of gas dehydration is sometimes economical if the gas temperature is unusually high. It is often a good practice that cooling is used in conjunction with other dehydration processes.

Gas compressors can be used partially as dehydrators. Because the saturation water content of gases decreases at higher pressure, some water is condensed and removed from gas at compressor stations by the compressor discharge coolers. Modern lean oil absorption gas plants use mechanical refrigeration to chill the inlet gas stream. Ethylene glycol is usually injected into the gas chilling section of the plant, which simultaneously dehydrates the gas and recovers liquid hydrocarbons, in a manner similar to the low-temperature separators.

10.3.2.2 Dehydration by Adsorption

"Adsorption" is defined as the ability of a substance to hold gases or liquids on its surface. In adsorption dehydration, the water vapor from the gas is concentrated and held at the surface of the solid desiccant by forces caused by residual valiancy. Solid desiccants have very large surface areas per unit weight to take advantage of these surface forces. The most common solid adsorbents used today are silica, alumina, and certain silicates known as *molecular sieves*. Dehydration plants can remove practically all water from natural gas using solid desiccants. Because of their great drying ability, solid desiccants are employed where higher efficiencies are required.

Depicted in Fig. 10.7 is a typical solid desiccant dehydration plant. The incoming wet gas should be cleaned preferably by a filter separator to remove solid and liquid contaminants in the gas. The filtered gas flows downward during dehydration through one adsorber containing a desiccant bed. The down-flow arrangement reduces disturbance of the bed caused by the high gas velocity during the adsorption. While one adsorber is dehydrating, the other adsorber is being regenerated by a hot stream of inlet gas from the regeneration gas heater. A direct-fired heater, hot oil, steam, or an indirect heater can supply the necessary regeneration heat. The regeneration gas usually flows upward through the bed to ensure thorough regeneration of the bottom of the bed, which is the last area contacted by the gas being dehydrated. The hot regenerated bed is cooled by shutting off or bypassing the heater. The cooling gas then flows downward through the bed so that any water adsorbed from the cooling gas will be at the top of the bed and will not be desorbed into the gas during the dehydration step. The still-hot regeneration gas and the cooling gas flow through the regeneration gas cooler to condense the desorbed water. Power-operated valves activated by a timing device switch the adsorbers between the dehydration, regeneration, and cooling steps.

Under normal operating conditions, the usable life of a desiccant ranges from 1 to 4 years. Solid desiccants become less effective in normal use because of loss of effective surface area as they age. Abnormally fast degradation occurs through blockage of the small pores and capillary openings lubricating oils, amines, glycols, corrosion inhibitors, and other contaminants, which cannot be removed during the regeneration cycle. Hydrogen sulfide can also damage the desiccant and reduce its capacity.

The advantages of solid-desiccant dehydration include the following:

- Lower dew point, essentially dry gas (water content <1.0 lb/MMcf) can be produced
- Higher contact temperatures can be tolerated with some adsorbents
- Higher tolerance to sudden load changes, especially on startup
- Quick start up after a shutdown
- High adaptability for recovery of certain liquid hydrocarbons in addition to dehydration functions

Operating problems with the solid-desiccant dehydration include the following:

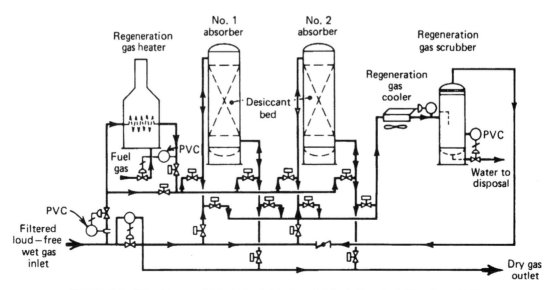

Figure 10.7 Flow diagram of a typical solid desiccant dehydration plant (Guenther, 1979).

- Space adsorbents degenerate with use and require replacement

Dehydrating tower must be regenerated and cooled for operation before another tower approaches exhaustion. The maximum allowable time on dehydration gradually shortens because desiccant loses capacity with use.

Although this type of dehydrator has high adaptability to sudden load changes, sudden pressure surges should be avoided because they may upset the desiccant bed and channel the gas stream resulting in poor dehydration. If a plant is operated above its rated capacity, high-pressure loss may introduce some attrition to occur. Attrition causes fines, which may in turn cause excessive pressure loss and result in loss of capacity.

Replacing the desiccant should be scheduled and completed ahead of the operating season. To maintain continuous operation, this may require discarding the desiccant before its normal operating life is reached. To cut operating costs, the inlet part of the tower can be recharged and the remainder of the desiccant retained because it may still possess some useful life. Additional service life of the desiccant may be obtained if the direction of gas flow is reversed at a time when the tower would normally be recharged.

10.3.2.3 Dehydration by Absorption

Water vapor is removed from the gas by intimate contact with a hygroscopic liquid desiccant in absorption dehydration. The contacting is usually achieved in packed or trayed towers. Glycols have been widely used as effective liquid desiccants. Dehydration by absorption with glycol is usually economically more attractive than dehydration by solid desiccant when both processes are capable of meeting the required dew point.

Glycols used for dehydrating natural gas are ethylene glycol (EG), diethylene glycol (DEG), triethylene glycol (TEG), and tetraethylene glycol (T_4EG). Normally a single type of pure glycol is used in a dehydrator, but sometimes a glycol blend is economically attractive. TEG has gained nearly universal acceptance as the most cost effective of the glycols because of its superior dew-point depression, operating cost, and operation reliability. TEG has been successfully used to dehydrate sweet and sour natural gases over wide ranges of operating conditions. Dew-point depression of 40–140°F can be achieved at a gas pressure ranging from 25 to 2,500 psig and gas temperature between 40 and 160°F. The dew-point depression obtained depends on the equilibrium dew-point temperature for a given TEG concentration and contact temperature. Increased glycol viscosity may cause problems at lower contact temperature. Thus, heating of the natural gas may be desirable. Very hot gas streams are often cooled before dehydration to prevent vaporization of TEG.

The feeding-in gas must be cleaned to remove all liquid water and hydrocarbons, wax, sand, drilling muds, and other impurities. These substances can cause severe foaming, flooding, higher glycol losses, poor efficiency, and increased maintenance in the dehydration tower or absorber. These impurities can be removed using an efficient scrubber, separator, or even a filter separator for very contaminated gases. Methanol, injected at the wellhead as hydrate inhibitor, can cause several problems for glycol dehydration plants. It increases the heat requirements of the glycol regeneration system. Slugs of liquid methanol can cause flooding in the absorber. Methanol vapor vented to the atmosphere with the water vapor from the regeneration system is hazardous and should be recovered or vented at nonhazardous concentrations.

10.3.2.3.1 Glycol Dehydration Process Illustrated in Fig. 10.8 shows the process and flow through a typical glycol dehydrator. The dehydration process can be described as follows:

1. The feeding-in gas stream first enters the unit through an inlet gas scrubber to remove liquid accumulations. A two-phase inlet scrubber is normally required.
2. The wet gas is then introduced to the bottom of the glycol-gas contactor and allowed to flow upward through the trays, while glycol flows downward through the column. The gas contacts the glycol on

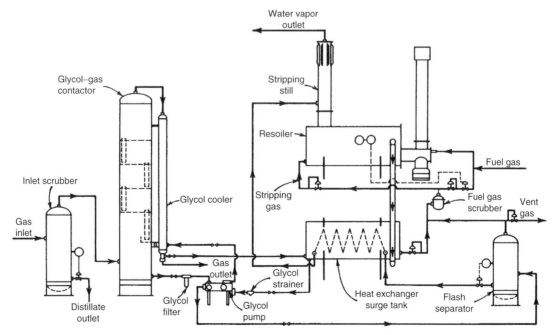

Figure 10.8 Flow diagram of a typical glycol dehydrator (Sivalls, 1977).

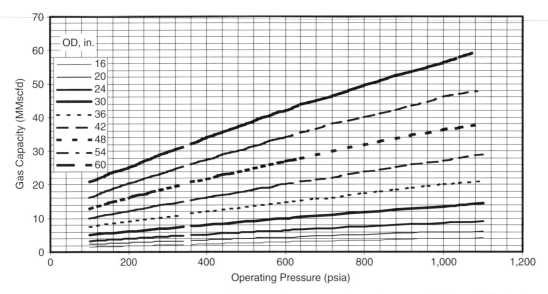

Figure 10.9 Gas capacity of vertical inlet scrubbers based on 0.7-specific gravity at 100 °F (Guo and Ghalambor, 2005).

each tray and the glycol absorbs the water vapor from the gas steam.
3. The gas then flows down through a vertical glycol cooler, usually fabricated in the form of a concentric pipe heat exchanger, where the outlet dry gas aids in cooling the hot regenerated glycol before it enters the contactor. The dry gas then leaves the unit from the bottom of the glycol cooler.
4. The dry glycol enters the top of the glycol-gas contactor from the glycol cooler and is injected onto the top tray. The glycol flows across each tray and down through a downcomer pipe onto the next tray. The bottom tray downcomer is fitted with a seal pot to hold a liquid seal on the trays.
5. The wet glycol, which has now absorbed the water vapor from the gas stream, leaves the bottom of the glycol-gas contactor column, passes through a high-pressure glycol filter, which removes any foreign solid particles that may have been picked up from the gas stream, and enters the power side of the glycol pump.
6. In the glycol pump, the wet high-pressure glycol from the contactor column pumps the dry regenerated glycol into the column. The wet glycol stream flows from the glycol pump to the flash separator, which allows for the release of the entrained solution gas.
7. The gas separated in the flash separator leaves the top of the flash separator vessel and can be used to supplement the fuel gas required for the reboiler. Any excess vent gas is discharged through a backpressure valve. The flash separator is equipped with a liquid level control and diaphragm motor valve that discharges the wet glycol stream through a heat exchange coil in the surge tank to preheat the wet glycol stream.
8. The wet glycol stream leaves the heat exchange coil in the surge tank and enters the stripping still mounted on top of the reboiler at the feed point in the still. The stripping still is packed with a ceramic intalox saddle-type packing, and the glycol flows downward through the column and enters the reboiler. The wet glycol passing downward through the still is contacted by hot rising glycol and water vapors passing upward through the column. The water vapors released in the reboiler and stripped from the glycol in the stripping still pass upward through the still column through an atmospheric reflux condenser that provides a partial reflux for the column. The water vapor then leaves the top of the stripping still column and is released to the atmosphere.
9. The glycol flows through the reboiler in essentially a horizontal path from the stripping still column to the opposite end. In the reboiler, the glycol is heated to approximately 350–400 °F to remove enough water vapor to re-concentrate it to 99.5% or higher. In field dehydration units, the reboiler is generally equipped with a direct-fired firebox, using a portion of the natural gas stream for fuel.
10. The re-concentrated glycol leaves the reboiler through an overflow pipe and passes into the shell side of the heat exchanger/surge tank. In the surge tank, the hot re-concentrated glycol is cooled by exchanging heat with the wet glycol stream passing through the coil. The surge tank also acts as a liquid accumulator for feed for the glycol pump. The re-concentrated glycol flows from the surge tank through a strainer and into the glycol pump. From the pump, it passes into the shell side of the glycol cooler mounted on the glycol-gas contactor. It then flows upward through the glycol cooler where it is further cooled and enters the column on the top tray.

10.3.2.3.2 Advantages and Limitations Glycol dehydrators have several advantages including the following:

- Low initial equipment cost
- Low pressure drop across absorption towers
- Continuous operation
- Makeup requirements may be added readily
- Recharging of towers presents no problems
- Plant may be used satisfactorily in the presence of materials that would cause fouling of some solid adsorbents

Glycol dehydrators also present several operating problems including the following:

- Suspended matter, such as dirt, scale, and iron oxide, may contaminate glycol solutions.

- Overheating of solution may produce both low and high boiling decomposition products.
- The resultant sludge may collect on heating surfaces, causing some loss in efficiency, or in severe cases, complete flow stoppage.
- When both oxygen and hydrogen sulfide are present, corrosion may become a problem because of the formation of acid material in glycol solution.
- Liquids (e.g., water, light hydrocarbons, or lubrication oils) in inlet gas may require installation of an efficient separator ahead of the absorber. Highly mineralized water entering the system with inlet gas may, over long periods, crystallize and fill the reboiler with solid salts.
- Foaming of solution may occur with a resultant carry over of liquid. The addition of a small quantity of antifoam compound usually remedies this problem.
- Some leakage around the packing glands of pumps may be permitted because excessive tightening of packing may result in the scouring of rods. This leakage is collected and periodically returned to the system.
- Highly concentrated glycol solutions tend to become viscous at low temperatures and, therefore, are hard to pump. Glycol lines may solidify completely at low temperatures when the plant is not operating. In cold weather, continuous circulation of part of the solution through the heater may be advisable. This practice can also prevent freezing in water coolers.
- To start a plant, all absorber trays must be filled with glycol before good contact of gas and liquid can be expected. This may also become a problem at low circulation rates because weep holes on trays may drain solution as rapidly as it is introduced.
- Sudden surges should be avoided in starting and shutting down a plant. Otherwise, large carryover losses of solution may occur.

10.3.2.3.3 Sizing Glycol Dehydrator Unit
Dehydrators with TEG in trays or packed-column contactors can be sized from standard models by using the following information:

- Gas flow rate
- Specific gravity of gas
- Operating pressure
- Maximum working pressure of contact
- Gas inlet temperature
- Outlet gas water content required

One of the following two design criteria can be employed:

1. Glycol/water ratio (GWR): A value of 2–6 gal TEG/lb$_m$ H$_2$O removed is adequate for most glycol dehydration requirements. Very often 2.5–4.0 gal TEG/lb$_m$ H$_2$O is used for field dehydrators.
2. Lean TEG concentration from re-concentrator. Most glycol re-concentrators can output 99.0–99.9% lean TEG. A value of 99.5% lean TEG is used in most designs.

Inlet Scrubber. It is essential to have a good inlet scrubber for efficient operation of a glycol dehydrator unit. Two-phase inlet scrubbers are generally constructed with $7\frac{1}{2}$-ft shell heights. The required minimum diameter of a vertical inlet scrubber can be determined based on the operating pressure and required gas capacity using Fig. 10.9, which was prepared by Guo and Ghalambor (2005) based on Sivalls's data (1977).

Glycol-Gas Contactor. Glycol contactors are generally constructed with a standard height of $7\frac{1}{2}$ ft. The minimum required diameter of the contactor can be determined based on the gas capacity of the contactor for standard gas of 0.7 specific gravity at standard temperature 100 °F. If the gas is not the standard gas and/or the operating temperature is different from the standard temperature, a correction should be first made using the following relation:

$$q_s = \frac{q}{C_t C_g}, \qquad (10.7)$$

where

q = gas capacity of contactor at operating conditions, MMscfd
q_s = gas capacity of contactor for standard gas (0.7 specific gravity) at standard temperature (100 °F), MMscfd
C_t = correction factor for operating temperature
C_g = correction factor for gas-specific gravity

The temperature and gas-specific gravity correction factors for trayed glycol contactors are given in Tables 10.9 and 10.10, respectively. The temperature and specific gravity factors for packed glycol contactors are contained in Tables 10.11 and 10.12, respectively.

Once the gas capacity of the contactor for standard gas at standard temperature is calculated, the required minimum diameter of a trayed glycol contactor can be calculated using Fig. 10.10. The required minimum diameter of a packed glycol contactor can be determined based on Fig. 10.11.

Table 10.9 Temperature Correction Factors for Trayed Glycol Contactors

Operating temperature (°F)	Correction factor (C_t)
40	1.07
50	1.06
60	1.05
70	1.04
80	1.02
90	1.01
100	1.00
110	0.99
120	0.98

Source: Used, with permission, from Sivalls, 1977.

Table 10.10 Specific Gravity Correction Factors for Trayed Glycol Contactors

Gas-specific gravity (air = 1)	Correction factor (C_g)
0.55	1.14
0.60	1.08
0.65	1.04
0.70	1.00
0.75	0.97
0.80	0.93
0.85	0.90
0.90	0.88

Source: Used, with permission, from Sivalls, 1977.

Table 10.11 Temperature Correction Factors for Packed Glycol Contactors

Operating temperature (°F)	Correction factor (C_t)
50	0.93
60	0.94
70	0.96
80	0.97
90	0.99
100	1.00
110	1.01
120	1.02

Source: Used, with permission, from Sivalls, 1977.

Table 10.12 Specific Gravity Correction Factors for Packed Glycol Contactors

Gas-specific gravity (air =1)	Correction Factor (C_g)
0.55	1.13
0.60	1.08
0.65	1.04
0.70	1.00
0.75	0.97
0.80	0.94
0.85	0.91
0.90	0.88

Source: Used, with permission, from Sivalls, 1977.

The required minimum height of packing of a packed contactor, or the minimum number of trays of a trayed contactor, can be determined based on Fig. 10.12.

Example Problem 10.2 Size a trayed-type glycol contactor for a field installation to meet the following requirements:

Gas flow rate:	12 MMscfd
Gas specific gravity:	0.75
Operating line pressure:	900 psig
Maximum working pressure of contactor:	1,440 psig
Gas inlet temperature:	90 °F
Outlet gas water content:	6 lb H_2O/MMscf
Design criteria:	GWR = 3 gal TEG/lb$_m$ H_2O with 99.5% TEG

Solution Because the given gas is not a standard gas and the inlet temperature is not the standard temperature, corrections need to be made. Tables 10.9 and 10.10 give $C_t = 1.01$ and $C_g = 0.97$. The gas capacity of contactor is calculated with Eq. (10.7):

$$q_s = \frac{12}{(1.01)(0.97)} = 12.25 \, \text{MMscfd}.$$

Figure 10.10 gives contactor diameter $D_C = 30$ in.
Figure 10.6 gives water content of inlet gas: $C_{wi} = 50 \, \text{lb}_m/\text{MMscf}$.
The required water content of outlet gas determines the dew-point temperature of the outlet gas through Fig. 10.6: $t_{do} = 28 \, °F$.
Therefore, the dew-point depression is $\Delta t_d = 90 - 28 = 62 \, °F$.
Based on GWR = 3 gal TEG/lb$_m$ H_2O and $\Delta t_d = 62 \, °F$, Fig. 10.12 gives the number of trays rounded off to be four.

Glycol Re-concentrator: Sizing the various components of a glycol re-concentrator starts from calculating the required glycol circulation rate:

$$q_G = \frac{(GWR)C_{wi}q}{24}, \qquad (10.8)$$

where

q_G = glycol circulation rate, gal/hr
GWR = GWR, gal TEG/lb$_m$ H_2O
C_{wi} = water content of inlet gas, lb$_m$ H_2O/MMscf
q = gas flow rate, MMscfd

Reboiler: The required heat load for the reboiler can be approximately estimated from the following equation:

$$H_t = 2,000 q_G, \qquad (10.9)$$

where

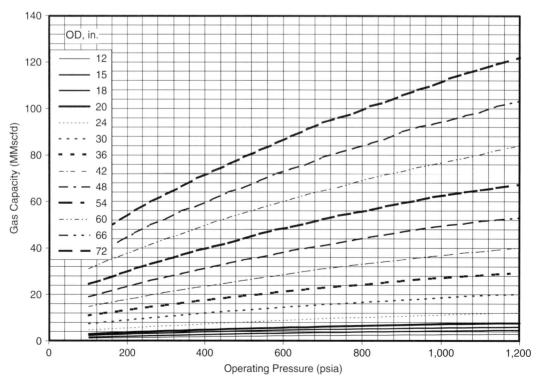

Figure 10.10 Gas capacity for trayed glycol contactors based on 0.7-specific gravity at 100 °F (Sivalls, 1977).

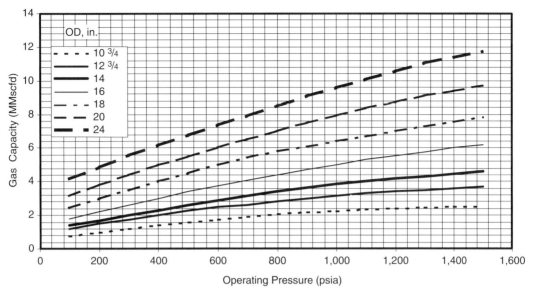

Figure 10.11 *Gas capacity for packed glycol contactors based on 0.7-specific gravity at 100 °F (Sivalls, 1977).*

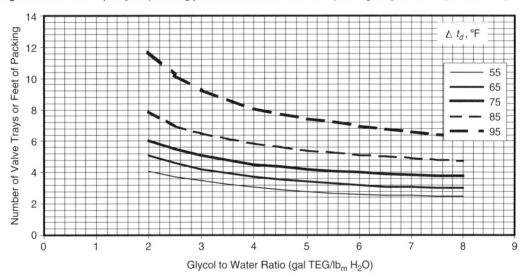

Figure 10.12 *The required minimum height of packing of a packed contactor, or the minimum number of trays of a trayed contactor (Sivalls, 1977).*

H_t = total heat load on reboiler, Btu/hr

Equation (10.9) is accurate enough for most high-pressure glycol dehydrator sizing. A more detailed procedure for determination of the required reboiler heat load can be found from Ikoku (1984). The general overall size of the reboiler can be determined as follows:

$$A_{fb} = \frac{H_t}{7,000}, \qquad (10.10)$$

where A_{fb} is the total firebox surface area in squared feet.

Glycol Circulating Pump: The glycol circulating pump can be sized using the glycol circulation rate and the maximum operating pressure of the contactor. Commonly used glycol-powered pumps use the rich glycol from the bottom of the contactor to power the pump and pump the lean glycol to the top of the contactor. The manufacturers of these pumps should be consulted to meet the specific needs of the glycol dehydrator.

Glycol Flash Separator: A glycol flash separator is usually installed downstream from the glycol pump to remove any entrained hydrocarbons from the rich glycol. A small 125-psi vertical two-phase separator is usually adequate for this purpose. The separator should be sized based on a liquid retention time in the vessel of at least 5 minutes.

$$V_s = \frac{q_G t_r}{60}, \qquad (10.11)$$

where

V_s = required settling volume in separator, gal
q_G = glycol circulation rate, gph
t_r = retention time approximately 5 minute

Liquid hydrocarbon is not allowed to enter the glycol-gas contactor. If this is a problem, a three-phase glycol flash separator should be used to keep these liquid hydrocarbons out of the reboiler and stripping still. Three-phase flash separators should be sized with a liquid retention time of 20–30 minutes. The hydrocarbon gas released from the flash separator can be piped to the reboiler to use as fuel gas and stripping gas. Based on the glycol circulation rate and the operating pressure of the contactor, the amount of gas available from the glycol pump can be determined.

Stripping Still: The size of the packed stripping still for the glycol re-concentrator can be determined based on the glycol-to-water circulation rate (gas TEG/lb$_m$ H$_2$O) and the glycol circulation rate (gph). The required diameter for the stripping still is normally based on the required diameter at the base of the still using the vapor and liquid loading conditions at the base point. The vapor load consists of the water vapor and stripping gas flowing up through the still. The liquid load consists of the rich glycol stream and reflux flowing downward through the still column. One tray is normally sufficient for most stripping still requirements for TEG dehydration units. The amount of stripping gas required to re-concentrate the glycol is approximately 2–10 ft^3 per gal of glycol circulated.

Summary

This chapter gives a brief introduction to fluid separation and gas dehydration systems. A guideline to selection of system components is also presented. Operators need to consult with equipment providers in designing their separation systems.

References

AHMED, T. *Hydrocarbon Phase Behavior*. Houston: Gulf Publishing Company, 1989.

CAMPBELL, J.M. *Gas Conditioning and Processing*. Norman, OK: Campbell Petroleum Services, 1976.

GUENTHER, J.D. Natural gas dehydration. Paper presented at the Seminar on Process Equipment and Systems on Treatment Platforms, April 26, 1979, Taastrup, Denmark.

GUO, B. and GHALAMBOR, A. *Natural Gas Engineering Handbook*. Houston: Gulf Publishing Company, 2005.

IKOKU, C.U. *Natural Gas Production Engineering*. New York: John Wiley & Sons, 1984.

MCCARTHY, E.L., BOYD, W.L., and REID, L.S. The water vapor content of essentially nitrogen-free natural gas saturated at various conditions of temperature and pressure. *Trans. AIME* 1950;189:241–243.

MCKETTA, J.J. and WEHE, W.L. Use this chart for water content of natural gases. *Petroleum Refinery* 1958;37:153–154.

SIVALLS, C.R. Fundamentals of oil and gas separation. Proceedings of the Gas Conditioning Conference, University of Oklahoma, Norman, Oklahoma, 1977.

Problems

10.1 Calculate the minimum required size of a standard oil/gas separator for the following conditions (consider vertical, horizontal, and spherical separators):

Gas flow rate:	4.0 MMscfd
Gas-specific gravity:	0.7
Condensate-gas ratio (CGR):	15 bbl/MMscf
Condensate gravity:	65 °API
Operating pressure:	600 psig
Operating temperature:	70 °F

10.2 A three-stage separation is proposed to treat a well stream at a flowline pressure of 1,000 psia. Calculate pressures at each stage of separation.

10.3 Estimate water contents of a natural gas at a pressure of 2,000 psia and temperatures of 40, 80, 120, 160, 200, and 240 °F.

10.4 Design a glycol contactor for a field dehydration installation to meet the following requirements. Consider both trayed-type and packed-type contactors.

Gas flow rate:	10 MMscfd
Gas-specific gravity:	0.65
Operating line pressure:	1,000 psig
Maximum working pressure of contactor:	1,440 psig
Gas inlet temperature:	90 °F
Outlet gas water content:	7 lb H$_2$O/MMscf
Design criteria with 99.5% TEG:	GWR = 3 gal TEG/lb$_m$ H$_2$O

11 Transportation Systems

Contents
11.1 Introduction 11/134
11.2 Pumps 11/134
11.3 Compressors 11/136
11.4 Pipelines 11/143
Summary 11/156
References 11/157
Problems 11/157

11.1 Introduction

Crude oil and natural gas are transmitted over short and long distances mainly through pipelines. Pumps and compressors are used for providing pressures required for the transportation. This chapter presents principles of pumps and compressors and techniques that are used for selecting these equipments. Pipeline design criteria and fluid flow in pipelines are also discussed. Flow assurance issues are addressed.

11.2 Pumps

Reciprocating piston pumps (also called "slush pumps" or "power pumps") are widely used for transporting crude oil through pipelines. There are two types of piston strokes: the single-action piston stroke and the double-action piston stroke. These are graphically shown in Figs. 11.1 and 11.2. The double-action stroke is used for duplex (two pistons) pumps. The single-action stroke is used for pumps with three or more pistons (e.g., triplex pump). Normally, duplex pumps can handle higher flow rate and triplex pumps can provide higher pressure.

11.2.1 Triplex Pumps

The work per stroke for a single piston is expressed as

$$\overline{W_1} = P\left(\frac{\pi D^2}{4}\right)L \,(\text{ft} - \text{lbs}).$$

The work per one rotation of crank is

$$\overline{W_2} = P\left(\frac{\pi D^2}{4}\right)L(1) \,(\text{ft} - \text{lbs})/\text{rotation},$$

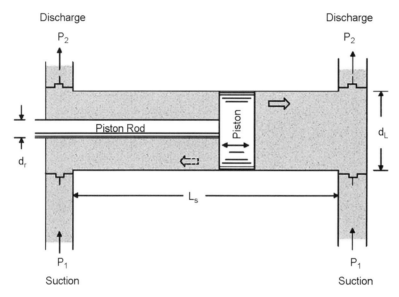

Figure 11.1 Double-action stroke in a duplex pump.

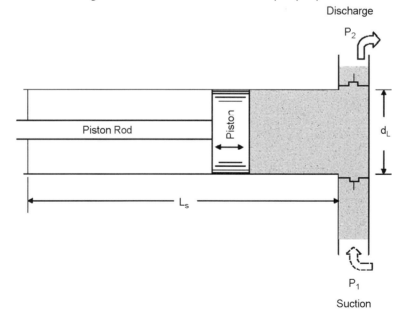

Figure 11.2 Single-action stroke in a triplex pump.

where

$P = $ pressure, lb/ft^2
$L = $ stroke length, ft
$D = $ piston diameter, ft.

Thus, for a triplex pump, the theoretical power is

$$\text{Power} = 3P\left(\frac{\pi D^2}{4}\right) LN \left(\frac{\text{ft} - \text{lb}}{\text{min}}\right), \quad (11.1)$$

where N is pumping speed in strokes per minute. The theoretical horsepower is

$$HP_{th} = \frac{3P\left(\frac{\pi D^2}{4}\right)}{550(60)} LN \text{ (hp)} \quad (11.2)$$

or

$$HP_{th} = \frac{3P\left(\frac{\pi D^2}{4}\right)}{33,000} LN \text{ (hp)}. \quad (11.3)$$

The input horsepower needed from the prime mover is

$$HP_i = \frac{3P\left(\frac{\pi D^2}{4}\right)}{33,000 e_m} LN \text{ (hp)}, \quad (11.4)$$

where e_m is the mechanical efficiency of the mechanical system transferring power from the prime mover to the fluid in the pump. Usually e_m is taken to be about 0.85.

The theoretical volume output from a triplex pump per revolution is

$$Q_{th} = 3\left(\frac{\pi D^2}{4}\right) \frac{LN}{60} \text{ (ft}^3/\text{sec}). \quad (11.5)$$

The theoretical output in bbl/day is thus

$$q_{th} = 604 LND^2 \left(\frac{\text{bbl}}{\text{day}}\right). \quad (11.6)$$

If we use inches (i.e., d [in.] and l [in.]) for D and L, then

$$q_{th} = 0.35 lNd^2 \left(\frac{\text{bbl}}{\text{day}}\right). \quad (11.7)$$

The real output of the pump is dependent on how efficiently the pump can fill the chambers of the pistons. Using the volumetric efficiency e_v in Eq. 11.7 gives

$$q_r = 0.35 e_v lNd^2 \left(\frac{\text{bbl}}{\text{day}}\right) \quad (11.8)$$

or

$$q_r = 0.01 e_v d^2 lN \text{ (gal/min)}, \quad (11.9)$$

where e_v is usually taken to be 0.88–0.98.

As the above volumetric equation can be written in d and l, then the horsepower equation can be written in d, l, and p (psi). Thus,

$$HP_i = \frac{3p\left(\frac{\pi d^2}{4}\right)\frac{l}{12} N}{33,000 e_m} \quad (11.10)$$

reduces to

$$HP_i = \frac{pd^2 lN}{168,067 e_m}. \quad (11.11)$$

11.2.2 Duplex Pumps

The work per stroke cycle is expressed as

$$W_1 = P\left(\frac{\pi D_1^2}{4}\right)L + P\left[\frac{\pi D_1^2}{4} - \frac{\pi D_2^2}{4}\right]L \text{(ft} - \text{lbs)}. \quad (11.12)$$

The work per one rotation of crank is

$$W_2 = \left\{P\left(\frac{\pi D_1^2}{4}\right)L + P\left[\frac{\pi D_1^2}{4} - \frac{\pi D_2^2}{4}\right]L\right\}(1)\left(\frac{\text{ft} - \text{lbs}}{\text{rotation}}\right). \quad (11.13)$$

Thus, for a duplex pump, the theoretical power is

$$\text{Power} = 2$$
$$\times \left\{P\left(\frac{\pi D_1^2}{4}\right)L + P\left[\frac{\pi D_1^2}{4} - \frac{\pi D_2^2}{4}\right]L\right\} N \left(\frac{\text{ft} - \text{lbs}}{\min}\right). \quad (11.14)$$

The theoretical horsepower is

$$HP_{th} = \frac{2\left\{P\left(\frac{\pi D_1^2}{4}\right)L + P\left[\frac{\pi D_1^2}{4} - \frac{\pi D_2^2}{4}\right]L\right\} N}{550(60)} \text{ (hp)}$$

or

$$HP_{th} = \frac{2\left\{P\left(\frac{\pi}{4} D_1^2\right)L + P\left[\frac{\pi D_1^2}{4} - \frac{\pi D_2^2}{4}\right]L\right\} N}{33,000}. \quad (11.15)$$

The input horsepower needed from the prime mover is

$$HP_i = \frac{2\left\{P\left(\frac{\pi}{4} D_1^2\right)L + P\left[\frac{\pi D_1^2}{4} - \frac{\pi D_2^2}{4}\right]L\right\} N}{33,000 e_m} \text{ (hp)}. \quad (11.16)$$

The theoretical volume output from the double-acting duplex pump per revolution is

$$Q_{th} = 2\left\{\frac{\pi D_1^2}{4}L + \left[\frac{\pi D_1^2}{4} - \frac{\pi D_2^2}{4}\right]L\right\} \frac{N}{60} \text{ (ft}^3/\text{sec)}. \quad (11.17)$$

The theoretical output in gals/min is thus

$$q_{th} = 2\left\{\frac{\pi D_1^2}{4}L + \left[\frac{\pi D_1^2}{4} - \frac{\pi D_2^2}{4}\right]L\right\}$$
$$\times \frac{N}{0.1337} \text{(gal/min)}. \quad (11.18)$$

If we use inches (i.e., d [in.] and l [in.]), for D and L, then

$$q_{th} = 2\left\{\frac{\pi d_1^2}{4}l + \left[\frac{\pi d_1^2}{4} - \frac{\pi d_2^2}{4}\right]l\right\} \frac{N}{231} \text{(gal/min)}. \quad (11.19)$$

The real output of the pump is

$$q_r = 2\left\{\frac{\pi d_1^2}{4}l + \left[\frac{\pi d_1^2}{4} - \frac{\pi d_2^2}{4}\right]l\right\} \frac{N}{231} e_v \text{(gal/min)}$$

or

$$q_r = 0.0068 (2d_1^2 - d_2^2) lN e_v \text{ (gal/min)}, \quad (11.20)$$

that is,

$$q_r = 0.233 (2d_1^2 - d_2^2) lN e_v \text{ (bbl/day)}. \quad (11.21)$$

As in the volumetric output, the horsepower equation can also be reduced to a form with p, d_1, d_2, and l

$$HP_i = \frac{p(2d_1^2 - d_2^2) lN}{252,101 e_m}. \quad (11.22)$$

Returning to Eq. (11.16) for the duplex double-action pump, let us derive a simplified pump equation. Rewriting Eq. (11.16), we have

$$HP_i = \frac{2\left\{P\left(\frac{\pi}{4} D_1^2\right)L + P\left[\frac{\pi D_1^2}{4} - \frac{\pi D_2^2}{4}\right]L\right\} N}{33,000 e_m}. \quad (11.23)$$

The flow rate is

$$Q_{th} = 2\left\{\frac{\pi D_1^2}{4}L + \left[\frac{\pi D_1^2}{4} - \frac{\pi D_2^2}{4}\right]L\right\} N \text{ (ft}^3/\text{min)}, \quad (11.24)$$

11/136 EQUIPMENT DESIGN AND SELECTION

so

$$HP_i = \frac{PQ_{th}}{33,000e_m}. \tag{11.25}$$

The usual form of this equation is in p (psi) and q (gal/min):

$$HP_i = \frac{[p(12)^2][q(0.1337)]}{33,000e_m}, \tag{11.26}$$

that is,

$$HP_i = \frac{pq}{1714e_m}. \tag{11.27}$$

The other form of this equation is in p (psi) and q_o (bbl/day) for oil transportation:

$$HP_i = \frac{pq_o}{58,766e_m}. \tag{11.28}$$

Equations (11.27) and (11.28) are valid for any type of pump.

Example Problem 11.1 A pipeline transporting 5,000 bbl/day of oil requires a pump with a minimum output pressure of 1,000 psi. The available suction pressure is 300 psi. Select a triplex pump for this operation.

Solution Assuming a mechanical efficient of 0.85, the horsepower requirement is

$$HP_i = \frac{pq_o}{58,766e_m} = \frac{(1,000)(5,000)}{58,766(0.85)} = 100\,\text{hp}.$$

According to a product sheet of the Oilwell Plunger Pumps, the Model 336-ST Triplex with forged steel fluid end has a rated brake horsepower of 160 hp at 320 rpm. The maximum working pressure is 3,180 psi with the minimum plunger (piston) size of $1\tfrac{3}{4}$ in. It requires a suction pressure of 275 psi. With 3-in. plungers, the pump displacement is 0.5508 gal/rpm, and it can deliver liquid flow rates in the range of 1,889 bbl/day (55.08 gpm) at 100 rpm to 6,046 bbl/day (176.26 gpm) at 320 rpm, allowing a maximum pressure of 1,420 psi. This pump can be selected for the operation. The required operating rpm is

$$\text{RPM} = \frac{(5,000)(42)}{(24)(60)(0.5508)} = 265\,\text{rpm}.$$

11.3 Compressors

When natural gas does not have sufficient potential energy to flow, a compressor station is needed. Five types of compressor stations are generally used in the natural gas production industry:

- Field gas-gathering stations to gather gas from wells in which pressure is insufficient to produce at a desired rate of flow into a transmission or distribution system. These stations generally handle suction pressures from below atmospheric pressure to 750 psig and volumes from a few thousand to many million cubic feet per day.
- Relay or main-line stations to boost pressure in transmission lines compress generally large volumes of gas at a pressure range between 200 and 1,300 psig.
- Re-pressuring or recycling stations to provide gas pressures as high as 6,000 psig for processing or secondary oil recovery projects.
- Storage field stations to compress trunk line gas for injection into storage wells at pressures up to 4,000 psig.
- Distribution plant stations to pump gas from holder supply to medium- or high-pressure distribution lines at about 20–100 psig, or pump into bottle storage up to 2,500 psig.

11.3.1 Types of Compressors

The compressors used in today's natural gas production industry fall into two distinct types: reciprocating and rotary compressors. Reciprocating compressors are most commonly used in the natural gas industry. They are built for practically all pressures and volumetric capacities. As shown in Fig. 11.3, reciprocating compressors have more moving parts and, therefore, lower mechanical efficiencies than rotary compressors. Each cylinder assembly of a reciprocation compressor consists of a piston, cylinder, cylinder heads, suction and discharge valves, and other parts necessary to convert rotary motion to reciprocation motion. A reciprocating compressor is designed for a certain range of compression ratios through the selection of proper piston displacement and clearance volume within the cylinder. This clearance volume can be either fixed or variable, depending on the extent of the operation range and the percent of load variation desired. A typical reciprocating compressor can deliver a volumetric gas flow rate up to 30,000 cubic feet per minute (cfm) at a discharge pressure up to 10,000 psig.

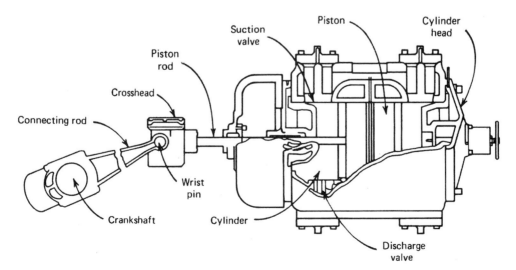

Figure 11.3 Elements of a typical reciprocating compressor (courtesy of Petroleum Extension Services).

Rotary compressors are divided into two classes: the centrifugal compressor and the rotary blower. A centrifugal compressor (Fig. 11.4) consists of a housing with flow passages, a rotating shaft on which the impeller is mounted, bearings, and seals to prevent gas from escaping along the shaft. Centrifugal compressors have few moving parts because only the impeller and shaft rotate. Thus, its efficiency is high and lubrication oil consumption and maintenance costs are low. Cooling water is normally unnecessary because of lower compression ratio and lower friction loss. Compression rates of centrifugal compressors are lower because of the absence of positive displacement. Centrifugal compressors compress gas using centrifugal force. In this type of compressor, work is done on the gas by an impeller. Gas is then discharged at a high velocity into a diffuser where the velocity is reduced and its kinetic energy is converted to static pressure. Unlike reciprocating compressors, all this is done without confinement and physical squeezing. Centrifugal compressors with relatively unrestricted passages and continuous flow are inherently high-capacity, low-pressure ratio machines that adapt easily to series arrangements within a station. In this way, each compressor is required to develop only part of the station compression ratio. Typically, the volume is more than 100,000 cfm and discharge pressure is up to 100 psig.

A rotary blower is built of a casing in which one or more impellers rotate in opposite directions. Rotary blowers are primarily used in distribution systems where the pressure differential between suction and discharge is less than 15 psi. They are also used for refrigeration and closed regeneration of adsorption plants. The rotary blower has several advantages: large quantities of low-pressure gas can be handled at comparatively low horsepower, it has small initial cost and low maintenance cost, it is simple to install and easy to operate and attend, it requires minimum floor space for the quantity of gas removed, and it has almost pulsation-less flow. As its disadvantages, it cannot withstand high pressures, it has noisy operation because of gear noise and clattering impellers, it improperly seals the clearance between the impellers and the casing, and it overheats if operated above safe pressures. Typically, rotary blowers deliver a volumetric gas flow rate of up to 17,000 cfm and have a maximum intake pressure of 10 psig and a differential pressure of 10 psi.

When selecting a compressor, the pressure–volume characteristics and the type of driver must be considered. Small rotary compressors (vane or impeller type) are generally driven by electric motors. Large-volume positive compressors operate at lower speeds and are usually driven by steam or gas engines. They may be driven through reduction gearing by steam turbines or an electric motor. Reciprocation compressors driven by steam turbines or electric motors are most widely used in the natural gas industry as the conventional high-speed compression machine. Selection of compressors requires considerations of volumetric gas deliverability, pressure, compression ratio, and horsepower.

The following are important characteristics of the two types of compressors:

- Reciprocating piston compressors can adjust pressure output to backpressure.
- Reciprocating compressors can vary their volumetric flow-rate output (within certain limits).
- Reciprocating compressors have a volumetric efficiency, which is related to the relative clearance volume of the compressor design.
- Rotary compressors have a fixed pressure ratio, so they have a constant pressure output.
- Rotary compressors can vary their volumetric flow-rate output (within certain limits).

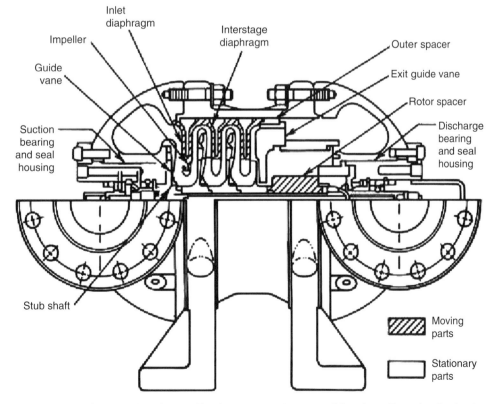

Figure 11.4 Cross-section of a centrifugal compressor (courtesy of Petroleum Extension Services).

11.3.2 Reciprocating Compressors

Figure 11.5 shows a diagram volume relation during gas compression. The shaft work put into the gas is expressed as

$$W_s = \frac{V_2^2}{2g} - \frac{V_1^2}{2g} + \left(P_2 v_2 - \int_1^2 P dv - P_1 v_1\right), \quad (11.29)$$

where

- W_s = mechanical shaft work into the system, ft-lbs per lb of fluid
- V_1 = inlet velocity of fluid to be compressed, ft/sec
- V_2 = outlet velocity of compressed fluid, ft/sec
- P_1 = inlet pressure, lb/ft² abs
- P_2 = outlet pressure, lb/ft² abs
- v_1 = specific volume at inlet, ft³/lb
- v_2 = specific volume at outlet, ft³/lb.

Note that the mechanical kinetic energy term $\frac{V^2}{2g}$ is in $ft \cdot \left(\frac{lb}{lb}\right)$ to get ft-lbs per lb.

Rewriting Eq. (11.29), we can get

$$W_s + \frac{V_1^2}{2g} - \frac{V_2^2}{2g} + P_1 v_1 - P_2 v_2 = -\int_1^2 P dv. \quad (11.30)$$

An isentropic process is usually assumed for reciprocating compression, that is, $P_1 v_1^k = P_2 v_2^k = P v^k =$ constant, where $k = \frac{c_p}{c_v}$. Because $P = \frac{P_1 v_1^k}{v^k}$, the right-hand side of Eq. (11.30) is formulated as

$$-\int_1^2 P dv = -\int_1^2 \frac{P_1 v_1^k}{v^k} dv = -P_1 v_1^k \int_1^2 \frac{dv}{v^k}$$

$$= -P_1 v_1^k \left[\frac{v^{1-k}}{1-k}\right]_1^2 = -\frac{P_1 v_1^k}{1-k}[v_2^{1-k} - v_1^{1-k}]$$

$$= \frac{P_1 v_1^k}{1-k}\left(\frac{v_1^{1-k}}{v_1^{1-k}}\right)[v_2^{1-k} - v_1^{1-k}]$$

$$= \frac{P_1 v_1}{1-k}\left[\frac{v_2^{1-k}}{v_1^{1-k}} - 1\right]$$

$$= \frac{P_1 v_1}{1-k}\left[\left(\frac{v_1}{v_2}\right)^{k-1} - 1\right]. \quad (11.31)$$

Using the ideal gas law

$$\frac{P}{\gamma} = RT, \quad (11.32)$$

where γ (lb/ft³) is the specific weight of the gas and T (°R) is the temperature and $R = 53.36$ (lb-ft/lb-°R) is the gas constant, and $v = \frac{1}{\gamma}$, we can write Eq. (11.32) as

$$Pv = RT \quad (11.33)$$

or

$$P_1 v_1 = RT_1. \quad (11.34)$$

Using $P_1 v_1^k = P_2 v_2^k = P v^k =$ constant, which gives

$$\left(\frac{v_1}{v_2}\right)^k = \frac{P_2}{P_1}$$

or

$$\frac{v_1}{v_2} = \left(\frac{P_2}{P_1}\right)^{\frac{1}{k}}. \quad (11.35)$$

Substituting Eqs. (11.35) and (11.34) into Eq. (11.31) gives

$$-\int_1^2 P dv = \frac{RT_1}{k-1}\left[\left(\frac{P_2}{P_1}\right)^{\frac{k-1}{k}} - 1\right]. \quad (11.36)$$

We multiply Eq. (11.33) by v^{k-1}, which gives

$$Pv(v^{k-1}) = RT(v^{k-1})$$
$$Pv^k = RTv^{k-1} = C_1$$
$$\frac{Pv^k}{R} = Tv^{k-1} = \frac{C_1}{R} = C_1'$$

Thus,

$$Tv^{k-1} = C_1'. \quad (11.37)$$

Also we can rise $Pv^k =$ constant to the $\left(\frac{k-1}{k}\right)$ power. This is

$$(Pv^k)^{\frac{k-1}{k}} = C_1'^{\frac{k-1}{k}}$$
$$P^{\frac{k-1}{k}}v^{k-1} = C_1'^{\frac{k-1}{k}}$$

or

$$v^{k-1} = \frac{C_1'^{\frac{k-1}{k}}}{P^{\frac{k-1}{k}}} = \frac{C_1''}{P^{\frac{k-1}{k}}}. \quad (11.38)$$

Substituting Eq. (3.38) into (3.37) gives

$$T\frac{C_1''}{P^{\frac{k-1}{k}}} = C_1'$$

or

$$\frac{T}{P^{\frac{k-1}{k}}} = \frac{C_1'}{C_1''} = C_1''' = \text{constant}. \quad (11.39)$$

Thus, Eq. (11.39) can be written as

$$\frac{T_1}{P_1^{\frac{k-1}{k}}} = \frac{T_2}{P_2^{\frac{k-1}{k}}}. \quad (11.40)$$

Thus, Eq. (11.40) is written

$$\left(\frac{P_2}{P_1}\right)^{\frac{k-1}{k}} = \frac{T_2}{T_1}. \quad (11.41)$$

Substituting Eq. (11.41) into (11.36) gives

$$-\int_1^2 P dv = \frac{RT_1}{k-1}\left[\frac{T_2}{T_1} - 1\right]$$

$$-\int_1^2 P dv = \frac{R}{k-1}(T_2 - T_1). \quad (11.42)$$

Therefore, our original expression, Eq. (11.30), can be written as

$$W_s + \frac{V_1^2}{2g} - \frac{V_2^2}{2g} + P_1 v_1 - P_2 v_2 = \frac{R}{k-1}(T_2 - T_1)$$

or

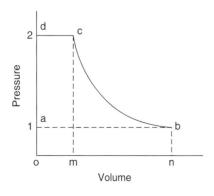

Figure 11.5 Basic pressure–volume diagram.

$$W_s = \frac{R}{k-1}(T_2 - T_1) + P_2v_2 - P_1v_1$$
$$+ \frac{(V_2^2 - V_1^2)}{2g}. \qquad (11.43)$$

And because
$$P_1v_1 = RT_1 \qquad (11.44)$$
and
$$P_2v_2 = RT_2, \qquad (11.45)$$

Eq. (11.43) becomes
$$W_s = \frac{R}{k-1}(T_2 - T_1) + R(T_2 - T_1)$$
$$+ \frac{(V_2^2 - V_1^2)}{2g}, \qquad (11.46)$$

but rearranging Eq. (11.46) gives
$$W_s = \frac{k}{k-1} RT_1 \left(\frac{T_2}{T_1} - 1\right) + \frac{(V_2^2 - V_1^2)}{2g}.$$

Substituting Eq. (11.41) and (11.44) into the above gives
$$W_s = \frac{k}{k-1} P_1v_1 \left[\left(\frac{P_2}{P_1}\right)^{\frac{k-1}{k}} - 1\right] + \frac{(V_2^2 - V_1^2)}{2g}. \qquad (11.47)$$

Neglecting the kinetic energy term, we arrive at
$$W_s = \frac{k}{k-1} P_1v_1 \left[\left(\frac{P_2}{P_1}\right)^{\frac{k-1}{k}} - 1\right], \qquad (11.48)$$

where W_s is ft-lb/lb, that is, work done per lb.

It is convenient to obtain an expression for power under conditions of steady state gas flow. Substituting Eq. (11.44) into (11.48) yields
$$W_s = \frac{k}{k-1} RT_1 \left[\left(\frac{P_2}{P_1}\right)^{\frac{k-1}{k}} - 1\right]. \qquad (11.49)$$

If we multiply both sides of Eq. (11.49) by the weight rate of flow, w_t (lb/sec), through the system, we get
$$P_s = \frac{k}{k-1} w_t RT_1 \left[\left(\frac{P_2}{P_1}\right)^{\frac{k-1}{k}} - 1\right], \qquad (11.50)$$

where $P_s = W_s w_t \frac{\text{ft-lb}}{\text{sec}}$ and is shaft power. However, the term w_t is
$$w_t = \gamma_1 Q_1 = \gamma_2 Q_2, \qquad (11.51)$$

where Q_1 (ft^3/sec) is the volumetric flow rate into the compressor and Q_2 (ft^3/sec) would be the compressed volumetric flow rate out of the compressor. Substituting Eq. (11.32) and (11.51) into (11.50) yields
$$P_s = \frac{k}{k-1} P_1 Q_1 \left[\left(\frac{P_2}{P_1}\right)^{\frac{k-1}{k}} - 1\right]. \qquad (11.52)$$

If we use more conventional field terms such as
$$P_1 = p_1(144) \text{ where } p_1 \text{ is in psia}$$
$$P_2 = p_2(144) \text{ where } p_2 \text{ is in psia}$$
and
$$Q_1 = \frac{q_1}{60} \text{ where } q_1 \text{ is in cfm,}$$

and knowing that 1 horsepower = 550 ft-lb/sec, then Eq. (11.52) becomes
$$HP = \frac{k}{(k-1)} \frac{p_1(144)q_1}{550(60)} \left[\left(\frac{p_2}{p_1}\right)^{\frac{k-1}{k}} - 1\right],$$

which yields
$$HP = \frac{k}{(k-1)} \frac{p_1 q_1}{229.2} \left[\left(\frac{p_2}{p_1}\right)^{\frac{k-1}{k}} - 1\right]. \qquad (11.53)$$

If the gas flow rate is given in Q_{MM} (MMscf/day) in a standard base condition at base pressure p_b (e.g., 14.7 psia) and base temperature T_b (e.g., 520 °R), since
$$q_1 = \frac{p_b T_1 Q_{MM}(1,000,000)}{p_1 T_b(24)}, \qquad (11.54)$$

Eq. (11.53) becomes
$$HP = \frac{181.79 p_b T_1 Q_{MM}}{T_b} \frac{k}{(k-1)} \left[\left(\frac{p_2}{p_1}\right)^{\frac{k-1}{k}} - 1\right]. \qquad (11.55)$$

It will be shown later that the efficiency of compression drops with increased compression ratio p_2/p_1. Most field applications require multistage compressors (two, three, and sometimes four stages) to reduce compression ratio in each stage. Figure 11.6 shows a two-stage compression unit. Using compressor stages with perfect intercooling between stages gives a theoretical minimum power for gas compression. To obtain this minimum power, the compression ratio in each stage must be the same and the cooling between each stage must bring the gas entering each stage to the same temperature.

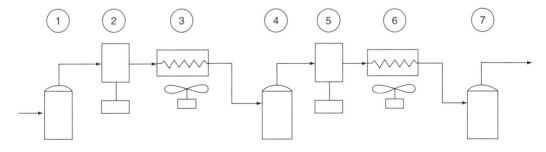

① ④ ⑦ Knockout drums (to remove condensed liquids)

② ⑤ Compressors (first and second stages)

③ Interstage cooler/intercooler (air–type)

⑥ Aftercooler (air–type)

Figure 11.6 Flow diagram of a two-stage compression unit.

The compression ratio in each stage should be less than six to increase compression efficiency. The equation to calculate stage-compression ratio is

$$r_s = \left(\frac{P_{dis}}{P_{in}}\right)^{1/n_s}, \tag{11.56}$$

where P_{dis}, P_{in}, and n_s are final discharge pressure, inlet pressure, and number of stages, respectively.

For a two-stage compression, the compression ratio for each stage should be

$$r_s = \sqrt{\frac{P_{dis}}{P_{in}}}. \tag{11.57}$$

Using Eq. (11.50), we can write the total power requirement for the two-stage compressor as

$$P_{total} = \frac{k}{k-1} w_t R T_{in1} \left[\left(\frac{P_{dis1}}{P_{in1}}\right)^{\frac{k-1}{k}} - 1\right]$$

$$+ \frac{k}{k-1} w_t R T_{in2} \left[\left(\frac{P_{dis2}}{P_{in2}}\right)^{\frac{k-1}{k}} - 1\right]. \tag{11.58}$$

The ideal intercooler will cool the gas flow stage one to stage two to the temperature entering the compressor. Thus, we have $T_{in1} = T_{in2}$. Also, the pressure $P_{in2} = P_{dis1}$. Equation (11.58) may be written as

$$P_{total} = \frac{k}{k-1} w_t R T_{in1} \left[\left(\frac{P_{dis1}}{P_{in1}}\right)^{\frac{k-1}{k}} - 1\right]$$

$$+ \frac{k}{k-1} w_t R T_{in1} \left[\left(\frac{P_{dis2}}{P_{dis1}}\right)^{\frac{k-1}{k}} - 1\right]. \tag{11.59}$$

We can find the value of P_{dis1} that will minimize the power required, P_{total}. We take the derivative of Eq. (11.59) with respect to P_{dis1} and set this equal to zero and solve for P_{dis1}. This gives

$$P_{dis1} = \sqrt{P_{in1} P_{dis2}},$$

which proves Eq. (11.57).

For the two-stage compressor, Eq. (11.59) can be rewritten as

$$P_{total} = 2\frac{k}{k-1} w_t R T_1 \left[\left(\frac{P_{dis2}}{P_{in1}}\right)^{\frac{k-1}{2k}} - 1\right]. \tag{11.60}$$

The ideal intercooling does not extend to the gas exiting the compressor. Gas exiting the compressor is governed by Eq. (11.41). Usually there is an adjustable after-cooler on a compressor that allows the operators to control the temperature of the exiting flow of gas. For greater number of stages, Eq. (11.60) can be written in field units as

$$HP_t = \frac{n_s p_1 q_1}{229.2} \frac{k}{(k-1)} \left[\left(\frac{p_2}{p_1}\right)^{\frac{k-1}{n_s k}} - 1\right] \tag{11.61}$$

or

$$HP_t = \frac{181.79 n_s p_b T_1 Q_{MM}}{T_b} \frac{k}{(k-1)} \left[\left(\frac{p_2}{p_1}\right)^{\frac{k-1}{n_s k}} - 1\right]. \tag{11.62}$$

In the above, p_1 (psia) is the intake pressure of the gas and p_2 (psia) is the outlet pressure of the compressor after the final stage, q_1 is the actual cfm of gas into the compressor, HP_t is the theoretical horsepower needed to compress the gas. This HP_t value has to be matched with a prime mover motor. The proceeding equations have been coded in the spreadsheet *ReciprocatingCompressorPower.xls* for quick calculations.

Reciprocating compressors have a clearance at the end of the piston. This clearance produces a volumetric efficiency e_v. The relation is given by

$$e_v = 0.96\left\{1 - \varepsilon\left[r_s^{\frac{1}{k}} - 1\right]\right\}, \tag{11.63}$$

where ε is the clearance ratio defined as the clearance volume at the end of the piston stroke divided by the entire volume of the chamber (volume contacted by the gas in the cylinder). In addition, there is a mechanical efficiency e_m of the compressor and its prime mover. This results in two separate expressions for calculating the required HP_t for reciprocating compressors and rotary compressors. The required minimum input prime mover motor to practically operate the compressor (either reciprocating or rotary) is

$$HP_{in} = \frac{HP_t}{e_v e_m}, \tag{11.64}$$

where $e_v \approx 0.80 - 0.99$ and $e_m \approx 0.80$ to 0.95 for reciprocating compressors, and $e_v = 1.0$ and $e_m \approx 0.70$ to 0.75 for rotary compressors.

Equation (11.64) stands for the input power required by the compressor, which is the minimum power to be provided by the prime mover. The prime movers usually have fixed power HP_p under normal operating conditions. The usable prime mover power ratio is

$$PR = \frac{HP_{in}}{HP_p}. \tag{11.65}$$

If the prime mover is not fully loaded by the compressor, its rotary speed increases and fuel consumption thus increases. Figure 11.7 shows fuel consumption curves for prime movers using gasoline, propane/butane, and diesel as fuel. Figure 11.8 presents fuel consumption curve for prime movers using natural gas as fuel. It is also important to know that the prime mover power drops with surface location elevation (Fig. 11.9).

Example Problem 11.2 Consider a three-stage reciprocating compressor that is rated at $q = 900$ scfm and a maximum pressure capability of $p_{max} = 240$ psig (standard conditions at sea level). The diesel prime mover is a diesel motor (naturally aspirated) rated at 300 horsepower (at sea-level conditions). The reciprocating compressor has a clearance ratio of $\varepsilon = 0.06$ and $e_m \approx 0.90$. Determine the gallons/hr of fuel consumption if the working backpressure is 150 psig, and do for

1. operating at sea level
2. operating at 6,000 ft.

Solution

1. Operating at sea level:

$$r_s = \sqrt[3]{\frac{P_{dis}}{p_{in}}} = \sqrt[3]{\frac{150 + 14.7}{14.7}} = \sqrt[3]{\frac{164.7}{14.7}} = 2.24$$

$$e_v = 0.96\left\{1 - 0.06\left[(2.24)^{\frac{1}{1.4}} - 1\right]\right\} = 0.9151$$

Required theoretical power to compress the gas:

$$HP_t = (3)\frac{14.7(900)}{229.2}\left(\frac{1.4}{0.4}\right)\left[\left(\frac{164.7}{14.7}\right)^{\frac{0.4}{3(1.4)}} - 1\right] = 156.8\,\text{hp}$$

Required input power to the compressor:

$$HP_r = \frac{HP_t}{e_m e_v} = \frac{156.8}{0.90(0.9151)} = 190.3\,\text{hp}$$

Since the available power from the prime mover is 300 hp, which is greater than HP_r, the prime mover is okay. The power ratio is

$$PR = \frac{190.3}{300.0} = 0.634 \text{ or } 63.4\%.$$

From Fig. 11.7, fuel usage is approximately 0.56 lb/hp-hr. The weight of fuel requirement is, therefore,

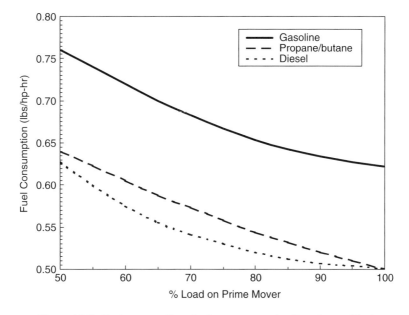

Figure 11.7 Fuel consumption of prime movers using three types of fuel.

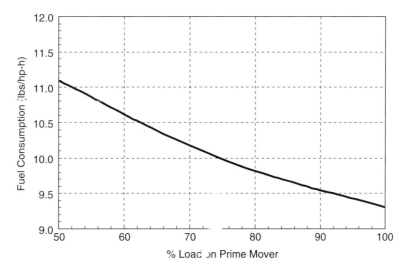

Figure 11.8 Fuel consumption of prime movers using natural gas as fuel.

$$w_f(\text{lb/hr}) \approx 0.56(190.3) = 106.6 \,\text{lb/hr}.$$

The volumetric fuel requirement is

$$q_f(\text{gallons/hr}) \approx \frac{106.6}{6.9} = 15.4 \,\text{gallons/hr}.$$

2. Operating at 6,000 ft,

the atmospheric pressure at an elevation of 6,000 is about 11.8 psia (Lyons et al., 2001). Figure 11.9 shows a power reduction of 22%.

$$r_s = \sqrt[3]{\frac{150 + 11.8}{11.8}} = 2.39$$

$$e_v = 0.96\{1 - 0.06[(2.39)^{0.714} - 1]\} = 0.9013$$

$$HP_t = (3)\frac{11.8(900)}{229.2}\left(\frac{1.4}{0.4}\right)\left[\left(\frac{161.8}{11.8}\right)^{0.0952} - 1\right] = 137.7 \,\text{hp}$$

$$HP_r = \frac{137.7}{e_m e_v} = \frac{137.7}{0.90(0.9103)} = 168.1 \,\text{hp}$$

$$HP_{in} = 300(1 - 0.22) = 234 \,\text{hp} > 168.1 \,\text{hp, so okay.}$$

$$\text{PR} = \frac{161.8}{234} = 0.718 \text{ or } 71.8\%$$

Figure 11.7 shows that a fuel usage of 0.54 lb/hp-hr at 71.8% power ratio. Thus,

$$w_f(\text{lbs/hr}) \approx 0.54(168.1) = 90.8 \,\text{lbs/hr}$$

$$q_f(\text{gallons/hr}) \approx \frac{90.8}{6.9} = 13.2 \,\text{gallons/hr}.$$

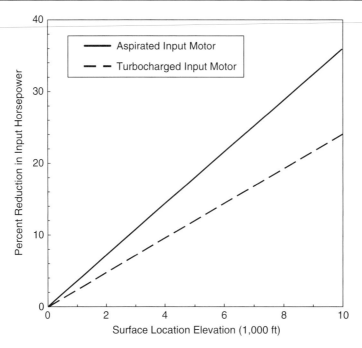

Figure 11.9 Effect of elevation on prime mover power.

11.3.3 Centrifugal Compressors

Although the adiabatic compression process can be assumed in centrifugal compression, polytropic compression process is commonly considered as the basis for comparing centrifugal compressor performance. The process is expressed as

$$pV^n = \text{constant}, \tag{11.66}$$

where n denotes the polytropic exponent. The isentropic exponent k applies to the ideal frictionless adiabatic process, while the polytropic exponent n applies to the actual process with heat transfer and friction. The n is related to k through polytropic efficiency E_p:

$$\frac{n-1}{n} = \frac{k-1}{k} \times \frac{1}{E_p}. \tag{11.67}$$

The polytropic efficiency of centrifugal compressors is nearly proportional to the logarithm of gas flow rate in the range of efficiency between 0.7 and 0.75. The polytropic efficiency chart presented by Rollins (1973) can be represented by the following correlation:

$$E_p = 0.61 + 0.03 \log(q_1), \tag{11.68}$$

where

q_1 – gas capacity at the inlet condition, cfm.

There is a lower limit of gas flow rate, below which severe gas surge occurs in the compressor. This limit is called "surge limit." The upper limit of gas flow rate is called "stone-wall limit," which is controlled by compressor horsepower.

The procedure of preliminary calculations for selection of centrifugal compressors is summarized as follows:

1. Calculate compression ratio based on the inlet and discharge pressures:

$$r = \frac{p_2}{p_1} \tag{11.69}$$

2. Based on the required gas flow rate under standard condition (q), estimate the gas capacity at inlet condition (q_1) by ideal gas law:

$$q_1 = \frac{p_b}{p_1} \frac{T_1}{T_b} q \tag{11.70}$$

3. Find a value for the polytropic efficiency E_p from the manufacturer's manual based on q_1.
4. Calculate polytropic ratio $(n-1)/n$ using Eq. (11.67):

$$R_p = \frac{n-1}{n} = \frac{k-1}{k} \times \frac{1}{E_p} \tag{11.71}$$

5. Calculate discharge temperature by

$$T_2 = T_1 r^{R_p}. \tag{11.72}$$

6. Estimate gas compressibility factor values at inlet and discharge conditions.
7. Calculate gas capacity at the inlet condition (q_1) by real gas law:

$$q_1 = \frac{z_1 p_b}{z_2 p_1} \frac{T_1}{T_b} q \tag{11.73}$$

8. Repeat Steps 2–7 until the value of q_1 converges within an acceptable deviation.
9. Calculate gas horsepower by

$$Hp_g = \frac{q_1 p_1}{229 E_p} \left(\frac{z_1 + z_2}{2 z_1}\right) \left(\frac{r^{R_p} - 1}{R_p}\right). \tag{11.74}$$

Some manufacturers present compressor specifications using polytropic head in $lb_f\text{-}ft/lb_m$ defined as

$$H_g = RT_1 \left(\frac{z_1 + z_2}{2}\right) \left(\frac{r^{R_p} - 1}{R_p}\right), \tag{11.75}$$

where R is the gas constant given by $1,544/MW_a$ in psia-ft^3/lb$_m$-°R. The polytropic head relates to the gas horsepower by

$$Hp_g = \frac{m_t H_g}{33,000 E_p}, \tag{11.76}$$

where m_t is mass flow rate in lb$_m$/min.

10. Calculate gas horsepower by:

$$Hp_b = Hp_g + \Delta Hp_m, \tag{11.77}$$

where ΔHp_m is mechanical power losses, which is usually taken as 20 horsepower for bearing and 30 horsepower for seals.

The proceeding equations have been coded in the spreadsheet *CentrifugalCompressorPower.xls* for quick calculations.

Example Problem 11.3 Size a centrifugal compressor for the following given data:

Gas-specific gravity: 0.68
Gas-specific heat ratio: 1.24
Gas flow rate: 144 MMscfd at 14.7 psia and 60 °F
Inlet pressure: 250 psia
Inlet temperature: 100 °F
Discharge pressure: 600 psia
Polytropic efficiency: $E_p = 0.61 + 0.03 \log(q_1)$

Solution Calculate compression ratio based on the inlet and discharge pressures:

$$r = \frac{600}{250} = 2.4$$

Calculate gas flow rate in scfm:

$$q = \frac{144,000,000}{(24)(60)} = 100,000 \text{ scfm}$$

Based on the required gas flow rate under standard condition (q), estimate the gas capacity at inlet condition (q_1) by ideal gas law:

$$q_1 = \frac{(14.7)}{(250)} \frac{(560)}{(520)} (100,000) = 6,332 \text{ cfm}$$

Find a value for the polytropic efficiency based on q_1:

$$E_p = 0.61 + 0.03 \log(6,332) = 0.724$$

Calculate polytropic ratio $(n-1)/n$:

$$R_p = \frac{n-1}{n} = \frac{1.24-1}{1.24} \times \frac{1}{0.724} = 0.2673$$

Calculate discharge temperature:

$$T_2 = (560)(2.4)^{0.2673} = 707.7\,°R = 247.7\,°F$$

Estimate gas compressibility factor values at inlet and discharge conditions (spreadsheet program *Hall-Yaborough-z.xls* can be used):

$$z_1 = 0.97 \text{ at } 250 \text{ psia and } 100\,°F$$

$$z_2 = 0.77 \text{ at } 600 \text{ psia and } 247.7\,°F$$

Calculate gas capacity at the inlet condition (q_1) by real gas law:

$$q_1 = \frac{(0.97)(14.7)}{(0.77)(250)} \frac{(560)}{(520)} (100,000) = 7,977 \text{ cfm}$$

Use the new value of q_1 to calculate E_p:

$$E_p = 0.61 + 0.03 \log(7,977) = 0.727$$

Calculate the new polytropic ratio $(n-1)/n$:

$$R_p = \frac{n-1}{n} = \frac{1.24-1}{1.24} \times \frac{1}{0.727} = 0.2662$$

Calculate the new discharge temperature:

$$T_2 = (560)(2.4)^{0.2662} = 707\,°R = 247\,°F$$

Estimate the new gas compressibility factor value:

$$z_2 = 0.77 \text{ at } 600 \text{ psia and } 247\,°F$$

Because z_2 did not change, q_1 remains the same value of 7,977 cfm.

Calculate gas horsepower:

$$Hp_g = \frac{(7,977)(250)}{(229)(0.727)} \left(\frac{0.97+0.77}{2(0.97)}\right) \left(\frac{2.4^{0.2662}-1}{0.2662}\right)$$

$$= 10,592 \text{ hp}$$

Calculate gas apparent molecular weight:

$$MW_a = (0.68)(29) = 19.72$$

Calculated gas constant:

$$R = \frac{1,544}{19.72} = 78.3 \text{ psia-ft}^3/\text{lb}_m\text{-}°R$$

Calculate polytropic head:

$$H_g = (78.3)(560)\left(\frac{0.97+0.77}{2}\right)\left(\frac{2.4^{0.2662}-1}{0.2662}\right)$$

$$= 37,610 \text{ lb}_f\text{-ft}/\text{lb}_m$$

Calculate gas horsepower requirement:

$$Hp_b = 10,592 + 50 = 10,642 \text{ hp}.$$

11.4 Pipelines

Transporting petroleum fluids with pipelines is a continuous and reliable operation. Pipelines have demonstrated an ability to adapt to a wide variety of environments including remote areas and hostile environments. With very minor exceptions, largely due to local peculiarities, most refineries are served by one or more pipelines, because of their superior flexibility to the alternatives.

Pipelines can be divided into different categories, including the following:

- Flowlines transporting oil and/or gas from satellite wells to manifolds
- Flowlines transporting oil and/or gas from manifolds to production facility
- Infield flowlines transporting oil and/or gas from between production facilities
- Export pipelines transporting oil and/or gas from production facilities to refineries/users

The pipelines are sized to handle the expected pressure and fluid flow on the basis of flow assurance analysis. This section covers the following topics:

1. Flow in oil and gas pipelines
2. Design of pipelines
3. Operation of pipelines.

11.4.1 Flow in Pipelines

Designing a long-distance pipeline for transportation of crude oil and natural gas requires knowledge of flow formulas for calculating capacity and pressure requirements. Based on the first law of thermal dynamics, the total pressure gradient is made up of three distinct components:

$$\frac{dP}{dL} = \frac{g}{g_c} \rho \sin\theta + \frac{f_M \rho u^2}{2 g_c D} + \frac{\rho u \, du}{g_c \, dL}, \quad (11.78)$$

where

$\frac{g}{g_c} \rho \sin\theta$ = pressure gradient due to elevation or potential energy change

$\frac{f_M \rho u^2}{2 g_c D}$ = pressure gradient due to frictional losses

$\frac{\rho u du}{g_c dL}$ = pressure gradient due to acceleration or kinetic energy change
P = pressure, lbf/ft^2
L = pipe length, ft
g = gravitational acceleration, ft/sec^2
g_c = 32.17, ft-lbm/lbf-sec^2
ρ = density lbm/ft^3
θ = dip angle from horizontal direction, degrees
f_M = Darcy–Wiesbach (Moody) friction factor
u = flow velocity, ft/sec
D = pipe inner diameter, ft

The elevation component is pipe-angle dependent. It is zero for horizontal flow. The friction loss component applies to any type of flow at any pipe angle and causes a pressure drop in the direction of flow. The acceleration component causes a pressure drop in the direction of velocity increase in any flow condition in which velocity changes occurs. It is zero for constant-area, incompressible flow. This term is normally negligible for both oil and gas pipelines.

The friction factor f_M in Eq. (11.78) can be determined based on flow regimes, that is, laminar flow or turbulent flow. Reynolds number (N_{Re}) is used as a parameter to distinguish between laminar and turbulent fluid flow. Reynolds number is defined as the ratio of fluid momentum force to viscous shear force. The Reynolds number can be expressed as a dimensionless group defined as

$$N_{Re} = \frac{D u \rho}{\mu}, \quad (11.79)$$

where

D = pipe ID, ft
u = fluid velocity, f/sec
ρ = fluid density, lb$_m$/ft^3
μ = fluid viscosity, lb$_m$/ft-sec.

The change from laminar to turbulent flow is usually assumed to occur at a Reynolds number of 2,100 for flow in a circular pipe. If U.S. field units of ft for diameter, ft/sec for velocity, lb$_m$/ft^3 for density and centipoises for viscosity are used, the Reynolds number equation becomes

$$N_{Re} = 1,488 \frac{D u \rho}{\mu}. \quad (11.80)$$

For a gas with specific gravity γ_g and viscosity μ_g (cp) flowing in a pipe with an inner diameter D (in.) at flow rate q (Mcfd) measured at base conditions of T_b (°R) and p_b (psia), the Reynolds number can be expressed as

$$N_{Re} = \frac{711 p_b q \gamma_g}{T_b D \mu_g}. \quad (11.81)$$

The Reynolds number usually takes values greater than 10,000 in gas pipelines. As T_b is 520 °R and p_b varies only from 14.4 to 15.025 psia in the United States, the value of $711 p_b/T_b$ varies between 19.69 and 20.54. For all practical purposes, the Reynolds number for natural gas flow problems may be expressed as

$$N_{Re} = \frac{20 q \gamma_g}{\mu_g d}, \quad (11.82)$$

where

q = gas flow rate at 60 °F and 14.73 psia, Mcfd
γ_g = gas-specific gravity (air = 1)
μ_g = gas viscosity at in-situ temperature and pressure, cp
d = pipe diameter, in.

The coefficient 20 becomes 0.48 if q is in scfh.

Figure 11.10 is a friction factor chart covering the full range of flow conditions. It is a log-log graph of $(\log f_M)$ versus $(\log N_{Re})$. Because of the complex nature of the curves, the equation for the friction factor in terms of the Reynolds number and relative roughness varies in different regions.

In the laminar flow region, the friction factor can be determined analytically. The Hagen–Poiseuille equation for laminar flow is

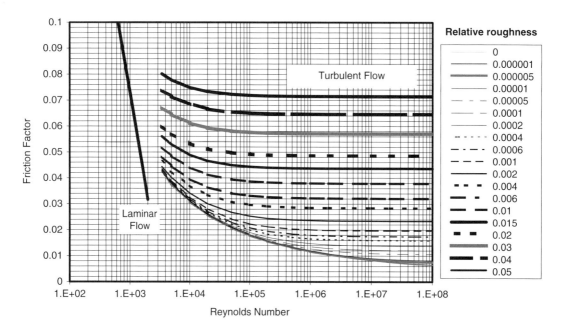

Figure 11.10 Darcy–Wiesbach friction factor chart (Moody, 1944).

$$\left(\frac{dp}{dL}\right)_f = \frac{32\mu u}{g_c D^2}. \quad (11.83)$$

Equating the frictional pressure gradients given by Eqs. (11.78) and (11.83) gives

$$\frac{f_M \rho u^2}{2g_c D} = \frac{32\mu u}{g_c D^2}, \quad (11.84)$$

which yields

$$f_M = \frac{64\mu}{du\rho} = \frac{64}{N_{Re}}. \quad (11.85)$$

In the turbulent flow region, a number of empirical correlations for friction factors are available. Only the most accurate ones are presented in this section.

For smooth wall pipes in the turbulent flow region, Drew et al. (1930) presented the most commonly used correlation:

$$f_M = 0.0056 + \frac{0.5}{N_{Re}^{0.32}}, \quad (11.86)$$

which is valid over a wide range of Reynolds numbers, $3 \times 10^3 < N_{Re} < 3 \times 10^6$.

For rough wall pipes in the turbulent flow region, the effect of wall roughness on friction factor depends on the relative roughness and Reynolds number. The Nikuradse (1933) friction factor correlation is still the best one available for fully developed turbulent flow in rough pipes:

$$\frac{1}{\sqrt{f_M}} = 1.74 - 2\log(2e_D) \quad (11.87)$$

This equation is valid for large values of the Reynolds number where the effect of relative roughness is dominant. The correlation that is used as the basis for modern friction factor charts was proposed by Colebrook (1938):

$$\frac{1}{\sqrt{f_M}} = 1.74 - 2\log\left(2e_D + \frac{18.7}{N_{Re}\sqrt{f_M}}\right), \quad (11.88)$$

which is applicable to smooth pipes and to flow in transition and fully rough zones of turbulent flow. It degenerates to the Nikuradse correlation at large values of the Reynolds number. Equation (11.88) is not explicit in f_M. However, values of f_M can be obtained by a numerical procedure such as Newton–Raphson iteration. An explicit correlation for friction factor was presented by Jain (1976):

$$\frac{1}{\sqrt{f_M}} = 1.14 - 2\log\left(e_D + \frac{21.25}{N_{Re}^{0.9}}\right). \quad (11.89)$$

This correlation is comparable to the Colebrook correlation. For relative roughness between 10^{-6} and 10^{-2} and the Reynolds number between 5×10^3 and 10^8, the errors were reported to be within $\pm 1\%$ when compared with the Colebrook correlation. Therefore, Eq. (11.89) is recommended for all calculations requiring friction factor determination of turbulent flow.

The wall roughness is a function of pipe material, method of manufacturing, and the environment to which it has been exposed. From a microscopic sense, wall roughness is not uniform, and thus, the distance from the peaks to valleys on the wall surface will vary greatly. The absolute roughness, ε, of a pipe wall is defined as the mean protruding height of relatively uniformly distributed and sized, tightly packed sand grains that would give the same pressure gradient behavior as the actual pipe wall. Analysis has suggested that the effect of roughness is not due to its absolute dimensions, but to its dimensions relative to the inside diameter of the pipe. Relative roughness, e_D, is defined as the ratio of the absolute roughness to the pipe internal diameter:

$$e_D = \frac{\varepsilon}{D}, \quad (11.90)$$

where ε and D have the same unit.

The absolute roughness is not a directly measurable property for a pipe, which makes the selection of value of pipe wall roughness difficult. The way to evaluate the absolute roughness is to compare the pressure gradients obtained from the pipe of interest with a pipe that is sand-roughened. If measured pressure gradients are available, the friction factor and Reynolds number can be calculated and an effective e_D obtained from the Moody diagram. This value of e_D should then be used for future predictions until updated. If no information is available on roughness, a value of $\varepsilon = 0.0006$ in. is recommended for tubing and line pipes.

11.4.1.1 Oil Flow

This section addresses flow of crude oil in pipelines. Flow of multiphase fluids is discussed in other literatures such as that of Guo et al. (2005).

Crude oil can be treated as an incompressible fluid. The relation between flow velocity and driving pressure differential for a given pipeline geometry and fluid properties is readily obtained by integration of Eq. (11.78) when the kinetic energy term is neglected:

$$P_1 - P_2 = \left(\frac{g}{g_c}\rho \sin\theta + \frac{f_M \rho u^2}{2g_c D}\right)L, \quad (11.91)$$

which can be written in flow rate as

$$P_1 - P_2 = \left(\frac{g}{g_c}\rho \sin\theta + \frac{f_M \rho q^2}{2g_c D A^2}\right)L, \quad (11.92)$$

where

q = liquid flow rate, ft^3/sec
A = inner cross-sectional area, ft^2

When changed to U.S. field units, Eq. (11.92) becomes

$$p_1 - p_2 = 0.433\gamma_o L \sin\theta + 1.15 \times 10^{-5}$$
$$\times \frac{f_M \gamma_o Q^2 L}{d^5}, \quad (11.93)$$

where

p_1 = inlet pressure, psi
p_2 = outlet pressure, psi
γ_o = oil specific gravity, water = 1.0
Q = oil flow rate, bbl/day
d = pipe inner diameter, in.

Example Problem 11.4 A 35 API gravity, 5 cp, oil is transported through a 6-in. (I.D.) pipeline with an uphill angle of 15 degrees across a distance of 5 miles at a flow rate of 5,000 bbl/day. Estimate the minimum required pump pressure to deliver oil at 50 psi pressure at the outlet. Assume $e = 0.0006$ in.

Solution

Pipe inner area:

$$A = \frac{\pi}{4}\left(\frac{6}{12}\right)^2 = 0.1963\,\text{ft}^2$$

The average oil velocity in pipe:

$$u = \frac{(5,000)(5.615)}{(24)(60)(60)(0.1963)} = 1.66\,\text{ft/sec}$$

Oil-specific gravity:

$$\gamma_o = \frac{141.5}{131.5 + 35} = 0.85$$

Reynolds number:

$$N_{Re} = 1,488 \frac{\left(\frac{6}{12}\right)(1.66)(0.85)(62.4)}{5} = 13,101$$

$> 2,100$ turbulent flow

Equation (11.89) gives

$$\frac{1}{\sqrt{f_M}} = 1.14 - 2\log\left(\left(\frac{0.0006}{6}\right) + \frac{21.25}{(13,101)^{0.9}}\right) = 5.8759,$$

which gives

$$f_M = 0.02896.$$

Equation (11.93) gives

$$p_1 = 50 + 0.433(0.85)(5)(5,280)\sin(15°) + 1.15 \times 10^{-5}$$

$$\times \frac{(0.02896)(0.85)(5,000)^2(5)(5,280)}{(6)^5}$$

$$= 2,590 \text{ psi}.$$

11.4.1.2 Gas Flow

Consider steady-state flow of dry gas in a constant-diameter, horizontal pipeline. The mechanical energy equation, Eq. (11.78), becomes

$$\frac{dp}{dL} = \frac{f_M \rho u^2}{2g_c D} = \frac{p(MW)_a}{zRT} \frac{fu^2}{2g_c D}, \quad (11.94)$$

which serves as a base for development of many pipeline equations. The difference in these equations originated from the methods used in handling the z-factor and friction factor. Integrating Eq. (11.94) gives

$$\int dp = \frac{(MW)_a f_M u^2}{2Rg_c D} \int \frac{p}{zT} dL. \quad (11.95)$$

If temperature is assumed constant at average value in a pipeline, $\bar{T}$, and gas deviation factor, $\bar{z}$, is evaluated at average temperature and average pressure, $\bar{p}$, Eq. (11.95) can be evaluated over a distance L between upstream pressure, p_1, and downstream pressure, p_2:

$$p_1^2 - p_2^2 = \frac{25\gamma_g Q^2 \bar{T} \bar{z} f_M L}{d^5}, \quad (11.96)$$

where

γ_g = gas gravity (air = 1)
Q = gas flow rate, MMscfd (at 14.7 psia, 60 °F)
$\bar{T}$ = average temperature, °R
$\bar{z}$ = gas deviation factor at $\bar{T}$ and $\bar{p}$
$\bar{p}$ = $(p_1 + p_2)/2$
L = pipe length, ft
d = pipe internal diameter, in.
F = Moody friction factor

Equation (11.96) may be written in terms of flow rate measured at arbitrary base conditions (T_b and p_b):

$$q = \frac{CT_b}{p_b}\sqrt{\frac{(p_1^2 - p_2^2)d^5}{\gamma_g \bar{T} \bar{z} f_M L}}, \quad (11.97)$$

where C is a constant with a numerical value that depends on the units used in the pipeline equation. If L is in miles and q is in scfd, $C = 77.54$.

The use of Eq. (11.97) involves an iterative procedure. The gas deviation factor depends on pressure and the friction factor depends on flow rate. This problem prompted several investigators to develop pipeline flow equations that are noniterative or explicit. This has involved substitutions for the friction factor f_M. The specific substitution used may be diameter-dependent only (Weymouth equation) or Reynolds number–dependent only (Panhandle equations).

11.4.1.2.1 Weymouth Equation for Horizontal Flow

Equation (11.97) takes the following form when the unit of scfh for gas flow rate is used:

$$q_h = \frac{3.23T_b}{p_b}\sqrt{\frac{1}{f_M}}\sqrt{\frac{(p_1^2 - p_2^2)d^5}{\gamma_g \bar{T} \bar{z} L}}, \quad (11.98)$$

where $\sqrt{\frac{1}{f_M}}$ is called the "transmission factor." The friction factor may be a function of flow rate and pipe roughness. If flow conditions are in the fully turbulent region, Eq. (11.89) degenerates to

$$f_M = \frac{1}{[1.14 - 2\log(e_D)]^2}, \quad (11.99)$$

where f_M depends only on the relative roughness, e_D. When flow conditions are not completely turbulent, f_M depends on the Reynolds number also.

Therefore, use of Eq. (11.98) requires a trial-and-error procedure to calculate q_h. To eliminate the trial-and-error procedure, Weymouth proposed that f vary as a function of diameter as follows:

$$f_M = \frac{0.032}{d^{1/3}} \quad (11.100)$$

With this simplification, Eq. (11.98) reduces to

$$q_h = \frac{18.062T_b}{p_b}\sqrt{\frac{(p_1^2 - p_2^2)D^{16/3}}{\gamma_g \bar{T} \bar{z} L}}, \quad (11.101)$$

which is the form of the Weymouth equation commonly used in the natural gas industry.

The use of the Weymouth equation for an existing transmission line or for the design of a new transmission line involves a few assumptions including no mechanical work, steady flow, isothermal flow, constant compressibility factor, horizontal flow, and no kinetic energy change. These assumptions can affect accuracy of calculation results.

In the study of an existing pipeline, the pressure-measuring stations should be placed so that no mechanical energy is added to the system between stations. No mechanical work is done on the fluid between the points at which the pressures are measured. Thus, the condition of no mechanical work can be fulfilled.

Steady flow in pipeline operation seldom, if ever, exists in actual practice because pulsations, liquid in the pipeline, and variations in input or output gas volumes cause deviations from steady-state conditions. Deviations from steady-state flow are the major cause of difficulties experienced in pipeline flow studies.

The heat of compression is usually dissipated into the ground along a pipeline within a few miles downstream from the compressor station. Otherwise, the temperature of the gas is very near that of the containing pipe, and because pipelines usually are buried, the temperature of the flowing gas is not influenced appreciably by rapid changes in atmospheric temperature. Therefore, the gas flow can be considered isothermal at an average effective temperature without causing significant error in long-pipeline calculations.

The compressibility of the fluid can be considered constant and an average effective gas deviation factor may be used. When the two pressures p_1 and p_2 lie in a region where z is essentially linear with pressure, it is accurate enough to evaluate $\bar{z}$ at the average pressure $\bar{p} = (p_1 + p_2)/2$. One can also use the arithmetic average of the z's with $\bar{z} = (z_1 + z_2)/2$, where z_1 and z_2 are obtained at p_1 and p_2, respectively. On the other hand, should p_1 and p_2 lie in the range where z is not linear with pressure (double-hatched lines), the proper average would result from determining the area under the z-curve and dividing it by the difference in pressure:

$$\bar{z} = \frac{\int_{p_1}^{p_2} z\,dp}{(p_1 - p_2)}, \qquad (11.102)$$

where the numerator can be evaluated numerically. Also, $\bar{z}$ can be evaluated at an average pressure given by

$$\bar{p} = \frac{2}{3}\left(\frac{p_1^3 - p_2^3}{p_1^2 - p_2^2}\right). \qquad (11.103)$$

Regarding the assumption of horizontal pipeline, in actual practice, transmission lines seldom, if ever, are horizontal, so that factors are needed in Eq. (11.101) to compensate for changes in elevation. With the trend to higher operating pressures in transmission lines, the need for these factors is greater than is generally realized. This issue of correction for change in elevation is addressed in the next section.

If the pipeline is long enough, the changes in the kinetic-energy term can be neglected. The assumption is justified for work with commercial transmission lines.

Example Problem 11.5 For the following data given for a horizontal pipeline, predict gas flow rate in ft^3/hr through the pipeline. Solve the problem using Eq. (11.101) with the trial-and-error method for friction factor and the Weymouth equation without the Reynolds number–dependent friction factor:

$d = 12.09$ in.
$L = 200$ mi
$e = 0.0006$ in.
$T = 80\,°\text{F}$
$\gamma_g = 0.70$
$T_b = 520\,°\text{R}$
$p_b = 14.7$ psia
$p_1 = 600$ psia
$p_2 = 200$ psia

Solution The average pressure is

$$\bar{p} = (200 + 600)/2 = 400\,\text{psia}.$$

With $\bar{p} = 400$ psia, $T = 540\,°\text{R}$ and $\gamma_g = 0.70$, *Brill-Beggs-Z.xls* gives

$$\bar{z} = 0.9188.$$

With $\bar{p} = 400$ psia, $T = 540\,°\text{R}$ and $\gamma_g = 0.70$, *Carr-Kobayashi-BurrowsViscosity.xls* gives

$$m = 0.0099\,\text{cp}.$$

Relative roughness:

$$e_D = 0.0006/12.09 = 0.00005$$

A. Trial-and-error calculation:

First trial:

$$q_h = 500{,}000\,\text{scfh}$$

$$N_{Re} = \frac{0.48(500{,}000)(0.7)}{(0.0099)(12.09)} = 1{,}403{,}733$$

$$\frac{1}{\sqrt{f_M}} = 1.14 - 2\log\left(0.00005 + \frac{21.25}{(1{,}403{,}733)^{0.9}}\right)$$

$$f_M = 0.01223$$

$$q_h = \frac{3.23(520)}{14.7}\sqrt{\frac{1}{0.01223}}\sqrt{\frac{(600^2 - 200^2)(12.09)^5}{(0.7)(540)(0.9188)(200)}}$$

$$= 1{,}148{,}450\,\text{scfh}$$

Second trial:

$$q_h = 1{,}148{,}450\,\text{cfh}$$

$$N_{Re} = \frac{0.48(1{,}148{,}450)(0.7)}{(0.0099)(12.09)} = 3{,}224{,}234$$

$$\frac{1}{\sqrt{f_M}} = 1.14 - 2\log\left(0.00005 + \frac{21.25}{(3{,}224{,}234)^{0.9}}\right)$$

$$f_M = 0.01145$$

$$q_h = \frac{3.23(520)}{14.7}\sqrt{\frac{1}{0.01145}}\sqrt{\frac{(600^2 - 200^2)(12.09)^5}{(0.7)(540)(0.9188)(200)}}$$

$$= 1{,}186{,}759\,\text{scfh}$$

Third trial:

$$q_h = 1{,}186{,}759\,\text{scfh}$$

$$N_{Re} = \frac{0.48(1{,}186{,}759)(0.7)}{(0.0099)(12.09)} = 3{,}331{,}786$$

$$\frac{1}{\sqrt{f_M}} = 1.14 - 2\log\left(0.00005 + \frac{21.25}{(3{,}331{,}786)^{0.9}}\right)$$

$$f_M = 0.01143$$

$$q_h = \frac{3.23(520)}{14.7}\sqrt{\frac{1}{0.01143}}\sqrt{\frac{(600^2 - 200^2)(12.09)^5}{(0.7)(540)(0.9188)(200)}}$$

$$= 1{,}187{,}962\,\text{scfh},$$

which is close to the assumed 1,186,759 scfh.

B. Using the Weymouth equation:

$$q_h = \frac{18.062(520)}{14.7}\sqrt{\frac{(600^2 - 200^2)(12.09)^{16/3}}{(0.7)(540)(0.9188)(200)}}$$

$$= 1{,}076{,}035\,\text{scfh}$$

Problems similar to this one can be quickly solved with the spreadsheet program *PipeCapacity.xls*.

11.4.1.2.2 Weymouth Equation for Non-horizontal Flow
Gas transmission pipelines are often nonhorizontal. Account should be taken of substantial pipeline elevation changes. Considering gas flow from point 1 to point 2 in a nonhorizontal pipe, the first law of thermal dynamics gives

$$\int_1^2 v\,dP + \frac{g}{g_c}\Delta z + \int_1^2 \frac{f_M u^2}{2g_c D}\,dL = 0. \qquad (11.104)$$

Based on the pressure gradient due to the weight of gas column,

$$\frac{dP}{dz} = \frac{\rho_g}{144}, \qquad (11.105)$$

and real gas law, $\rho_g = \frac{p(MW)_a}{zRT} = \frac{29\gamma_g p}{zRT}$, Weymouth (1912) developed the following equation:

$$q_h = \frac{3.23 T_b}{p_b}\sqrt{\frac{(p_1^2 - e^s p_2^2) d^5}{f_M \gamma_g \bar{T}\bar{z} L}}, \qquad (11.106)$$

where

$e = 2.718$ and

$$s = \frac{0.0375\gamma_g \Delta z}{\bar{T}\bar{z}}, \quad (11.107)$$

and Δz is equal to outlet elevation minus inlet elevation (note that Δz is positive when outlet is higher than inlet). A general and more rigorous form of the Weymouth equation with compensation for elevation is

$$q_h = \frac{3.23T_b}{p_b}\sqrt{\frac{(p_1^2 - e^s p_2^2)d^5}{f_M \gamma_g \bar{T}\bar{z}L_e}}, \quad (11.108)$$

where L_e is the effective length of the pipeline. For a uniform slope, L_e is defined as $L_e = \frac{(e^s-1)L}{s}$.

For a non-uniform slope (where elevation change cannot be simplified to a single section of constant gradient), an approach in steps to any number of sections, n, will yield

$$L_e = \frac{(e^{s_1}-1)}{s_1}L_1 + \frac{e^{s_1}(e^{s_2}-1)}{s_2}L_2$$
$$+ \frac{e^{s_1+s_2}(e^{s_3}-1)}{s_3}L_3 + \ldots\ldots + \sum_{i=1}^{n}$$
$$\times \frac{e^{\sum_{j=1}^{i-1}s_j}(e^{s_i}-1)}{s_i}L_i, \quad (11.109)$$

where

$$s_i = \frac{0.0375\gamma_g \Delta z_i}{\bar{T}\bar{z}}. \quad (11.110)$$

11.4.1.2.3 Panhandle-A Equation for Horizontal Flow
The Panhandle-A pipeline flow equation assumes the following Reynolds number–dependent friction factor:

$$f_M = \frac{0.085}{N_{Re}^{0.147}} \quad (11.111)$$

The resultant pipeline flow equation is, thus,

$$q = 435.87\frac{d^{2.6182}}{\gamma_g^{0.4604}}\left(\frac{T_b}{p_b}\right)^{1.07881}\left[\frac{(p_1^2-p_2^2)}{\bar{T}\bar{z}L}\right]^{0.5394}, \quad (11.112)$$

where q is the gas flow rate in scfd measured at T_b and p_b, and other terms are the same as in the Weymouth equation.

11.4.1.2.4 Panhandle-B Equation for Horizontal Flow (Modified Panhandle)
The Panhandle-B equation is the most widely used equation for long transmission and delivery lines. It assumes that f_M varies as

$$f_M = \frac{0.015}{N_{Re}^{0.0392}}, \quad (11.113)$$

and it takes the following resultant form:

$$q = 737d^{2.530}\left(\frac{T_b}{p_b}\right)^{1.02}\left[\frac{(p_1^2-p_2^2)}{\bar{T}\bar{z}L\gamma_g^{0.961}}\right]^{0.510} \quad (11.114)$$

11.4.1.2.5 Clinedinst Equation for Horizontal Flow
The Clinedinst equation rigorously considers the deviation of natural gas from ideal gas through integration. It takes the following form:

$$q = 3973.0\frac{z_b p_b p_{pc}}{p_b}$$
$$\times \sqrt{\frac{d^5}{\bar{T}f_M L\gamma_g}\left(\int_0^{p_{r1}}\frac{p_r}{z}dp_r - \int_0^{p_{r2}}\frac{p_r}{z}dp_r\right)}, \quad (11.115)$$

where

- q = volumetric flow rate, Mcfd
- p_{pc} = pseudocritical pressure, psia
- d = pipe internal diameter, in.
- L = pipe length, ft
- p_r = pseudo-reduced pressure
- $\bar{T}$ = average flowing temperature, °R
- γ_g = gas gravity, air = 1.0
- z_b = gas deviation factor at T_b and p_b, normally accepted as 1.0.

Based on Eqs. (2.29), (2.30), and (2.51), Guo and Ghalambor (2005) generated curves of the integral function $\int_0^{p_r}\frac{p_r}{z}dp_r$ for various gas-specific gravity values.

11.4.1.2.6 Pipeline Efficiency
All pipeline flow equations were developed for perfectly clean lines filled with gas. In actual pipelines, water, condensates, sometimes crude oil accumulates in low spots in the line. There are often scales and even "junk" left in the line. The net result is that the flow rates calculated for the 100% efficient cases are often modified by multiplying them by an efficiency factor E. The efficiency factor expresses the actual flow capacity as a fraction of the theoretical flow rate. An efficiency factor ranging from 0.85 to 0.95 would represent a "clean" line. Table 11.1 presents typical values of efficiency factors.

Table 11.1 Typical Values of Pipeline Efficiency Factors

Type of line	Liquid content (gal/MMcf)	Efficiency E
Dry-gas field	0.1	0.92
Casing-head gas	7.2	0.77
Gas and condensate	800	0.6

11.4.2 Design of Pipelines
Pipeline design includes determination of material, diameter, wall thickness, insulation, and corrosion protection measure. For offshore pipelines, it also includes weight coating and trenching for stability control. Bai (2001) provides a detailed description on the analysis–analysis-based approach to designing offshore pipelines. Guo et al. (2005) presents a simplified approach to the pipeline design.

The diameter of pipeline should be determined based on flow capacity calculations presented in the previous section. This section focuses on the calculations to design wall thickness and insulation.

11.4.2.1 Wall Thickness Design
Wall thickness design for steel pipelines is governed by U.S. Code ASME/ANSI B32.8. Other codes such as Z187 (Canada), DnV (Norway), and IP6 (UK) have essentially the same requirements but should be checked by the readers.

Except for large-diameter pipes (>30 in.), material grade is usually taken as X-60 or X-65 (414 or 448 MPa) for high-pressure pipelines or on deepwater. Higher grades can be selected in special cases. Lower grades such as X-42, X-52, or X-56 can be selected in shallow water or for low-pressure, large-diameter pipelines to reduce material cost or in cases in which high ductility is required for improved impact resistance. Pipe types include

- Seamless
- Submerged arc welded (SAW or DSAW)

- Electric resistance welded (ERW)
- Spiral weld.

Except in specific cases, only seamless or SAW pipes are to be used, with seamless being the preference for diameters of 12 in. or less. If ERW pipe is used, special inspection provisions such as full-body ultrasonic testing are required. Spiral weld pipe is very unusual for oil/gas pipelines and should be used only for low-pressure water or outfall lines.

11.4.2.1.1 Design Procedure Determination of pipeline wall thickness is based on the design internal pressure or the external hydrostatic pressure. Maximum longitudinal stresses and combined stresses are sometimes limited by applicable codes and must be checked for installation and operation. However, these criteria are not normally used for wall thickness determination. Increasing wall thickness can sometimes ensure hydrodynamic stability in lieu of other stabilization methods (such as weight coating). This is not normally economical, except in deepwater where the presence of concrete may interfere with the preferred installation method. We recommend the following procedure for designing pipeline wall thickness:

Step 1: Calculate the minimum wall thickness required for the design internal pressure.
Step 2: Calculate the minimum wall thickness required to withstand external pressure.
Step 3: Add wall thickness allowance for corrosion if applicable to the maximum of the above.
Step 4: Select next highest nominal wall thickness.
Step 5: Check selected wall thickness for hydrotest condition.
Step 6: Check for handling practice, that is, pipeline handling is difficult for D/t larger than 50; welding of wall thickness less than 0.3 in (7.6 mm) requires special provisions.

Note that in certain cases, it may be desirable to order a nonstandard wall. This can be done for large orders.

Pipelines are sized on the basis of the maximum expected stresses in the pipeline under operating conditions. The stress calculation methods are different for thin-wall and thick-wall pipes. A thin-wall pipe is defined as a pipe with D/t greater than or equal to 20. Figure 11.11 shows stresses in a thin-wall pipe. A pipe with D/t less than 20 is considered a thick-wall pipe. Figure 11.12 illustrates stresses in a thick-wall pipe.

11.4.2.1.2 Design for Internal Pressure Three pipeline codes typically used for design are ASME B31.4 (ASME, 1989), ASME B31.8 (ASME, 1990), and DnV 1981 (DnV, 1981). ASME B31.4 is for all oil lines in North America. ASME B31.8 is for all gas lines and two-phase flow pipelines in North America. DnV 1981 is for oil, gas, and two-phase flow pipelines in North Sea. All these codes can be used in other areas when no other code is available.

The nominal pipeline wall thickness (t_{NOM}) can be calculated as follows:

$$t_{NOM} = \frac{P_d D}{2E_w \eta \sigma_y F_t} + t_a, \qquad (11.116)$$

where P_d is the design internal pressure defined as the difference between the internal pressure (P_i) and external pressure (P_e), D is nominal outside diameter, t_a is thickness allowance for corrosion, and σ_y is the specified

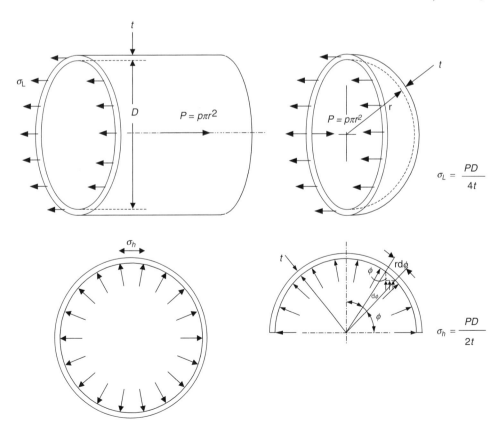

Figure 11.11 *Stresses generated by internal pressure p in a thin-wall pipe, D/t > 20.*

Table 11.2 Design and Hydrostatic Pressure Definitions and Usage Factors for Oil Lines

Parameter	ASME B31.4, 1989 Edition	Dnv (Veritas, 1981)
Design internal pressure P_d^a	$P_i - P_e$ [401.2.2]	$P_i - P_e$ [4.2.2.2]
Usage factor η	0.72 [402.3.1(a)]	0.72 [4.2.2.1]
Hydrotest pressure P_h	1.25 P_i^b [437.4.1(a)]	1.25 P_d [8.8.4.3]

[a] Credit can be taken for external pressure for gathering lines or flowlines when the MAOP (P_i) is applied at the wellhead or at the seabed. For export lines, when P_i is applied on a platform deck, the head fluid shall be added to P_i for the pipeline section on the seabed.

[b] If hoop stress exceeds 90% of yield stress based on nominal wall thickness, special care should be taken to prevent overstrain of the pipe.

minimum yield strength. Equation (11.116) is valid for any consistent units.

Most codes allow credit for external pressure. This credit should be used whenever possible, although care should be exercised for oil export lines to account for head of fluid and for lines that traverse from deep to shallow water.

ASME B31.4 and DnV 1981 define P_i as the maximum allowable operating pressure (MAOP) under normal conditions, indicating that surge pressures up to 110% MAOP is acceptable. In some cases, P_i is defined as wellhead shut-in pressure (WSIP) for flowlines or specified by the operators.

In Eq. (11.116), the weld efficiency factor (E_w) is 1.0 for seamless, ERW, and DSAW pipes. The temperature derating factor (F_t) is equal to 1.0 for temperatures under 250 °F. The usage factor (η) is defined in Tables 11.2 and 11.3 for oil and gas lines, respectively.

The underthickness due to manufacturing tolerance is taken into account in the design factor. There is no need to add any allowance for fabrication to the wall thickness calculated with Eq. (11.116).

11.4.2.1.3 Design for External Pressure Different practices can be found in the industry using different external pressure criteria. As a rule of thumb, or unless qualified thereafter, it is recommended to use propagation criterion for pipeline diameters under 16-in. and collapse criterion for pipeline diameters more than or equal to 16-in.

Propagation Criterion: The propagation criterion is more conservative and should be used where optimization of the wall thickness is not required or for pipeline installation methods not compatible with the use of buckle arrestors such as reel and tow methods. It is generally economical to design for propagation pressure for diameters less than 16-in. For greater diameters, the wall thickness penalty is too high. When a pipeline is designed based on the collapse criterion, buckle arrestors are recommended. The external pressure criterion should be based on nominal wall thickness, as the safety factors included below account for wall variations.

Although a large number of empirical relationships have been published, the recommended formula is the latest given by AGA.PRC (AGA, 1990):

$$P_P = 33 S_y \left(\frac{t_{NOM}}{D}\right)^{2.46}, \tag{11.117}$$

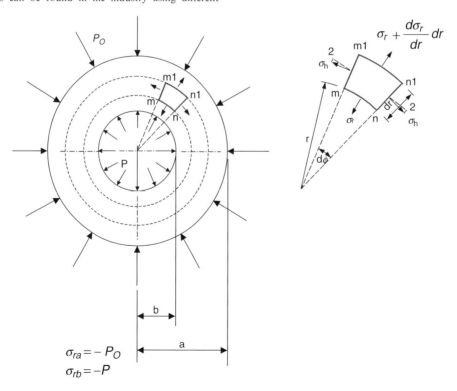

Figure 11.12 Stresses generated by internal pressure p in a thick-wall pipe, D/t < 20.

Table 11.3 Design and Hydrostatic Pressure Definitions and Usage Factors for Gas Lines

Parameter	ASME B31.8, 1989 Edition, 1990 Addendum	DnV (Veritas, 1981)
P_d^a	$P_i - P_e$ [A842.221]	$P_i - P_e$ [4.2.2.2]
Usage factor η	0.72 [A842.221]	0.72 [4.2.2.1]
Hydrotest pressure P_h	1.25 P_i^b [A847.2]	1.25P_d [8.8.4.3]

[a] Credit can be taken for external pressure for gathering lines or flowlines when the MAOP (P_i) is applied at wellhead or at the seabed. For export lines, when P_i is applied on a platform deck, the head of fluid shall be added to P_i for the pipeline section on the seabed (particularly for two-phase flow).
[b] ASME B31.8 imposes $P_h = 1.4P_i$ for offshore risers but allows onshore testing of prefabricated portions.

which is valid for any consistent units. The nominal wall thickness should be determined such that $P_p > 1.3\,P_e$. The safety factor of 1.3 is recommended to account for uncertainty in the envelope of data points used to derive Eq. (11.117). It can be rewritten as

$$t_{NOM} \geq D\left(\frac{1.3P_p}{33S_y}\right)^{\frac{1}{2.46}}. \tag{11.118}$$

For the reel barge method, the preferred pipeline grade is below X-60. However, X-65 steel can be used if the ductility is kept high by selecting the proper steel chemistry and microalloying. For deepwater pipelines, D/t ratios of less than 30 are recommended. It has been noted that bending loads have no demonstrated influence on the propagation pressure.

Collapse Criterion: The mode of collapse is a function of D/t ratio, pipeline imperfections, and load conditions. The theoretical background is not given in this book. An empirical general formulation that applies to all situations is provided. It corresponds to the transition mode of collapse under external pressure (P_e), axial tension (T_a), and bending strain (σ_b), as detailed elsewhere (Murphey and Langner, 1985; AGA, 1990).

The nominal wall thickness should be determined such that

$$\frac{1.3P_p}{P_C} + \frac{\varepsilon_b}{\varepsilon_B} \leq g_p, \tag{11.119}$$

where 1.3 is the recommended safety factor on collapse, ε_B is the bending strain of buckling failure due to pure bending, and g is an imperfection parameter defined below.

The safety factor on collapse is calculated for D/t ratios along with the loads (P_e, ε_b, T_a) and initial pipeline out-of-roundness (δ_o). The equations are

$$P_C = \frac{P_{el}P_y'}{\sqrt{P_{el}^2 + P_y'^2}}, \tag{11.120}$$

$$P_y' = P_y\left[\sqrt{1 - 0.75\left(\frac{T_a}{T_y}\right)^2} - \frac{T_a}{2T_y}\right], \tag{11.121}$$

$$P_{el} = \frac{2E}{1-\nu^2}\left(\frac{t}{D}\right)^3, \tag{11.122}$$

$$P_y = 2S_y\left(\frac{t}{D}\right), \tag{11.123}$$

$$T_y = AS_y, \tag{11.124}$$

where g_p is based on pipeline imperfections such as initial out-of-roundness (δ_o), eccentricity (usually neglected), and residual stress (usually neglected). Hence,

$$g_p = \sqrt{\frac{1+p^2}{p^2 - \frac{1}{f_p^2}}}, \tag{11.125}$$

where

$$p = \frac{P_y'}{P_{el}}, \tag{11.126}$$

$$f_p = \sqrt{1 + \left(\delta_o\frac{D}{t}\right)^2} - \delta_o\frac{D}{t}, \tag{11.127}$$

$$\varepsilon_B = \frac{t}{2D}, \tag{11.128}$$

and

$$\delta_o = \frac{D_{\max} - D_{\min}}{D_{\max} + D_{\min}}. \tag{11.129}$$

When a pipeline is designed using the collapse criterion, a good knowledge of the loading conditions is required (T_a and ε_b). An upper conservative limit is necessary and must often be estimated.

Under high bending loads, care should be taken in estimating ε_b using an appropriate moment-curvature relationship. A Ramberg Osgood relationship can be used as

$$K^* = M^* + AM^{*B}, \tag{11.130}$$

where $K^* = K/K_y$ and $M^* = M/M_y$ with $K_y = 2S_y/ED$ is the yield curvature and $M_y = 2IS_y/D$ is the yield moment. The coefficients A and B are calculated from the two data points on stress–strain curve generated during a tensile test.

11.4.2.1.4 Corrosion Allowance To account for corrosion when water is present in a fluid along with contaminants such as oxygen, hydrogen sulfide (H_2S), and carbon dioxide (CO_2), extra wall thickness is added. A review of standards, rules, and codes of practices (Hill and Warwick, 1986) shows that wall allowance is only one of several methods available to prevent corrosion, and it is often the least recommended.

For H_2S and CO_2 contaminants, corrosion is often localized (pitting) and the rate of corrosion allowance ineffective. Corrosion allowance is made to account for damage during fabrication, transportation, and storage. A value of $1/16$ in. may be appropriate. A thorough assessment of the internal corrosion mechanism and rate is necessary before any corrosion allowance is taken.

11.4.2.1.5 Check for Hydrotest Condition The minimum hydrotest pressure for oil and gas lines is given in Tables 11.2 and 11.3, respectively, and is equal to 1.25 times the design pressure for pipelines. Codes do not require that the pipeline be designed for hydrotest conditions but sometimes give a tensile hoop stress limit 90% SMYS, which is always satisfied if credit has not been taken for external pressure. For cases where the wall thickness is based on $P_d = P_i - P_e$, codes recommend not to overstrain the pipe. Some of the codes are ASME B31.4 (Clause 437.4.1), ASME B31.8 (no limit on hoop stress during hydrotest), and DnV (Clause 8.8.4.3).

For design purposes, condition $\sigma_h \leq \sigma_y$ should be confirmed, and increasing wall thickness or reducing test pressure should be considered in other cases. For offshore pipelines connected to riser sections requiring $P_h = 1.4P_i$, it is recommended to consider testing the riser separately (for prefabricated sections) or to determine the hydrotest pressure based on the actual internal pressure experienced by the pipeline section. It is important to note that most pressure testing of subsea pipelines is done with water, but on occasion, nitrogen or air has been used. For low D/t ratios (<20), the actual hoop stress in a pipeline tested from the surface is overestimated when using the thin wall equations provided in this chapter. Credit for this effect is allowed by DnV Clause 4.2.2.2 but is not normally taken into account.

Example Problem 11.6 Calculate the required wall thickness for the pipeline in Example Problem 11.4 assuming a seamless still pipe of X-60 grade and onshore gas field (external pressure $P_e = 14.65$ psia).

Solution The wall thickness can be designed based on the hoop stress generated by the internal pressure $Pi = 2,590$ psia. The design pressure is

$$P_d = P_i - P_e = 2,590 - 14.65 = 2,575.35 \text{ psi}.$$

The weld efficiency factor is $E_w = 1.0$. The temperature de-rating factor $F_t = 1.0$. Table 11.3 gives $\eta = 0.72$. The yield stress is $\sigma_y = 60,000$ psi. A corrosion allowance $1/16$ in. is considered. The nominal pipeline wall thickness can be calculated using Eq. (11.116) as

$$t_{NOM} = \frac{(2,574.3)(6)}{2(1.0)(0.72)(60,000)(1.0)} + \frac{1}{16} = 0.2413 \text{ in}.$$

Considering that welding of wall thickness less than 0.3 in. requires special provisions, the minimum wall thickness is taken, 0.3 in.

11.4.2.2 Insulation Design

Oil and gas field pipelines are insulated mainly to conserve heat. The need to keep the product fluids in the pipeline at a temperature higher than the ambient temperature could exist, for reasons including the following:

- Preventing formation of gas hydrates
- Preventing formation of wax or asphaltenes
- Enhancing product flow properties
- Increasing cool-down time after shutting down

In liquefied gas pipelines, such as liquefied natural gas, insulation is required to maintain the cold temperature of the gas to keep it in a liquid state.

Designing pipeline insulation requires thorough knowledge of insulation materials and heat transfer mechanisms across the insulation. Accurate predictions of heat loss and temperature profile in oil- and gas-production pipelines are essential to designing and evaluating pipeline operations.

11.4.2.2.1 Insulation Materials Polypropylene, polyethylene, and polyurethane are three base materials widely used in the petroleum industry for pipeline insulation. Their thermal conductivities are given in Table 11.4 (Carter et al., 2002). Depending on applications, these base materials are used in different forms, resulting in different overall conductivities. A three-layer polypropylene applied to pipe surface has a conductivity of 0.225 W/M-°C (0.13 btu/hr-ft-°F), while a four-layer polypropylene has a conductivity of 0.173 W/M-°C (0.10 btu/hr-ft-°F). Solid polypropylene has higher conductivity than polypropylene foam. Polymer syntactic polyurethane has a conductivity of 0.121 W/M-°C (0.07 btu/hr-ft-°F), while glass syntactic polyurethane has a conductivity of 0.156 W/M-°C (0.09 btu/hr-ft-°F). These materials have lower conductivities in dry conditions such as that in pipe-in-pipe (PIP) applications.

Because of their low thermal conductivities, more and more polyurethane foams are used in deepwater pipeline applications. Physical properties of polyurethane foams include density, compressive strength, thermal conductivity, closed-cell content, leachable halides, flammability, tensile strength, tensile modulus, and water absorption. Typical values of these properties are available elsewhere (Guo et al., 2005).

In steady-state flow conditions in an insulated pipeline segment, the heat flow through the pipe wall is given by

$$Q_r = UA_r \Delta T, \qquad (11.131)$$

where Q_r is heat-transfer rate; U is overall heat-transfer coefficient (OHTC) at the reference radius; A_r is area of the pipeline at the reference radius; ΔT is the difference in temperature between the pipeline product and the ambient temperature outside.

The OHTC, U, for a system is the sum of the thermal resistances and is given by (Holman, 1981):

$$U = \frac{1}{A_r \left[\frac{1}{A_i h_i} + \sum_{m=1}^{n} \frac{\ln(r_{m+1}/r_m)}{2\pi L k_m} + \frac{1}{A_o h_o} \right]}, \qquad (11.132)$$

Table 11.4 Thermal Conductivities of Materials Used in Pipeline Insulation

Material name	Thermal conductivity	
	W/M-°C	Btu/hr-ft-°F
Polyethylene	0.35	0.20
Polypropylene	0.22	0.13
Polyurethane	0.12	0.07

Table 11.5 Typical Performance of Insulated Pipelines

Insulation type	U-Value		Water depth (M)	
	$(Btu/hr - ft^2 - °F)$	$W/M^2 - K$	Field proven	Potential
Solid polypropylene	0.50	2.84	1,600	4,000
Polypropylene foam	0.28	1.59	700	2,000
Syntactic polyurethane	0.32	1.81	1,200	3,300
Syntactic polyurethane foam	0.30	1.70	2,000	3,300
Pipe-in-pipe syntactic polyurethane foam	0.17	0.96	3,100	4,000
Composite	0.12	0.68	1,000	3,000
Pipe-in-pipe high efficiency	0.05	0.28	1,700	3,000
Glass syntactic polyurethane	0.03	0.17	2,300	3,000

where h_i is film coefficient of pipeline inner surface; h_o is film coefficient of pipeline outer surface; A_i is area of pipeline inner surface; A_o is area of pipeline outer surface; r_m is radius of layer m; and k_m is thermal conductivity of layer m.

Similar equations exist for transient-heat flow, giving an instantaneous rate for heat flow. Typically required insulation performance, in terms of OHTC (U value) of steel pipelines in water, is summarized in Table 11.5.

Pipeline insulation comes in two main types: dry insulation and wet insulation. The dry insulations require an outer barrier to prevent water ingress (PIP). The most common types of this include the following:

- Closed-cell polyurethane foam
- Open-cell polyurethane foam
- Poly-isocyanurate foam
- Extruded polystyrene
- Fiber glass
- Mineral wool
- Vacuum-insulation panels

Under certain conditions, PIP systems may be considered over conventional single-pipe systems. PIP insulation may be required to produce fluids from high-pressure/high-temperature (>150 °C) reservoirs in deepwater (Carmichael et al., 1999). The annulus between pipes can be filled with different types of insulation materials such as foam, granular particles, gel, and inert gas or vacuum.

A pipeline-bundled system—a special configuration of PIP insulation—can be used to group individual flowlines together to form a bundle (McKelvie, 2000); heat-up lines can be included in the bundle, if necessary. The complete bundle may be transported to site and installed with a considerable cost savings relative to other methods. The extra steel required for the carrier pipe and spacers can sometimes be justified (Bai, 2001).

Wet-pipeline insulations are those materials that do not need an exterior steel barrier to prevent water ingress, or the water ingress is negligible and does not degrade the insulation properties. The most common types of this are as follows:

- Polyurethane
- Polypropylene
- Syntactic polyurethane
- Syntactic polypropylene
- Multilayered

The main materials that have been used for deepwater insulations have been polyurethane and polypropylene based. Syntactic versions use plastic or glass matrix to improve insulation with greater depth capabilities. Insulation coatings with combinations of the two materials have also been used. Guo et al. (2005) gives the properties of these wet insulations. Because the insulation is buoyant, this effect must be compensated by the steel pipe weight to obtain lateral stability of the deepwater pipeline on the seabed.

11.4.2.2.2 Heat Transfer Models

Heat transfer across the insulation of pipelines presents a unique problem affecting flow efficiency. Although sophisticated computer packages are available for predicting fluid temperatures, their accuracies suffer from numerical treatments because long pipe segments have to be used to save computing time. This is especially true for transient fluid-flow analyses in which a very large number of numerical iterations are performed.

Ramey (1962) was among the first investigators who studied radial-heat transfer across a well casing with no insulation. He derived a mathematical heat-transfer model for an outer medium that is infinitely large. Miller (1980) analyzed heat transfer around a geothermal wellbore without insulation. Winterfeld (1989) and Almehaideb (1989) considered temperature effect on pressure-transient analyses in well testing. Stone et al. (1989) developed a numerical simulator to couple fluid flow and heat flow in a wellbore and reservoir. More advanced studies on the wellbore heat-transfer problem were conducted by Hasan and Kabir (1994, 2002), Hasan et al. (1997, 1998), and Kabir et al. (1996). Although multilayers of materials have been considered in these studies, the external temperature gradient in the longitudinal direction has not been systematically taken into account. Traditionally, if the outer temperature changes with length, the pipe must be divided into segments, with assumed constant outer temperature in each segment, and numerical algorithms are required for heat-transfer computation. The accuracy of the computation depends on the number of segments used. Fine segments can be employed to ensure accuracy with computing time sacrificed.

Guo et al. (2006) presented three analytical heat-transfer solutions. They are the transient-flow solution for startup mode, steady-flow solution for normal operation mode, and transient-flow solution for flow rate change mode (shutting down is a special mode in which the flow rate changes to zero).

Temperature and Heat Transfer for Steady Fluid Flow. The internal temperature profile under steady fluid-flow conditions is expressed as

$$T = \frac{1}{\alpha^2}\left[\beta - \alpha\beta L - \alpha\gamma - e^{-\alpha(L+C)}\right], \quad (11.133)$$

where the constant groups are defined as

$$\alpha = \frac{2\pi Rk}{v\rho C_p sA}, \quad (11.134)$$

$$\beta = \alpha G \cos(\theta), \quad (11.135)$$

$$\gamma = -\alpha T_0, \quad (11.136)$$

and

$$C = -\frac{1}{\alpha}\ln(\beta - \alpha^2 T_s - \alpha\gamma), \quad (11.137)$$

where T is temperature inside the pipe, L is longitudinal distance from the fluid entry point, R is inner radius of insulation layer, k is the thermal conductivity of the insulation material, v is the average flow velocity of fluid in the pipe, ρ is fluid density, C_p is heat capacity of fluid at constant pressure, s is thickness of the insulation layer, A is the inner cross-sectional area of pipe, G is principal thermal-gradient outside the insulation, θ is the angle between the principal thermal gradient and pipe orientation, T_0 is temperature of outer medium at the fluid entry location, and T_s is temperature of fluid at the fluid entry point.

The rate of heat transfer across the insulation layer over the whole length of the pipeline is expressed as

$$q = -\frac{2\pi Rk}{s}$$
$$\times \left(T_0 L - \frac{G\cos(\theta)}{2}L^2 - \frac{1}{\alpha^2}\left\{(\beta - \alpha\gamma)L - \frac{\alpha\beta}{2}L^2\right.\right.$$
$$\left.\left. + \frac{1}{\alpha}\left[e^{-\alpha(L+C)} - e^{-\alpha C}\right]\right\}\right), \quad (11.138)$$

where q is the rate of heat transfer (heat loss).

Transient Temperature During Startup. The internal temperature profile after starting up a fluid flow is expressed as follows:

$$T = \frac{1}{\alpha^2}\{\beta - \alpha\beta L - \alpha\gamma - e^{-\alpha[L+f(L-vt)]}\}, \quad (11.139)$$

11/154 EQUIPMENT DESIGN AND SELECTION

Table 11.6 Base Data for Pipeline Insulation Design

Length of pipeline:	8,047	M
Outer diameter of pipe:	0.2032	M
Wall thickness:	0.00635	M
Fluid density:	881	kg/M^3
Fluid specific heat:	2,012	J/kg-°C
Average external temperature:	10	°C
Fluid temperature at entry point:	28	°C
Fluid flow rate:	7,950	M^3/day

where the function f is given by

$$f(L - vt) = -(L - vt) - \frac{1}{\alpha} \ln\{\beta - \alpha\beta(L - vt)$$
$$- \alpha\gamma - \alpha^2[T_s - G\cos(\theta)(L - vt)]\} \quad (11.140)$$

and t is time.

Transient Temperature During Flow Rate Change.
Suppose that after increasing or decreasing the flow rate, the fluid has a new velocity v' in the pipe. The internal temperature profile is expressed as follows:

$$T = \frac{1}{\alpha'^2}\{\beta' - \alpha'\beta' L - \alpha'\gamma' - e^{-\alpha'[L+f(L-v't)]}\}, \quad (11.141)$$

where

$$\alpha' = \frac{2\pi Rk}{v'\rho C_p sA}, \quad (11.142)$$

$$\beta' = \alpha' G \cos(\theta), \quad (11.143)$$

$$\gamma' = -\alpha' T_0, \quad (11.144)$$

and the function f is given by

$$f(L - v't) = -(L - v't) - \frac{1}{\alpha'} \ln(\beta' - \alpha'\beta'(L$$
$$- v't) - \alpha'\gamma' - \left(\frac{\alpha'}{\alpha}\right)^2 \{\beta - \alpha\beta(L$$
$$- v't) - \alpha\gamma - e^{-\alpha[(L-v't)+C]}\}). \quad (11.145)$$

Example Problem 11.7 A design case is shown in this example. Design base for a pipeline insulation is presented in Table 11.6. The design criterion is to ensure that the temperature at any point in the pipeline will not drop to less than 25 °C, as required by flow assurance. Insulation materials considered for the project were polyethylene, polypropylene, and polyurethane.

Solution A polyethylene layer of 0.0254 M (1 in.) was first considered as the insulation. Figure 11.13 shows the temperature profiles calculated using Eqs. (11.133) and (11.139). It indicates that at approximately 40 minutes after startup, the transient-temperature profile in the pipeline will approach the steady-flow temperature profile. The temperature at the end of the pipeline will be slightly lower than 20 °C under normal operating conditions. Obviously, this insulation option does not meet design criterion of 25 °C in the pipeline.

Figure 11.14 presents the steady-flow temperature profiles calculated using Eq. (11.133) with polyethylene layers of four thicknesses. It shows that even a polyethylene layer 0.0635-M (2.5-in.) thick will still not give a pipeline temperature higher than 25 °C; therefore, polyethylene should not be considered in this project.

A polypropylene layer of 0.0254 M (1 in.) was then considered as the insulation. Figure 11.15 illustrates the temperature profiles calculated using Eq. (11.133) and (11.139). It again indicates that at approximately 40 minutes after startup, the transient-temperature profile in the pipe will approach the steady-flow temperature profile. The temperature at the end of the pipeline will be approximately 22.5 °C under normal operating conditions. Obviously, this insulation option, again, does not meet design criterion of 25 °C in the pipeline.

Figure 11.16 demonstrates the steady-flow temperature profiles calculated using Eq. (11.133) with polypropylene layers of four thicknesses. It shows that a polypropylene layer of 0.0508 M (2.0 in.) or thicker will give a pipeline temperature of higher than 25 °C.

A polyurethane layer of 0.0254 M (1 in.) was also considered as the insulation. Figure 11.17 shows the temperature profiles calculated using Eqs. (11.133) and (11.139). It indicates that the temperature at the end of pipeline will drop to slightly lower than 25 °C under normal operating conditions. Figure 11.18 presents the steady-flow temperature profiles calculated using Eq. (11.133) with polyurethane layers of four thicknesses. It shows that a polyurethane layer of 0.0381 M (1.5 in.)

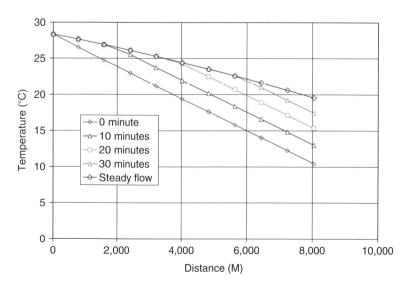

Figure 11.13 Calculated temperature profiles with a polyethylene layer of 0.0254 M (1 in.).

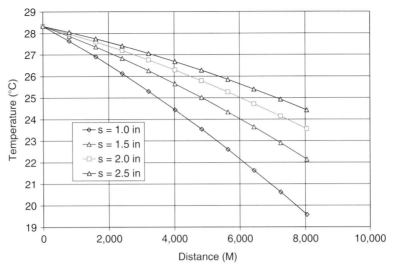

Figure 11.14 Calculated steady-flow temperature profiles with polyethylene layers of various thicknesses.

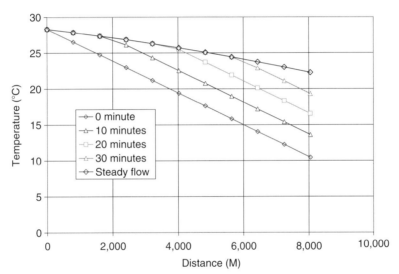

Figure 11.15 Calculated temperature profiles with a polypropylene layer of 0.0254 M (1 in.).

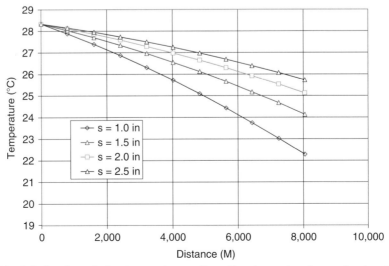

Figure 11.16 Calculated steady-flow temperature profiles with polypropylene layers of various thicknesses.

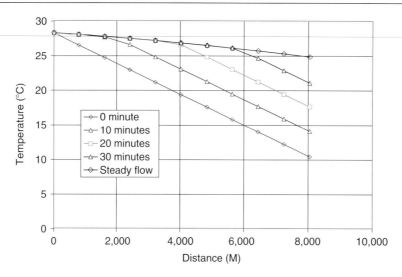

Figure 11.17 Calculated temperature profiles with a polyurethane layer of 0.0254 M (1 in.).

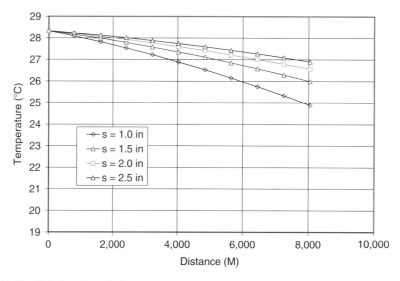

Figure 11.18 Calculated steady-flow temperature profiles with polyurethane layers of four thicknesses.

is required to keep pipeline temperatures higher than 25 °C under normal operating conditions.

Therefore, either a polypropylene layer of 0.0508 M (2.0 in.) or a polyurethane layer of 0.0381 M (1.5 in.) should be chosen for insulation of the pipeline. Cost analyses can justify one of the options, which is beyond the scope of this example.

The total heat losses for all the steady-flow cases were calculated with Eq. (11.138). The results are summarized in Table 11.7. These data may be used for sizing heaters for the pipeline if heating of the product fluid is necessary.

Summary

This chapter described oil and gas transportation systems. The procedure for selection of pumps and gas compressors were presented and demonstrated. Theory and applications of pipeline design were illustrated.

Table 11.7 Calculated Total Heat Losses for the Insulated Pipelines (kW)

Material name	Insulation thickness (M) 0.0254	0.0381	0.0508	0.0635
Polyethylene	1,430	1,011	781	636
Polypropylene	989	685	524	424
Polyurethane	562	383	290	234

References

ALMEHAIDEB, R.A., AZIZ, K., and PEDROSA, O.A., JR. A reservoir/wellbore model for multiphase injection and pressure transient analysis. Presented at the 1989 SPE Middle East Oil Show, Bahrain, 11–14 March. Paper SPE 17941.

American Gas Association. Collapse of Offshore Pipelines. Pipeline Research Committee, Seminar held 20 February, 1990, in Houston, Texas.

American Society of Mechanical Engineers. Liquid transportation systems for hydrocarbons, liquid petroleum gas, anhydrous ammonia and alcohols. *ASME B31.4* 1989. Washington.

American Society of Mechanical Engineers. Gas Transmission and Distribution Piping Systems," ASME Code for Pressure Piping, B31.8–(1989 Edition and 1990 Addendum). Washington.

BAI, Y. *Pipelines and Risers*, Vol. 3, *Ocean Engineering Book Series*. Amsterdam: Elsevier, 2001.

BROWN, G.G. A series of enthalpy-entropy charts for natural gases. *Trans. AIME* 1945;60:65.

CARMICHAEL, R., FANG, J., and TAM, C. Pipe-in-pipe systems for deepwater developments. Proceedings of the Deepwater Pipeline Technology Conference, in New Orleans, 1999.

CARTER, B., GRAY, C., and CAI, J. 2002 survey of offshore non-chemical flow assurance solutions. Poster published by *Offshore Magazine*, Houston, 2003.

Det norske Veritas. *Rules for Submarine Pipeline Systems*. 1981.

GUO, B. et al. *Offshore Pipelines*. Burlington: Gulf Professional Publishing, 2005.

GUO, B., DUAN S., and GHALAMBOR, A. A simple model for predicting heat loss and temperature profiles in insulated pipelines. *SPE Prod. Operations J.* February 2006.

GUO, B. and GHALAMBOR, A. *Natural Gas Engineering Handbook*. Houston, TX: Gulf Publishing Company, 2005.

HASAN, A.R. and KABIR, C.S. Aspects of wellbore heat transfer during two-phase flow. *SPEPF* 1994;9(2):211–218.

HASAN, A.R., KABIR, C.S., and WANG, X. Wellbore two-phase flow and heat transfer during transient testing. *SPEJ* 1998:174–181.

HASAN, A.R., KABIR, C.S., and WANG, X. Development and application of a wellbore/reservoir simulator for testing oil wells. *SPEFE* 1997;12(3)182–189.

HASAN, A.R. and KABIR, C.S. *Fluid Flow and Heat Transfer in Wellbores*. Richardson, TX: SPE, 2002.

HILL, R.T. and WARWICK, P.C. Internal Corrosion Allowance for Marine Pipelines: A Question of Validity. OTC paper No. 5268, 1986.

HOLMAN, J.P. *Heat Transfer*. New York: McGraw-Hill Book Co., 1981.

IKOKU, C.U. *Natural Gas Production Engineering*. New York: John Wiley & Sons, 1984.

KABIR, C.S. et al. A wellbore/reservoir simulator for testing gas well in high-temperature reservoirs. *SPEFE* 1996;11(2):128–135.

KATZ, D.L. and LEE, R.L. *Natural Gas Engineering—Production and Storage*. New York: McGraw-Hill Publishing Co., 1990.

LYONS, W.C. *Air and Gas Drilling Manual*. New York: McGraw-Hill, 2001:4–5.

MCKELVIE, M. "Bundles—Design and Construction," *Integrated Graduate Development Scheme*, Heriot-Watt U., 2000.

MILLER, C.W. Wellbore storage effect in geothermal wells. *SPEJ* 1980:555.

MURPHEY, C.E. and LANGNER, C.G. Ultimate pipe strength under bending, collapse, and fatigue. Proceedings of the OMAE Conference, 1985.

RAMEY, H.J., JR. Wellbore heat transmission. *JPT* April 1962;427, *Trans. AIME* 14.

ROLLINS, J.P. *Compressed Air and Gas Handbook*. New York: Compressed Air and Gas Institute, 1973.

STONE, T.W., EDMUNDS, N.R., and KRISTOFF, B.J. A comprehensive wellbore/reservoir simulator. Presented at the 1989 SPE Reservoir Simulation Symposium, 6–8 February, in Houston. Paper SPE 18419.

WINTERFELD, P.H. Simulation of pressure buildup in a multiphase wellbore/reservoir system. *SPEFE* 1989; 4(2):247–252.

Problems

11.1 A pipeline transporting 10,000 bbl/day of oil requires a pump with a minimum output pressure of 500 psi. The available suction pressure is 300 psi. Select a triplex pump for this operation.

11.2 A pipeline transporting 8,000 bbl/day of oil requires a pump with a minimum output pressure of 400 psi. The available suction pressure is 300 psi. Select a duplex pump for this operation.

11.3 For a reciprocating compressor, calculate the theoretical and brake horsepower required to compress 30 MMcfd of a 0.65 specific gravity natural gas from 100 psia and 70 °F to 2,000 psia. If intercoolers and end-coolers cool the gas to 90 °F, what is the heat load on the coolers? Assuming the overall efficiency is 0.80.

11.4 For a centrifugal compressor, use the following data to calculate required input horsepower and polytropic head:

Gas-specific gravity:	0.70
Gas-specific heat ratio:	1.30
Gas flow rate:	50 MMscfd at 14.7 psia and 60 °F
Inlet pressure:	200 psia Inlet temperature: 70 °F
Discharge pressure:	500 psia
Polytropic efficiency:	$E_p = 061 + 003 \log(q_1)$

11.5 For the data given in Problem 11.4, calculate the required brake horsepower if a reciprocating compressor is used.

11.6 A 40-API gravity, 3-cp oil is transported through an 8-in. (I.D.) pipeline with a downhill angle of 5 degrees across a distance of 10 miles at a flow rate of 5,000 bbl/day. Estimate the minimum required pump pressure to deliver oil at 100 psi pressure at the outlet. Assume $e = 0.0006$ in.

11.7 For the following data given for a horizontal pipeline, predict gas flow rate in cubic feet per hour through the pipeline. Solve the problem using Eq. (11.101) with the trial-and-error method for friction factor and the Weymouth equation without the Reynolds number–dependent friction factor:

$d = 6$ in.
$L = 100$ mi
$e = 0.0006$ in.
$T = 70$ °F
$\gamma_g = 0.70$
$T_b = 520$ °R

$p_b = 14.65$ psia
$p_1 = 800$ psia
$p_2 = 200$ psia

11.8 Solve Problem 11.7 using
   a. Panhandle-A Equation
   b. Panhandle-B Equation

11.9 Assuming a 10-degree uphill angle, solve Problem 11.7 using the Weymouth equation.

11.10 Calculate the required wall thickness for a pipeline using the following data:
   Water depth 2,000 ft offshore oil field
   Water temperature 45 °F
   12.09 in. pipe inner diameter
   Seamless still pipe of X-65 grade
   Maximum pipeline pressure 3,000 psia

11.11 Design insulation for a pipeline with the following given data:

Length of pipeline:	7,000 M
Outer diameter of pipe:	0.254 M
Wall thickness:	0.0127 M
Fluid density:	800 kg/M^3
Fluid specific heat:	2,000 J/kg- °C
Average external temperature:	15 °C
Fluid temperature at entry point:	30 °C
Fluid flow rate:	5,000 M^3/day

Part III Artificial Lift Methods

Most oil reservoirs are of the volumetric type where the driving mechanism is the expansion of solution gas when reservoir pressure declines because of fluid production. Oil reservoirs will eventually not be able to produce fluids at economical rates unless natural driving mechanisms (e.g., aquifer and/or gas cap) or pressure maintenance mechanisms (e.g., water flooding or gas injection) are present to maintain reservoir energy. The only way to obtain a high production rate of a well is to increase production pressure drawdown by reducing the bottom-hole pressure with artificial lift methods.

Approximately 50% of wells worldwide need artificial lift systems. The commonly used artificial lift methods include the following:

- Sucker rod pumping
- Gas lift
- Electrical submersible pumping
- Hydraulic piston pumping
- Hydraulic jet pumping
- Plunger lift
- Progressing cavity pumping

Each method has applications for which it is the optimum installation. Proper selection of an artificial lift method for a given production system (reservoir and fluid properties, wellbore configuration, and surface facility restraints) requires a thorough understanding of the system. Economics analysis is always performed. Relative advantages and disadvantages of artificial lift systems are discussed in the beginning of each chapter in this part of this book. The chapters in this part provide production engineers with fundamentals of sucker rod pumping and gas lifts, as well as an introduction to other artificial lift systems. The following three chapters are included in this part of the book:

Chapter 12: Sucker Rod Pumping
Chapter 13: Gas Lift
Chapter 14: Other Artificial Lift Methods

12 Sucker Rod Pumping

Contents
12.1 Introduction 12/162
12.2 Pumping System 12/162
12.3 Polished Rod Motion 12/165
12.4 Load to the Pumping Unit 12/168
12.5 Pump Deliverability and Power Requirements 12/170
12.6 Procedure for Pumping Unit Selection 12/172
12.7 Principles of Pump Performance Analysis 12/174
Summary 12/179
References 12/179
Problems 12/179

12.1 Introduction

Sucker rod pumping is also referred to as "beam pumping." It provides mechanical energy to lift oil from bottom hole to surface. It is efficient, simple, and easy for field people to operate. It can pump a well down to very low pressure to maximize oil production rate. It is applicable to slim holes, multiple completions, and high-temperature and viscous oils. The system is also easy to change to other wells with minimum cost. The major disadvantages of beam pumping include excessive friction in crooked/deviated holes, solid-sensitive problems, low efficiency in gassy wells, limited depth due to rod capacity, and bulky in offshore operations. Beam pumping trends include improved pump-off controllers, better gas separation, gas handling pumps, and optimization using surface and bottom-hole cards.

12.2 Pumping System

As shown in Fig. 12.1, a sucker rod pumping system consists of a pumping unit at surface and a plunger pump submerged in the production liquid in the well.

The *prime mover* is either an electric motor or an internal combustion engine. The modern method is to supply each well with its own motor or engine. Electric motors are most desirable because they can easily be automated. The power from the prime mover is transmitted to the input shaft of a gear reducer by a *V-belt drive*. The output shaft of the gear reducer drives the *crank arm* at a lower speed (~4–40 revolutions per minute [rpm] depending on well characteristics and fluid properties). The rotary motion of the crank arm is converted to an oscillatory motion by means of the *walking beam* through a *pitman arm*. The *horse's head* and the *hanger cable* arrangement is used to ensure that the upward pull on the sucker rod string is vertical at all times (thus, no bending moment is applied to the *stuffing box*). The *polished rod* and stuffing box combine to maintain a good liquid seal at the surface and, thus, force fluid to flow into the "T" connection just below the stuffing box.

Conventional pumping units are available in a wide range of sizes, with stroke lengths varying from 12 to almost 200 in. The strokes for any pumping unit type are available in increments (unit size). Within each unit size, the stroke length can be varied within limits (about six different lengths being possible). These different lengths are achieved by varying the position of the pitman arm connection on the crank arm.

Walking beam ratings are expressed in allowable polished rod loads (PRLs) and vary from approximately 3,000 to 35,000 lb. Counterbalance for conventional pumping units is accomplished by placing weights directly on the beam (in smaller units) or by attaching weights to the rotating crank arm (or a combination of the two methods for larger units). In more recent designs, the rotary counterbalance can be adjusted by shifting the position of the weight on the crank by a jackscrew or rack and pinion mechanism.

There are two other major types of pumping units. These are the Lufkin Mark II and the Air-Balanced Units (Fig. 12.2). The pitman arm and horse's head are in the same side of the walking beam in these two types of units (Class III lever system). Instead of using counter-weights in Lufkin Mark II type units, air cylinders are used in the air-balanced units to balance the torque on the crankshaft.

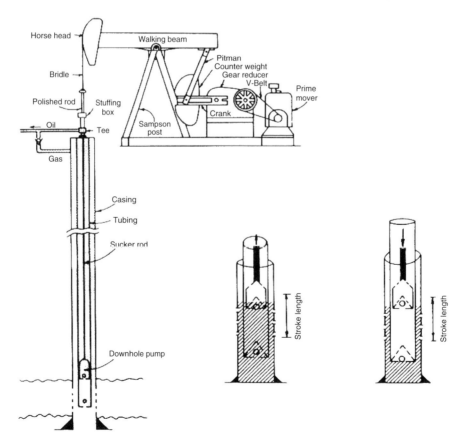

Figure 12.1 A diagrammatic drawing of a sucker rod pumping system (Golan and Whitson, 1991).

The American Petroleum Institute (API) has established designations for sucker rod pumping units using a string of characters containing four fields. For example,

C-228D-200-74.

The first field is the code for type of pumping unit. C is for conventional units, A is for air-balanced units, B is for beam counterbalance units, and M is for Mark II units. The second field is the code for peak torque rating in thousands of inch-pounds and gear reducer. D stands for double-reduction gear reducer. The third field is the code for PRL rating in hundreds of pounds. The last field is the code for stroke length in inches.

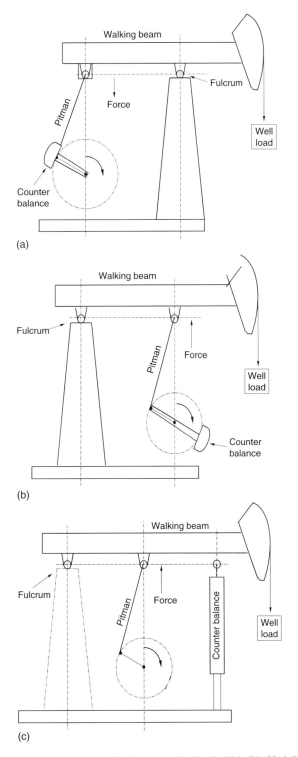

Figure 12.2 *Sketch of three types of pumping units: (a) conventional unit; (b) Lufkin Mark II Unit; (c) air-balanced unit.*

12/164 ARTIFICIAL LIFT METHODS

Figure 12.3 illustrates the working principle of a plunger pump. The pump is installed in the tubing string below the dynamic liquid level. It consists of a *working barrel* and *liner*, *standing valve* (SV), and *traveling valve* (TV) at the bottom of the *plunger*, which is connected to *sucker rods*.

As the plunger is moved downward by the sucker rod string, the TV is open, which allows the fluid to pass through the valve, which lets the plunger move to a position just above the SV. During this downward motion of the plunger, the SV is closed; thus, the fluid is forced to pass through the TV.

When the plunger is at the bottom of the stroke and starts an upward stroke, the TV closes and the SV opens. As upward motion continues, the fluid in the well below the SV is drawn into the volume above the SV (fluid

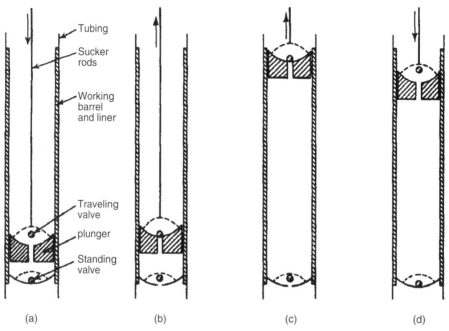

Figure 12.3 The pumping cycle: (a) plunger moving down, near the bottom of the stroke; (b) plunger moving up, near the bottom of the stroke; (c) plunger moving up, near the top of the stroke; (d) plunger moving down, near the top of the stroke (Nind, 1964).

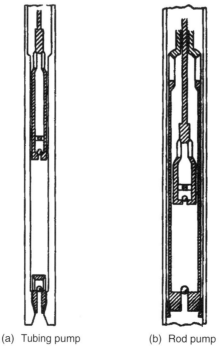

(a) Tubing pump   (b) Rod pump

Figure 12.4 Two types of plunger pumps (Nind, 1964).

passing through the open SV). The fluid continues to fill the volume above the SV until the plunger reaches the top of its stroke.

There are two basic types of plunger pumps: tubing pump and rod pump (Fig. 12.4). For the tubing pump, the working barrel or liner (with the SV) is made up (i.e., attached) to the bottom of the production tubing string and must be run into the well with the tubing. The plunger (with the TV) is run into the well (inside the tubing) on the sucker rod string. Once the plunger is seated in the working barrel, pumping can be initiated. A rod pump (both working barrel and plunger) is run into the well on the sucker rod string and is seated on a wedged type seat that is fixed to the bottom joint of the production tubing. Plunger diameters vary from $\frac{5}{8}$ to $4\frac{5}{8}$ in. Plunger area varies from 0.307 in.^{2} to 17.721 in.^{2}.

12.3 Polished Rod Motion

The theory of polished rod motion has been established since 1950s (Nind, 1964). Figure 12.5 shows the cyclic motion of a polished rod in its movements through the stuffing box of the conventional pumping unit and the air-balanced pumping unit.

Conventional Pumping Unit. For this type of unit, the acceleration at the bottom of the stroke is somewhat greater than true simple harmonic acceleration. At the top of the stroke, it is less. This is a major drawback for the conventional unit. Just at the time the TV is closing and the fluid load is being transferred to the rods, the acceleration for the rods is at its maximum. These two factors combine to create a maximum stress on the rods that becomes one of the limiting factors in designing an installation. Table 12.1 shows dimensions of some API conventional pumping units. Parameters are defined in Fig. 12.6.

Air-Balanced Pumping Unit. For this type of unit, the maximum acceleration occurs at the top of the stroke (the acceleration at the bottom of the stroke is less than simple harmonic motion). Thus, a lower maximum stress is set up in the rod system during transfer of the fluid load to the rods.

The following analyses of polished rod motion apply to conventional units. Figure 12.7 illustrates an approximate motion of the connection point between pitman arm and walking beam.

If x denotes the distance of B below its top position C and is measured from the instant at which the crank arm and pitman arm are in the vertical position with the crank arm vertically upward, the law of cosine gives

$$(AB)^2 = (OA)^2 + (OB)^2 - 2(OA)(OB)\cos AOB,$$

that is,

$$h^2 = c^2 + (h + c - x)^2 - 2c(h + c - x)\cos \omega t,$$

where ω is the angular velocity of the crank. The equation reduces to

$$x^2 - 2x[h + c(1 - \cos \omega t)] + 2c(h + c)(1 - \cos \omega t) = 0$$

so that

$$x = h + c(1 - \cos \omega t) \pm \sqrt{c^2 \cos^2 \omega t + (h^2 - c^2)}.$$

When ωt is zero, x is also zero, which means that the negative root sign must be taken. Therefore,

$$x = h + c(1 - \cos \omega t) - \sqrt{c^2 \cos^2 \omega t + (h^2 + c^2)}.$$

Acceleration is

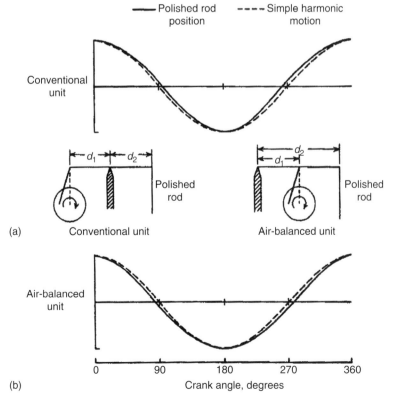

Figure 12.5 Polished rod motion for (a) conventional pumping unit and (b) air-balanced unit (Nind, 1964).

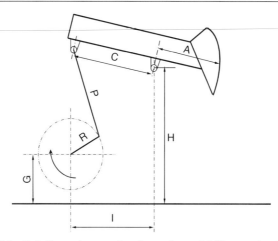

Figure 12.6 Definitions of conventional pumping unit API geometry dimensions.

Table 12.1 Conventional Pumping Unit API Geometry Dimensions

API Unit designation	A (in.)	C (in.)	I (in.)	P (in.)	H (in.)	G (in.)	R1, R2, R3 (in.)	C_s (lb)	Torque factor
C-912D-365-168	210	120.03	120	148.5	237.88	86.88	47, 41, 35	−1,500	80.32
C-912D-305-168	210	120.03	120	148.5	237.88	86.88	47, 41, 35	−1,500	80.32
C-640D-365-168	210	120.03	120	148.5	237.88	86.88	47, 41, 35	−1,500	80.32
C-640D-305-168	210	120.03	120	148.5	237.88	86.88	47, 41, 35	−1,500	80.32
C-456D-305-168	210	120.03	120	148.5	237.88	86.88	47, 41, 35	−1,500	80.32
C-912D-427-144	180	120.03	120	148.5	237.88	86.88	47, 41, 35	−650	68.82
C-912D-365-144	180	120.03	120	148.5	237.88	86.88	47, 41, 35	−650	68.82
C-640D-365-144	180	120.03	120	148.5	238.88	89.88	47, 41, 35	−650	68.82
C-640D-305-144	180	120.08	120	144.5	238.88	89.88	47, 41, 35	−520	68.45
C-456D-305-144	180	120.08	120	144.5	238.88	89.88	47, 41, 35	−520	68.45
C-640D-256-144	180	120.08	120	144.5	238.88	89.88	47, 41, 35	−400	68.45
C-456D-256-144	180	120.08	120	144.5	238.88	89.88	47, 41, 35	−400	68.45
C-320D-256-144	180	120.08	120	144.5	238.88	89.88	47, 41, 35	−400	68.45
C-456D-365-120	152	120.03	120	148.5	238.88	89.88	47, 41, 35	570	58.12
C-640D-305-120	155	111.09	111	133.5	213	75	42, 36, 30	−120	57.02
C-456D-305-120	155	111.09	111	133.5	213	75	42, 36, 30	−120	57.02
C-320D-256-120	155	111.07	111	132	211	75	42, 36, 30	55	57.05
C-456D-256-120	155	111.07	111	132	211	75	42, 36, 30	55	57.05
C-456D-213-120	155	111.07	111	132	211	75	42, 36, 30	0	57.05
C-320D-213-120	155	111.07	111	132	211	75	42, 36, 30	0	57.05
C-228D-213-120	155	111.07	111	132	211	75	42, 36, 30	0	57.05
C-456D-265-100	129	111.07	111	132	211	75	42, 36, 30	550	47.48
C-320D-265-100	129	111.07	111	132	211	75	42, 36, 30	550	47.48
C-320D-305-100	129	111.07	111	132	211	75	42, 36, 30	550	47.48
C-228D-213-100	129	96.08	96	113	180	63	37, 32, 27	0	48.37
C-228D-173-100	129	96.05	96	114	180	63	37, 32, 27	0	48.37
C-160D-173-100	129	96.05	96	114	180	63	37, 32, 27	0	48.37
C-320D-246-86	111	111.04	111	133	211	75	42, 36, 30	800	40.96
C-228D-246-86	111	111.04	111	133	211	75	42, 36, 30	800	40.96
C-320D-213-86	111	96.05	96	114	180	63	37, 32, 27	450	41.61
C-228D-213-86	111	96.05	96	114	180	63	37, 32, 27	450	41.61
C-160D-173-86	111	96.05	96	114	180	63	37, 32, 27	450	41.61
C-114D-119-86	111	84.05	84	93.75	150.13	53.38	32, 27, 22	115	40.98
C-320D-245-74	96	96.05	96	114	180	63	37, 32, 27	800	35.99
C-228D-200-74	96	96.05	96	114	180	63	37, 32, 27	800	35.99
C-160D-200-74	96	96.05	96	114	180	63	37, 32, 27	800	35.99
C-228D-173-74	96	84.05	84	96	152.38	53.38	32, 27, 22	450	35.49
C-160D-173-74	96	84.05	84	96	152.38	53.38	32, 27, 22	450	35.49
C-160D-143-74	96	84.05	84	93.75	150.13	53.38	32, 27, 22	300	35.49
C-114D-143-74	96	84.05	84	93.75	150.13	53.38	32, 27, 22	300	35.49
C-160D-173-64	84	84.05	84	93.75	150.13	53.38	32, 27, 22	550	31.02
C-114D-173-64	84	84.05	84	93.75	150.13	53.38	32, 27, 22	550	31.02
C-160D-143-64	84	72.06	72	84	132	45	27, 22, 17	360	30.59
C-114D-143-64	84	72.06	72	84	132	45	27, 22, 17	360	30.59
C-80D-119-64	84	64	64	74.5	116	41	24, 20, 16	0	30.85
C-160D-173-54	72	72.06	72	84	132	45	27, 22, 17	500	26.22

(*Continued*)

Table 12.1 Conventional Pumping Unit API Geometry Dimensions (Continued)

API Unit designation	A (in.)	C (in.)	I (in.)	P (in.)	H (in.)	G (in.)	R1, R2, R3 (in.)	C_s (lb)	Torque factor
C-114D-133-54	72	64	64	74.5	116	41	24, 20, 16	330	26.45
C-80D-133-54	72	64	64	74.5	116	41	24, 20, 16	330	26.45
C-80D-119-54	72	64	64	74.5	116	41	24, 20, 16	330	26.45
C-P57D-76-54	64	51	51	64	103	39	21, 16, 11	105	25.8
C-P57D-89-54	64	51	51	64	103	39	21, 16, 11	105	25.8
C-80D-133-48	64	64	64	74.5	116	41	24, 20, 16	440	23.51
C-80D-109-48	64	56.05	56	65.63	105	37	21, 16, 11	320	23.3
C-57D-109-48	64	56.05	56	65.63	105	37	21, 16, 11	320	23.3
C-57D-95-48	64	56.05	56	65.63	105	37	21, 16, 11	320	23.3
C-P57D-109-48	57	51	51	64	103	39	21, 16, 11	180	22.98
C-P57D-95-48	57	51	51	64	103	39	21, 16, 11	180	22.98
C-40D-76-48	64	48.17	48	57.5	98.5	37	18, 14, 10	0	23.1
C-P40D-76-48	61	47	47	56	95	39	18, 14, 10	190	22.92
C-P57D-89-42	51	51	51	64	103	39	21, 16, 11	280	20.56
C-P57D-76-42	51	51	51	64	103	39	21, 16, 11	280	20.56
C-P40D-89-42	53	47	47	56	95	39	18, 14, 10	280	19.92
C-P40D-76-42	53	47	47	56	95	39	18, 14, 10	280	19.92
C-57D-89-42	56	48.17	48	57.5	98.5	37	18, 14, 10	150	20.27
C-57D-76-42	56	48.17	48	57.5	98.5	37	18, 14, 10	150	20.27
C-40D-89-42	56	48.17	48	57.5	98.5	37	18, 14, 10	150	20.27
C-40D-76-42	56	48.17	48	57.5	98.5	37	18, 14, 10	150	20.27
C-40D-89-36	48	48.17	48	57.5	98.5	37	18, 14, 10	275	17.37
C-P40D-89-36	47	47	47	56	95	39	18, 14, 10	375	17.66
C-25D-67-36	48	48.17	48	57.5	98.5	37	18, 14, 10	275	17.37
C-25D-56-36	48	48.17	48	57.5	98.5	37	18, 14, 10	275	17.37
C-25D-67-30	45	36.22	36	49.5	84.5	31	12, 8	150	14.53
C-25D-53-30	45	36.22	36	49.5	84.5	31	12, 9	150	14.53

$$a = \frac{d^2x}{dt^2}.$$

Carrying out the differentiation for acceleration, it is found that the maximum acceleration occurs when ωt is equal to zero (or an even multiple of π radians) and that this maximum value is

$$a_{\max} = \omega^2 c (1 + \frac{c}{h}). \tag{12.1}$$

It also appears that the minimum value of acceleration is

$$a_{\min} = \omega^2 c (1 - \frac{c}{h}). \tag{12.2}$$

If N is the number of pumping strokes per minute, then

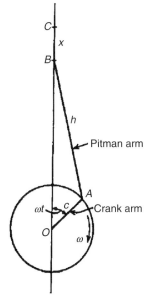

AB = length of pitman arm (h)
OA = length of crank arm (c)
OB = distance from center O to pitman arm-walking beam connection at B

Figure 12.7 Approximate motion of connection point between pitman arm and walking beam (Nind, 1964).

$$\omega = \frac{2\pi N}{60}\,(\text{rad/sec}). \tag{12.3}$$

The maximum downward acceleration of point B (which occurs when the crank arm is vertically upward) is

$$a_{\max} = \frac{cN^2}{91.2}\left(1+\frac{c}{h}\right)(\text{ft/sec}^2) \tag{12.4}$$

or

$$a_{\max} = \frac{cN^2 g}{2936.3}\left(1+\frac{c}{h}\right)\,(\text{ft/sec}^2). \tag{12.5}$$

Likewise the minimum upward ($a_{\min}$) acceleration of point B (which occurs when the crank arm is vertically downward) is

$$a_{\min} = \frac{cN^2 g}{2936.3}\left(1-\frac{c}{h}\right)\,(\text{ft/sec}^2). \tag{12.6}$$

It follows that in a conventional pumping unit, the maximum upward acceleration of the horse's head occurs at the bottom of the stroke (polished rod) and is equal to

$$a_{\max} = \frac{d_1}{d_2}\frac{cN^2 g}{2936.3}\left(1+\frac{c}{h}\right)\,(\text{ft/sec}^2), \tag{12.7}$$

where d_1 and d_2 are shown in Fig. 12.5. However,

$$\frac{2cd_2}{d_1} = S,$$

where S is the polished rod stroke length. So if S is measured in inches, then

$$\frac{2cd_2}{d_1} = \frac{S}{12}$$

or

$$\frac{cd_2}{d_1} = \frac{S}{24}. \tag{12.8}$$

So substituting Eq. (12.8) into Eq. (12.7) yields

$$a_{\max} = \frac{SN^2 g}{70471.2}\left(1+\frac{c}{h}\right)(\text{ft/sec}^2), \tag{12.9}$$

or we can write Eq. (12.9) as

$$a_{\max} = \frac{SN^2 g}{70{,}471.2}M(\text{ft/sec}^2), \tag{12.10}$$

where M is the machinery factor and is defined as

$$M = 1 + \frac{c}{h}. \tag{12.11}$$

Similarly,

$$a_{\min} = \frac{SN^2 g}{70471.2}\left(1-\frac{c}{h}\right)(\text{ft/sec}^2). \tag{12.12}$$

For air-balanced units, because of the arrangements of the levers, the acceleration defined in Eq. (12.12) occurs at the bottom of the stroke, and the acceleration defined in Eq. (12.9) occurs at the top. With the lever system of an air-balanced unit, the polished rod is at the top of its stroke when the crank arm is vertically upward (Fig. 12.5b).

12.4 Load to the Pumping Unit

The load exerted to the pumping unit depends on well depth, rod size, fluid properties, and system dynamics. The maximum PRL and peak torque are major concerns for pumping unit.

12.4.1 Maximum PRL

The PRL is the sum of weight of fluid being lifted, weight of plunger, weight of sucker rods string, dynamic load due to acceleration, friction force, and the up-thrust from below on plunger. In practice, no force attributable to fluid acceleration is required, so the acceleration term involves only acceleration of the rods. Also, the friction term and the weight of the plunger are neglected. We ignore the reflective forces, which will tend to underestimate the maximum PRL. To compensate for this, we set the up-thrust force to zero. Also, we assume the TV is closed at the instant at which the acceleration term reaches its maximum. With these assumptions, the $\text{PRL}_{\max}$ becomes

$$\text{PRL}_{\max} = S_f(62.4)D\frac{(A_p - A_r)}{144} + \frac{\gamma_s DA_r}{144}$$
$$+ \frac{\gamma_s DA_r}{144}\left(\frac{SN^2 M}{70{,}471.2}\right), \tag{12.13}$$

where

S_f = specific gravity of fluid in tubing
D = length of sucker rod string (ft)
A_p = gross plunger cross-sectional area (in.^{2})
A_r = sucker rod cross-sectional area (in.^{2})
γ_s = specific weight of steel (490 lb/ft^3)
M = Eq. (12.11).

Note that for the air-balanced unit, M in Eq. (12.13) is replaced by $1-c/h$.

Equation (12.13) can be rewritten as

$$\text{PRL}_{\max} = S_f(62.4)\frac{DA_p}{144} - S_f(62.4)\frac{DA_r}{144} + \frac{\gamma_s DA_r}{144}$$
$$+ \frac{\gamma_s DA_r}{144}\left(\frac{SN^2 M}{70{,}471.2}\right). \tag{12.14}$$

If the weight of the rod string in air is

$$W_r = \frac{\gamma_s DA_r}{144}, \tag{12.15}$$

which can be solved for A_r, which is

$$A_r = \frac{144 W_r}{\gamma_s D}. \tag{12.16}$$

Substituting Eq. (12.16) into Eq. (12.14) yields

$$\text{PRL}_{\max} = S_f(62.4)\frac{DA_p}{144} - S_f(62.4)\frac{W_r}{\gamma_s} + W_r$$
$$+ W_r\left(\frac{SN^2 M}{70{,}471.2}\right). \tag{12.17}$$

The above equation is often further reduced by taking the fluid in the second term (the subtractive term) as an 50°API with $S_f = 0.78$. Thus, Eq. (12.17) becomes (where $\gamma_s = 490$)

$$\text{PRL}_{\max} = S_f(62.4)\frac{DA_p}{144} - 0.1W_r + W_r + W_r\left(\frac{SN^2 M}{70{,}471.2}\right)$$

or

$$\text{PRL}_{\max} = W_f + 0.9W_r + W_r\left(\frac{SN^2 M}{70{,}471.2}\right), \tag{12.18}$$

where $W_f = S_f(62.4)\frac{DA_p}{144}$ and is called the fluid load (not to be confused with the actual fluid weight on the rod string).

Thus, Eq. (12.18) can be rewritten as

$$\text{PRL}_{\max} = W_f + (0.9 + F_1)W_r, \tag{12.19}$$

where for conventional units

$$F_1 = \frac{SN^2(1+\frac{c}{h})}{70{,}471.2} \tag{12.20}$$

and for air-balanced units

$$F_1 = \frac{SN^2(1-\frac{c}{h})}{70{,}471.2}. \tag{12.21}$$

12.4.2 Minimum PRL

The minimum PRL occurs while the TV is open so that the fluid column weight is carried by the tubing and not the rods. The minimum load is at or near the top of the stroke. Neglecting the weight of the plunger and friction term, the minimum PRL is

$$\text{PRL}_{\min} = -S_f(62.4)\frac{W_r}{\gamma_s} + W_r - W_r F_2,$$

which, for 50°API oil, reduces to

$$\text{PRL}_{\min} = 0.9W_r - F_2 W_r = (0.9 - F_2)W_r, \quad (12.22)$$

where for the conventional units

$$F_2 = \frac{SN^2(1-\frac{c}{h})}{70,471.2} \quad (12.23)$$

and for air-balanced units

$$F_2 = \frac{SN^2(1+\frac{c}{h})}{70,471.2}. \quad (12.24)$$

12.4.3 Counterweights

To reduce the power requirements for the prime mover, a counterbalance load is used on the walking beam (small units) or the rotary crank. The ideal counterbalance load C is the average PRL. Therefore,

$$C = \frac{1}{2}(\text{PRL}_{\max} + \text{PRL}_{\min}).$$

Using Eqs. (12.19) and (12.22) in the above, we get

$$C = \frac{1}{2}W_f + 0.9W_r + \frac{1}{2}(F_1 - F_2)W_r \quad (12.25)$$

or for conventional units

$$C = \frac{1}{2}W_f + W_r\left(0.9 + \frac{SN^2}{70,471.2}\frac{c}{h}\right) \quad (12.26)$$

and for air-balanced units

$$C = \frac{1}{2}W_f + W_r\left(0.9 - \frac{SN^2}{70,471.2}\frac{c}{h}\right). \quad (12.27)$$

The counterbalance load should be provided by structure unbalance and counterweights placed at walking beam (small units) or the rotary crank. The counterweights can be selected from manufacturer's catalog based on the calculated C value. The relationship between the counterbalance load C and the total weight of the counterweights is

$$C = C_s + W_c \frac{r}{c}\frac{d_1}{d_2},$$

where

C_s = structure unbalance, lb
W_c = total weight of counterweights, lb
r = distance between the mass center of counterweights and the crank shaft center, in.

12.4.4 Peak Torque and Speed Limit

The peak torque exerted is usually calculated on the most severe possible assumption, which is that the peak load (polished rod less counterbalance) occurs when the effective crank length is also a maximum (when the crank arm is horizontal). Thus, peak torque T is (Fig. 12.5)

$$T = c[C - (0.9 - F_2)W_r]\frac{d_2}{d_1}. \quad (12.28)$$

Substituting Eq. (12.25) into Eq. (12.28) gives

$$T = \frac{1}{2}S[C - (0.9 - F_2)W_r] \quad (12.29)$$

or

$$T = \frac{1}{2}S\left[\frac{1}{2}W_f + \frac{1}{2}(F_1 + F_2)W_r\right]$$

or

$$T = \frac{1}{4}S\left(W_f + \frac{2SN^2 W_r}{70,471.2}\right)(\text{in.-lb}). \quad (12.30)$$

Because the pumping unit itself is usually not perfectly balanced ($C_s \neq 0$), the peak torque is also affected by structure unbalance. Torque factors are used for correction:

$$T = \frac{\frac{1}{2}[\text{PRL}_{\max}(TF_1) + \text{PRL}_{\min}(TF_2)]}{0.93}, \quad (12.31)$$

where

TF_1 = maximum upstroke torque factor
TF_2 = maximum downstroke torque factor
0.93 = system efficiency.

For symmetrical conventional and air-balanced units, $TF = TF_1 = TF_2$.

There is a limiting relationship between stroke length and cycles per minute. As given earlier, the maximum value of the downward acceleration (which occurs at the top of the stroke) is equal to

$$a_{\max/\min} = \frac{SN^2 g(1 \pm \frac{c}{h})}{70,471.2}, \quad (12.32)$$

(the $\pm$ refers to conventional units or air-balanced units, see Eqs. [12.9] and [12.12]). If this maximum acceleration divided by g exceeds unity, the downward acceleration of the hanger is greater than the free-fall acceleration of the rods at the top of the stroke. This leads to severe pounding when the polished rod shoulder falls onto the hanger (leading to failure of the rod at the shoulder). Thus, a limit of the above downward acceleration term divided by g is limited to approximately 0.5 (or where L is determined by experience in a particular field). Thus,

$$\frac{SN^2(1 \pm \frac{c}{h})}{70,471.2} \leq L \quad (12.33)$$

or

$$N_{\text{limit}} = \sqrt{\frac{70,471.2 L}{S(1 \mp \frac{c}{h})}}. \quad (12.34)$$

For $L = 0.5$,

$$N_{\text{limit}} = \frac{187.7}{\sqrt{S(1 \mp \frac{c}{h})}}. \quad (12.35)$$

The minus sign is for conventional units and the plus sign for air-balanced units.

12.4.5 Tapered Rod Strings

For deep well applications, it is necessary to use a tapered sucker rod strings to reduce the PRL at the surface. The larger diameter rod is placed at the top of the rod string, then the next largest, and then the least largest. Usually these are in sequences up to four different rod sizes. The tapered rod strings are designated by 1/8-in. (in diameter) increments. Tapered rod strings can be identified by their numbers such as

a. No. 88 is a nontapered $^8/_8$- or 1-in. diameter rod string
b. No. 76 is a tapered string with $^7/_8$-in. diameter rod at the top, then a $^6/_8$-in. diameter rod at the bottom.
c. No. 75 is a three-way tapered string consisting of
 $^7/_8$-in. diameter rod at top
 $^6/_8$-in. diameter rod at middle
 $^5/_8$-in. diameter rod at bottom
d. No. 107 is a four-way tapered string consisting of
 $^{10}/_8$-in. (or $1^1/_4$-in.) diameter rod at top
 $^9/_8$-in. (or $1^1/_8$-in.) diameter rod below $^{10}/_8$-in. diameter rod
 $^8/_8$-in. (or 1-in.) diameter rod below $^9/_8$-in. diameter rod
 $^7/_8$-in. diameter rod below $^8/_8$-in. diameter rod

Tapered rod strings are designed for static (quasi-static) lads with a sufficient factor of safety to allow for random low level dynamic loads. Two criteria are used in the design of tapered rod strings:

1. Stress at the top rod of each rod size is the same throughout the string.
2. Stress in the top rod of the smallest (deepest) set of rods should be the highest (~30,000 psi) and the stress progressively decreases in the top rods of the higher sets of rods.

The reason for the second criterion is that it is preferable that any rod breaks occur near the bottom of the string (otherwise macaroni).

Example Problem 12.1 The following geometric dimensions are for the pumping unit C–320D–213–86:

$d_1 = 96.05$ in.
$d_2 = 111$ in.
$c = 37$ in.
$c/h = 0.33$.

If this unit is used with a $2\frac{1}{2}$-in. plunger and $\frac{7}{8}$-in. rods to lift 25 °API gravity crude (formation volume factor 1.2 rb/stb) at depth of 3,000 ft, answer the following questions:

a. What is the maximum allowable pumping speed if $L = 0.4$ is used?
b. What is the expected maximum polished rod load?
c. What is the expected peak torque?
d. What is the desired counterbalance weight to be placed at the maximum position on the crank?

Solution The pumping unit C–320D–213–86 has a peak torque of gearbox rating of 320,000 in.-lb, a polished rod rating of 21,300 lb, and a maximum polished rod stroke of 86 in.

a. Based on the configuration for conventional unit shown in Fig. 12.5a and Table 12.1, the polished rod stroke length can be estimated as

$$S = 2c\frac{d_2}{d_1} = (2)(37)\frac{111}{96.05} = 85.52 \text{ in.}$$

The maximum allowable pumping speed is

$$N = \sqrt{\frac{70,471.2L}{S(1-\frac{c}{h})}} = \sqrt{\frac{(70,471.2)(0.4)}{(85.52)(1-0.33)}} = 22 \text{ SPM}.$$

b. The maximum PRL can be calculated with Eq. (12.17). The 25 °API gravity has an $S_f = 0.9042$. The area of the $2\frac{1}{2}$-in. plunger is $A_p = 4.91 \text{ in.}^2$. The area of the $\frac{7}{8}$-in. rod is $A_r = 0.60 \text{ in.}^2$. Then

$$W_f = S_f(62.4)\frac{DA_p}{144} = (0.9042)(62.4)\frac{(3,000)(4.91)}{144}$$
$$= 5,770 \text{ lbs}$$

$$W_r = \frac{\gamma_s DA_r}{144} = \frac{(490)(3,000)(0.60)}{144} = 6,138 \text{ lbs}$$

$$F_1 = \frac{SN^2(1+\frac{c}{h})}{70,471.2} = \frac{(85.52)(22)^2(1+0.33)}{70,471.2} = 0.7940.$$

Then the expected maximum PRL is

$$\text{PRL}_{\max} = W_f - S_f(62.4)\frac{W_r}{\gamma_s} + W_r + W_rF_1$$
$$= 5,770 - (0.9042)(62.4)(6,138)/(490)$$
$$\quad + 6,138 + (6,138)(0.794)$$
$$= 16,076 \text{ lbs} < 21,300 \text{ lb}$$

c. The peak torque is calculated by Eq. (12.30):

$$T = \frac{1}{4}S\left(W_f + \frac{2SN^2 W_r}{70,471.2}\right)$$
$$= \frac{1}{4}(85.52)\left(5,770 + \frac{2(85.52)(22)^2(6,138)}{70,471.2}\right)$$
$$= 280,056 \text{ lb-in.} < 320,000 \text{ lb-in.}$$

d. Accurate calculation of counterbalance load requires the minimum PRL:

$$F_2 = \frac{SN^2(1-\frac{c}{h})}{70,471.2} = \frac{(85.52)(22)^2(1-0.33)}{70,471.2} = 0.4$$

$$\text{PRL}_{\min} = -S_f(62.4)\frac{W_r}{\gamma_s} + W_r - W_rF_2$$
$$= -(0.9042)(62.4)\frac{6,138}{490} + 6,138 - (6,138)(0.4)$$
$$= 2,976 \text{ lb}$$

$$C = \frac{1}{2}(\text{PRL}_{\max} + \text{PRL}_{\min}) = \frac{1}{2}(16,076 + 2,976) = 9,526 \text{ lb}.$$

A product catalog of LUFKIN Industries indicates that the structure unbalance is 450 lb and 4 No. 5ARO counterweights placed at the maximum position (c in this case) on the crank will produce an effective counterbalance load of 10,160 lb, that is,

$$W_c\frac{(37)}{(37)}\frac{(96.05)}{(111)} + 450 = 10,160,$$

which gives $W_c = 11,221$ lb. To generate the ideal counterbalance load of $C = 9,526$ lb, the counterweights should be placed on the crank at

$$r = \frac{(9,526)(111)}{(11,221)(96.05)}(37) = 36.30 \text{ in.}$$

The computer program *SuckerRodPumpingLoad.xls* can be used for quickly seeking solutions to similar problems. It is available from the publisher with this book. The solution is shown in Table 12.2.

12.5 Pump Deliverability and Power Requirements

Liquid flow rate delivered by the plunger pump can be expressed as

$$q = \frac{A_p}{144}N\frac{S_p}{12}\frac{E_v}{B_o}\frac{(24)(60)}{5.615}(\text{bbl/day})$$

or

$$q = 0.1484\frac{A_pNS_pE_v}{B_o}(\text{stb/day}),$$

where S_p is the effective plunger stroke length (in.), E_v is the volumetric efficiency of the plunger, and B_o formation volume factor of the fluid.

12.5.1 Effective Plunger Stroke Length

The motion of the plunger at the pump-setting depth and the motion of the polished rod do not coincide in time and in magnitude because sucker rods and tubing strings are elastic. Plunger motion depends on a number of factors including polished rod motion, sucker rod stretch, and

Table 12.2 *Solution Given by Computer Program* SuckerRodPumpingLoad.xls

SuckerRodPumpingLoad.xls
Description: This spreadsheet calculates the maximum allowable pumping speed, the maximum PRL, the minimum PRL, peak torque, and counterbalance load.
Instruction: (1) Update parameter values in the Input section; and (2) view result in the Solution section.

Input data

Pump setting depth (D):	3,000 ft
Plunger diameter (d_p):	2.5 in.
Rod section 1, diameter (d_{r1}):	1 in.
Length (L_1):	0 ft
Rod section 2, diameter (d_{r2}):	0.875 in.
Length (L_2):	3,000 ft
Rod section 3, diameter (d_{r3}):	0.75 in.
Length (L_3):	0 ft
Rod section 4, diameter (d_{r4}):	0.5 in.
Length (L_4):	0 ft
Type of pumping unit (1 = conventional; -1 = Mark II or Air-balanced):	1
Beam dimension 1 (d_1):	96.05 in.
Beam dimension 2 (d_2):	111 in.
Crank length (c):	37 in.
Crank to pitman ratio (c/h):	0.33
Oil gravity (API):	25 °API
Maximum allowable acceleration factor (L):	0.4

Solution

$S = 2c\frac{d_2}{d_1}$	= 85.52 in.
$N = \sqrt{\frac{70471.2L}{S(1-\frac{c}{h})}}$	= 22 SPM
$A_p = \frac{\pi d_p^2}{4}$	= 4.91 in.^{2}
$A_r = \frac{\pi d_r^2}{4}$	= 0.60 in.
$W_f = S_f(62.4)\frac{DA_p}{144}$	= 5,770 lb
$W_r = \frac{\gamma_s DA_r}{144}$	= 6,138 lb
$F_1 = \frac{SN^2(1\pm\frac{c}{h})}{70,471.2}$	= 0.7940 °
$PRL_{max} = W_f - S_f(62.4)\frac{W_r}{\gamma_s} + W_r + W_r F_1$	= 16,076 lb
$T = \frac{1}{4}S\left(W_f + \frac{2SN^2 W_r}{70,471.2}\right)$	= 280,056 lb
$F_2 = \frac{SN^2(1\mp\frac{c}{h})}{70,471.2}$	= 0.40
$PRL_{min} = -S_f(62.4)\frac{W_r}{\gamma_s} + W_r - W_r F_2$	= 2,976 lb
$C = \frac{1}{2}(PRL_{max} + PRL_{min})$	= 9,526 lb

tubing stretch. The theory in this subject has been well established (Nind, 1964).

Two major sources of difference in the motion of the polished rod and the plunger are elastic stretch (elongation) of the rod string and overtravel. Stretch is caused by the periodic transfer of the fluid load from the SV to the TV and back again. The result is a function of the stretch of the rod string and the tubing string. Rod string stretch is caused by the weight of the fluid column in the tubing coming on to the rod string at the bottom of the stroke when the TV closes (this load is removed from the rod string at the top of the stroke when the TV opens). It is apparent that the plunger stroke will be less than the polished rod stroke length S by an amount equal to the rod stretch. The magnitude of the rod stretch is

$$\delta l_r = \frac{W_f D_r}{A_r E}, \quad (12.36)$$

where

W_f = weight of fluid (lb)
D_r = length of rod string (ft)
A_r = cross-sectional area of rods (in.^{2})
E = modulus of elasticity of steel (30×10^6 lb/in.^{2}).

Tubing stretch can be expressed by a similar equation:

$$\delta l_t = \frac{W_f D_t}{A_t E} \quad (12.37)$$

But because the tubing cross-sectional area A_t is greater than the rod cross-sectional area A_r, the stretch of the tubing is small and is usually neglected. However, the tubing stretch can cause problems with wear on the casing. Thus, for this reason a tubing anchor is almost always used.

Plunger overtravel at the bottom of the stroke is a result of the upward acceleration imposed on the downward-moving sucker rod elastic system. An approximation to the extent of the overtravel may be obtained by considering a sucker rod string being accelerated vertically upward at a rate n times the acceleration of gravity. The vertical force required to supply this acceleration is nW_r. The magnitude of the rod stretch due to this force is

$$\delta l_o = n\frac{W_r D_r}{A_r E} \text{ (ft).} \quad (12.38)$$

But the maximum acceleration term n can be written as

$$n = \frac{SN^2\left(1\pm\frac{c}{h}\right)}{70,471.2}$$

so that Eq. (12.38) becomes

$$\delta l_o = \frac{W_r D_r}{A_r E}\frac{SN^2\left(1\pm\frac{c}{h}\right)}{70,471.2}\text{ (ft),} \quad (12.39)$$

where again the plus sign applies to conventional units and the minus sign to air-balanced units.

Let us restrict our discussion to conventional units. Then Eq. (12.39) becomes

$$\delta l_o = \frac{W_r D_r}{A_r E} \frac{SN^2 M}{70{,}471.2} \text{ (ft).} \qquad (12.40)$$

Equation (12.40) can be rewritten to yield δl_o in inches. W_r is

$$W_r = \gamma_s A_r D_r$$

and $\gamma_S = 490 \text{ lb/ft}^3$ with $E = 30 \times 10^6 \text{ lb/m}^2$. Eq. (12.40) becomes

$$\delta l_o = 1.93 \times 10^{-11} D_r^2 SN^2 M \text{(in.)}, \qquad (12.41)$$

which is the familiar Coberly expression for overtravel (Coberly, 1938).

Plunger stroke is approximated using the above expressions as

$$S_p = S - \delta l_r - \delta l_t + \delta l_o$$

or

$$S_p = S - \frac{12D}{E}$$
$$\times \left[W_f \left(\frac{1}{A_r} + \frac{1}{A_t} \right) - \frac{SN^2 M}{70{,}471.2} \frac{W_r}{A_r} \right] \text{(in.).} \qquad (12.42)$$

If pumping is carried out at the maximum permissible speed limited by Eq. (12.34), the plunger stroke becomes

$$S_p = S - \frac{12D}{E}$$
$$\times \left[W_f \left(\frac{1}{A_r} + \frac{1}{A_t} \right) - \frac{1 + \frac{c}{h}}{1 - \frac{c}{h}} \frac{LW_r}{A_r} \right] \text{(in.).} \qquad (12.43)$$

For the air-balanced unit, the term $\frac{1+\frac{c}{h}}{1-\frac{c}{h}}$ is replaced by its reciprocal.

12.5.2 Volumetric Efficiency

Volumetric efficiency of the plunger mainly depends on the rate of slippage of oil past the pump plunger and the solution–gas ratio under pump condition.

Metal-to-metal plungers are commonly available with plunger-to-barrel clearance on the diameter of -0.001, -0.002, -0.003, -0.004, and -0.005 in. Such fits are referred to as -1, -2, -3, -4, and -5, meaning the plunger outside diameter is 0.001 in. smaller than the barrel inside diameter. In selecting a plunger, one must consider the viscosity of the oil to be pumped. A loose fit may be acceptable for a well with high viscosity oil (low °API gravity). But such a loose fit in a well with low viscosity oil may be very inefficient. Guidelines are as follows:

a. Low-viscosity oils (1–20 cps) can be pumped with a plunger to barrel fit of -0.001 in.
b. High-viscosity oils (7,400 cps) will probably carry sand in suspension so a plunger-to-barrel fit or approximately 0.005 in. can be used.

An empirical formula has been developed that can be used to calculate the slippage rate, q_s (bbl/day), through the annulus between the plunger and the barrel:

$$q_s = \frac{k_p}{\mu} \frac{(d_b - d_p)^{2.9} (d_b + d_p)}{d_b^{0.1}} \frac{\Delta p}{L_p}, \qquad (12.44)$$

where

k_p = a constant
d_p = plunger outside diameter (in.)
d_b = barrel inside diameter (in.)
Δp = differential pressure drop across plunger (psi)
L_p = length of plunger (in.)
μ = viscosity of oil (cp).

The value of k_p is 2.77×10^6 to 6.36×10^6 depending on field conditions. An average value is 4.17×10^6. The value of Δp may be estimated on the basis of well productivity index and production rate. A reasonable estimate may be a value that is twice the production drawdown.

Volumetric efficiency can decrease significantly due to the presence of free gas below the plunger. As the fluid is elevated and gas breaks out of solution, there is a significant difference between the volumetric displacement of the bottom-hole pump and the volume of the fluid delivered to the surface. This effect is denoted by the shrinkage factor greater than 1.0, indicating that the bottom-hole pump must displace more fluid by some additional percentage than the volume delivered to the surface (Brown, 1980). The effect of gas on volumetric efficiency depends on solution–gas ratio and bottom-hole pressure. Down-hole devices, called "gas anchors," are usually installed on pumps to separate the gas from the liquid.

In summary, volumetric efficiency is mainly affected by the slippage of oil and free gas volume below plunger. Both effects are difficult to quantify. Pump efficiency can vary over a wide range but are commonly 70–80%.

12.5.3 Power Requirements

The prime mover should be properly sized to provide adequate power to lift the production fluid, to overcome friction loss in the pump, in the rod string and polished rod, and in the pumping unit. The power required for lifting fluid is called "hydraulic power." It is usually expressed in terms of net lift:

$$P_h = 7.36 \times 10^{-6} q \gamma_l L_N, \qquad (12.45)$$

where

P_h = hydraulic power, hp
q = liquid production rate, bbl/day
γ_l = liquid specific gravity, water = 1
L_N = net lift, ft,

and

$$L_N = H + \frac{p_{tf}}{0.433 \gamma_l}, \qquad (12.46)$$

where

H = depth to the average fluid level in the annulus, ft
p_{tf} = flowing tubing head pressure, psig.

The power required to overcome friction losses can be empirically estimated as

$$P_f = 6.31 \times 10^{-7} W_r SN. \qquad (12.47)$$

Thus, the required prime mover power can be expressed as

$$P_{pm} = F_s (P_h + P_f), \qquad (12.48)$$

where F_s is a safety factor of 1.25–1.50.

Example Problem 12.2 A well is pumped off (fluid level is the pump depth) with a rod pump described in Example Problem 12.1. A 3-in. tubing string (3.5-in. OD, 2.995 ID) in the well is not anchored. Calculate (a) expected liquid production rate (use pump volumetric efficiency 0.8), and (b) required prime mover power (use safety factor 1.35).

Solution This problem can be quickly solved using the program *SuckerRodPumpingFlowrate&Power.xls*. The solution is shown in Table 12.3.

12.6 Procedure for Pumping Unit Selection

The following procedure can be used for selecting a pumping unit:

Table 12.3 Solution Given by SuckerRodPumpingFlowrate&Power.xls

SuckerRodPumpingFlowRate&Power.xls
Description: This spreadsheet calculates expected deliverability and required prime mover power for a given sucker rod pumping system.
Instruction: (1) Update parameter values in the Input section; and (2) view result in the Solution section.

Input data

Pump setting depth (D):	4,000 ft
Depth to the liquid level in annulus (H):	4,000 ft
Flowing tubing head pressure (p_{tf}):	100 ft
Tubing outer diameter (d_{to}):	3.5 in.
Tubing inner diameter (d_{ti}):	2.995 in.
Tubing anchor (1 = yes; 0 = no):	0
Plunger diameter (d_p):	2.5 in.
Rod section 1, diameter (d_{r1}):	1 in.
Length (L_1):	0 ft
Rod section 2, diameter (d_{r2}):	0.875 in.
Length (L_2):	0 ft
Rod section 3, diameter (d_{r3}):	0.75 in.
Length (L_3):	4,000 ft
Rod section 4, diameter (d_{r4}):	0.5 in.
Length (L_4):	0 ft
Type of pumping unit (1 = conventional; −1 = Mark II or Air-balanced):	1
Polished rod stroke length (S)	86 in.
Pumping speed (N)	22 spm
Crank to pitman ratio (c/h):	0.33 °
Oil gravity (API):	25 °API
Fluid formation volume factor (B_o):	1.2 rb/stb
Pump volumetric efficiency (E_v):	0.8
Safety factor to prime mover power (F_s):	1.35

Solution

$A_t = \frac{\pi d_t^2}{4}$	= 2.58 in.^{2}
$A_p = \frac{\pi d_p^2}{4}$	= 4.91 in.^{2}
$A_r = \frac{\pi d_r^2}{4}$	= 0.44 in.
$W_f = S_f (62.4) \frac{DA_p}{144}$	= 7,693 lb
$W_r = \frac{\gamma_s DA_r}{144}$	= 6,013 lb
$M = 1 \pm \frac{c}{h}$	= 1.33
$S_p = S - \frac{12D}{E}\left[W_f\left(\frac{1}{A_r}+\frac{1}{A_t}\right) - \frac{SN^2 M}{70471.2}\frac{W_r}{A_r}\right]$	= 70 in.
$q = 0.1484 \frac{A_p N S_p E_v}{B_o}$	= 753 sbt/day
$L_N = H + \frac{p_{tf}}{0.433 \gamma_l}$	= 4,255 ft
$P_h = 7.36 \times 10^{-6} q \gamma_l L_N$	= 25.58 hp
$P_f = 6.31 \times 10^{-7} W_r S N$	= 7.2 hp
$P_{pm} = F_s(P_h + P_f)$	= 44.2 hp

1. From the maximum anticipated fluid production (based on IPR) and estimated volumetric efficiency, calculate required pump displacement.
2. Based on well depth and pump displacement, determine API rating and stroke length of the pumping unit to be used. This can be done using either Fig. 12.8 or Table 12.4.
3. Select tubing size, plunger size, rod sizes, and pumping speed from Table 12.4.
4. Calculate the fractional length of each section of the rod string.
5. Calculate the length of each section of the rod string to the nearest 25 ft.
6. Calculate the acceleration factor.
7. Determine the effective plunger stroke length.
8. Using the estimated volumetric efficiency, determine the probable production rate and check it against the desired production rate.
9. Calculate the dead weight of the rod string.
10. Calculate the fluid load.
11. Determine peak polished rod load and check it against the maximum beam load for the unit selected.
12. Calculate the maximum stress at the top of each rod size and check it against the maximum permissible working stress for the rods to be used.
13. Calculate the ideal counterbalance effect and check it against the counterbalance available for the unit selected.
14. From the manufacturer's literature, determine the position of the counterweight to obtain the ideal counterbalance effect.
15. On the assumption that the unit will be no more than 5% out of counterbalance, calculate the peak torque on the gear reducer and check it against the API rating of the unit selected.
16. Calculate hydraulic horsepower, friction horsepower, and brake horsepower of the prime mover. Select the prime mover.
17. From the manufacturer's literature, obtain the gear reduction ratio and unit sheave size for the unit selected, and the speed of the prime mover. From this, determine the engine sheave size to obtain the desired pumping speed.

Example Problem 12.3 A well is to be put on a sucker rod pump. The proposed pump setting depth is 3,500 ft. The anticipated production rate is 600 bbl/day oil of 0.8 specific gravity against wellhead pressure 100 psig. It is assumed that the working liquid level is low, and a sucker rod string having a working stress of 30,000 psi is to be used. Select surface and subsurface equipment for the installation. Use a safety factor of 1.35 for the prime mover power.

Solution

1. Assuming volumetric efficiency of 0.8, the required pump displacement is

$$(600)/(0.8) = 750 \, \text{bbl/day}.$$

2. Based on well depth 3,500 ft and pump displacement 750 bbl/day, Fig. 12.8 suggests API pump size 320 unit with 84 in. stroke, that is, a pump is selected with the following designation:

$$C\text{-}320D\text{-}213\text{-}86$$

3. Table 12.4 g suggests the following:

   Tubing size: 3 in. OD, 2.992 in. ID
   Plunger size: $2\tfrac{1}{2}$ in.
   Rod size: $\tfrac{7}{8}$ in.
   Pumping speed: 18 spm

4. Table 12.1 gives $d_1 = 96.05$ in., $d_2 = 111$ in., $c = 37$ in., and $h = 114$ in., thus $c/h = 0.3246$. The spreadsheet program *SuckerRodPumpingFlowRate&Power.xls* gives

   $q_o = 687 \, \text{bbl/day} > 600 \, \text{bbl/day}$

   $P_{pm} = 30.2 \, \text{hp}$

5. The spreadsheet program *SuckerRodPumpingLoad.xls* gives

   $PRL_{\max} = 16{,}121 \, \text{lb}$
   $PRL_{\min} = 4{,}533 \, \text{lb}$
   $T = 247{,}755 \, \text{lb} < 320{,}000 \, \text{in.-lb}$
   $C = 10{,}327 \, \text{lb}$

6. The cross-sectional area of the $\tfrac{7}{8}$-in. rod is 0.60 in.^{2}. Thus, the maximum possible stress in the sucker rod is

$$\sigma_{\max} = (16{,}121)/(0.60) = 26{,}809 \, \text{psi} < 30{,}000 \, \text{psi}.$$

Therefore, the selected pumping unit and rod meet well load and volume requirements.

7. If a LUFKIN Industries C-320D-213-86 unit is chosen, the structure unbalance is 450 lb and 4 No. 5 ARO counterweights placed at the maximum position (*c* in this case) on the crank will produce an effective counterbalance load of 12,630 lb, that is,

$$W_c \frac{(37)}{(37)} \frac{(96.05)}{(111)} + 450 = 12{,}630 \, \text{lb},$$

which gives $W_c = 14{,}075 \, \text{lb}$. To generate the ideal counterbalance load of $C = 10{,}327 \, \text{lb}$, the counterweights should be placed on the crank at

$$r = \frac{(10{,}327)(111)}{(14{,}076)(96.05)}(37) = 31.4 \, \text{in}.$$

8. The LUFKIN Industries C-320D-213-86 unit has a gear ratio of 30.12 and unit sheave sizes of 24, 30, and 44 in. are available. If a 24-in. unit sheave and a 750-rpm electric motor are chosen, the diameter of the motor sheave is

$$d_e = \frac{(18)(30.12)(24)}{(750)} = 17.3 \, \text{in}.$$

12.7 Principles of Pump Performance Analysis

The efficiency of sucker rod pumping units is usually analyzed using the information from pump dynagraph and polisher rod dynamometer cards. Figure 12.9 shows a schematic of a pump dynagraph. This instrument is installed immediately above the plunger to record the plunger stroke and the loads carried by the plunger during the pump cycle.

The relative motion between the cover tube (which is attached to the pump barrel and hence anchored to the tubing) and the calibrated rod (which is an integral part of the sucker rod string) is recorded as a horizontal line on the recording tube. This is achieved by having the recording tube mounted on a winged nut threaded onto the calibrated rod and prevented from rotating by means of

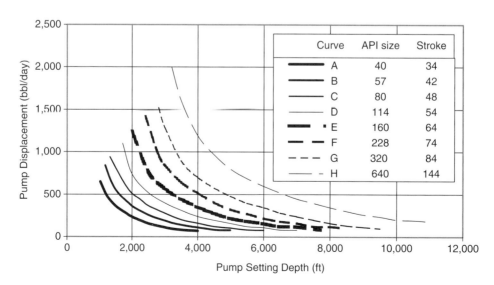

Figure 12.8 Sucker rod pumping unit selection chart (Kelley and Willis, 1954).

Table 12.4 Design Data for API Sucker Rod Pumping Units

(a) Size 40 unit with 34-in. stroke

Pump depth (ft)	Plunger size (in.)	Tubing size (in.)	Rod sizes (in.)	Pumping speed (stroke/min)
1,000–1,100	$2\tfrac{3}{4}$	3	$\tfrac{7}{8}$	24–19
1,100–1,250	$2\tfrac{1}{2}$	3	$\tfrac{7}{8}$	24–19
1,250–1,650	$2\tfrac{1}{4}$	$2\tfrac{1}{2}$	$\tfrac{3}{4}$	24–19
1,650–1,900	2	$2\tfrac{1}{2}$	$\tfrac{3}{4}$	24–19
1,900–2,150	$1\tfrac{3}{4}$	$2\tfrac{1}{2}$	$\tfrac{3}{4}$	24–19
2,150–3,000	$1\tfrac{1}{2}$	2	$\tfrac{5}{8}-\tfrac{3}{4}$	24–19
3,000–3,700	$1\tfrac{1}{4}$	2	$\tfrac{5}{8}-\tfrac{3}{5}$	22–18
3,700–4,000	1	2	$\tfrac{5}{8}-\tfrac{3}{6}$	21–18

(b) Size 57 unit with 42-in. stroke

Pump depth (ft)	Plunger size (in.)	Tubing size (in.)	Rod sizes (in.)	Pumping speed (stroke/min)
1,150–1,300	$2\tfrac{3}{4}$	3	$\tfrac{7}{8}$	24–19
1,300–1,450	$2\tfrac{1}{2}$	3	$\tfrac{7}{8}$	24–19
1,450–1,850	$2\tfrac{1}{4}$	$2\tfrac{1}{2}$	$\tfrac{3}{4}$	24–19
1,850–2,200	2	$2\tfrac{1}{2}$	$\tfrac{3}{4}$	24–19
2,200–2,500	$1\tfrac{3}{4}$	$2\tfrac{1}{2}$	$\tfrac{3}{4}$	24–19
2,500–3,400	$1\tfrac{1}{2}$	2	$\tfrac{5}{8}-\tfrac{3}{4}$	23–18
3,400–4,200	$1\tfrac{1}{4}$	2	$\tfrac{5}{8}-\tfrac{3}{5}$	22–17
4,200–5,000	1	2	$\tfrac{5}{8}-\tfrac{3}{6}$	21–17

(c) Size 80 unit with 48-in. stroke

Pump depth (ft)	Plunger size (in.)	Tubing size (in.)	Rod sizes (in.)	Pumping speed (stroke/min)
1,400–1,500	$2\tfrac{3}{4}$	3	$\tfrac{7}{8}$	24–19
1,550–1,700	$2\tfrac{1}{2}$	3	$\tfrac{7}{8}$	24–19
1,700–2,200	$2\tfrac{1}{4}$	$2\tfrac{1}{2}$	$\tfrac{3}{4}$	24–19
2,200–2,600	2	$2\tfrac{1}{2}$	$\tfrac{3}{4}$	24–19
2,600–3,000	$1\tfrac{3}{4}$	$2\tfrac{1}{2}$	$\tfrac{3}{4}$	23–18
3,000–4,100	$1\tfrac{1}{2}$	2	$\tfrac{5}{8}-\tfrac{3}{4}$	23–19
4,100–5,000	$1\tfrac{1}{4}$	2	$\tfrac{5}{8}-\tfrac{3}{5}$	21–17
5,000–6,000	1	2	$\tfrac{5}{8}-\tfrac{3}{6}$	19–17

(d) Size 114 unit with 54-in. stroke

Pump depth (ft)	Plunger size (in.)	Tubing size (in.)	Rod sizes (in.)	Pumping speed (stroke/min)
1,700–1,900	$2\tfrac{3}{4}$	3	$\tfrac{7}{8}$	24–19
1,900–2,100	$2\tfrac{1}{2}$	3	$\tfrac{7}{8}$	24–19
2,100–2,700	$2\tfrac{1}{4}$	$2\tfrac{1}{2}$	$\tfrac{3}{4}$	24–19
2,700–3,300	2	$2\tfrac{1}{2}$	$\tfrac{3}{4}$	23–18
3,300–3,900	$1\tfrac{3}{4}$	$2\tfrac{1}{2}$	$\tfrac{3}{4}$	22–17
3,900–5,100	$1\tfrac{1}{2}$	2	$\tfrac{5}{8}-\tfrac{3}{4}$	21–17
5,100–6,300	$1\tfrac{1}{4}$	2	$\tfrac{5}{8}-\tfrac{3}{5}$	19–16
6,300–7,000	1	2	$\tfrac{5}{8}-\tfrac{3}{6}$	17–16

(e) Size 160 unit with 64-in. stroke

Pump depth (ft)	Plunger size (in.)	Tubing size (in.)	Rod sizes (in.)	Pumping speed (stroke/min)
2,000–2,200	$2\tfrac{3}{4}$	3	$\tfrac{7}{8}$	24–19
2,200–2,400	$2\tfrac{1}{2}$	3	$\tfrac{7}{8}$	24–19
2,400–3,000	$2\tfrac{1}{4}$	$2\tfrac{1}{2}$	$\tfrac{3}{4}-\tfrac{7}{8}$	24–19
3,000–3,600	2	$2\tfrac{1}{2}$	$\tfrac{3}{4}-\tfrac{7}{8}$	23–18
3,600–4,200	$1\tfrac{3}{4}$	$2\tfrac{1}{2}$	$\tfrac{3}{4}-\tfrac{7}{8}$	22–17
4,200–5,400	$1\tfrac{1}{2}$	2	$\tfrac{5}{8}-\tfrac{3}{4}-\tfrac{7}{8}$	21–17
5,400–6,700	$1\tfrac{1}{4}$	2	$\tfrac{5}{8}-\tfrac{3}{4}-\tfrac{7}{8}$	19–15
6,700–7,700	1	2	$\tfrac{5}{8}-\tfrac{3}{4}-\tfrac{7}{8}$	17–15

(f) Size 228 unit with 74-in. stroke

Pump depth (ft)	Plunger size (in.)	Tubing size (in.)	Rod sizes (in.)	Pumping speed (stroke/min)
2,400–2,600	$2\tfrac{3}{4}$	3	$\tfrac{7}{8}$	24–20
2,600–3,000	$2\tfrac{1}{2}$	3	$\tfrac{7}{8}$	23–18
3,000–3,700	$2\tfrac{1}{4}$	$2\tfrac{1}{2}$	$\tfrac{3}{4}-\tfrac{7}{8}$	22–17
3,700–4,500	2	$2\tfrac{1}{2}$	$\tfrac{3}{4}-\tfrac{7}{8}$	21–16
4,500–5,200	$1\tfrac{3}{4}$	$2\tfrac{1}{2}$	$\tfrac{3}{4}-\tfrac{7}{8}$	19–15
5,200–6,800	$1\tfrac{1}{2}$	2	$5/8-\tfrac{3}{4}-\tfrac{7}{8}$	18–14
6,800–8,000	$1\tfrac{1}{4}$	2	$5/8-\tfrac{3}{4}-\tfrac{7}{8}$	16–13
8,000–8,500	11/16	2	$5/8-\tfrac{3}{4}-\tfrac{7}{8}$	14–13

(g) Size 320 unit with 84-in. stroke

Pump depth (ft)	Plunger size (in.)	Tubing size (in.)	Rod sizes (in.)	Pumping speed (stroke/min)
2,800–3,200	$2\tfrac{3}{4}$	3	$\tfrac{7}{8}$	23–18
3,200–3,600	$2\tfrac{1}{2}$	3	$\tfrac{7}{8}$	21–17
3,600–4,100	$2\tfrac{1}{4}$	$2\tfrac{1}{2}$	$\tfrac{3}{4}-\tfrac{7}{8}-1$	21–17
4,100–4,800	2	$2\tfrac{1}{2}$	$\tfrac{3}{4}-\tfrac{7}{8}-1$	20–16
4,800–5,600	$1\tfrac{3}{4}$	$2\tfrac{1}{2}$	$\tfrac{3}{4}-\tfrac{7}{8}-1$	19–16
5,600–6,700	$1\tfrac{1}{2}$	$2\tfrac{1}{2}$	$\tfrac{3}{4}-\tfrac{7}{8}-1$	18–15
6,700–8,000	$1\tfrac{1}{4}$	$2\tfrac{1}{2}$	$\tfrac{3}{4}-\tfrac{7}{8}-1$	17–13
8,000–9,500	11/16	$2\tfrac{1}{2}$	$\tfrac{3}{4}-\tfrac{7}{8}-1$	14–11

(h) Size 640 unit with 144-in. stroke

Pump depth (ft)	Plunger size (in.)	Tubing size (in.)	Rod sizes (in.)	Pumping speed (stroke/min)
3,200–3,500	$2\tfrac{3}{4}$	3	$\tfrac{7}{8}-1$	18–14
3,500–4,000	$2\tfrac{1}{2}$	3	$\tfrac{7}{8}-1$	17–13
4,000–4,700	$2\tfrac{1}{4}$	$2\tfrac{1}{2}$	$\tfrac{3}{4}-\tfrac{7}{8}-1$	16–13
4,700–5,700	2	$2\tfrac{1}{2}$	$\tfrac{3}{4}-\tfrac{7}{8}-1$	15–12
5,700–6,600	$1\tfrac{3}{4}$	$2\tfrac{1}{2}$	$\tfrac{3}{4}-\tfrac{7}{8}-1$	14–12
6,600–8,000	$1\tfrac{1}{2}$	$2\tfrac{1}{2}$	$\tfrac{3}{4}-\tfrac{7}{8}-1$	14–11
8,000–9,600	$1\tfrac{1}{4}$	$2\tfrac{1}{2}$	$\tfrac{3}{4}-\tfrac{7}{8}-1$	13–10
9,600–11,000	11/16	$2\tfrac{1}{2}$	$\tfrac{3}{4}-\tfrac{7}{8}-1$	12–10

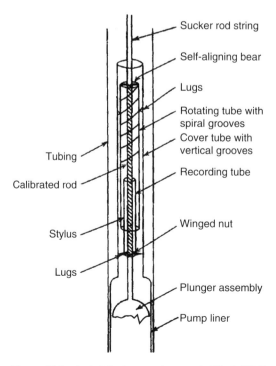

Figure 12.9 A sketch of pump dynagraph (Nind, 1964).

two lugs, which are attached to the winged nut, which run in vertical grooves in the cover tube. The stylus is mounted on a third tube, which is free to rotate and is connected by a self-aligning bearing to the upper end of the calibrated rod. Lugs attached to the cover tube run in spiral grooves cut in the outer surface of the rotating tube. Consequently, vertical motion of the plunger assembly relative to the barrel results in rotation of the third tube, and the stylus cuts a horizontal line on a recording tube.

Any change in plunger loading causes a change in length of the section of the calibrated rod between the winged nut supporting the recording tube and the self-aligning bearing supporting the rotating tube (so that a vertical line is cut on the recording tube by the stylus). When the pump is in operation, the stylus traces a series of cards, one on top of the other. To obtain a new series of cards, the polished rod at the well head is rotated. This rotation is transmitted to the plunger in a few pump strokes. Because the recording tube is prevented from rotating by the winged nut lugs that run in the cover tube grooves, the rotation of the sucker rod string causes the winged nut to travel—upward or downward depending on the direction of rotation—on the threaded calibrated rod. Upon the completion of a series of tests, the recording tube (which is 36 in. long) is removed.

It is important to note that although the bottom-hole dynagraph records the plunger stroke and variations in plunger loading, no zero line is obtained. Thus, quantitative interpretation of the cards becomes somewhat speculative unless a pressure element is run with the dynagraph.

Figure 12.10 shows some typical dynagraph card results. Card (a) shows an ideal case where instantaneous valve actions at the top and bottom of the stroke are indicated. In general, however, some free gas is drawn into the pump on the upstroke, so a period of gas compression can occur on the down-stroke before the TV opens. This is shown in card (b). Card (c) shows gas expansion during the upstroke giving a rounding of the card just as the upstroke begins. Card (d) shows fluid pounding that occurs when the well is almost pumped off (the pump displacement rate is higher than the formation of potential liquid production rate). This fluid pounding results in a rapid fall off in stress in the rod string and the sudden imposed shock to the system. Card (e) shows that the fluid pounding has progressed so that the mechanical shock causes oscillations in the system. Card (f) shows that the pump is operating at a very low volumetric efficiency where almost all the pump stroke is being lost in gas compression and expansion (no liquid is being pumped). This results in no valve action and the area between the card nearly disappears (thus, is gas locked). Usually, this gas-locked condition is only temporary, and as liquid leaks past the plunger, the volume of liquid in the pump barrel increases until the TV opens and pumping recommences.

The use of the pump dynagraph involves pulling the rods and pump from the well bath to install the instrument and to recover the recording tube. Also, the dynagraph cannot be used in a well equipped with a tubing pump. Thus, the dynagraph is more a research instrument than an operational device. Once there is knowledge from a dynagraph, surface dynamometer cards can be interpreted.

The surface, or polished rod, dynamometer is a device that records the motion of (and its history) the polished rod during the pumping cycle. The rod string is forced by the pumping unit to follow a regular time versus position pattern. However, the polished rod reacts with the loadings (on the rod string) that are imposed by the well.

The surface dynamometer cards record the history of the variations in loading on the polished rod during a cycle. The cards have three principal uses:

a. To obtain information that can be used to determine load, torque, and horsepower changes required of the pump equipment
b. To improve pump operating conditions such as pump speed and stroke length
c. To check well conditions after installation of equipment to prevent or diagnose various operating problems (like pounding, etc.)

Surface instruments can be mechanical, hydraulic, and electrical. One of the most common mechanical instruments is a ring dynamometer installed between the hanger bar and the polished rod clamp in such a manner as the ring may carry the entire well load. The deflection of the ring is proportional to the load, and this deflection is amplified and transmitted to the recording arm by a series of levers. A stylus on the recording arm traces a record of the imposed loads on a waxed (or via an ink pen) paper card located on a drum. The loads are obtained in terms of polished rod displacements by having the drum oscillate back and forth to reflect the polished rod motion. Correct interpretation of surface dynamometer card leads to estimate of various parameter values.

- Maximum and minimum PRLs can be read directly from the surface card (with the use of instrument calibration). These data then allow for the determination of the torque, counterbalance, and horsepower requirements for the surface unit.
- Rod stretch and contraction is shown on the surface dynamometer card. This phenomenon is reflected in the surface unit dynamometer card and is shown in Fig. 12.11a for an ideal case.
- Acceleration forces cause the ideal card to rotate clockwise. The PRL is higher at the bottom of the stroke and lower at the top of the stroke. Thus, in Fig. 12.11b, Point A is at the bottom of the stroke.

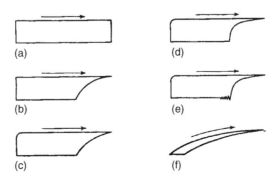

Figure 12.10 Pump dynagraph cards: (a) ideal card, (b) gas compression on down-stroke, (c) gas expansion on upstroke, (d) fluid pound, (e) vibration due to fluid pound, (f) gas lock (Nind, 1964).

- Rod vibration causes a serious complication in the interpretation of the surface card. This is result of the closing of the TV and the "pickup" of the fluid load by the rod string. This is, of course, the fluid pounding. This phenomenon sets up damped oscillation (longitudinal and bending) in the rod string. These oscillations result in waves moving from one end of the rod string to the other. Because the polished rod moves slower near the top and bottom of the strokes, these stress (or load) fluctuations due to vibrations tend to show up more prominently at those locations on the cards. Figure 12.11c shows typical dynamometer card with vibrations of the rod string.

Figure 12.12 presents a typical chart from a strain-gage type of dynamometer measured for a conventional unit operated with a 74-in. stroke at 15.4 strokes per minute. It shows the history of the load on the polished rod as a function of time (this is for a well 825 ft in depth with a No. 86 three-tapered rod string). Figure 12.13 reproduces the data in Fig. 12.12 in a load versus displacement diagram. In the surface chart, we can see the peak load of 22,649 lb (which is 28,800 psi at the top of the 1-in. rod) in Fig. 12.13a. In Fig. 12.13b, we see the peak load of 17,800 lb (which is 29,600 psi at the top of the $7/8$-in. rod). In Fig. 12.13c, we see the peak load of 13,400 lb (which is 30,300 psi at the top of the $3/4$-in. rod). In Fig. 12.13d is

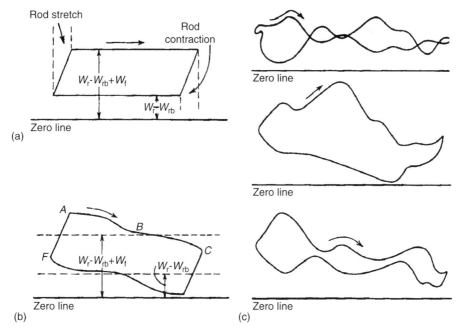

Figure 12.11 Surface dynamometer card: (a) ideal card (stretch and contraction), (b) ideal card (acceleration), (c) three typical cards (Nind, 1964).

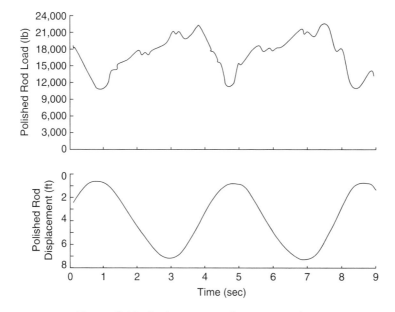

Figure 12.12 Strain-gage–type dynamometer chart.

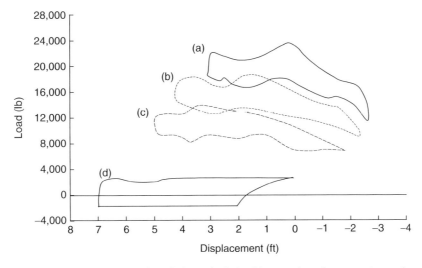

Figure 12.13 Surface to down hole cards derived from surface dynamometer card.

the dynagraph card at the plunger itself. This card indicates gross pump stroke of 7.1 ft, a net liquid stroke of 4.6 ft, and a fluid load of $W_f = 3,200$ lb. The shape of the pump card, Fig. 12.13d, indicates some down-hole gas compression. The shape also indicates that the tubing anchor is holding properly. A liquid displacement rate of 200 bbl/day is calculated and, compared to the surface measured production of 184 bbl/day, indicated no serious tubing flowing leak. The negative in Fig. 12.13d is the buoyancy of the rod string.

The information derived from the dynamometer card (dynagraph) can be used for evaluation of pump performance and troubleshooting of pumping systems. This subject is thoroughly addressed by Brown (1980).\

Summary

This chapter presents the principles of sucker rod pumping systems and illustrates a procedure for selecting components of rod pumping systems. Major tasks include calculations of polished rod load, peak torque, stresses in the rod string, pump deliverability, and counterweight placement. Optimization of existing pumping systems is left to Chapter 18.

References

BROWN, K.E. *The Technology of Artificial Lift Methods*, Vol. 2a. Tulsa, OK: Petroleum Publishing Co., 1980.

COBERLY, C.J. Problems in modern deep-well pumping. *Oil Gas J.* May 12, 1938.

GOLAN, M. and WHITSON, C.H. *Well Performance*, 2nd edition. Englewood Cliffs: Prentice Hall, 1991.

NIND, T.E.W. *Principles of Oil Well Production.* New York: McGraw-Hill Book Co., New York, 1964.

Problems

12.1 If the dimensions d_1, d_2, and c take the same values for both conventional unit (Class I lever system) and air-balanced unit (Class III lever system), how different will their polished rod strokes length be?

12.2 What are the advantages of the Lufkin Mark II and air-balanced units in comparison with conventional units?

12.3 Use your knowledge of kinematics to prove that for Class I lever systems,

a. the polished rod will travel faster in down stroke than in upstroke if the distance between crankshaft and the center of Sampson post is less than dimension d_1.

b. the polished rod will travel faster in up stroke than in down stroke if the distance between crankshaft and the center of Sampson post is greater than dimension d_1.

12.4 Derive a formula for calculating the effective diameter of a tapered rod string.

12.5 Derive formulas for calculating length fractions of equal-top-rod-stress tapered rod strings for (a) two-sized rod strings, (b) three-sized rod strings, and (c) four-sized rod strings. Plot size fractions for each case as a function of plunger area.

12.6 A tapered rod string consists of sections of $5/8$- and $1/2$-in. rods and a 2-in. plunger. Use the formulas from Problem 12.5 to calculate length fraction of each size of rod.

12.7 A tapered rod string consists of sections of $3/4$-, $5/8$-, and $1/2$-in. rods and a $1\tfrac{3}{4}$-in. plunger. Use the formulas from Problem 12.5 to calculate length fraction of each size of rod.

12.8 The following geometry dimensions are for the pumping unit C–80D–133–48:

$d_1 = 64$ in.
$d_2 = 64$ in.
$c = 24$ in.
$h = 74.5$ in.

Can this unit be used with a 2-in. plunger and $3/4$-in. rods to lift 30 °API gravity crude (formation volume factor 1.25 rb/stb) at depth of 2,000 ft? If yes, what is the required counter-balance load?

12.9 The following geometry dimensions are for the pumping unit C–320D–256–120:

$d_1 = 111.07$ in.
$d_2 = 155$ in.
$c = 42$ in.
$h = 132$ in.

Can this unit be used with a $2\tfrac{1}{2}$-in. plunger and $3/4$-, $7/8$-, 1-in. tapered rod string to lift 22 °API gravity crude (formation volume factor 1.22 rb/stb) at a depth of 3,000 ft? If yes, what is the required counter-balance load?

12.10 A well is pumped off with a rod pump described in Problem 12.8. A 2½-in. tubing string (2.875-in. OD, 2.441 ID) in the well is not anchored. Calculate (a) expected liquid production rate (use pump volumetric efficiency 0.80) and (b) required prime mover power (use safety factor 1.3).

12.11 A well is pumped with a rod pump described in Problem 12.9 to a liquid level of 2,800 ft. A 3-in. tubing string (3½-in. OD, 2.995-in. ID) in the well is anchored. Calculate (a) expected liquid production rate (use pump volumetric efficiency 0.85) and (b) required prime mover power (use safety factor 1.4).

12.12 A well is to be put on a sucker rod pump. The proposed pump setting depth is 4,500 ft. The anticipated production rate is 500 bbl/day oil of 40 °API gravity against wellhead pressure 150 psig. It is assumed that the working liquid level is low, and a sucker rod string having a working stress of 30,000 psi is to be used. Select surface and subsurface equipment for the installation. Use a safety factor of 1.40 for prime mover power.

12.13 A well is to be put on a sucker rod pump. The proposed pump setting depth is 4,000 ft. The anticipated production rate is 550 bbl/day oil of 35 °API gravity against wellhead pressure 120 psig. It is assumed that working liquid level will be about 3,000 ft, and a sucker rod string having a working stress of 30,000 psi is to be used. Select surface and subsurface equipment for the installation. Use a safety factor of 1.30 for prime mover power.

13 Gas Lift

Contents
13.1 Introduction 13/182
13.2 Gas Lift System 13/182
13.3 Evaluation of Gas Lift Potential 13/183
13.4 Gas Lift Gas Compression Requirements 13/185
13.5 Selection of Gas Lift Valves 13/192
13.6 Special Issues in Intermittent-Flow Gas Lift 13/201
13.7 Design of Gas Lift Installations 13/203
Summary 13/205
References 13/205
Problems 13/205

13.1 Introduction

Gas lift technology increases oil production rate by injection of compressed gas into the lower section of tubing through the casing–tubing annulus and an orifice installed in the tubing string. Upon entering the tubing, the compressed gas affects liquid flow in two ways: (a) the energy of expansion propels (pushes) the oil to the surface and (b) the gas aerates the oil so that the effective density of the fluid is less and, thus, easier to get to the surface.

There are four categories of wells in which a gas lift can be considered:

1. High productivity index (PI), high bottom-hole pressure wells
2. High PI, low bottom-hole pressure wells
3. Low PI, high bottom-hole pressure wells
4. Low PI, low bottom-hole pressure wells

Wells having a PI of 0.50 or less are classified as low productivity wells. Wells having a PI greater than 0.50 are classified as high productivity wells. High bottom-hole pressures will support a fluid column equal to 70% of the well depth. Low bottom-hole pressures will support a fluid column less than 40% of the well depth.

Gas lift technology has been widely used in the oil fields that produce sandy and gassy oils. Crooked/deviated holes present no problem. Well depth is not a limitation. It is also applicable to offshore operations. Lifting costs for a large number of wells are generally very low. However, it requires lift gas within or near the oil fields. It is usually not efficient in lifting small fields with a small number of wells if gas compression equipment is required. Gas lift advancements in pressure control and automation systems have enabled the optimization of individual wells and gas lift systems.

13.2 Gas Lift System

A complete gas lift system consists of a gas compression station, a gas injection manifold with injection chokes and time cycle surface controllers, a tubing string with installations of unloading valves and operating valve, and a down-hole chamber.

Figure 13.1 depicts a configuration of a gas-lifted well with installations of unloading valves and operating valve on the tubing string. There are four principal advantages to be gained by the use of multiple valves in a well:

1. Deeper gas injection depths can be achieved by using valves for wells with fixed surface injection pressures.
2. Variation in the well's productivity can be obtained by selectively injecting gas valves set at depths "higher" or "lower" in the tubing string.
3. Gas volumes injected into the well can be "metered" into the well by the valves.
4. Intermittent gas injection at progressively deeper set valves can be carried out to "kick off" a well to either continuous or intermittent flow.

A continuous gas lift operation is a steady-state flow of the aerated fluid from the bottom (or near bottom) of the well to the surface. Intermittent gas lift operation is characterized by a start-and-stop flow from the bottom (or near bottom) of the well to the surface. This is *unsteady state flow*.

In continuous gas lift, a small volume of high-pressure gas is introduced into the tubing to aerate or lighten the fluid column. This allows the flowing bottom-hole pressure with the aid of the expanding injection gas to deliver liquid to the surface. To accomplish this efficiently, it is desirable to design a system that will permit injection through a single valve at the greatest depth possible with the available injection pressure.

Continuous gas lift method is used in wells with a high PI (≥ 0.5 stb/day/psi) and a reasonably high reservoir pressure relative to well depth. Intermittent gas lift method is suitable to wells with (1) high PI and low reservoir pressure or (2) low PI and low reservoir pressure.

The type of gas lift operation used, continuous or intermittent, is also governed by the volume of fluids to be produced, the available lift gas as to both volume and pressure, and the well reservoir's conditions such as the case when the high instantaneous BHP drawdown encountered with intermittent flow would cause excessive sand production, or coning, and/or gas into the wellbore.

Figure 13.2 illustrates a simplified flow diagram of a closed rotary gas lift system for a single well in an intermittent gas lift operation. The time cycle surface controller regulates the start-and-stop injection of lift gas to the well.

For proper selection, installation, and operations of gas lift systems, the operator must know the equipment and the fundamentals of gas lift technology. The basic equipment for gas lift technology includes the following:

a. Main operating valves
b. Wire-line adaptations
c. Check valves
d. Mandrels
e. Surface control equipment
f. Compressors

This chapter covers basic system engineering design fundamentals for gas lift operations. Relevant topics include the following:

1. Liquid flow analysis for evaluation of gas lift potential
2. Gas flow analysis for determination of lift gas compression requirements
3. Unloading process analysis for spacing subsurface valves
4. Valve characteristics analysis for subsurface valve selection
5. Installation design for continuous and intermittent lift systems.

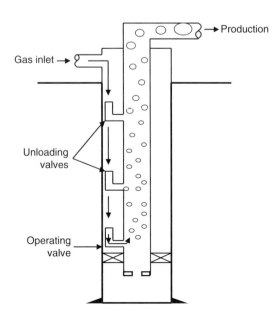

Figure 13.1 Configuration of a typical gas lift well.

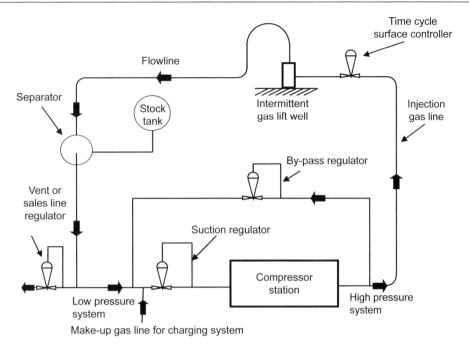

Figure 13.2 *A simplified flow diagram of a closed rotary gas lift system for single intermittent well.*

13.3 Evaluation of Gas Lift Potential

Continuous gas lift can be satisfactorily applied to most wells having a reasonable degree of bottom-hole maintenance and a PI of approximately 0.5 bbl/day/psi or greater. A PI as low as 0.2 bbl/day/psi can be used for a continuous gas lift operation if injection gas is available at a sufficiently high pressure. An intermittent gas lift is usually applied to wells having a PI less than 0.5 bbl/day/psi.

Continuous gas lift wells are changed to intermittent gas lift wells after reservoir pressures drop to below a certain level. Therefore, intermittent gas lift wells usually give lower production rates than continuous gas lift wells. The decision of whether to use gas lift technology for oil well production starts from evaluating gas lift potential with continuous gas injection.

Evaluation of gas lift potential requires system analyses to determine well operating points for various lift gas availabilities. The principle is based on the fact that there is only one pressure at a given point (node) in any system; no matter, the pressure is estimated based on the information from upstream (inflow) or downstream (outflow). The node of analysis is usually chosen to be the gas injection point inside the tubing, although bottom hole is often used as a solution node.

The potential of gas lift wells is controlled by gas injection rate or gas liquid ratio (GLR). Four gas injection rates are significant in the operation of gas lift installations:

1. Injection rates of gas that result in no liquid (oil or water) flow up the tubing. The gas amount is insufficient to lift the liquid. If the gas enters the tubing at an extremely low rate, it will rise to the surface in small semi-spheres (bubbly flow).
2. Injection rates of maximum efficiency where a minimum volume of gas is required to lift a given amount of liquid.
3. Injection rate for maximum liquid flow rate at the "optimum GLR."
4. Injection rate of no liquid flow because of excessive gas injection. This occurs when the friction (pipe) produced by the gas prevents liquid from entering the tubing.

Figure 13.3 depicts a continuous gas lift operation. The tubing is filled with reservoir fluid below the injection point and with the mixture of reservoir fluid and injected gas above the injection point. The pressure relationship is shown in Fig. 13.4.

The inflow performance curve for the node at the gas injection point inside the tubing is well IPR curve minus the pressure drop from bottom hole to the node. The outflow performance curve is the vertical lift performance curve, with total GLR being the sum of formation GLR and injected GLR. Intersection of the two curves defines the operation point, that is, the well production potential.

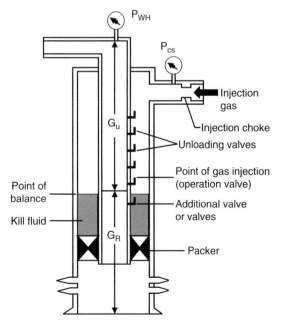

Figure 13.3 *A sketch of continuous gas lift.*

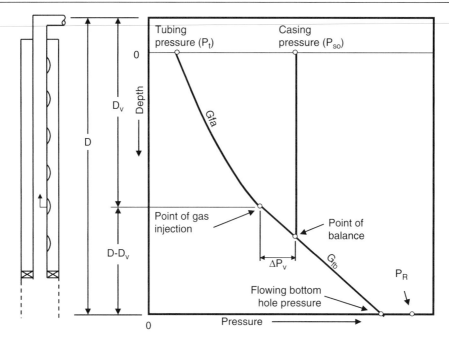

Figure 13.4 Pressure relationship in a continuous gas lift.

In a field-scale evaluation, if an unlimited amount of lift gas is available for a given gas lift project, the injection rate of gas to individual wells should be optimized to maximize oil production of each well. If only a limited amount of gas is available for the gas lift, the gas should be distributed to individual wells based on predicted well lifting performance, that is, the wells that will produce oil at higher rates at a given amount of lift gas are preferably chosen to receive more lift gas.

If an unlimited amount of gas lift gas is available for a well, the well should receive a lift gas injection rate that yields the optimum GLR in the tubing so that the flowing bottom-hole pressure is minimized, and thus, oil production is maximized. The optimum GLR is liquid flow rate dependent and can be found from traditional gradient curves such as those generated by Gilbert (Gilbert, 1954). Similar curves can be generated with modern computer programs using various multiphase correlations. The computer program *OptimumGLR.xls* in the CD attached to this book was developed based on modified Hagedorn and Brown method (Brown, 1977) for multiphase flow calculations and the Chen method (1979) for friction factor determination. It can be used for predicting the optimum GLR in tubing at a given tubing head pressure and liquid flow rate.

After the system analysis is completed with the optimum GLRs in the tubing above the injection point, the expected liquid production rate (well potential) is known. The required injection GLR to the well can be calculated by

$$GLR_{inj} = GLR_{opt,o} - GLR_{fm}, \quad (13.1)$$

where

GLR_{inj} = injection GLR, scf/stb
$GLR_{opt,o}$ = optimum GLR at operating flow rate, scf/stb
GLR_{fm} = formation oil GLR, scf/stb.

Then the required gas injection rate to the well can be calculated by

$$q_{g,inj} = GLR_{inj}q_o, \quad (13.2)$$

where q_o is the expected operating liquid flow rate.

If a limited amount of gas lift gas is available for a well, the well potential should be estimated based on GLR expressed as

$$GLR = GLR_{fm} + \frac{q_{g,inj}}{q}, \quad (13.3)$$

where q_g is the lift gas injection rate (scf/day) available to the well.

Example Problem 13.1 An oil well has a pay zone around the mid-perf depth of 5,200 ft. The formation oil has a gravity of 26 °API and GLR of 300 scf/stb. Water cut remains 0%. The IPR of the well is expressed as

$$q = q_{max}\left[1 - 0.2\frac{p_{wf}}{\bar{p}} - 0.8\left(\frac{p_{wf}}{\bar{p}}\right)^2\right],$$

where

$q_{max} = 1500\,\text{stb/day}$
$\bar{p} = 2,000\,\text{psia}.$

A $2\frac{1}{2}$-in. tubing (2.259 in. inside diameter [ID]) can be set with a packer at 200 ft above the mid-perf. What is the maximum expected oil production rate from the well with continuous gas lift at a wellhead pressure of 200 psia if

a. an unlimited amount of lift gas is available for the well?
b. only 1 MMscf/day of lift gas is available for the well?

Solution The maximum oil production rate is expected when the gas injection point is set right above the packer. Assuming that the pressure losses due to friction below the injection point are negligible, the inflow-performance curve for the gas injection point (inside tubing) can be expressed as

$$p_{vf} = 0.125\bar{p}[\sqrt{81 - 80(q/q_{max})} - 1] - G_R(D - D_v),$$

where p_{vf} is the pressure at the gas injection point, G_R is the pressure gradient of the reservoir fluid, D is the pay zone depth, and D_v is the depth of the gas injection point. Based on the oil gravity of 26 °API, G_R is calculated to be 0.39 psi/ft. D and D_v are equal to 5,200 ft and 5,000 ft, respectively in this problem.

The outflow performance curve for the gas injection point can be determined based on 2.259-in. tubing ID, 200 psia wellhead pressure, and the GLRs.

a. Spreadsheet *OptimumGLR.xls* gives the following:

q (stb/d)	GLR_{opt} (scf/stb)
400	4,500
600	3,200
800	2,400

Using these data to run computer program *HagedornBrown-Correlation.xls* (on the CD attached to this book) gives

q (stb/day)	p_t (psia)
400	603
600	676
800	752

Figure 13.5 shows the system analysis plot given by the computer program *GasLiftPotential.xls*. It indicates an operating point of $q = 632$ stb/day and $p_{t,v} = 698$ psia tubing pressure at the depth of injection.

The optimum GLR at the operating point is calculated with interpolation as

$$GLR_{opt,o} = 2,400 + \frac{3,200 - 2,400}{800 - 600}(800 - 632)$$
$$= 3,072 \text{ scf/stb}.$$

The injection GLR is

$$GLR_{inj} = 3,072 - 300 = 2,772 \text{ scf/stb}.$$

Then the required gas injection rate to the well can be calculated:

$$q_{g,inj} = (2,772)(632) = 1,720,000 \text{ scf/day}$$

b. For a given amount of lift gas 1 MMscf/day, the GLR can be calculated with Eq. (13.3) as

q (stb/day)	GLR (scf/stb)
400	2,800
600	1,967
800	1,550

Using these data to run computer program *Hagedorn-BrownCorrelation.xls* gives

q (stb/day)	p_t (psia)
400	614
600	694
800	774

Figure 13.6 shows the system analysis plot given by the computer program *GasLiftPotential.xls*. It indicates an operating point of $q = 620$ stb/day and $p_t = 702$ psia tubing pressure at the depth of injection.

This example shows that increasing the gas injection rate from 1 MMscf/day to 1.58 MMscf/day will not make a significant difference in the oil production rate.

13.4 Gas Lift Gas Compression Requirements

The gas compression station should be designed to provide an adequate gas lift gas flow rate at sufficiently high pressure. These gas flow rates and output pressures determine the required power of the compression station.

13.4.1 Gas Flow Rate Requirement

The total gas flow rate of the compression station should be designed on the basis of gas lift at peak operating condition for all the wells with a safety factor for system leak consideration, that is,

$$q_{g,total} = S_f \sum_{i=1}^{N_w} (q_{g,inj})_i, \quad (13.4)$$

where

q_g = total output gas flow rate of the compression station, scf/day
S_f = safety factor, 1.05 or higher
N_w = number of wells.

The procedure for determination of lift gas injection rate $q_{g,inj}$ to each well has been illustrated in Example Problem 13.1.

13.4.2 Output Gas Pressure Requirement

Kickoff of a dead well (non-natural flowing) requires much higher compressor output pressures than the ultimate goal of steady production (either by continuous gas lift or by intermittent gas lift operations). Mobil compressor trailers are used for the kickoff operations. The output pressure of the compression station should be designed on the basis of the gas distribution pressure under normal flow conditions, not the kickoff conditions. It can be expressed as

$$p_{out} = S_f p_L, \quad (13.5)$$

where

p_{out} = output pressure of the compression station, psia
S_f = safety factor
p_L = pressure at the inlet of the gas distribution line, psia.

Starting from the tubing pressure at the valve ($p_{t,v}$), the pressure at the inlet of the gas distribution line can be estimated based on the relationships of pressures along the injection path. These relationships are discussed in the following subsections.

13.4.2.1 Injection Pressure at Valve Depth

The injection pressure at valve depth in the casing side can be expressed as

$$p_{c,v} = p_{t,v} + \Delta p_v, \quad (13.6)$$

where

$p_{c,v}$ = casing pressure at valve depth, psia
Δp_v = pressure differential across the operating valve (orifice).

It is a common practice to use $\Delta p_v = 100$ psi. The required size of the orifice can be determined using the choke-flow equations presented in Subsection 13.4.2.3.

13.4.2.2 Injection Pressure at Surface

Accurate determination of the surface injection pressure $p_{c,s}$ requires rigorous methods such as the Cullender and

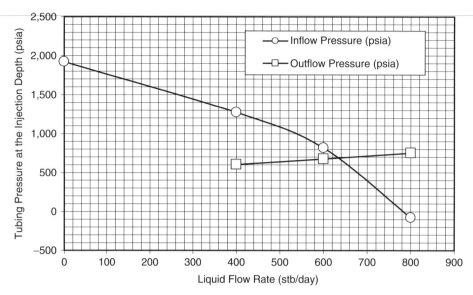

Figure 13.5 System analysis plot given by GasLiftPotential.xls for the unlimited gas injection case.

Smith method (Katz et al., 1959). The average temperature and compressibility factor method also gives results with acceptable accuracy. In both methods, the frictional pressure losses in the annulus are considered. However, because of the large cross-sectional area of the annular space, the frictional pressure losses are often negligible. Then the average temperature and compressibility factor model degenerates to (Economides et al., 1994)

$$p_{c,v} = p_{c,s}\, e^{0.01875\frac{\gamma_g D_v}{\bar{z}\bar{T}}}, \qquad (13.7)$$

where
$p_{c,v}$ = casing pressure at valve depth, psia
$p_{c,s}$ = casing pressure at surface, psia
γ_g = gas specific gravity, air = 1.0
$\bar{z}$ = the average gas compressibility factor
$\bar{T}$ = the average temperature, °R.

Equation (13.7) can be rearranged to be

$$p_{c,s} = p_{c,v} e^{-0.01875\frac{\gamma_g D_v}{\bar{z}\bar{T}}}. \qquad (13.8)$$

Since the z factor also depends on $p_{c,s}$, this equation can be solved for $p_{c,s}$ with a trial-and-error approach. Because Eq. (13.8) involves exponential function that is difficult to handle without a calculator, an approximation to the equation has been used traditionally. In fact, when Eq. (13.7) is expended as a Taylor series, and if common fluid properties for a natural gas and reservoir are considered such as $\gamma_g = 0.7$, $\bar{z} = 0.9$, and $\bar{T} = 600\,°R$, it can be approximated as

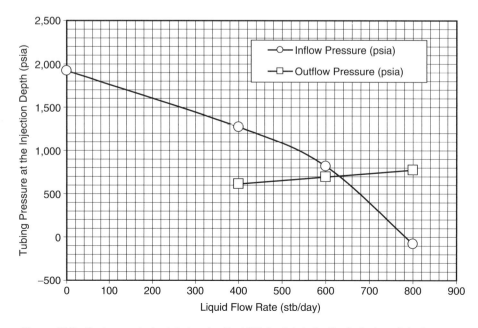

Figure 13.6 System analysis plot given by GasLiftPotential.xls for the limited gas injection case.

$$p_{c,v} = p_{c,s}\left(1 + \frac{D_v}{40{,}000}\right), \quad (13.9)$$

which gives

$$p_{c,s} = \frac{p_{c,v}}{1 + \dfrac{D_v}{40{,}000}}. \quad (13.10)$$

Neglecting the pressure losses between injection choke and the casing head, the pressure downstream of the choke (p_{dn}) can be assumed to be the casing surface injection pressure, that is,

$$p_{dn} = p_{c,s}.$$

13.4.2.3 Pressure Upstream of the Choke

The pressure upstream of the injection choke depends on flow condition at the choke, that is, sonic or subsonic flow. Whether a sonic flow exists depends on a downstream-to-upstream pressure ratio. If this pressure ratio is less than a critical pressure ratio, sonic (critical) flow exists. If this pressure ratio is greater than or equal to the critical pressure ratio, subsonic (subcritical) flow exists. The critical pressure ratio through chokes is expressed as

$$R_c = \left(\frac{2}{k+1}\right)^{\frac{k}{k-1}}, \quad (13.11)$$

where $k = C_p/C_v$ is the gas-specific heat ratio. The value of the k is about 1.28 for natural gas. Thus, the critical pressure ratio is about 0.55.

Pressure equations for choke flow are derived based on an isentropic process. This is because there is no time for heat to transfer (adiabatic) and the friction loss is negligible (assuming reversible) at choke.

13.4.2.3.1 Sonic Flow
Under sonic flow conditions, the gas passage rate reaches and remains its maximum value. The gas passage rate is expressed in the following equation for ideal gases:

$$q_{gM} = 879 C_c A p_{up} \sqrt{\left(\frac{k}{\gamma_g T_{up}}\right)\left(\frac{2}{k+1}\right)^{\frac{k+1}{k-1}}}, \quad (13.12)$$

where

q_{gM} = gas flow rate, Mscf/day
p_{up} = pressure upstream the choke, psia
A = cross-sectional area of choke, in.^{2}
T_{up} = upstream temperature, °R
γ_g = gas specific gravity related to air
C_c = choke flow coefficient.

The choke flow coefficient C_c can be determined using charts in Figs. 5.2 and 5.3 (Chapter 5) for nozzle- and orifice-type chokes, respectively. The following correlation has been found to give reasonable accuracy for Reynolds numbers between 10^4 and 10^6 for nozzle-type chokes (Guo and Ghalambor, 2005):

$$C = \frac{d}{D} + \frac{0.3167}{\left(\dfrac{d}{D}\right)^{0.6}} + 0.025[\log(N_{Re}) - 4], \quad (13.13)$$

where

d = choke diameter, inch
D = pipe diameter, in.
N_{Re} = Reynolds number

and the Reynolds number is given by

$$N_{Re} = \frac{20 q_{gM} \gamma_g}{\mu d}, \quad (13.14)$$

where
μ = gas viscosity at *in situ* temperature and pressure, cp.

Equation (13.12) indicates that the upstream pressure is independent of downstream pressure under sonic flow conditions. If it is desirable to make a choke work under sonic flow conditions, the upstream pressure should meet the following condition:

$$p_{up} \geq \frac{p_{dn}}{0.55} = 1.82 p_{dn} \quad (13.15)$$

Once the pressure upstream of the choke/orifice is determined by Eq. (13.15), the required choke/orifice diameter can be calculated with Eq. (13.12) using a trial-and-error approach.

13.4.2.3.2 Subsonic Flow
Under subsonic flow conditions, gas passage through a choke can be expressed as

$$q_{gM} = 1{,}248 C_c A p_{up}$$

$$\times \sqrt{\frac{k}{(k-1)\gamma_g T_{up}}\left[\left(\frac{p_{dn}}{p_{up}}\right)^{\frac{2}{k}} - \left(\frac{p_{dn}}{p_{up}}\right)^{\frac{k+1}{k}}\right]}. \quad (13.16)$$

If it is desirable to make a choke work under subsonic flow conditions, the upstream pressure should be determined from Eq. (13.16) with a trial-and-error method.

13.4.2.4 Pressure of the Gas Distribution Line

The pressure at the inlet of gas distribution line can be calculated using the Weymouth equation for horizontal flow (Weymouth, 1912):

$$q_{gM} = \frac{0.433 T_b}{p_b} \sqrt{\frac{\left(p_L^2 - p_{up}^2\right) D^{16/3}}{\gamma_g \bar{T} \bar{z} L_g}}, \quad (13.17)$$

where

T_b = base temperature, °R
p_b = base pressure, psi
p_L = pressure at the inlet of gas distribution line, psia
L_g = length of distribution line, mile

Equation (13.17) can be rearranged to solve for pressure:

$$p_L = \sqrt{p_{up}^2 + \left(\frac{q_{gM} p_b}{0.433 T_b}\right)^2 \frac{\gamma_g \bar{T} \bar{z} L_g}{D^{16/3}}} \quad (13.18)$$

Example Problem 13.2 An oil field has 16 oil wells to be gas lifted. The gas lift gas at the central compressor station is first pumped to two injection manifolds with 4-in. ID, 1-mile lines and then is distributed to the wellheads with 4-in. ID, 0.2-mile lines. Given the following data, calculate the required output pressure of compression station:

Gas-specific gravity (γ_g):	0.65
Valve depth (D_v):	5,000 ft
Maximum tubing pressure at valve depth (p_t):	500 psia
Required lift gas injection rate per well:	2 MMscf/day
Pressure safety factor (S_f):	1.1
Base temperature (T_b):	60 °F
Base pressure (p_b):	14.7 psia

Solution Using $\Delta p_v = 100$ psi, the injection pressure at valve depth is then 600 psia. Equation (13.10) gives

$$p_{c,s} = \frac{p_{c,v}}{1 + \dfrac{D_v}{40{,}000}} = \frac{600}{1 + \dfrac{5{,}000}{40{,}000}} = 533 \text{ psia}.$$

Neglecting the pressure losses between the injection choke and the casing head, pressure downstream of the choke

(p_{dn}) can be assumed to be the surface injection pressure, that is,

$$p_{dn} = p_{c,s} = 533 \text{ psia}.$$

Assuming minimum sonic flow at the injection choke, the pressure upstream of the choke is calculated as

$$p_{up} \geq \frac{p_{dn}}{0.55} = 1.82 p_{dn} = (1.82)(533) = 972 \text{ psia}.$$

The gas flow rate in each of the two gas distribution lines is (2)(16)/(2), or 16 MMscf/day. Using the trial-and-error method, Eq. (13.18) gives

$$p_L = \sqrt{(972)^2 + \left(\frac{(16{,}000)(14.7)}{0.433(60+460)}\right)^2 \frac{(0.65)(530)(0.79)(1)}{(4)^{16/3}}}$$

$$= 1{,}056 \text{ psia}.$$

The required output pressure of the compressor is determined to be

$$p_{out} = S_f p_L = (1.1)(1{,}056) = 1{,}162 \text{ psia}.$$

The computer program *CompressorPressure.xls* can be used for solving similar problems. The solution given by the program to this example problem is shown in Table 13.1.

13.4.3 Compression Power Requirement

The compressors used in the petroleum industry fall into two distinct categories: reciprocating and rotary compressors. Reciprocating compressors are built for practically all pressures and volumetric capacities. Reciprocating compressors have more moving parts and, therefore, lower mechanical efficiencies than rotary compressors. Each cylinder assembly of a reciprocation compressor consists of a piston, cylinder, cylinder heads, suction and discharge valves, and other parts necessary to convert rotary motion to reciprocation motion. A reciprocating compressor is designed for a certain range of compression ratios through the selection of proper piston displacement and clearance volume within the cylinder. This clearance volume can be either fixed or variable, depending on the extent of the operation range and the percent of load variation desired. A typical reciprocating compressor can deliver a volumetric gas flow rate up to 30,000 cubic feet per minute (cfm) at a discharge pressure up to 10,000 psig.

Rotary compressors are divided into two classes: the centrifugal compressor and the rotary blower. A centrifugal compressor consists of a housing with flow passages, a rotating shaft on which the impeller is mounted, bearings, and seals to prevent gas from escaping along the shaft. Centrifugal compressors have few moving parts because only the impeller and shaft rotate. Thus, its efficiency is high and lubrication oil consumption and maintenance costs are low. Cooling water is normally unnecessary because of lower compression ratio and lower friction loss. Compression rates of centrifugal compressors are lower because of the absence of positive displacement. Centrifugal compressors compress gas using centrifugal force. Work is done on the gas by an impeller. Gas is then discharged at a high velocity into a diffuser where the velocity is reduced and its kinetic energy is converted to static pressure. Unlike reciprocating compressors, all this is done without confinement and physical squeezing. Centrifugal compressors with relatively unrestricted passages and continuous flow are inherently high-capacity, low-pressure ratio machines that adapt easily to series arrangements within a station. In this way, each compressor is required to develop only part of the station compression ratio. Typically, the volume is more than 100,000 cfm and discharge pressure is up to 100 psig.

When selecting a compressor, the pressure-volume characteristics and the type of driver must be considered. Small

Table 13.1 Result Given by Computer Program CompressorPressure.xls

CompressorPressure.xls
Description: This spreadsheet calculates required pressure from compressor.
Instruction: (1) Select a unit system; (2) click "Solution" button; and (3) view result.

Input data	U.S. units	SI units 1
Depth of operating valve (D_v):	5,000 ft	
Length of the main distribution line (L_g):	1 mi	
ID of the main distribution line (D):	4.00 in.	
Gas flow rate in main distribution line ($q_{g,l}$):	16 MMscf/day	
Surface temperature (T_s):	70 °F	
Temperature at valve depth (T_v):	120 °F	
Gas-specific gravity (γ_g):	0.65 (air = 1)	
Gas-specific heat ratio (k):	1.25	
Tubing pressure at valve depth (p_t):	500 psia	
Valve pressure differential (Δp_v):	100 psia	
Base temperature (T_b):	60 °F	
Base pressure (p_b):	14.7 psia	
Pressure safety factor (S_f):	1.1	
Solution		
$p_{c,v} = p_{t,v} + \Delta p_v$	600 psia	
Average z-factor in annulus:	0.9189?	
$p_{c,s} - p_{c,v} e^{-0.01875 \frac{\gamma_g D_v}{zT}} = 0$ gives $p_{c,s}$	532 psia	
$p_{dn} = p_{c,s}$	532 psia	
$p_{up} \geq \frac{p_{dn}}{0.55} = 1.82 p_{dn}$	969 psia	
Average z-factor at surface:	0.8278	
$p_L - \sqrt{p_{up}^2 + \left(\frac{q_{gM} p_b}{0.433 T_b}\right)^2 \frac{\gamma_g \bar{T} z L_g}{D^{16/3}}} = 0$ gives p_L	1,063 psia	
$p_{out} = S_f p_L$	1,170 psia	

rotary compressors (vane or impeller type) are generally driven by electric motors. Large-volume positive compressors operate at lower speeds and are usually driven by steam or gas engines. They may be driven through reduction gearing by steam turbines or an electric motor. Reciprocation compressors driven by steam turbines or electric motors are most widely used in the petroleum industry as the conventional high-speed compression machine. Selection of compressors requires considerations of volumetric gas deliverability, pressure, compression ratio, and horsepower.

13.4.3.1 Reciprocating Compressors

Two basic approaches are used to calculate the horsepower theoretically required to compress natural gas. One is to use analytical expressions. In the case of adiabatic compression, the relationships are complicated and are usually based on the ideal-gas equation. When used for real gases where deviation from ideal-gas law is appreciable, they are empirically modified to take into consideration the gas deviation factor. The second approach is the enthalpy-entropy or Mollier diagram for real gases. This diagram provides a simple, direct, and rigorous procedure for determining the horsepower theoretically necessary to compress the gas.

Even though in practice the cylinders in the reciprocating compressors may be water-cooled, it is customary to consider the compression process as fundamentally adiabatic—that is, to idealize the compression as one in which there is no cooling of the gas. Furthermore, the process is usually considered to be essentially a perfectly reversible adiabatic, that is, an isentropic process. Thus, in analyzing the performance of a typical reciprocating compressor, one may look upon the compression path following the general law

$$pV^k = \text{a constant.} \tag{13.19}$$

For real natural gases in the gravity range $0.55 < \gamma_g < 1$, the following relationship can be used at approximately 150 °F:

$$k^{150\,°F} \approx \frac{2.738 - \log \gamma_g}{2.328} \tag{13.20}$$

When a real gas is compressed in a single-stage compression, the compression is polytropic tending to approach adiabatic or constant-entropy conditions. Adiabatic compression calculations give the maximum theoretical work or *horsepower* necessary to compress a gas between any two pressure limits, whereas isothermal compression calculations give the minimum theoretical work or horsepower necessary to compress a gas. Adiabatic and isothermal work of compression, thus, give the upper and lower limits, respectively, of work or horsepower requirements to compress a gas. One purpose of intercoolers between multistage compressors is to reduce the horsepower necessary to compress the gas. The more intercoolers and stages, the closer the horsepower requirement approaches the isothermal value.

13.4.3.1.1 Volumetric Efficiency
The volumetric efficiency represents the efficiency of a compressor cylinder to compress gas. It may be defined as the ratio of the volume of gas actually delivered to the piston displacement, corrected to suction temperature and pressure. The principal reasons that the cylinder will not deliver the piston displacement capacity are wire-drawing, a throttling effect on the valves; heating of the gas during admission to the cylinder; leakage past valves and piston rings; and re-expansion of the gas trapped in the clearance-volume space from the previous stroke. Re-expansion has by far the greatest effect on volumetric efficiency.

The theoretical formula for volumetric efficiency is

$$E_v = 1 - (r^{1/k} - 1)\, C_l, \tag{13.21}$$

where

E_v = volumetric efficiency, fraction
r = cylinder compression ratio
C_l = clearance, fraction.

In practice, adjustments are made to the theoretical formula in computing compressor performance:

$$E_v = 0.97 - \left[\left(\frac{z_s}{z_d}\right) r^{1/k} - 1\right] C_l - e_v, \tag{13.22}$$

where

z_s = gas deviation factor at suction of the cylinder
z_d = gas deviation factor at discharge of the cylinder
e_v = correction factor.

In this equation, the constant 0.97 is a reduction of 1 to correct for minor inefficiencies such as incomplete filling of the cylinder during the intake stroke. The correction factor e_v is to correct for the conditions in a particular application that affect the volumetric efficiency and for which the theoretical formula is inadequate.

13.4.3.1.2 Stage Compression
The ratio of the discharge pressure to the inlet pressure is called the *pressure ratio*. The volumetric efficiency becomes less, and mechanical stress limitation becomes more, pronounced as pressure ratio increases. Natural gas is usually compressed in stages, with the pressure ratio per stage being less than 6. In field practice, the pressure ratio seldom exceeds 4 when boosting gas from low pressure for processing or sale. When the total compression ratio is greater than this, more stages of compression are used to reach high pressures.

The total power requirement is a minimum when the pressure ratio in each stage is the same. This may be expressed in equation form as

$$r = \left(\frac{p_d}{p_s}\right)^{1/N_s}, \tag{13.23}$$

where

p_d = final discharge pressure, absolute
p_s = suction pressure, absolute
N_s = number of stages required.

As large compression ratios result in gas being heated to undesirably high temperatures, it is common practice to cool the gas between stages and, if possible, after the final stage of compression.

13.4.3.1.3 Isentropic Horsepower
The computation is based on the assumption that the process is ideal isentropic or perfectly reversible adiabatic. The total ideal horsepower for a given compression is the sum of the ideal work computed for each stage of compression. The ideal isentropic work can be determined for each stage of compression in a number of ways. One way to solve a compression problem is by using the Mollier diagram. This method is not used in this book because it is not easily computerized. Another approach commonly used is to calculate the horsepower for each stage from the isentropic work formula:

$$w = \frac{k}{k-1} \frac{53.241\, T_1}{\gamma_g} \left[\left(\frac{p_2}{p_1}\right)^{(k-1)/k} - 1\right], \tag{13.24}$$

where

w = theoretical shaft work required to compress the gas, ft-lb$_f$/lb$_m$
T_1 = suction temperature of the gas, °R

γ_g = gas-specific gravity, air = 1
p_1 = suction pressure of the gas, psia
p_2 = pressure of the gas at discharge point, psia.

When the deviation from ideal gas behavior is appreciable, Eq. (13.24) is empirically modified. One such modification is

$$w = \frac{k}{k-1} \frac{53.241 T_1}{\gamma_g} \left[\left(\frac{p_2}{p_1}\right)^{z_1(k-1)/k} - 1 \right] \quad (13.25)$$

or, in terms of power,

$$Hp_{MM} = \frac{k}{k-1} \frac{3.027 p_b}{T_b} T_1 \left[\left(\frac{p_2}{p_1}\right)^{z_1(k-1)/k} - 1 \right], \quad (13.26)$$

where

Hp_{MM} = required theoretical compression power, hp/MMcfd
z_1 = compressibility factor at suction conditions.

The theoretical adiabatic horsepower obtained by the proceeding equations can be converted to brake horsepower (Hp_b) required at the end of prime mover of the compressor using an overall efficiency factor, E_o. The brake horsepower is the horsepower input into the compressor. The efficiency factor E_o consists of two components: compression efficiency (compressor-valve losses) and the mechanical efficiency of the compressor. The overall efficiency of a compressor depends on a number of factors, including design details of the compressor, suction pressure, speed of the compressor, compression ratio, loading, and general mechanical condition of the unit. In most modern compressors, the compression efficiency ranges from 83 to 93%. The mechanical efficiency of most modern compressors ranges from 88 to 95%. Thus, most modern compressors have an overall efficiency ranging from 75 to 85%, based on the ideal isentropic compression process as a standard. The actual efficiency curves can be obtained from the manufacturer. Applying these factors to the theoretical horsepower gives

$$Hp_b = \frac{q_{MM} Hp_{MM}}{E_o}, \quad (13.27)$$

where q_{MM} is the gas flow rate in MMscfd.

The discharge temperature for real gases can be calculated by

$$T_2 = T_1 \left(\frac{p_2}{p_1}\right)^{z_1(k-1)/k}. \quad (13.28)$$

Calculation of the heat removed by intercoolers and aftercoolers can be accomplished using constant pressure-specific heat data:

$$\Delta H = n_G \overline{C}_p \Delta T, \quad (13.29)$$

where

n_G = number of lb-mole of gas
$\overline{C}_p$ = specific heat under constant pressure evaluated at cooler operating pressure and the average temperature, btu/lb-mol-°F.

Example Problem 13.3 For data given in Example Problem 13.2, assuming the overall efficiency is 0.80, calculate the theoretical and brake horsepower required to compress the 32 MMcfd of a 0.65-specific gravity natural gas from 100 psia and 70 °F to 1,165 psia. If intercoolers cool the gas to 70 °F, what is the heat load on the intercoolers and what is the final gas temperature?

Solution The overall compression ratio is

$$r_{ov} = \frac{1,165}{100} = 11.65.$$

Because this is greater than 6, more than one-stage compression is required. Using two stages of compression gives

$$r = \left(\frac{1,165}{100}\right)^{1/2} = 3.41.$$

The gas is compressed from 100 to 341 psia in the first stage, and from 341 to 1,165 psia in the second stage. Based on gas-specific gravity, the following gas property data can be obtained:

$T_c = 358\,°R$
$p_c = 671$ psia
$T_r = 1.42$
$p_{r,1} = 0.149$ at 100 psia
$p_{r,2} = 0.595$ at 341 psia
$z_1 = 0.97$ at 70 °F and 100 psia
$z_2 = 0.95$ at 70 °F and 341 psia.

First stage:

$$Hp_{MM} = \frac{1.25}{0.25} \left(3.027 \times \frac{14.7}{520}\right) 530 \left[(3.41)^{0.97(0.25/1.25)} - 1\right]$$

$$= 61\,\text{hp/MMcfd}$$

Second stage:

$$Hp_{MM} = \frac{1.25}{0.25} \left(3.027 \times \frac{14.7}{520}\right) 530 \left[(3.41)^{0.95(0.25/1.25)} - 1\right]$$

$$= 59\,\text{hp/MMcfd}$$

Total theoretical compression work = 61 + 59 = 120 hp/MMcfd.

Required brake horsepower is

$$Hp_b = \frac{(32)(120)}{(0.8)} = 4,800\,\text{hp}.$$

Number of moles of gas is

$$n_G = \frac{1,000,000}{378.6}(32) = 2.640 \times 10^3 (32)$$

$$= 84 \times 10^6\,\text{lb-mole/day}.$$

Gas temperature after the first stage of compression is

$$T_2 = (530)(3.41)^{0.97(0.25/1.25)} = 670\,°R = 210\,°F.$$

The average cooler temperature is $\frac{210 + 70}{2} = 140\,°F$.

$\overline{C}_p$ at 140 °F and 341 psia $= 9.5 \frac{btu}{lb-mol\cdot°F}$.

Intercooler load $= 2.640 \times 10^3 (32)(9.5)(210 - 70)$

$$= 55.67 \times 10^6\,btu/day.$$

Final gas temperature:

$$T_d = (530)(3.41)^{0.95(0.25/1.25)} = 669\,°R = 209\,°F$$

It can be shown that the results obtained using the analytical expressions compare very well to those obtained from the Mollier diagram.

The computer program *ReciprocatingCompressorPower.xls* can be used for computing power requirement of each stage of compression. The solution given by the program for the first stage of compression in this example problem is shown in Table 13.2.

13.4.3.2 Centrifugal Compressors
Although the adiabatic compression process can be assumed in centrifugal compression, polytropic compression process is commonly considered as the basis for comparing centrifugal compressor performance. The process is expressed as

Table 13.2 Result Given by Computer Program ReciprocatingCompressorPower.xls for the First-Stage Compression

ReciprocatingCompressorPower.xls
Description: This spreadsheet calculates stage power of reciprocating compressor.
Instruction: (1) Update parameter valves in the "Input data" in blue; (2) click "Solution" button; (3) view result in the Solution section.

Input data
Gas flow rate (q_g):	32 MMscf/day
Stage inlet temperature (T_1):	70 °F
Stage inlet pressure (p_1):	100 psia
Gas-specific gravity (γ_g):	0.65(air = 1)
Stage outlet pressure (p_2):	341 psia
Gas-specific heat ratio (k):	1.25
Overall efficiency (E_o):	0.8
Base temperature (T_b):	60 °F
Base pressure (p_b):	14.7 psia

Solution

z = Hall–Yarborogh Method $= 0.9574$

$r = \dfrac{p_2}{p_1}$ $= 3.41$

$Hp_{MM} = \dfrac{k}{k-1} \dfrac{3.027 p_b}{T_b} T_1 \left[\left(\dfrac{p_2}{p_1}\right)^{z_1(k-1)/k} - 1 \right]$ $= 60$ hp

$Hp_b = \dfrac{q_{MM} Hp_{MM}}{E_o}$ $= 2{,}401$ hp

$T_2 = T_1 \left(\dfrac{p_2}{p_1}\right)^{z_1(k-1)/k}$ $= 210.33\,°F$

$T_{avg} = \dfrac{T_1 + T_2}{2}$ $= 140.16\,°F$

$\bar{C}_p$ $= 9.50$ btu/lbm-mol °F

Cooler load $= 2.640 \times 10^3 q_{MM} \bar{C}_p (T_{avg} - T_1)$ $= 56{,}319{,}606$ btu/day

---

$$pV^n = \text{constant}, \tag{13.30}$$

where n denotes the polytropic exponent. The isentropic exponent k applies to the ideal frictionless adiabatic process, while the polytropic exponent n applies to the actual process with heat transfer and friction. The n is related to k through polytropic efficiency E_p:

$$\dfrac{n-1}{n} = \dfrac{k-1}{k} \times \dfrac{1}{E_p} \tag{13.31}$$

The polytropic efficiency of centrifugal compressors is nearly proportional to the logarithm of gas flow rate in the range of efficiency between 0.7 and 0.75. The polytropic efficiency chart presented by Rollins (1973) can be represented by the following correlation (Guo and Ghalambor, 2005):

$$E_p = 0.61 + 0.03 \log(q_1), \tag{13.32}$$

where q_1 = gas capacity at the inlet condition, cfm.

There is a lower limit of gas flow rate below which severe gas surge occurs in the compressor. This limit is called *surge limit*. The upper limit of gas flow rate is called *stone-wall limit*, which is controlled by compressor horsepower.

The procedure of preliminary calculations for selection of centrifugal compressors is summarized as follows:

1. Calculate compression ratio based on the inlet and discharge pressures:

$$r = \dfrac{p_2}{p_1} \tag{13.33}$$

2. Based on the required gas flow rate under standard condition (q), estimate the gas capacity at inlet condition (q_1) by ideal gas law:

$$q_1 = \dfrac{p_b}{p_1} \dfrac{T_1}{T_b} q \tag{13.34}$$

3. Find a value for the polytropic efficiency E_p from the manufacturer's manual based on q_1.

4. Calculate polytropic ratio $(n-1)/n$:

$$R_p = \dfrac{n-1}{n} = \dfrac{k-1}{k} \times \dfrac{1}{E_p} \tag{13.35}$$

5. Calculate discharge temperature by

$$T_2 = T_1 \, r^{R_p}. \tag{13.36}$$

6. Estimate gas compressibility factor values at inlet and discharge conditions.

7. Calculate gas capacity at the inlet condition (q_1) by real gas law:

$$q_1 = \dfrac{z_1 p_b}{z_2 p_1} \dfrac{T_1}{T_b} q \tag{13.37}$$

8. Repeat steps 2 through 7 until the value of q_1 converges within an acceptable deviation.

9. Calculate gas horsepower by

$$Hp_g = \dfrac{q_1 p_1}{229 E_p} \left(\dfrac{z_1 + z_2}{2 z_1}\right) \left(\dfrac{r^{R_p} - 1}{R_p}\right). \tag{13.38}$$

Some manufacturers present compressor specifications using polytropic head in lb$_f$-ft/lb$_m$ defined as

$$H_g = RT_1 \left(\dfrac{z_1 + z_2}{2}\right) \left(\dfrac{r^{R_p} - 1}{R_p}\right), \tag{13.39}$$

where R is the gas constant given by $1{,}544/MW_a$ in psia-ft^3/lb$_m$-°R. The polytropic head relates to the gas horsepower by

$$Hp_g = \dfrac{M_F H_g}{33{,}000 E_p}, \tag{13.40}$$

where M_F is mass flow rate in lb$_m$/min.

10. Calculate gas horsepower by

$$Hp_b = Hp_g + \Delta Hp_m, \tag{13.41}$$

where ΔHp_m is mechanical power losses, which is usually taken as 20 horsepower for bearing and 30 horsepower for seals.

The proceeding equations have been coded in the computer program *CnetriComp.xls* (on the CD attached to this book) for quick calculation.

Example Problem 13.4 Assuming two centrifugal compressors in series are used to compress gas for a gas lift operation. Size the first compressor using the formation given in Example Problem 13.3.

Solution Calculate compression ratio based on the inlet and discharge pressures:

$$r = \sqrt{\frac{1,165}{100}} = 3.41$$

Calculate gas flow rate in scfm:

$$q = \frac{32,000,000}{(24)(60)} = 22,222 \text{ scfm}$$

Based on the required gas flow rate under standard condition (q), estimate the gas capacity at inlet condition (q_1) by ideal gas law:

$$q_1 = \frac{(14.7)}{(250)} \frac{(560)}{(520)} (22,222) = 3,329 \text{ cfm}$$

Find a value for the polytropic efficiency based on q_1:

$$E_p = 0.61 + 0.03 \log(3,329) = 0.719$$

Calculate polytropic ratio $(n-1)/n$:

$$R_p = \frac{1.25 - 1}{1.25} \times \frac{1}{0.719} = 0.278$$

Calculate discharge temperature by

$$T_2 = (530)(3.41)^{0.278} = 745\,°R = 285\,°F.$$

Estimate gas compressibility factor values at inlet and discharge conditions:

$z_1 = 1.09$ at 100 psia and 70 °F
$z_2 = 0.99$ at 341 psia and 590 °F

Calculate gas capacity at the inlet condition (q_1) by real gas law:

$$q_1 = \frac{(1.09)(14.7)}{(0.99)(100)} \frac{(530)}{(520)} (22,222) = 3,674 \text{ cfm}$$

Use the new value of q_1 to calculate E_p:

$$E_p = 0.61 + 0.03 \log(3,674) = 0.721$$

Calculate the new polytropic ratio $(n-1)/n$:

$$R_p = \frac{1.25 - 1}{1.25} \times \frac{1}{0.721} = 0.277$$

Calculate the new discharge temperature:

$$T_2 = (530)(3.41)^{0.277} = 746\,°R = 286\,°F$$

Estimate the new gas compressibility factor value:

$z_2 = 0.99$ at 341 psia and 286 °F

Because z_2 did not change, q_1 remains the same value of 3,674 cfm.

Calculate gas horsepower:

$$Hp_g = \frac{(3,674)(100)}{(229)(0.721)} \left(\frac{1.09 + 0.99}{2(1.09)}\right) \left(\frac{3.41^{0.277} - 1}{0.277}\right)$$

$$= 3,100 \text{ hp}$$

Calculate gas apparent molecular weight:

$$MW_a = (0.65)(29) = 18.85$$

Calculated gas constant:

$$R = \frac{1,544}{18.85} = 81.91 \text{ psia-ft}^3/\text{lb}_m\text{-°R}$$

Calculate polytropic head:

$$H_g = (81.91)(530) \left(\frac{1.09 + 0.99}{2}\right) \left(\frac{3.41^{0.277} - 1}{0.277}\right)$$

$$= 65,850 \text{ lb}_f\text{-ft/lb}_m$$

Calculate gas horsepower:

$$Hp_b = 3,100 + 50 = 3,150 \text{ hp}$$

The computer program *CentrifugalCompressorPower.xls* can be used for solving similar problems. The solution given by the program to this example problem is shown in Table 13.3.

13.5 Selection of Gas Lift Valves

Kickoff of a dead well requires a much higher gas pressure than the ultimate operating pressure. Because of the kickoff problem, gas lift valves have been developed and are run as part of the overall tubing string. These valves permit the introduction of gas (which is usually injected down the annulus) into the fluid column in tubing at intermediate depths to unload the well and initiate well flow. Proper design of these valve depths to unsure unloading requires a thorough understanding of the unloading process and valve characteristics.

13.5.1 Unloading Sequence

Figure 13.7 shows a well unloading process. Usually all valves are open at the initial condition, as depicted in Fig. 13.7a, due to high tubing pressures. The fluid in tubing has a pressure gradient G_s of static liquid column. When the gas enters the first (top) valve as shown in Fig. 13.7b, it creates a slug of liquid–gas mixture of less-density in the tubing above the valve depth. Expansion of the slug pushes the liquid column above it to flow to the surface. It can also cause the liquid in the bottom hole to flow back to reservoir if no check valve is installed at the end of the tubing string. However, as the length of the light slug grows due to gas injection, the bottom-hole pressure will eventually decrease to below reservoir pressure, which causes inflow of reservoir fluid. When the tubing pressure at the depth of the first valve is low enough, the first valve should begin to close and the gas should be forced to the second valve as shown in Fig. 13.7c. Gas injection to the second valve will gasify the liquid in the tubing between the first and the second valve. This will further reduce bottom-hole pressure and cause more inflow. By the time the slug reaches the depth of the first valve, the first valve should be closed, allowing more gas to be injected to the second valve. The same process should occur until the gas enters the main valve (Fig. 13.7d). The main valve (sometimes called the *master valve* or *operating valve*) is usually the lower most valve in the tubing string. It is an orifice type of valve that never closes. In continuous gas lift operations, once the well is fully unloaded and a steady-state flow is established, the main valve is the only valve open and in operation (Fig. 13.7e).

13.5.2 Valve Characteristics

Equations (13.12) and (13.16) describing choke flow are also applicable to the main valve of orifice type. Flow characteristics of this type of valve are depicted in Fig. 13.8. Under sonic flow conditions, the gas passage is independent of tubing pressure but not casing pressure.

Table 13.3 *Result Given by the Computer Program* CentrifugalCompressorPower.xls

CentrifugalCompressorPower.xls
Description: This spreadsheet calculates stage power of reciprocating compressor.
Instruction: (1) Update parameter valves in the "Input data" in blue; (2) click "Solution" button; (3) view result in the Solution section.

Input data

Gas flow rate (q_g):	32 MMscf/day
Inlet temperature (T_1):	70 °F
Inlet pressure (p_1):	100 psia
Gas-specific gravity (γ_g):	0.65 (air = 1)
Discharge pressure (p_2):	341 psia
Gas-specific heat ratio (k):	1.25
Base temperature (T_b):	60 °F
Base pressure (p_b):	14.7 psia

Solution

$r = \dfrac{p_2}{p_1}$	$= 3.41$
$q = \dfrac{q_{MM}}{(24)(60)}$	$= 22{,}222$ scfm
$q_1 = \dfrac{p_b}{p_1}\dfrac{T_1}{T_b} q$	$= 3{,}329$ scfm
$E_p = 0.61 + 0.03 \log(q_1)$	$= 0.7192$
$R_p = \dfrac{n-1}{n} = \dfrac{k-1}{k} \times \dfrac{1}{E_p}$	$= 0.2781$
$T_2 = T_1 r^{R_p}$	$= 285$ °F
z_1 by Hall–Yarborogh Method	$= 1.0891$
z_2 by Hall–Yarborogh Method	$= 0.9869$
$q_1 = \dfrac{z_1 p_b}{z_2 p_1}\dfrac{T_1}{T_b} q$	$= 3{,}674$
$E_p = 0.61 + 0.03 \log(q_1)$	$= 0.7205$
$R_p = \dfrac{n-1}{n} = \dfrac{k-1}{k} \times \dfrac{1}{E_p}$	$= 0.2776$
$T_2 = T_1 r^{R_p}$	$= 285$ °F
$Hp_g = \dfrac{q_1 p_1}{229 E_p}\left(\dfrac{z_1 + z_2}{2 z_1}\right)\left(\dfrac{r^{R_p}-1}{R_p}\right)$	$= 3{,}102$ hp
$Hp_b = Hp_g + 50$	$= 3{,}152$ hp
$MW_a = 29 \gamma_g$	$= 18.85$
$R = \dfrac{1{,}544}{MW_a}$	$= 81.91$
$H_g = RT_1 \left(\dfrac{z_1+z_2}{2}\right)\left(\dfrac{r^{R_p}-1}{R_p}\right)$	$= 65{,}853$ lbf-ft/lbm

There are different types of unloading valves, namely casing pressure-operated valve (usually called a *pressure valve*), throttling pressure valve (also called a *proportional valve* or *continuous flow valve*), fluid-operated valve (also called a *fluid valve*), and combination valve (also called a *fluid open-pressure closed valve*). Different gas lift design methods have been developed and used in the oil industry for applications of these valves.

13.5.2.1 Pressure Valve

Pressure valves are further classified as unbalanced bellow valves, balanced pressure valves, and pilot valves. Tubing pressure affects the opening action of the unbalanced valves, but it does not affect the opening or closing of balanced valves. Pilot valves were developed for intermittent gas lift with large ports.

13.5.2.1.1 Unbalanced Bellow Valve
As shown in Fig. 13.9, an unbalanced bellow valve has a pressure-charged nitrogen dome and an optional spring loading element. While the forces from the dome pressure and spring act to cause closing of the valve, the forces due to casing and tubing pressures act to cause opening of the valve. Detailed discussions of valve mechanics can be found in Brown (1980). When a valve is at its closed condition (as shown in Fig. 13.9), the minimum casing pressure required to open the valve is called the *valve opening pressure* and is expressed as

$$P_{vo} = \dfrac{1}{1-R} P_d + S_t - \dfrac{R}{1-R} P_t, \quad (13.42)$$

where

P_{vo} = valve opening pressure, psig
P_d = pressure in the dome, psig
S_t = equivalent pressure caused by spring tension, psig
P_t = tubing pressure at valve depth when the valve opens, psi
R = area ratio A_p/A_b
A_p = valve seat area, in.^{2}
A_b = total effective bellows area, in.^{2}.

The term $\dfrac{R}{1-R} P_t$ is called *tubing effect* (T.E.) and $\dfrac{R}{1-R}$ is called *tubing effect factor* (T.E.F.). With other parameters given, Eq. (13.42) is used for determining the required dome pressure at depth, that is, $P_d = (1-R)P_{vo} - S_t + RP_t$, in valve selection.

When a valve is at its open condition (as shown in Fig. 13.10), the maximum pressure under the ball (assumed to be casing pressure) required to close the valve is called the *valve closing pressure* and is expressed as

$$P_{vc} = P_d + S_t(1-R), \quad (13.43)$$

where P_{vc} = valve closing pressure, psig.

The difference between the valve opening and closing pressures, $P_{vo} - P_{vc}$, is called *spread*. Spread can be important

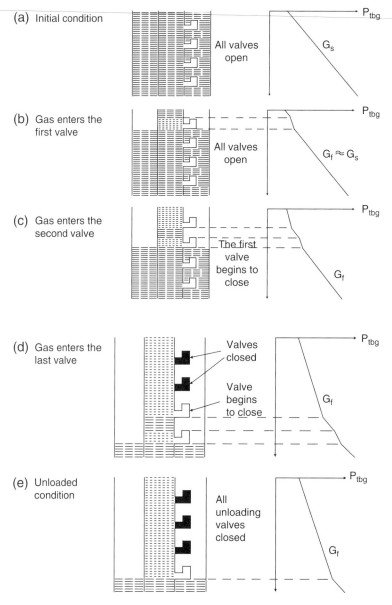

Figure 13.7 Well unloading sequence.

in continuous flow installations but is particularly important in intermittent gas lift installations where unbalanced valves are used. The spread controls the minimum amount of gas used for each cycle. As the spread increases, the amount of gas injected during the cycle increases.

Gas passage of unbalanced valves are tubing-pressure dependent due to partial travel of the valve stem. Figure 13.11 illustrates flow characteristics of unbalanced valves.

13.5.2.1.2 Balanced Pressure Valve Figure 13.12 depicts a balanced pressure valve. Tubing pressure does not influence valve status when in the closed or open condition. The valve opens and closes at the same pressure—dome pressure. Balanced pressure valves act as expanding orifice regulators, opening to pass any amount of gas injected from the surface and partial closing to control the lower gas flow rate.

13.5.2.1.3 Pilot Valve Figure 13.13 shows a sketch of a pilot valve used for intermittent gas lift where a large port

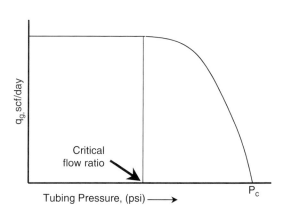

Figure 13.8 Flow characteristics of orifice-type valves.

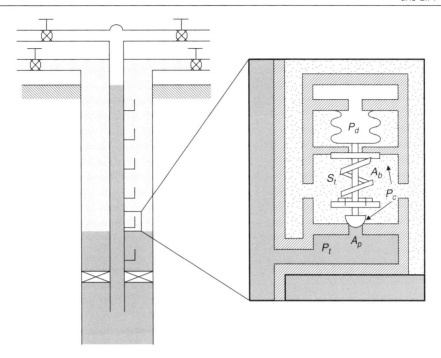

Figure 13.9 Unbalanced bellow valve at its closed condition.

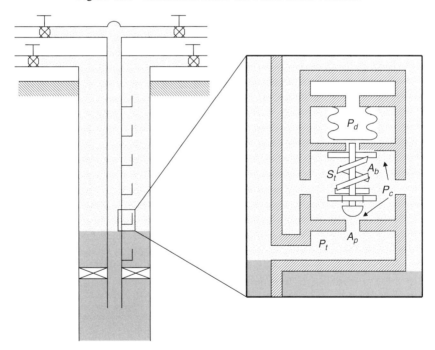

Figure 13.10 Unbalanced bellow valve at its open condition.

for gas passage and a close control over the spread characteristics are desirable. It has two ports. The smaller port (control port) is used for opening calculations and the large port (power port) is used for gas passage calculations. The equations derived from unbalanced valves are also valid for pilot valves.

13.5.2.2 Throttling Pressure Valve

Throttling pressure valves are also called *continuous flow valves*. As shown in Fig. 13.14, the basic elements of a throttling valve are the same as the pressure-operated valve except that the entrance port of the valve is choked to drop the casing pressure to tubing pressure by using a tapered stem or seat, which allows the port area to sense tubing pressure when the valve is open. Unlike pressure-operated valves where the casing pressure must drop to a pressure set by dome pressure and spring for the valve to close, a throttling pressure valve will close on a reduction in tubing pressure with the casing pressure held constant. The equations derived from pressure-operated

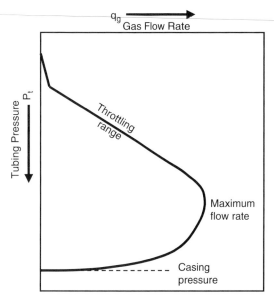

Figure 13.11 Flow characteristics of unbalanced valves.

valves are also to be applied to throttling valves for opening pressure calculations.

13.5.2.3 Fluid-Operated Valve
As shown in Fig. 13.15, the basic elements of a fluid-operated valve are identical to those in a pressure-operated valve except that tubing pressure now acts on the larger area of the bellows and casing pressure acts on the area of the port. This configuration makes the valve mostly sensitive to the tubing fluid pressure. Therefore, the opening pressure is defined as the tubing pressure required to open the valve under actual operating conditions. Force balance gives

$$P_{vo} = \frac{1}{1-R}P_d + S_t - \frac{R}{1-R}P_c, \quad (13.44)$$

where P_c = casing pressure, psig.

The term $\frac{R}{1-R}P_c$ is called the C.E. and $\frac{R}{1-R}$ is called T.E.F. for fluid valves. With other parameters given, Eq. (13.44) is used for determining required dome pressure at depth, that is, $P_d = (1-R)P_{vo} - S_t + RP_c$, in valve selection.

When a fluid valve is in its open position under operating conditions, the maximum pressure under the ball (assumed to be tubing pressure) required to close the valve is called the *valve closing pressure* and is expressed as

$$P_{vc} = P_d + S_t(1-R), \quad (13.45)$$

which is identical to that for a pressure-operated valve.

The first generation of fluid valves is a differential valve. As illustrated in Fig. 13.16, a differential valve relies on the difference between the casing pressure and the spring pressure effect to open and close. The opening and closing pressures are the same tubing pressure defined as

$$P_{vo} = P_{vc} = P_c - S_t. \quad (13.46)$$

13.5.2.4 Combination Valves
Figure 13.17 shows that a combination valve consists of two portions. The upper portion is essentially the same as that found in pressure-operated valves, and the lower portion is a fluid pilot, or a differential pressure device incorporating a stem and a spring. Holes in the pilot housing allow the casing pressure to act on the area of the stem at the upper end. The spring acts to hold the stem

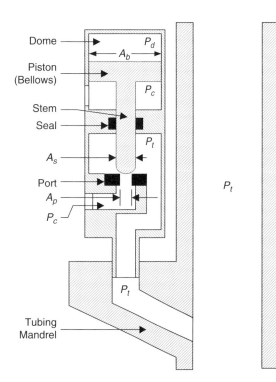

Figure 13.12 A sketch of a balanced pressure valve.

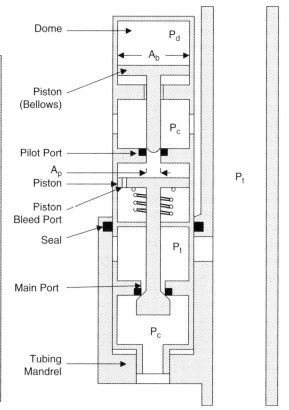

Figure 13.13 A sketch of a pilot valve.

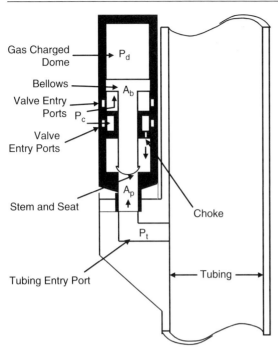

Figure 13.14 A sketch of a throttling pressure valve.

in the upward position. This is the open position for the pilot. The casing pressure acts to move the stem to the closed position. The fluid pilot will only open when tubing pressure acting on the pilot area is sufficient to overcome the casing pressure force and move the stem up to the open position. At the instant of opening, the pilot opens completely, providing instantaneous operation for intermittent lift.

13.5.3 Valve Spacing

Various methods are being used in the industry for designing depths of valves of different types. They are the universal design method, the API-recommended method, the fallback method, and the percent load method. However, the basic objective should be the same:

1. To be able to open unloading valves with kickoff and injection operating pressures
2. To ensure single-point injection during unloading and normal operating conditions
3. To inject gas as deep as possible

No matter which method is used, the following principles apply:

- The design tubing pressure at valve depth is between gas injection pressure (loaded condition) and the minimum tubing pressure (fully unloaded condition).
- Depth of the first valve is designed on the basis of kickoff pressure from a special compressor for well kickoff operations.
- Depths of other valves are designed on the basis of injection operating pressure.
- Kickoff casing pressure margin, injection operating casing pressure margin, and tubing transfer pressure margin are used to consider the following effects:
  ○ Pressure drop across the valve
  ○ Tubing pressure effect of the upper valve
  ○ Nonlinearity of the tubing flow gradient curve.

The universal design method explained in this section is valid for all types of continuous-flow gas lift valves. Still, different procedures are used with the universal design method, including the following:

a. Design procedure using constant surface opening pressure for pressure-operated valves.
b. Design procedure using 10- to 20-psi drop in surface closing pressures between valves for pressure-operated valves.

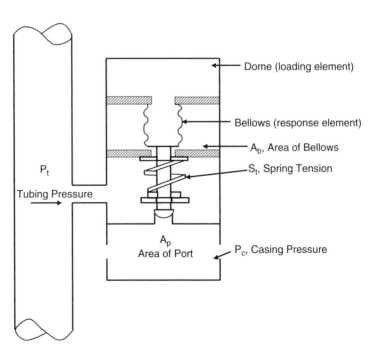

Figure 13.15 A sketch of a fluid-operated valve.

13/198 ARTIFICIAL LIFT METHODS

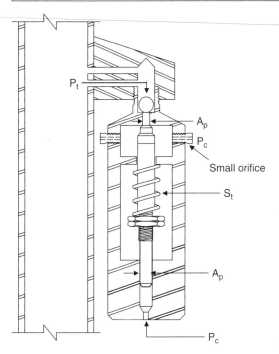

Figure 13.16 A sketch of a differential valve.

c. Design procedure for fluid-operated valves.
d. Design procedure for combination of pressure-closed fluid-opened values.

Detailed descriptions of these procedures are given by Brown (1980). Only the design procedure using constant surface opening pressure for pressure-operated valves is illustrated in this section.

Figure 13.18 illustrates a graphical solution procedure of valve spacing using constant surface opening pressure for pressure-operated valves. The arrows in the figure depict the sequence of line drawing.

For a continuous-flow gas lift, the analytical solution procedure is outlined as follows:

1. Starting from a desired wellhead pressure p_{hf} at surface, compute a flowing tubing-pressure traverse under fully unloaded condition. This can be done using various two-phase flow correlations such as the modified Hagedorn–Brown correlation (*HagedornBrownCorrelation.xls*).
2. Starting from a design wellhead pressure $p_{hf,d} = p_{hf} + \Delta p_{hf,d}$ at surface, where Δp_{hf} can be taken as $0.25 p_{c,s}$, establish a design tubing line meeting the flowing tubing-pressure traverse at tubing shoe. Pressures in this line, denoted by p_{td}, represent tubing pressure after adjustment for tubing pressure margin. Gradient of this line is denoted by G_{fd}. Set $\Delta p_{hf} = 0$ if tubing pressure margin is not required.
3. Starting from a desired injection operating pressure p_c at surface, compute a injection operating pressure line. This can be done using Eq. (13.7) or Eq. (13.9).
4. Starting from $p_{cs} - \Delta p_{cm}$ at surface, where the casing pressure margin Δp_{cm} can be taken as 50 psi, establish a design casing line parallel to the injection operating pressure line. Pressures in this line, denoted by p_{cd}, represent injection pressure after adjustment for casing pressure margin. Set $\Delta p_{cm} = 0$ if the casing pressure margin is not required as in the case of using the universal design method.

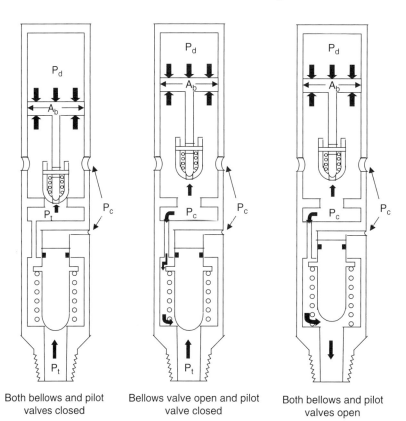

Figure 13.17 A sketch of combination valve.

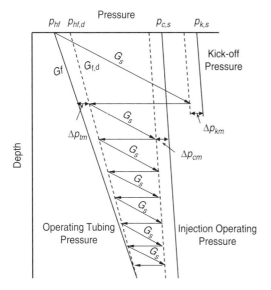

Figure 13.18 A flow diagram to illustrate procedure of valve spacing.

5. Starting from available kickoff surface pressure $p_{k,s}$, establish kickoff casing pressure line. This can be done using Eq. (13.7) or Eq. (13.9).
6. Starting from $p_k - \Delta p_{km}$ at surface, where the kickoff pressure margin Δp_{km} can be taken as 50 psi, establish a design kickoff line parallel to the kickoff casing pressure line. Pressures in this line, denoted p_{kd}, represent kickoff pressure after adjustment for kickoff pressure margin. Set $\Delta p_{km} = 0$ if kickoff casing pressure margin is not required.
7. Calculate depth of the first valve. Based on the fact that $p_{hf} + G_s D_1 = p_{kd1}$, the depth of the top valve is expressed as

$$D_1 = \frac{p_{kd1} - p_{hf}}{G_s}, \quad (13.47)$$

where

p_{kd1} = kickoff pressure opposite the first valve (psia)
G_s = static (dead liquid) gradient; psi/ft

Applying Eq. (13.9) gives

$$p_{kd1} = (p_{k,s} - \Delta p_{km})\left(1 + \frac{D_1}{40,000}\right). \quad (13.48)$$

Solving Eqs. (13.47) and (13.48) yields

$$D_1 = \frac{p_{k,s} - \Delta p_{km} - p_{hf}}{G_s - \frac{p_k - \Delta p_{km}}{40,000}}. \quad (13.49)$$

When the static liquid level is below the depth calculated by use of Eq. (13.49), the first valve is placed at a depth slightly deeper than the static level. If the static liquid level is known, then

$$D_1 = D_s + S_1, \quad (13.50)$$

where D_s is the static level and S_1 is the submergence of the valve below the static level.

8. Calculate the depths to other valves. Based on the fact that $p_{hf} + G_{fd} D_2 + G_s(D_2 - D_1) = p_{cd2}$, the depth of valve 2 is expressed as

$$D_2 = \frac{p_{cd2} - G_{fd} D_1 - p_{hf}}{G_s} + D_1, \quad (13.51)$$

where

p_{cd2} = design injection pressure at valve 2, psig
G_{fd} = design unloading gradient, psi/ft

Applying Eq. (13.9) gives

$$p_{cd2} = (p_{c,s} - \Delta p_{cm})\left(1 + \frac{D_2}{40,000}\right). \quad (13.52)$$

Solving Eqs. (13.51) and (13.52) yields

$$D_2 = \frac{p_{c,s} - \Delta p_{cm} - p_{hf,d} + (G_s - G_{fd})D_1}{G_s - \frac{p_c - \Delta p_{cm}}{40,000}}. \quad (13.53)$$

Similarly, the depth to the third valve is

$$D_3 = \frac{p_{c,s} - \Delta p_{cm} - p_{hf,d} + (G_s - G_{fd})D_2}{G_s - \frac{p_c - \Delta p_{cm}}{40,000}}. \quad (13.54)$$

Thus, a general equation for depth of valve i is

$$D_i = \frac{p_{c,s} - \Delta p_{cm} - p_{hf,d} + (G_s - G_{fd})D_{i-1}}{G_s - \frac{p_c - \Delta p_{cm}}{40,000}}. \quad (13.55)$$

Depths of all valves can be calculated in a similar manner until the minimum valve spacing (~ 400 ft) is reached.

Example Problem 13.5 Only 1 MMscf/day of lift gas is available for the well described in the Example Problem 13.1. If 1,000 psia is available to kick off the well and then a steady injection pressure of 800 psia is maintained for gas lift operation against a wellhead pressure of 130 psia, design locations of unloading and operating valves. Assume a casing pressure margin of 50 psi.

Solution The hydrostatic pressure of well fluid (26 °API oil) is (0.39 psi/ft) (5,200 ft), or 2,028 psig, which is greater than the given reservoir pressure of 2,000 psia. Therefore, the well does not flow naturally. The static liquid level depth is estimated to be

$$5,200 - (2,000 - 14.7)/(0.39) = 110 \text{ ft}.$$

Depth of the top valve is calculated with Eq. (13.49):

$$D_1 = \frac{1,000 - 50 - 130}{0.39 - \frac{1,000 - 50}{40,000}} = 2,245 \text{ ft} > 110 \text{ ft}$$

Tubing pressure margin at surface is (0.25)(800), or 200 psi. The modified Hagedorn–Brown correlation gives tubing pressure of 591 psia at depth of 5,000 ft. The design tubing flowing gradient is $G_{fd} = [591 - (130 + 200)]/(5,000)$ or 0.052 psi/ft. Depth of the second valve is calculated with Eq. (13.53):

$$D_2 = \frac{1,000 - 50 - 330 + (0.39 - 0.052)(2,245)}{0.39 - \frac{1,000 - 50}{40,000}} = 3,004 \text{ ft}$$

Similarly,

$$D_3 = \frac{1,000 - 50 - 330 + (0.39 - 0.052)(3,004)}{0.39 - \frac{1,000 - 50}{40,000}} = 3,676 \text{ ft}$$

$$D_4 = \frac{1,000 - 50 - 330 + (0.39 - 0.052)(3,676)}{0.39 - \frac{1,000 - 50}{40,000}} = 4,269 \text{ ft}$$

$$D_5 = \frac{1,000 - 50 - 330 + (0.39 - 0.052)(4,269)}{0.39 - \frac{1,000 - 50}{40,000}} = 4,792 \text{ ft},$$

which is the depth of the operating valve.

Similar problems can be quickly solved with the computer spreadsheet *GasLiftValveSpacing.xls*.

13.5.4 Valve selection and testing
Valve selection starts from sizing of valves to determine required proper port size A_p and area ratio R. Valve testing sets dome pressure P_d and/or string load S_t. Both of the processes are valve-type dependent.

13.5.4.1 Valve Sizing
Gas lift valves are sized on the basis of required gas passage through the valve. All the equations presented in Section 13.4.2.3 for choke flow are applicable to valve port area calculations. Unloading and operating valves (orifices) are sized on the basis of subcritical (subsonic flow) that occurs when the pressure ratio P_t/P_c is greater than the critical pressure ratio defined in the right-hand side of Eq. (13.11). The value of the k is about 1.28 for natural gas. Thus, the critical pressure ratio is about 0.55. Rearranging Eq. (13.12) gives

$$A_p = \frac{q_{gM}}{1{,}248 C p_{up} \sqrt{\frac{k}{(k-1)\gamma_g T_{up}} \left[\left(\frac{p_{dn}}{p_{up}}\right)^{\frac{2}{k}} - \left(\frac{p_{dn}}{p_{up}}\right)^{\frac{k+1}{k}} \right]}}. \quad (13.56)$$

Since the flow coefficient C is port-diameter dependent, a trial-and-error method is required to get a solution. A conservative C value is 0.6 for orifice-type valve ports. Once the required port area is determined, the port diameter can then be calculated by $d_p = 1.1284\sqrt{A_p}$ and up-rounded off to the nearest $\frac{1}{16}$ in.

The values of the port area to bellows area ratio R are fixed for given valve sizes and port diameters by valve manufacturers. Table 13.4 presents R values for Otis Spreadmaster Valves.

Example Problem 13.6 Size port for the data given below:

Upstream pressure:	900 psia
Downstream pressure for subsonic flow:	600 psia
Tubing ID:	2.259 in.
Gas rate:	2,500 Mscf/day
Gas-specific gravity:	0.75 (1 for air)
Gas-specific heat ratio:	1.3
Upstream temperature:	110 °F
Gas viscosity:	0.02 cp
Choke discharge coefficient:	0.6
Use Otis Spreadmaster Valve	

Solution

$$A_p = \frac{2{,}500}{1{,}248(0.6)(900)\sqrt{\frac{1.3}{(1.3-1)(0.75)(110+460)} \left[\left(\frac{600}{900}\right)^{\frac{2}{1.3}} - \left(\frac{600}{900}\right)^{\frac{1.3+1}{1.3}} \right]}}$$

$A_p = 0.1684$ in.^{2}

$d_p = 1.1284\sqrt{1.684} = 0.4631$ in.

Table 13.1 shows that an Otis $1\frac{1}{2}$-in. outside diameter (OD) valve with $\frac{1}{2}$-in. diameter seat will meet the requirement. It has an R value of 0.2562.

13.5.4.2 Valve Testing
Before sending to field for installation, every gas lift valve should be set and tested at an opening pressure in the shop that corresponds to the desired opening pressure in the well. The pressure is called *test rack opening pressure* (P_{tro}). The test is run with zero tubing pressure for pressure-operated valves and zero casing pressure for fluid-operated valves at a standard temperature (60 °F in the U.S. petroleum industry). For pressure-operated unbalanced bellow valves at zero tubing pressure, Eq. (13.42) becomes

$$P_{tro} = \frac{P_d \text{ at } 60\,°F}{1 - R} + S_t. \quad (13.57)$$

For fluid-operated valves at zero casing pressure, Eq. (13.44) also reduces to Eq. (13.57) at zero casing pressure and 60 °F.

To set P_d at 60 °F to a value representing P_d at valve depth condition, real gas law must be used for correction:

$$P_d \text{ at } 60\,°F = \frac{520 z_{60°F} P_d}{T_d z_d}, \quad (13.58)$$

where
T_d = temperature at valve depth, °R
z_d = gas compressibility factor at valve depth condition.

The z factors in Eq. (13.58) can be determined using the Hall–Yarborogh correlation. Computer spreadsheet *Hall-Yarborogh-z.xls* is for this purpose.

Table 13.4 R Values for Otis Spreadmaster Valves

Port Diameter (in.)	$\frac{9}{16}$-in. OD Valves			1-in. OD Valves			$1\frac{1}{2}$-in. OD Valves		
	R	1 − R	T.E.F.	R	1 − R	T.E.F.	R	1 − R	T.E.F.
($\frac{1}{8}$) 0.1250	0.1016	0.8984	0.1130	0.0383	0.9617	0.0398			
0.1520	0.1508	0.8429	0.1775						
0.1730	0.1958	0.8042	0.2434						
($\frac{3}{16}$) 0.1875				0.0863	0.9137	0.0945	0.0359	0.9641	0.0372
0.1960	0.2508	0.7492	0.3347						
($\frac{13}{64}$) 0.2031				0.1013	0.8987	0.1127			
0.2130	0.2966	0.7034	0.4216						
0.2460	0.3958	0.6042	0.6550						
($\frac{1}{4}$) 0.2500				0.1534	0.8466	0.1812	0.0638	0.9362	0.0681
($\frac{9}{32}$) 0.2812				0.1942	0.8058	0.2410			
($\frac{5}{16}$) 0.3125				0.2397	0.7603	0.3153	0.0996	0.9004	0.1106
($1\frac{1}{32}$) 0.3437				0.2900	0.7100	0.4085			
($\frac{3}{8}$) 0.3750				0.3450	0.6550	0.5267	0.1434	0.8566	0.1674
($\frac{7}{16}$) 0.4375				0.4697	0.5303	0.8857	0.1952	0.8048	0.2425
($\frac{1}{2}$) 0.5000							0.2562	0.7438	0.3444
($\frac{9}{16}$) 0.5625							0.3227	0.6773	0.4765
($\frac{5}{8}$) 0.6250							0.3984	0.6016	0.6622
($\frac{3}{4}$) 0.7500							0.5738	0.4262	1.3463

Equation (13.57) indicates that the P_{tro} also depends on the optional string load S_t for double-element valves. The S_t value can be determined on the basis of manufacturer's literature.

The procedure for setting and testing valves in a shop is as follows:

- Install valve in test rack.
- Adjust spring setting until the valve opens with S_t psig applied pressure. This sets S_t value in the valve.
- Pressure up the dome with nitrogen gas. Cool valve to 60 °F.
- Bleed pressure off of dome until valve opens with P_{tro} psig applied pressure.

Example Problem 13.7 Design gas lift valves using the following data:

Pay zone depth:	6,500 ft
Casing size and weight:	7 in., 23 lb.
Tubing 2⅜ in., 4.7 lb. (1.995 in. ID)	
Liquid level surface:	
Kill fluid gradient:	0.4 psi/ft
Gas gravity:	0.75
Bottom-hole temperature:	170 °F
Temperature surface flowing:	100 °F
Injection depth:	6,300 ft
Minimum tubing pressure at injection point:	600 psi
Pressure kickoff:	1,000 psi
Pressure surface operating:	900 psi
Pressure of wellhead:	120 psi
Tubing pressure margin at surface:	200 psi
Casing pressure margin:	0 psi

Valve specifications given by Example Problem 13.6

Solution Design tubing pressure at surface ($p_{hf,d}$):

$$120 + 200 = 320 \text{ psia}$$

Design tubing pressure gradient (G_{fd}):

$$(600 - 320)/6,300 = 0.044 \text{ psi/ft}$$

Temperature gradient (G_t):

$$(170 - 100)/6,300 = 0.011 \text{ F/ft}$$

$1 - R$ $1.0 - 0.2562 = 0.7438$

T.E.F. = $R/(1 - R)$ $0.2562/0.7438 = 0.3444$

Depth of the top valve is calculated with Eq. (13.49):

$$D_1 = \frac{1,000 - 0 - 120}{0.40 - \frac{1,000 - 0}{40,000}} = 2,347 \text{ ft}$$

Temperature at the top valve: $100 + (0.011)(2,347) = 126 °F$

Design tubing pressure at the top valve: $320 + (0.044)(2,347) = 424 \text{ psia}$

For constant surface opening pressure of 900 psia, the valve opening pressure is calculated with Eq. (13.9):

$$p_{vo1} = (900)\left(1 + \frac{2,347}{40,000}\right) = 953 \text{ psia}$$

The dome pressure at the valve depth is calculated on the basis of Eq. (13.42):

$$P_d = 0.7438(953) - 0 + (0.2562)(424) = 817 \text{ psia}$$

The valve closing pressure at the valve depth is calculated with Eq. (13.43):

$$P_{vc} = 817 + (0)(0.7438) = 817 \text{ psia}$$

The dome pressure at 60 °F can be calculated with a trial-and-error method. The first estimate is given by idea gas law:

$$P_d \text{ at } 60° \text{ F} = \frac{520 P_d}{T_d} = \frac{(520)(817)}{(126 + 460)} = 725 \text{ psia}$$

Spreadsheet programs give $z_{60F} = 0.80$ at 725 psia and 60 °F. The same spreadsheet gives $z_d = 0.85$ at 817 psia and 126 °F. Then Eq. (13.58) gives

$$P_d \text{ at } 60°\text{F} = \frac{(520)(0.80)P_d}{(126 + 460)(0.85)}(817) = 683 \text{ psia.}$$

Test rack opening pressure is given by Eq. (13.57) as

$$P_{tro} = \frac{683}{0.7438} + 0 = 918 \text{ psia.}$$

Following the same procedure, parameters for other valves are calculated. The results are summarized in Table 13.5.

The spreadsheet program *GasLiftValveDesign.xls* can be used to seek solutions of similar problems.

13.6 Special Issues in Intermittent-Flow Gas Lift

The intermittent-flow mechanism is very different from that of the continuous-flow gas lift. It is normally applicable in either high-BHP–low PI or low-BHP–low PI reservoirs. In these two reservoir cases, an excessive high drawdown is needed, which results in a prohibitively high GLR to produce the desired quantity of oil (liquid) by continuous gas lift. In many instances, the reservoir simply is not capable of giving up the desired liquid regardless of drawdown.

The flow from a well using intermittent gas lift techniques is called "ballistic" or "slug" flow. Two major factors that define the intermittent-gas lift process must be understood:

1. Complex flowing gradient of the gas lifted liquids from the well.
2. Contribution of the PI of the well to the actual deliverability of liquid to the surface.

Figure 13.19 shows the BHP of a well being produced by intermittent-flow gas lift.

The BHP at the instant the valve opens is indicated by Point A. The pressure impulse results in an instantaneous pressure buildup at Point B, which reaches a maximum at C after the initial acceleration of the oil column.

Figure 13.20 shows the intermittent-flowing gradient, which is a summation of the gradient of gas above the slug, the gradient of the slug, and the gradient of the lift gas and entrained liquids below the slug.

Table 13.5 Summary of Results for Example Problem 13.7

Valve no.	Valve depth (ft)	Temperature (°F)	Design tubing pressure (psia)	Surface opening pressure (psia)	Valve opening pressure (psia)	Dome pressure at depth (psia)	Valve closing pressure (psia)	Dome pressure at 60 °F (psia)	Test rack opening (psia)
1	2,347	126	424	900	953	817	817	683	918
2	3,747	142	487	900	984	857	857	707	950
3	5,065	156	545	900	1,014	894	894	702	944
4	6,300	170	600	900	1,042	929	929	708	952

13/202 ARTIFICIAL LIFT METHODS

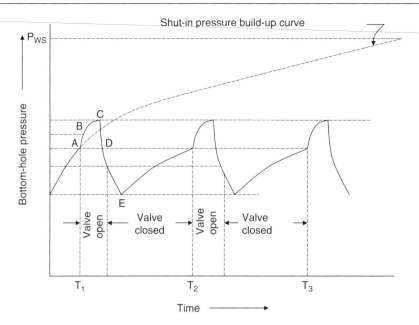

Figure 13.19 Illustrative plot of BHP of an intermittent flow.

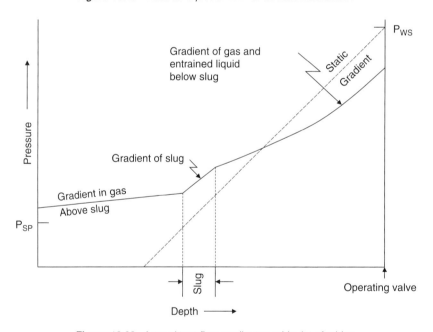

Figure 13.20 Intermittent flow gradient at midpoint of tubing.

Example Problem 13.8 Determine the depth to the operating (master) valve and the minimum GLR ratio for the following well data:

Depth = 8,000 ft
$p_{so} = 800$ psig
$2\frac{3}{8}$ -in. tubing = 1.995 in. ID
$5\frac{1}{2}$ -in., 20 lb/ft casing
No water production
$\gamma_o = 0.8762$, 30 °API
BHP (SI) = 2,000 psig
PI = 0.10 bbl/day/psi
$p_{tf} = 50$ psig
$t_{av} = 127$ °F
Cycle time: 45 minutes
Desired production: 100 bbl/day
$\gamma_g = 0.80$

Solution The static gradient is

$$G_s = 0.8762(0.433) = 0.379 \text{ psi/ft}.$$

Thus, the average flowing BHP is

$$P_{bhfave} = 2,000 - 1,000 = 1,000 \text{ psig}.$$

The depth to the static fluid level with the $p_{tf} = 50$ psig, is

$$D_s = 8,000 - \left(\frac{2,000 - 50}{0.379}\right) = 2,855 \text{ ft}.$$

The hydrostatic head after a 1,000 psi drawdown is

$$D_{dds} = \frac{1,000}{0.379} = 2,639 \text{ ft}.$$

Thus, the depth to the working fluid level is

$$WFL = D_s + D_{dds} = 2,855 + 2,639 = 5,494 \text{ ft}.$$

Figure 13.21 shows the example well and the WFL.

The number of cycles per day is approximately $\frac{24(60)}{45} = 32$ cycles/day.

The number of bbls per cycle is $\frac{100}{32} \approx 3$ bbls/cycle.

Intermittent-gas lift operating experience shows that depending on depth, 30–60% of the total liquid slug is lost due to slippage or fallback.

If a 40% loss of starting slug is assumed, the volume of the starting slug is $\frac{3}{0.60} \approx 5.0$ bbl/cycle.

Because the capacity of our tubing is 0.00387 bbl/ft, the length of the starting slug is $\frac{5.0}{0.00387} \approx 1{,}292$ ft.

This means that the operating valve should be located $\frac{1{,}292}{2} = 646$ ft below the working fluid level. Therefore, the depth to the operating valve is $5{,}494 + 646 = 6{,}140$ ft.

The pressure in the tubing opposite the operating valve with the 50 psig surface back-pressure (neglecting the weight of the gas column) is

$$p_t = 50 + (1{,}292)(0.379) = 540 \text{ psig}.$$

For minimum slippage and fallback, a minimum velocity of the slug up the tubing should be 100 ft/min. This is accomplished by having the pressure in the casing opposite the operating valve at the instant the valve opens to be at least 50% greater than the tubing pressure with a minimum differential of 200 psi. Therefore, for a tubing pressure at the valve depth of 540 psig, at the instant the valve opens, the minimum casing pressure at 6,140 ft is

$$p_{\min c} = 540 + 540/2 = 810 \text{ psig}.$$

Equation (13.10) gives a $p_{so} = 707$ psig.

The minimum volume of gas required to lift the slug to the surface will be that required to fill the tubing from injection depth to surface, less the volume occupied by the slug. Thus, this volume is $(6{,}140 + 1{,}292)\,0.00387 = 18.8$ bbls, which converts to 105.5 ft^3.

The approximate pressure in the tubing immediately under a liquid slug at the instant the slug surfaces is equal to the pressure due to the slug length plus the tubing backpressure. This is

$$p_{ts} = 50 + \left[\frac{3.0}{0.00387}\right](0.379) = 344 \text{ psig}.$$

Thus, the average pressure in the tubing is

$$p_{tave} = \frac{810 + 344}{2} = 577 \text{ psig} = 591.7 \text{ psia}.$$

The average temperature in the tubing is 127 °F or 587 °R. This gives $z = 0.886$. The volume of gas at standard conditions (API 60 °F, 14.695 psia) is

$$V_{sc} = 105.5 \left(\frac{591.7}{14.695}\right)\left(\frac{520}{587}\right)\frac{1}{0.886} = 4{,}246 \text{ scf/cycle}.$$

13.7 Design of Gas Lift Installations

Different types of gas lift installations are used in the industry depending on well conditions. They fall into four categories: (1) open installation, (2) semiclosed installation, (3) closed installation, and (4) chamber installation.

As shown in Fig. 13.22a, no packer is set in open installations. This type of installation is suitable for continuous flow gas lift in wells with good fluid seal. Although this type of installation is simple, it exposes all gas lift valves beneath the point of gas injection to severe fluid erosion due to the dynamic changing of liquid level in the annulus. Open installation is not recommended unless setting packer is not an option.

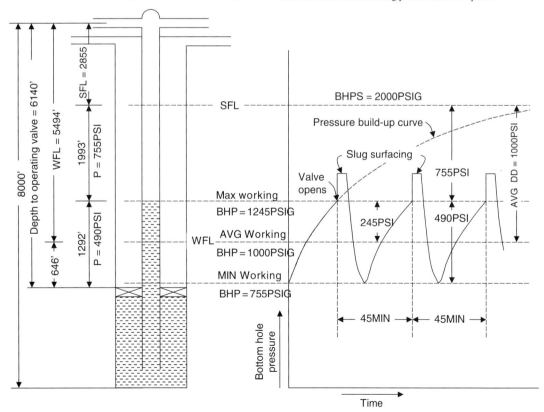

Figure 13.21 Example Problem 13.8 schematic and BHP buildup for slug flow.

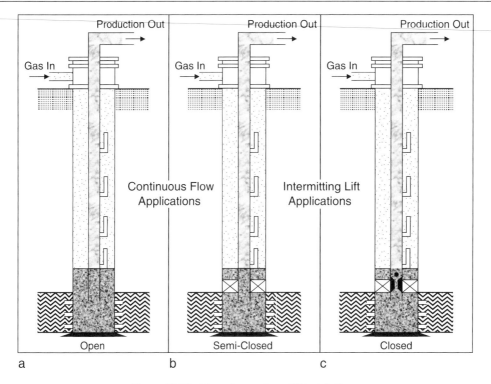

Figure 13.22 Three types of gas lift installations.

Figure 13.22b demonstrates a semiclosed installation. It is identical to the open installation except that a packer is set between the tubing and casing. This type of installation can be used for both continuous- and intermittent-flow gas lift operations. It avoids all the problems associated with the open installations. However, it still does not prevent flow of well fluids back to formation during unloading processes, which is especially important for intermittent operating.

Illustrated in Fig. 13.22c is a closed installation where a standing valve is placed in the tubing string or below the bottom gas lift valve. The standing valve effectively prevents the gas pressure from acting on the formation, which increases the daily production rate from a well of the intermittent type.

Chamber installations are used for accumulating liquid volume at bottom hole of intermittent-flow gas lift wells. A chamber is an ideal installation for a low BHP and high PI well. The chambers can be configured in various ways including using two packers, insert chamber, and reverse flow chamber. Figure 13.23 shows a standard two-packer chamber. This type of chamber is installed to ensure a large storage volume of liquids with a minimum amount of backpressure on the formation so that the liquid production rate is not hindered.

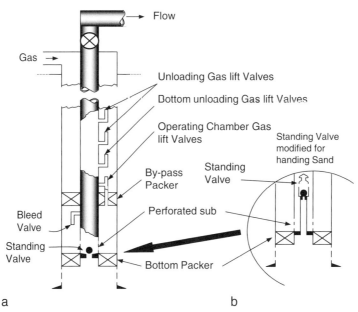

Figure 13.23 Sketch of a standard two-packer chamber.

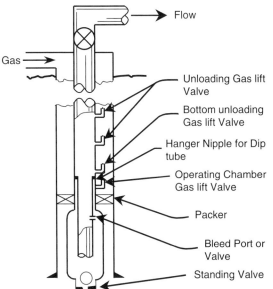

Figure 13.24 A sketch of an insert chamber.

Figure 13.24 illustrates an insert chamber. It is normally used in a long open hole or perforated interval where squeezing of fluids back to formation by gas pressure is a concern. It takes the advantage of existing bottom-hole pressure. The disadvantage of the installation is that the chamber size is limited by casing diameter.

Shown in Fig. 13.25 is a reverse flow chamber. It ensures venting of all formation gas into the tubing string to empty the chamber for liquid accumulation. For wells with high-formation GLR, this option appears to be an excellent choice.

Summary

This chapter presents the principles of gas lift systems and illustrates a procedure for designing gas lift operations. Major tasks include calculations of well deliverability, pressure and horsepower requirements for gas lift gas compression, gas lift valve selection and spacing, and selection of installation methods. Optimization of existing gas lift systems is left to Chapter 18.

References

BROWN, K.E. *The Technology of Artificial Lift Methods*, Vol. 1. Tulsa, OK: PennWell Books, 1977.

BROWN, K.E. *The Technology of Artificial Lift Methods*, Vol. 2a. Tulsa, OK: Petroleum Publishing Co., 1980.

ECONOMIDES, M.J., HILL, A.D., and EHIG-ECONOMIDES, C. *Petroleum Production Systems*. New Jersey: Prentice Hall PTR, 1994.

GILBERT, W.E. Flowing and gas-lift well performance. *API Drill. Prod. Practice* 1954.

GUO, B. and GHALAMBOR, A. *Natural Gas Engineering Handbook*. Houston, TX: Gulf Publishing Co., 2005.

KATZ, D.L., CORNELL, D., KOBAYASHI, R., POETTMANN, F.H., VARY, J.A., ELENBAAS, J.R., and WEINAUG, C.F. *Handbook of Natural Gas Engineering*. New York: McGraw-Hill Publishing Company, 1959.

WEYMOUTH, T.R. Problems in Natural Gas Engineering. *Trans. ASME* 1912;34:185.

Problems

13.1 An oil well has a pay zone around the mid-perf depth of 5,200 ft. The formation oil has a gravity of 30 °API and GLR of 500 scf/stb. Water cut remains 10%. The IPR of the well is expressed as

$$q = J\lfloor \bar{p} - p_{wf} \rfloor,$$

where
$J = 0.5\,\text{stb/day/psi}$
$\bar{p} = 2,000\,\text{psia}$.

A 2-in. tubing (1.995-in. ID) can be set with a packer at 200 ft above the mid-perf. What is the maximum expected oil production rate from the well with continuous gas lift at a wellhead pressure of 200 psia if

a. unlimited amount of lift gas is available for the well?
b. only 1.2 MMscf/day of lift gas is available for the well?

13.2 An oil well has a pay zone around the mid-perf depth of 6,200 ft. The formation oil has a gravity of 30 °API and GLR of 500 scf/stb. Water cut remains 10%. The IPR of the well is expressed as

$$q = q_{max}\left[1 - 0.2\frac{p_{wf}}{\bar{p}} - 0.80.2\left(\frac{p_{wf}}{\bar{p}}\right)^2\right],$$

where
$q_{max} = 2,000\,\text{stb/day}$
$\bar{p} = 2,500\,\text{psia}$.

A $2\frac{1}{2}$-in. tubing (2.259-in. ID) can be set with a packer at 200 ft above the mid-perf. What is the maximum expected oil production rate from the well with continuous gas lift at a wellhead pressure of 150 psia if

a. unlimited amount of lift gas is available for the well?
b. only 1.0 MMscf/day of lift gas is available for the well?

13.3 An oil field has 24 oil wells defined in Problem 13.1. The gas lift gas at the central compressor station is first pumped to three injection manifolds with 6-in. ID, 2-mile lines and then distributed to the well heads with 4 in. ID, 0.5-mile lines. Given the following

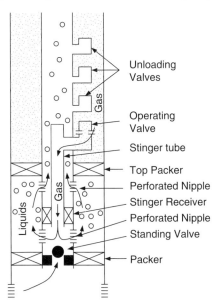

Figure 13.25 A sketch of a reserve flow chamber.

data, calculate the required output pressure of the compression station:

Gas-specific gravity (γ_g): 0.75
Base temperature (T_b): 60 °F
Base pressure (p_b): 14.7 psia.

13.4 An oil field has 32 oil wells defined in Problem 13.2. The gas lift gas at the central compressor station is first pumped to four injection manifolds with 4-in. ID, 1.5-mile lines and then distributed to the wellheads with 4-in. ID, 0.4-mile lines. Given the following data, calculate the required output pressure of compression station:

Gas-specific gravity (γ_g): 0.70
Base temperature (T_b): 60 °F
Base pressure (p_b): 14.7 psia

13.5 For a reciprocating compressor, calculate the theoretical and brake horsepower required to compress 50 MMcfd of a 0.7-gravity natural gas from 200 psia and 70 °F to 2,500 psia. If intercoolers cool the gas to 90 °F, what is the heat load on the intercoolers and what is the final gas temperature? Assuming the overall efficiency is 0.75.

13.6 For a reciprocating compressor, calculate the theoretical and brake horsepower required to compress 30 MMcfd of a 0.65-gravity natural gas from 100 psia and 70 °F to 2,000 psia. If intercoolers and endcoolers cool the gas to 90 °F, what is the heat load on the coolers? Assuming the overall efficiency is 0.80.

13.7 For a centrifugal compressor, use the following data to calculate required input horsepower and polytropic head:

Gas-specific gravity: 0.70
Gas-specific heat ratio: 1.30
Gas flow rate: 50 MMscfd at 14.7 psia and 60 °F
Inlet pressure: 200 psia
Inlet temperature: 70 °F
Discharge pressure: 500 psia
Polytropic efficiency: $E_p = 0.61 + 0.03 \log(q_1)$

13.8 For the data given in Problem 13.7, calculate the required brake horsepower if a reciprocating compressor is used.

13.9 Only 1 MMscf/day of lift gas is available for the well described in Problem 13.3. If 1,000 psia is available to kick off the well and then a steady injection pressure of 800 psia is maintained for gas lift operation against a wellhead pressure of 130 psia, design locations of unloading and operating valves. Assume a casing pressure margin of 0 psi.

13.10 An unlimited amount of lift gas is available for the well described in Problem 13.4. If 1,100 psia is available to kick off the well and then a steady injection pressure of 900 psia is maintained for gas lift operation against a wellhead pressure of 150 psia, design locations of unloading and operating valves. Assume a casing pressure margin of 50 psi.

13.11 Size port for the data given below:

Upstream pressure: 950 psia
Downstream pressure for subsonic flow: 650 psia
Tubing ID: 2.259 in.
Gas rate: 2,000 Mscf/day
Gas-specific gravity: 0.70 (1 for air)
Gas-specific heat ratio: 1.3
Upstream temperature: 100 °F
Gas viscosity: 0.02 cp
Choke discharge coefficient: 0.6
Use Otis Spreadmaster Valve

13.12 Size port for the data given below:

Upstream pressure: 950 psia
Downstream pressure for subsonic flow: 550 psia
Tubing ID: 1.995 in.
Gas rate: 1,500 Mscf/day
Gas specific gravity: 0.70 (1 for air)
Gas specific heat ratio: 1.3
Upstream temperature: 80 °F
Gas viscosity: 0.03 cp
Choke discharge coefficient: 0.6
Use Otis Spreadmaster Valve

13.13 Design gas lift valves using the following data:

Pay zone depth: 5,500 ft
Casing size and weight: 7 in., 23 lb
Tubing $2\frac{3}{8}$ in., 4.7 lb (1.995-in. ID):
Liquid level surface:
Kill fluid gradient: 0.4 psi/ft
Gas gravity: 0.65
Bottom-hole temperature: 150 °F
Temperature surface flowing: 80 °F
Injection depth: 5,300 ft
The minimum tubing pressure at injection point: 550 psi
Pressure kickoff: 950 psi
Pressure surface operating: 900 psi
Pressure of wellhead: 150 psi
Tubing pressure margin at surface: 200 psi
Casing pressure margin: 0 psi
Otis $1\frac{1}{2}$-in. OD valve with $\frac{1}{2}$-in. diameter seat: $R = 0.2562$

13.14 Design gas lift valves using the following data:

Pay zone depth: 7,500 ft
Casing size and weight: 7 in., 23 lb
Tubing $2\frac{3}{8}$-in., 4.7 lb (1.995 in. ID):
Liquid level surface:
Kill fluid gradient: 0.4 psi/ft
Gas gravity: 0.70
Bottom-hole temperature: 160 °F
Temperature surface flowing: 90 °F
Injection depth: 7,300 ft
The minimum tubing pressure at injection point: 650 psi
Pressure kickoff: 1,050 psi
Pressure surface operating: 950 psi
Pressure of wellhead: 150 psi
Tubing pressure margin at surface: 200 psi
Casing pressure margin: 10 psi
Otis 1-in. OD valve with $\frac{1}{2}$-in. diameter seat: $R = 0.1942$

13.15 Determine the gas lift gas requirement for the following well data:

Depth = 7,500 ft
p_{so} = 800 psig
$2\frac{3}{8}$-in. tubing = 1.995 in. ID
$5\frac{1}{2}$-in., 20-lb/ft casing
No water production
γ_o = 0.8762, 30 °API
BHP (SI) = 1,800 psig
PI = 0.125 bbl/day/psi
p_{tf} = 50 psig
t_{av} = 120 °F
Cycle time: 45 minutes
Desired production: 150 bbl/day
γ_g = 0.70

14

Other Artificial Lift Methods

Contents
14.1 Introduction 14/208
14.2 Electrical Submersible Pump 14/208
14.3 Hydraulic Piston Pumping 14/211
14.4 Progressive Cavity Pumping 14/213
14.5 Plunger Lift 14/215
14.6 Hydraulic Jet Pumping 14/220
Summary 14/222
References 14/222
Problems 14/223

14.1 Introduction

In addition to beam pumping and gas lift systems, other artificial lift systems are used in the oil industry. They are electrical submersible pumping, hydraulic piston pumping, hydraulic jet pumping, progressive cavity pumping, and plunger lift systems. All these systems are continuous pumping systems except the plunger lift, which is very similar to intermittent gas lift systems.

14.2 Electrical Submersible Pump

Electrical submersible pumps (ESPs) are easy to install and operate. They can lift extremely high volumes from highly productive oil reservoirs. Crooked/deviated holes present no problem. ESPs are applicable to offshore operations. Lifting costs for high volumes are generally very low. Limitations to ESP applications include high-voltage electricity availability, not applicable to multiple completions, not suitable to deep and high-temperature oil reservoirs, gas and solids production is troublesome, and costly to install and repair. ESP systems have higher horsepower, operate in hotter applications, are used in dual installations and as spare down-hole units, and include down-hole oil/water separation. Sand and gas problems have led to new products. Automation of the systems includes monitoring, analysis, and control.

The ESP is a relatively efficient artificial lift. Under certain conditions, it is even more efficient than sucker rod beam pumping. As shown in Fig. 14.1, an ESP consists of subsurface and surface components.

a. Subsurface components

- Pump
- Motor
- Seal electric cable
- Gas separator

b. Surface components

- Motor controller (or variable speed controller)
- Transformer
- Surface electric cable

The overall ESP system operates like any electric pump commonly used in other industrial applications. In ESP operations, electric energy is transported to the down-hole electric motor via the electric cables. These electric cables are run on the side of (and are attached to) the production tubing. The electric cable provides the electrical energy needed to actuate the down-hole electric motor. The electric motor drives the pump and the pump imparts energy to the fluid in the form of hydraulic power, which lifts the fluid to the surface.

14.2.1 Principle

ESPs are pumps made of dynamic pump stages or centrifugal pump stages. Figure 14.2 gives the internal schematic of a single-stage centrifugal pump. Figure 14.3 shows a cutaway of a multistage centrifugal pump.

The electric motor connects directly to the centrifugal pump module in an ESP. This means that the electric motor shaft connects directly to the pump shaft. Thus, the pump rotates at the same speed as the electric motor.

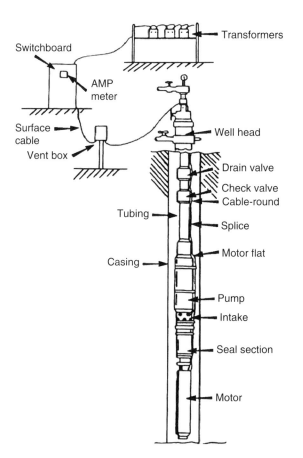

Figure 14.1 A sketch of an ESP installation (Centrilift-Hughes, Inc., 1998).

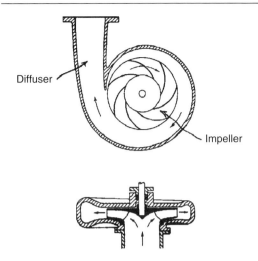

Figure 14.2 *An internal schematic of centrifugal pump.*

Like most down-hole tools in the oil field, ESPs are classified by their outside diameter (from 3.5 to 10.0 in.). The number of stages to be used in a particular outside diameter sized pump is determined by the volumetric flow rate and the lift (height) required. Thus, the length of a pump module can be 40–344 in. in length. Electric motors are three-phase (AC), squirrel cage, induction type. They can vary from 10 to 750 hp at 60 Hz or 50 Hz (and range from $3\frac{3}{4}$ to $7\frac{1}{4}$ in. in diameter). Their voltage requirements vary from 420–4,200 V.

The seal system (the protector) separates the well fluids from the electric motor lubrication fluids and the electrical wiring. The electric controller (surface) serves to energize the ESP, sensing such conditions as overload, well pump-off, short in cable, and so on. It also shuts down or starts up in response to down-hole pressure switches, tank levels, or remote commands. These controllers are available in conventional electromechanical or solid-state devices. Conventional electromechanical controllers give a fixed-speed, fixed flow rate pumping. To overcome this limitation, the variable speed controller has been developed (solid state). These controllers allow the frequency of the electric current to vary. This results in a variation in speed (rpm) and, thus, flow rate. Such a device allows changes to be made (on the fly) whenever a well changes volume (static level), pressure, GLR, or WOR. It also allows flexibility for operations in wells where the PI is not well known. The transformer (at surface) changes the voltage of the distribution system to a voltage required by the ESP system.

Unlike positive-displacement pumps, centrifugal pumps do not displace a fixed amount of fluid but create a relatively constant amount of pressure increase to the flow system. The output flow rate depends on backpressure. The pressure increase is usually expressed as pumping head, the equivalent height of freshwater that the pressure differential can support (pumps are tested with freshwater by the manufacturer). In U.S. field units, the pumping head is expressed as

$$h = \frac{\Delta p}{0.433}, \quad (14.1)$$

where
  $h =$ pumping head, ft
  $\Delta p =$ pump pressure differential, psi.

As the volumetric throughput increases, the pumping head of a centrifugal pump decreases and power slightly increases. However, there exists an optimal range of flow rate where the pump efficiency is maximal. A typical ESP characteristic chart is shown in Fig. 14.4.

ESPs can operate over a wide range of parameters (depths and volumes), to depths over 12,000 ft and volumetric flow rates of up to 45,000 bbl/day. Certain operating variables can severely limit ESP applications, including the following:

- Free gas in oil
- Temperature at depth
- Viscosity of oil
- Sand content fluid
- Paraffin content of fluid

Excessive free gas results in pump cavitation that leads to motor fluctuations that ultimately reduces run life and reliability. High temperature at depth will limit the life of the thrust bearing, the epoxy encapsulations (of electronics, etc.), insulation, and elastomers. Increased viscosity of the fluid to be pumped reduces the total head that the pump system can generate, which leads to an increased number of pump stages and increased horsepower requirements. Sand and paraffin content in the fluid will lead to wear and choking conditions inside the pump.

14.2.2 ESP Applications
The following factors are important in designing ESP applications:

- PI of the well
- Casing and tubing sizes
- Static liquid level

ESPs are usually for high PI wells. More and more ESP applications are found in offshore wells. The outside diameter of the ESP down-hole equipment is determined by the inside diameter (ID) of the borehole. There must be

Figure 14.3 *A sketch of a multistage centrifugal pump.*

clearance around the outside of the pump down-hole equipment to allow the free flow of oil/water to the pump intake. The desired flow rate and tubing size will determine the total dynamic head (TDH) requirements for the ESP system. The "TDH" is defined as the pressure head immediately above the pump (in the tubing). This is converted to feet of head (or meters of head). This TDH is usually given in water equivalent. Thus, TDH = static column of fluid (net) head + friction loss head + back-pressure head.

The following procedure can be used for selecting an ESP:

1. Starting from well inflow performance relationship (IPR), determine a desirable liquid production rate q_{Ld}. Then select a pump size from the manufacturer's specification that has a minimum delivering flow rate q_{Lp}, that is, $q_{Lp} > q_{Ld}$.
2. From the IPR, determine the flowing bottom-hole pressure p_{wf} at the pump-delivering flow rate q_{Lp}, not the q_{Ld}.
3. Assuming zero casing pressure and neglecting gas weight in the annulus, calculate the minimum pump depth by

$$D_{pump} = D - \frac{p_{wf} - p_{suction}}{0.433\gamma_L}, \quad (14.2)$$

where
D_{pump} = minimum pump depth, ft
D = depth of production interval, ft
p_{wf} = flowing bottom-hole pressure, psia
$p_{suction}$ = required suction pressure of pump, 150–300 psi
γ_L = specific gravity of production fluid, 1.0 for freshwater.

4. Determine the required pump discharge pressure based on wellhead pressure, tubing size, flow rate q_{Lp}, and fluid properties. This can be carried out quickly using the computer spreadsheet *HagedornBrownCorrelation.xls*.
5. Calculate the required pump pressure differential $\Delta p = p_{discharge} - p_{suction}$ and then required pumping head by Eq. (14.1).
6. From the manufacturer's pump characteristics curve, read pump head or head per stage. Then calculate the required number of stages.
7. Determine the total power required for the pump by multiplying the power per stage by the number of stages.

Example Problem 14.1 A 10,000-ft-deep well produces 32 °API oil with GOR 50 scf/stb and zero water cut through a 3-in. (2.992-in. ID) tubing in a 7-in. casing. The oil has a formation volume factor of 1.25 and average viscosity of 5 cp. Gas-specific gravity is 0.7. The surface and bottom-hole temperatures are 70 °F and 170 °F, respectively. The IPR of the well can be described by the Vogel model with a reservoir pressure 4,350 psia and AOF 15,000 stb/day. If the well is to be put in production with an ESP to produce liquid at 8,000 stb/day against a flowing wellhead pressure of 100 psia, determine the required specifications for an ESP for this application. Assume the minimum pump suction pressure is 200 psia.

Solution

1. Required liquid throughput at pump is

$$q_{Ld} = (1.25)(8,000) = 10,000 \text{ bbl/day}.$$

Select an ESP that delivers liquid flow rate $q_{Lp} = q_{Ld} = 10,000$ bbl/day in the neighborhood of its maximum efficiency (Fig. 14.4).

2. Well IPR gives

$$p_{wfd} = 0.125\bar{p}\left[\sqrt{81 - 80(q_{Ld}/q_{max})} - 1\right]$$
$$= 0.125(4,350)[\sqrt{81 - 80(8,000/15,000)} - 1]$$
$$= 2,823 \text{ psia}.$$

3. The minimum pump depth is

$$D_{pump} = D - \frac{p_{wf} - p_{suction}}{0.433\gamma_L}$$
$$= 10,000 - \frac{2,823 - 200}{0.433(0.865)}$$
$$= 2,997 \text{ ft}.$$

Use pump depth of $10,000 - 200 = 9,800$ ft. The pump suction pressure is

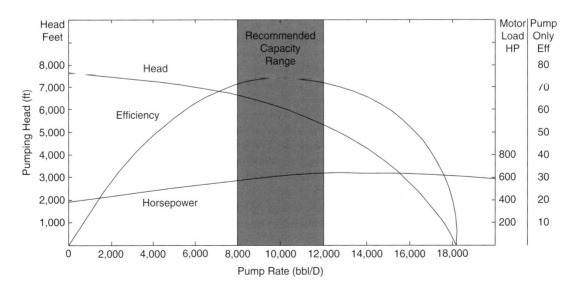

Figure 14.4 A typical characteristic chart for a 100-stage ESP.

$$p_{suction} = 2,823 - 0.433(0.865)(10,000 - 9,800)$$
$$= 2,748 \text{ psia.}$$

4. Computer spreadsheet *HagedornBrownCorrelation.xls* gives the required pump discharge pressure of 3,728 psia.
5. The required pump pressure differential is

$$\Delta p = p_{discharge} - p_{suction} = 3,728 - 2,748 = 980 \text{ psi.}$$

The required pumping head is

$$h = \frac{\Delta p}{0.433} = \frac{980}{0.433} = 2,263 \text{ feet of freshwater.}$$

6. At throughput 10,000 bbl/day, Fig. 14.4 gives a pumping head of 6,000 ft for the 100-stage pump, which yields 60 ft pumping head per stage. The required number of stages is $(2,263)/(60) = 38$ stages.
7. At throughput 10,000 bbl/day, Fig. 14.4 gives the power of the 100-stage pump of 600 hp, which yields 6 hp/stage. The required power for a 38-stage pump is then $(6)(38) = 226$ hp.

The solution given by the computer spreadsheet *ESPdesign.xls* is shown in Table 14.1.

14.3 Hydraulic Piston Pumping

Hydraulic piston pumping systems can lift large volumes of liquid from great depth by pumping wells down to fairly low pressures. Crooked holes present minimal problems. Both natural gas and electricity can be used as the power source. They are also applicable to multiple completions and offshore operations. Their major disadvantages include power oil systems being fire hazards and costly, power water treatment problems, and high solids production being troublesome.

As shown in Fig. 14.5, a hydraulic piston pump (HPP) consists of an engine with a reciprocating piston driven by a power fluid connected by a short shaft to a piston in the pump end. HPPs are usually double-acting, that is, fluid is being displaced from the pump on both the upstroke and the downstroke. The power fluid is injected down a tubing string from the surface and is either returned to the surface through another tubing (closed power fluid) or commingled with the produced fluid in the production string (open power fluid). Because the pump and engine pistons are directly connected, the volumetric flow rates in the pump and engine are related through a simple equation (Cholet, 2000):

$$q_{pump} = q_{eng} \frac{A_{pump}}{A_{eng}}, \qquad (14.3)$$

where

q_{pump} = flow rate of the produced fluid in the pump, bbl/day
q_{eng} = flow rate of the power fluid, bbl/day
A_{pump} = net cross-sectional area of pump piston, in.^{2}
A_{eng} = net cross-sectional area of engine piston, in.^{2}.

Equation (14.3) implies that liquid production rate is proportional to the power fluid injection rate. The proportionality factor A_{pump}/A_{eng} is called the "P/E ratio." By adjusting the power fluid injection rate, the liquid production rate can be proportionally changed. Although the P/E ratio magnifies production rate, a larger P/E ratio means higher injection pressure of the power fluid.

The following pressure relation can be derived from force balance in the HPP:

$$p_{eng,i} - p_{eng,d} = (p_{pump,d} - p_{pump,i})(P/E) + F_{pump}, \qquad (14.4)$$

where

$p_{eng,i}$ = pressure at engine inlet, psia
$p_{eng,d}$ = engine discharge pressure, psia
$p_{pump,d}$ = pump discharge pressure, psia

Table 14.1 Result Given by the Computer Spreadsheet ESPdesign.xls

ESPdesign.xls
Description: This spreadsheet calculates parameters for ESP selection.
Instruction: (1) Update parameter values in the Input data and Solution sections; and (2) view result in the Solution section.

Input data

Reservoir depth (D):	10,000 ft
Reservoir pressure (p_{bar}):	4,350 psia
AOF in Vogel equation for IPR (q_{max}):	15,000 stb/day
Production fluid gravity (γ_L):	0.865 1 for H_2O
Formation volume factor of production liquid (B_L):	1.25 rb/stb
Tubing inner diameter (d_{ti}):	2.992 in.
Well head pressure (p_{wh}):	100 psia
Required pump suction pressure ($p_{suction}$):	200 psia
Desired production rate (q_{Ld}):	8,000 stb/day

Solution

Desired bottom-hole pressure from IPR (p_{wfd})	= 2,823 psia
Desired production rate at pump (q_{Ld})	= 10,000 bbl/day
Input here the minimum capacity of selected pump (q_{Lp}):	10,000 bbl/day
Minimum pump setting depth (D_{pump})	= 2,997 ft
Input pump setting depth (D_{pump}):	9,800 ft
Pump suction pressure ($p_{suction}$)	= 2,748 psia
Input pump discharge pressure ($p_{discharge}$):	3,728 psia
Required pump pressure differential (Δp)	= 980 psia
Required pumping head (h)	= 2,263 ft H_2O
Input pumping head per stage of the selected pump (h_s):	60.00 ft/stage
Input horse power per stage of the selected pump (hp_s):	6.00 hp/stage
Input efficiency of the selected pump (E_p):	0.72
Required number of stages (N_s)	= 38
Total motor power requirement (hp_{motor})	= 226.35 hp

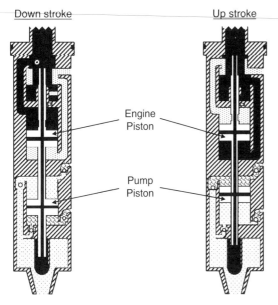

Figure 14.5 A sketch of a hydraulic piston pump.

$p_{pump,i}$ = pump intake pressure, psia
F_{pump} = pump friction-induced pressure loss, psia.

Equation (14.4) is also valid for open power fluid system where $p_{eng,d} = p_{pump,d}$.

The pump friction-induced pressure loss F_{pump} depends on pump type, pumping speed, and power fluid viscosity. Its value can be estimated with the following empirical equation:

$$F_{pump} = 50\gamma_L (0.99 + 0.01\nu_{pf})(7.1 e^{Bq_{total}})^{N/N_{max}}, \quad (14.5)$$

where

γ_L = specific gravity of production liquid, 1.0 for H_2O
ν_{pf} = viscosity of power fluid, centistokes
q_{total} = total liquid flow rate, bbl/day
N = pump speed, spm
N_{max} = maximum pump speed, spm
B = 0.000514 for $2\frac{3}{8}$-in. tubing
  = 0.000278 for $2\frac{7}{8}$-in. tubing
  = 0.000167 for $3\frac{1}{2}$-in. tubing
  = 0.000078 for $4\frac{1}{2}$-in. tubing.

The pump intake pressure $p_{pump,i}$ can be determined on the basis of well IPR and desired liquid production rate q_{Ld}. If the IPR follows Vogel's model, then for an HPP installed close to bottom hole, $p_{pump,i}$ can be estimated using

$$p_{pump,i} = 0.125\bar{p}\left[\sqrt{81 - 80(q_{Ld}/q_{max})} - 1\right] - G_b \\ \times (D - D_p), \quad (14.6)$$

where

G_b = pressure gradient below the pump, psi/ft
D = reservoir depth, ft
D_p = pump setting depth, ft.

The pump discharge pressure $p_{pump,d}$ can be calculated based on wellhead pressure and production tubing performance. The engine discharge pressure $p_{eng,d}$ can be calculated based on the flow performance of the power fluid returning tubing. With all these parameter values known, the engine inlet pressure $p_{eng,i}$ can be calculated by Eq. (14.6). Then the surface operating pressure can be estimated by

$$p_s = p_{eng,i} - p_h + p_f, \quad (14.7)$$

where

p_s = surface operating pressure, psia
p_h = hydrostatic pressure of the power fluid at pump depth, psia
p_f = frictional pressure loss in the power fluid injection tubing, psi.

The required input power can be estimated from the following equation:

$$HP = 1.7 \times 10^{-5} q_{eng} p_s \quad (14.8)$$

Selection of HPP is based on the net lift defined by

$$L_N = D_p - \frac{p_{pump,i}}{G_b} \quad (14.9)$$

and empirical value of P/E defined by

$$P/E = \frac{10,000}{L_N}. \quad (14.10)$$

The following procedure is used for selecting an HPP:

1. Starting from well IPR, determine a desirable liquid production rate q_{Ld}. Then calculate pump intake pressure with Eq. (14.6).
2. Calculate net lift with Eq. (14.9) and P/E ratio with Eq. (14.10).
3. Calculate flow rate at pump suction point by $q_{Ls} = B_o q_{Ld}$, where B_o is formation volume factor of oil. Then estimate pump efficiency E_p.
4. Select a pump rate ratio N/N_{max} between 0.2 and 0.8. Calculate the design flow rate of pump by

$$q_{pd} = \frac{q_{Ls}}{E_p(N/N_{max})}.$$

5. Based on q_{pd} and P/E values, select a pump from the manufacturer's literature and get rated displacement values q_{pump}, q_{eng}, and N_{max}. If not provided, calculate flow rates per stroke by

$$q'_{pump} = \frac{q_{pump}}{N_{max}}$$

and

$$q'_{eng} = \frac{q_{eng}}{N_{max}}.$$

6. Calculate pump speed by

$$N = \left(\frac{N}{N_{max}}\right) N_{max}.$$

7. Calculate power fluid rate by

$$q_{pf} = \left(\frac{N}{N_{max}}\right) \frac{q_{eng}}{E_{eng}}.$$

8. Determine the return production flow rate by

$$q_{total} = q_{pf} + q_{Ls}$$

for open power fluid system or

$$q_{total} = q_{Ls}$$

for closed power fluid system.

9. Calculate pump and engine discharge pressure $p_{pump,d}$ and $p_{eng,d}$ based on tubing performance.
10. Calculate pump friction-induced pressure loss using Eq. (14.5).
11. Calculate required engine pressure using Eq. (14.4).
12. Calculate pressure change Δp_{inj} from surface to engine depth in the power fluid injection tubing based on single-phase flow. It has two components:

$$\Delta p_{inj} = p_{potential} - p_{friction}$$

13. Calculate required surface operating pressure by

$$p_{so} = p_{eng,i} - \Delta p_{inj}.$$

14. Calculate required surface operating horsepower by

$$HP_{so} = 1.7 \times 10^{-5} \frac{q_{pf} p_{so}}{E_s},$$

where E_s is the efficiency of surface pump.

Example Problem 14.2 A 10,000-ft-deep well has a potential to produce 40 °API oil with GOR 150 scf/stb and 10% water cut through a 2-in. (1.995-in. ID) tubing in a 7-in. casing with a pump installation. The oil has a formation volume factor of 1.25 and average viscosity of 5 cp. Gas- and water-specific gravities are 0.7 and 1.05, respectively. The surface and bottom-hole temperatures are 80 and 180 °F, respectively. The IPR of the well can be described by Vogel's model with a reservoir pressure 2,000 psia and AOF 300 stb/day. If the well is to be put in production with an HPP at a depth of 9,700 ft in an open power fluid system to produce liquid at 200 stb/day against a flowing wellhead pressure of 75 psia, determine the required specifications for the HPP for this application. Assume the overall efficiencies of the engine, HHP, and surface pump to be 0.90, 0.80, and 0.85, respectively.

Solution This problem is solved by computer spreadsheet *HydraulicPistonPump.xls*, as shown in Table 14.2.

14.4 Progressive Cavity Pumping

The progressive cavity pump (PCP) is a positive displacement pump, using an eccentrically rotating single-helical rotor, turning inside a stator. The rotor is usually constructed of a high-strength steel rod, typically double-chrome plated. The stator is a resilient elastomer in a double-helical configuration molded inside a steel casing. A sketch of a PCP system is shown in Fig. 14.6.

Progressive cavity pumping systems can be used for lifting heavy oils at a variable flow rate. Solids and free gas production present minimal problems. They can be

Table 14.2 Solution Given by HydraulicPistonPump.xls

HydraulicPistonPump.xls
Description: This spreadsheet calculates parameters for HPP selection.
Instruction: (1) Update parameter values in the Input data and Solution sections; and (2) view result in the Solution section.

Input data

Reservoir depth (D):	10,000 ft
Reservoir pressure (p_{bar}):	2,000 psia
AOF in Vogel equation for IPR (q_{max}):	300 stb/day
Production fluid gravity (γ_L):	0.8251 1 for H_2O
Formation volume factor of production liquid (B_L):	1.25 rb/stb
Tubing inner diameter (d_{ti}):	1.995 in.
B value:	0.000514
Power fluid viscosity (v_{pf}):	1 cs
Well head pressure (p_{wh}):	100 psia
Pump setting depth (D_p):	9,700 ft
Desired production rate (q_{Ld}):	200 stb/day
HPP efficiency (E_p):	0.80
Surface pump efficiency (E_s):	0.85
Engine efficiency (E_e):	0.90
Pump speed ratio (N/N_{max}):	0.80
Power fluid flow system (1 = OPFS, 0 = CPFS):	1

Solution

Desired bottom-hole pressure from IPR (p_{wfd})	= 1,065 psia
Pump intake pressure (p_{pump})	= 958 psia
Net lift (L_N)	= 7,019 ft
Design pump to engine area ratio (P/E)	= 1.42
Flow rate at pump suction point (q_{Ls})	= 250 bbl/day
Design flow rate of pump (q_{pd})	= 391 bbl/day
Input from manufacturer's literature:	
Pump P/E:	1.13
$q_{p,max}$:	502 bbl/day
$q_{e,max}$:	572 bbl/day
N_{max}:	27
Flow rate per stroke/min in pump (q'_p)	= 18.59 bbl/day
Flow rate per stroke/min in engine (q'_e)	= 21.19 bbl/day
Pump speed (N)	= 21.60 spm
Power fluid rate (q_{pf})	= 508 bbl/day
Return production flow rate (q_{total})	= 758 bbl/day
Input pump discharge pressure by mHB correlation ($p_{pump,d}$):	2,914 psia
Input engine discharge pressure by mHB correlation ($p_{eng,d}$):	2,914 psia
Pump friction-induced pressure loss (F_{pump})	= 270 psi
Required engine pressure ($p_{eng,i}$)	= 5,395 psia
Input pressure change in the injection tubing (Δp_{inj}):	= 3,450 psi
Required surface operating pressure (p_{so})	= 1,945 psia
Required surface horsepower (HP_{so})	= 20 hp

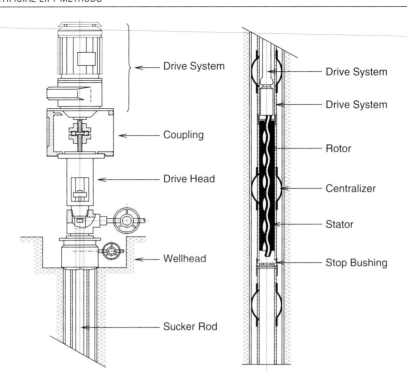

Figure 14.6 Sketch of a PCP system.

installed in deviated and horizontal wells. With its ability to move large volumes of water, the progressing cavity pump is also used for coal bed methane, dewatering, and water source wells. The PCP reduces overall operating costs by increasing operating efficiency while reducing energy requirements. The major disadvantages of PCPs include short operating life (2–5 years) and high cost.

14.4.1 Down-Hole PCP Characteristics
Proper selection of a PCP requires knowledge of PCP geometry, displacement, head, and torque requirements. Figure 14.7 (Cholet, 2000) illustrates rotor and stator geometry of PCP

where

D = rotor diameter, in.
E = rotor/stator eccentricity, in.
P_r = pitch length of rotor, ft
P_s = pitch length of stator, ft.

Two numbers define the geometry of the PCP: the number of lobes of rotor and the number of lobes of the stator. A pump with a single helical rotor and double helical stator is described as a "1-2 pump" where $P_s = 2P_r$. For a multilobe pump,

$$P_s = \frac{L_r + 1}{L_r} P_r, \quad (14.11)$$

where L_r is the number of rotor lobes. The ratio P_r/P_s is called the "kinematics ratio."

Pump displacement is defined by the fluid volume produced in one revolution of the rotor:

$$V_0 = 0.028 D E P_s, \quad (14.12)$$

where V_0 = pump displacement, ft³.

Pump flow rate is expressed as

$$Q_c = 7.12 D E P_s N - Q_s, \quad (14.13)$$

where

Q_c = pump flow rate, bbl/day

N = rotary speed, rpm
Q_s = leak rate, bbl/day.

The PCP head rating is defined by

$$\Delta P = (2n_p - 1)\delta p, \quad (14.14)$$

where

ΔP = pump head rating, psi
n_p = number of pitches of stator
δp = head rating developed into an elementary cavity, psi.

PCP mechanical resistant torque is expressed as

$$T_m = \frac{144 V_0 \Delta P}{e_p}, \quad (14.15)$$

where
T_m = mechanical resistant torque, lb$_f$-ft
e_p = efficiency.

The load on thrust bearing through the drive string is expressed as

$$F_b = \frac{\pi}{4}(2E + D)^2 \Delta P, \quad (14.16)$$

where F_b = axial load, lb$_f$.

14.4.2 Selection of Down-Hole PCP
The following procedure can be used in the selection of a PCP:

1. Starting from well IPR, select a desirable liquid flow rate q_{Lp} at pump depth and the corresponding pump intake pressure below the pump p_{pi}.
2. Based on manufacturer's literature, select a PCP that can deliver liquid rate Q_{Lp}, where $Q_{Lp} > q_{Lp}$. Obtain the value of head rating for an elementary cavity δp.
3. Determine the required pump discharge pressure p_{pd} based on wellhead pressure, tubing size, flow rate Q_{Lp},

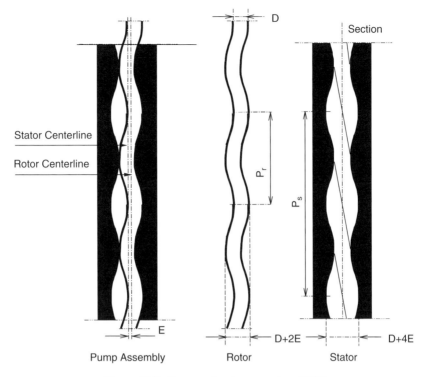

Figure 14.7 Rotor and stator geometry of PCP.

and fluid properties. This can be carried out quickly using the computer spreadsheet *HagedornBrownCorrelation.xls*.

4. Calculate required pump head by

$$\Delta P = p_{pd} - p_{pi}. \qquad (14.17)$$

5. Calculate the required number of pitches n_p using Eq. (14.14).
6. Calculate mechanical resistant torque with Eq. (14.15).
7. Calculate the load on thrust bearing with Eq. (14.16).

14.4.3 Selection of Drive String
Sucker rod strings used in beam pumping are also used in the PCP systems as drive strings. The string diameter should be properly chosen so that the tensile stress in the string times the rod cross-sectional area does not exceed the maximum allowable strength of the string. The following procedure can be used in selecting a drive string:

1. Calculate the weight of the selected rod string W_r in the effluent fluid (liquid level in annulus should be considered to adjust the effect of buoyancy).
2. Calculate the thrust generated by the head rating of the pump F_b with Eq. (14.16).
3. Calculate mechanical resistant torque T_m with Eq. (14.15).
4. Calculate the torque generated by the viscosity of the effluent in the tubing by

$$T_v = 2.4 \times 10^{-6} \mu_f L N \frac{d^3}{(D-d)} \frac{1}{\ln \frac{\mu_s}{\mu_f}} \left(\frac{\mu_s}{\mu_f} - 1 \right), \qquad (14.18)$$

where
T_v = viscosity-resistant torque, lb$_f$-ft
μ_f = viscosity of the effluent at the inlet temperature, cp
μ_s = viscosity of the effluent at the surface temperature, cp
L = depth of tubing, ft
d = drive string diameter, in.

5. Calculate total axial load to the drive string by

$$F = F_b + W_r. \qquad (14.19)$$

6. Calculate total torque by

$$T = T_m + T_v. \qquad (14.20)$$

7. Calculate the axial stress in the string by

$$\sigma_t = \frac{4}{\pi d^3} \sqrt{F^2 d^2 + 64 T^2 \times 144}, \qquad (14.21)$$

where the tensile stress σ_t is in pound per square inch. This stress value should be compared with the strength of the rod with a safety factor.

14.4.4 Selection of Surface Driver
The prime mover for PCP can be an electrical motor, hydraulic drive, or internal-combustion engine. The minimum required power from the driver depends on the total resistant torque requirement from the PCP, that is,

$$P_h = 1.92 \times 10^{-4} TN, \qquad (14.22)$$

where the hydraulic power P_h is in hp. Driver efficiency and a safety factor should be used in driver selection from manufacturer's literature.

14.5 Plunger Lift
Plunger lift systems are applicable to high gas–liquid ratio wells. They are very inexpensive installations. Plunger automatically keeps tubing clean of paraffin and scale. But they are good for low-rate wells normally less than 200 B/D. Listiak (2006) presents a thorough discussion of this technology.

Traditionally, plunger lift was used on oil wells. Recently, plunger lift has become more common on gas wells for de-watering purposes. As shown in Fig. 14.8, high-pressure gas wells produce gas carrying liquid water and/or condensate in the form of mist. As the gas flow velocity in the well drops as a result of the reservoir pressure depletion, the carrying capacity of the gas decreases. When the gas velocity drops to a critical level, liquid begins to accumulate in the well and the well flow can undergo annular flow regime followed by a slug flow regime. The accumulation of liquids (liquid loading) increases bottom-hole pressure that reduces gas production rate. Low gas production rate will cause gas velocity to drop further. Eventually the well will undergo bubbly flow regime and cease producing.

Liquid loading is not always obvious, and recognizing the liquid-loading problem is not an easy task. A thorough diagnostic analysis of well data needs to be performed. The symptoms to look for include onset of liquid slugs at the surface of well, increasing difference between the tubing and casing pressures with time, sharp changes in gradient on a flowing pressure survey, sharp drops in a production decline curve, and prediction with analytical methods.

Accurate prediction of the problem is vital for taking timely measures to solve the problem. Previous investigators have suggested several methods to predict the problem. Results from these methods often show discrepancies. Also, some of these methods are not easy to use because of the difficulties with prediction of bottom-hole pressure in multiphase flow.

Turner et al. (1969) were the pioneer investigators who analyzed and predicted the minimum gas flow rate capable of removing liquids from the gas production wells. They presented two mathematical models to describe the liquid-loading problem: the film movement model and entrained drop movement model. On the basis of analyses on field data they had, they concluded that the film movement model does not represent the controlling liquid transport mechanism.

The Turner et al. entrained drop movement model was derived on the basis of the terminal-free settling velocity of liquid drops and the maximum drop diameter corresponding to the critical Weber number of 30. According to Turner et al. (1969), gas will continuously remove liquids from the well until its velocity drops to below the terminal velocity. The minimum gas flow rate for a particular set of conditions (pressure and conduit geometry) can be calculated using a mathematical model. Turner et al. (1969) found that this entrained drop movement model gives underestimates of the minimum gas flow rates. They recommended the equation-derived values be adjusted upward by approximately 20% to ensure removal of all drops. Turner et al. (1969) believed that the discrepancy was attributed to several facts including the use of drag coefficients for solid spheres, the assumption of stagnation velocity, and the critical Weber number established for drops falling in air, not in compressed gas.

The main problem that hinders the application of the Turner et al. entrained drop model to gas wells comes from the difficulties of estimating the values of gas density and pressure. Using an average value of gas-specific gravity (0.6) and gas temperature (120 °F), Turner et al. derived an expression for gas density as 0.0031 times the pressure. However, they did not present a method for calculating the gas pressure in a multiphase flow wellbore.

Starting from the Turner et al. entrained drop model, Guo and Ghalambor (2005) determined the minimum kinetic energy of gas that is required to lift liquids. A four-phase (gas, oil, water, and solid particles) mist-flow model was developed. Applying the minimum kinetic energy criterion to the four-phase flow model resulted in a closed-form analytical equation for predicting the minimum gas flow rate. Through case studies, Guo and Ghalambor demonstrated that this new method is more conservative and accurate. Their analysis also indicates that the controlling conditions are bottom-hole conditions where gas has higher pressure and lower kinetic energy. This analysis is consistent with the observations from air-drilling operations where solid particles accumulate at bottom-hole rather than top-hole (Guo and Ghalambor, 2002). However, this analysis contradicts the results by Turner et al. (1969), that indicated that the wellhead conditions are, in most instances, controlling.

14.5.1 Working Principle

Figure 14.9 illustrates a plunger lift system. Plunger lift uses a free piston that travels up and down in the well's tubing string. It minimizes liquid fallback and uses the well's energy more efficiently than in slug or bubble flow.

The purpose of plunger lift is like that of other artificial lift methods: to remove liquids from the wellbore so that the well can be produced at the lowest bottom-hole pressures. Whether in a gas well, oil well, or gas lift well, the mechanics of a plunger lift system are the same. The plunger, a length of steel, is dropped down the tubing to the bottom of the well and allowed to travel back to the surface. It provides a piston-like interface between liquids and gas in the wellbore and prevents liquid fallback. By providing a "seal" between the liquid and gas, a well's own energy can be used to efficiently lift liquids out of the wellbore. A plunger changes the rules for liquid removal. However, in a well without a plunger, gas velocity must be high to remove liquids. With a plunger, gas velocity can be very low. Unloading relies much more on the well's ability to store enough gas pressure to lift the plunger and a liquid slug to surface, and less on critical flow rates.

Plunger operation consists of shut-in and flow periods. The flow period is further divided into an unloading period and flow after plunger arrival. Lengths of these periods will vary depending on the application, producing capability of the well, and pressures.

A plunger cycle starts with the shut-in period that allows the plunger to drop from the surface to the bottom of the well. At the same time, the well builds gas pressure stored either in the casing, in the fracture, or in the near wellbore region of the reservoir. The well must be shut in long enough to build reservoir pressure that will provide energy to lift both the plunger and the liquid slug to the surface against line pressure and friction. When this time and

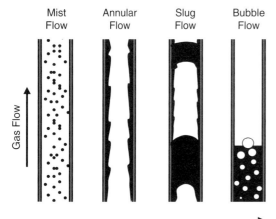

Figure 14.8 Four flow regimes commonly encountered in gas wells.

OTHER ARTIFICIAL LIFT METHODS 14/217

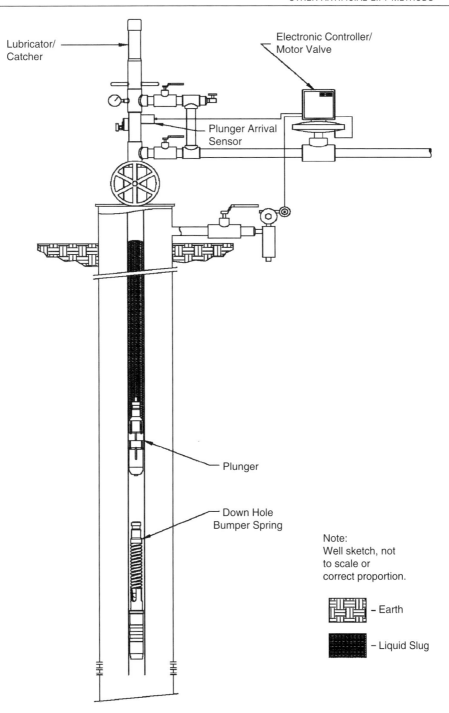

Figure 14.9 A sketch of a plunger lift system (courtesy Ferguson Beauregard).

pressure have been reached, the flow period is started and unloading begins. In the initial stages of the flow period, the plunger and liquid slug begin traveling to the surface. Gas above the plunger quickly flows from the tubing into the flowline, and the plunger and liquid slug follow up the hole. The plunger arrives at the surface, unloading the liquid. Initially, high rates prevail (often three to four times the average daily rate) while the stored pressure is blown down. The well can now produce free of liquids, while the plunger remains at the surface, held by the well's pressure and flow. As rates drop, velocities eventually drop below the critical rate, and liquids begin to accumulate in the tubing. The well is shut in and the plunger falls back to the bottom to repeat the cycle.

At the end of the shut-in period, the well has built pressure. The casing pressure is at its maximum, and the tubing pressure is lower than the casing pressure. The difference is equivalent to the hydrostatic pressure of the liquid in the tubing.

When the well is opened, the tubing pressure quickly drops down to line pressure, while the casing pressure slowly decreases until the plunger reaches the surface. As the plunger nears the surface, the liquid on top of the plunger may surge through the system, causing spikes in line pressure and flow rate. This continues until the plunger reaches the surface. After the plunger surfaces, a large increase in flow rate will produce higher tubing pressures and an increase in flowline pressure. Tubing pressure will then drop very close to line pressure. Casing pressure will reach its minimum either on plunger arrival or after, as the casing blows down and the well produces with minimal liquids in the tubing. If the well stays above the critical unloading rate, the casing pressure will remain fairly constant or may decrease further. As the gas rate drops, liquids become held up in the tubing and casing pressure will increase.

Upon shut in, the casing pressure builds more rapidly. How fast depends on the inflow performance and reservoir pressure of the well. The tubing pressure will increase quickly from line pressure, as the flowing gas friction ceases. It will eventually track casing pressure (less the liquid slug). Casing pressure will continue to increase to maximum pressure until the well is opened again.

As with most wells, maximum plunger lift production occurs when the well produces against the lowest possible bottom-hole pressure. On plunger lift, the lowest average bottom-hole pressures are almost always obtained by shutting the well in the minimum amount of time. Practical experience and plunger lift models demonstrate that lifting large liquid slugs requires higher average bottom-hole pressure. Lengthy shut-in periods also increase average bottom-hole pressure. So the goal of plunger lift should be to shut the well in the minimum amount of time and produce only enough liquids that can be lifted at this minimum buildup pressure.

What is the minimum shut-in time? The absolute minimum amount of time for shut-in is the time it takes the plunger to reach the bottom. The well must be shut-in in this length of time regardless of what other operating conditions exist. Plungers typically fall between 200 and 1,000 ft/min in dry gas and 20 and 250 ft/min in liquids. Total fall time varies and is affected by plunger type, amount of liquids in the tubing, the condition of the tubing (crimped, corkscrewed, corroded, etc.), and the deviation of the tubing or wellbore.

The flow period during and after plunger arrival is used to control liquid loads. In general, a short flow period brings in a small liquid load, and a long flow period brings in a larger liquid load. By controlling this flow time, the liquid load is controlled. So the well can be flowed until the desired liquid load has entered the tubing. A well with a high GLR may be capable of long flow periods without requiring more than minimum shut-in times. In this case, the plunger could operate as few as 1 or 2 cycles/day. Conversely, a well with a low GLR may never be able to flow after plunger arrival and may require 25 cycles/day or more. In practice, if the well is shutting in for only the minimum amount of time, it can be flowed as long as possible to maintain target plunger rise velocities. If the well is shutting in longer than the minimum shut-in time, there should be little or no flow after the plunger arrives at the surface.

14.5.2 Design Guideline

Plunger lift systems can be evaluated using rules of thumb in conjunction with historic well production or with a mathematical plunger model. Because plunger lift installations are typically inexpensive, easy to install, and easy to test, most evaluations are performed by rules of thumb.

14.5.2.1 Estimate of Production Rates with Plunger Lift
The simplest and sometimes most accurate method of determining production increases from plunger lift is from decline curve analysis. Gas and oil reservoirs typically have predictable declines, either exponential, harmonic, or hyperbolic. Initial production rates are usually high enough to produce the well above critical rates (unloaded) and establish a decline curve. When liquid loading occurs, a marked decrease and deviation from normal decline can be seen. By unloading the well with plunger lift, a normal decline can be reestablished. Production increases from plunger lift will be somewhere between the rates of the well when it started loading and the rate of an extended decline curve to the present time. Ideally, decline curves would be used in concert with critical velocity curves to predetermine when plunger lift should be installed. In this manner, plunger lift will maintain production on a steady decline and never allow the well to begin loading.

Another method to estimate production is to build an inflow performance curve based on the backpressure equation. This is especially helpful if the well has an open annulus and casing pressure is known. The casing pressure gives a good approximation of bottom-hole pressure. The IPR curve can be built based on the estimated reservoir pressure, casing pressure, and current flow rate. Because the job of plunger lift is to lower the bottom-hole pressure by removing liquids, the bottom-hole pressure can be estimated with no liquids. This new pressure can be used to estimate a production rate with lower bottom-hole pressures.

14.5.2.2 GLR and Buildup Pressure Requirements
There are two minimum requirements for plunger lift operation: minimum GLR and buildup pressure. For the plunger lift to operate, there must be available gas to provide the lifting force, in sufficient quantity per barrel of liquid for a given well depth.

14.5.2.2.1 Rules of Thumb As a rule of thumb, the minimum GLR requirement is considered to be about 400 scf/bbl/1,000 ft of well depth, that is,

$$GLR_{min} = 400 \frac{D}{1,000}, \tag{14.23}$$

where
GLR_{min} = minimum required GLR for plunger lift, scf/bbl
D = depth to plunger, ft.

Equation (14.23) is based on the energy stored in a compressed volume of 400 scf of gas expanding under the hydrostatic head of a barrel of liquid. The drawback is that no consideration is given to line pressures. Excessively high line pressures, relative to buildup pressure may increase the requirement. The rule of thumb also assumes that the gas expansion can be applied from a large open annulus without restriction. Slim-hole wells and wells with packers that require gas to travel through the reservoir or through small perforations in the tubing will cause a greater restriction and energy loss. This increases the minimum requirements to as much as 800–1,200 scf/bbl/1,000 ft.

Well buildup pressure is the second requirement for plunger operation. This buildup pressure is the bottom-hole pressure just before the plunger begins its ascent (equivalent to surface casing pressure in a well with an open annulus). In practice, the minimum shut-in pressure requirement for plunger lift is equivalent to 1½ times maximum sales line pressure. The actual requirement may be higher. The rule works well in intermediate-depth wells (2,000–8,000 ft) with slug sizes of 0.1–0.5 barrels/cycle. It breaks down for higher liquid volumes, deeper wells (due to increasing friction), and excessive pressure restrictions at the surface or in the wellbore.

An improved rule for minimum pressure is that a well can lift a slug of liquid equal to about 50–60% of the difference between shut-in casing pressure and maximum sales line pressure. This rule gives

$$p_c = p_{L\max} + \frac{p_{sh}}{f_{sl}}, \tag{14.24}$$

where

p_c = required casing pressure, psia
p_{Lmax} = maximum line pressure, psia
p_{sh} = slug hydrostatic pressure, psia
f_{sl} = slug factor, 0.5–0.6.

This rule takes liquid production into account and can be used for wells with higher liquid production that require more than 1–2 barrels/cycle. It is considered as a conservative estimate of minimum pressure requirements. To use Eq. (14.24), first the total liquid production on plunger lift and number of cycles possible per day should be estimated. Then the amount of liquid that can be lifted per cycle should be determined. That volume of liquid per cycle is converted into the slug hydrostatic pressure using the well tubing size. Finally, the equation is used to estimate required casing pressure to operate the system.

It should be noted that a well that does not meet minimum GLR and pressure requirements could still be plunger lifted with the addition of an external gas source. Design at this point becomes more a matter of the economics of providing the added gas to the well at desired pressures.

14.5.2.2.2 Analytical Method Analytical plunger lift design methods have been developed on the basis of force balance. Several studies in the literature address the addition of makeup gas to a plunger installation through either existing gas lift operations, the installation of a field gas supply system, or the use of wellhead compression. Some of the studies were presented by Beeson et al. (1955), Lebeaux and Sudduth (1955), Foss and Gaul (1965), Abercrombie (1980), Rosina (1983), Mower et al. (1985), and Lea (1981, 1999).

The forces acting on the plunger at any given point in the tubing include the following:

1. Stored casing pressure acting on the cross-section of the plunger
2. Stored reservoir pressure acting on the cross-section of the plunger
3. Weight of the fluid
4. Weight of the plunger
5. Friction of the fluid with the tubing
6. Friction of the plunger with the tubing
7. Gas friction in the tubing
8. Gas slippage upward past the plunger
9. Liquid slippage downward past the plunger
10. Surface pressure (line pressure and restrictions) acting against the plunger travel

Several publications have been written dealing with this approach. Beeson et al. (1955) first presented equations for high GLR wells based on an empirically derived analysis. Foss and Gaul (1965) derived a force balance equation for use on oil wells in the Ventura Avenue field. Mower et al. (1985) presented a dynamic analysis of plunger lift that added gas slippage and reservoir inflow and mathematically described the entire cycle (not just plunger ascent) for tight-gas/very high GLR wells.

The methodology used by Foss and Gaul (1965) was to calculate a casing pressure required to move the plunger and liquid slug just before it reached the surface, called Pc_{min}. Since Pc_{min} is at the end of the plunger cycle, the energy of the expanding gas from the casing to the tubing is at its minimum. Adjusting Pc_{min} for gas expansion from the casing to the tubing during the full plunger cycle results in the pressure required to start the plunger at the beginning of the plunger cycle, or Pc_{max}.

The equations below are essentially the same equations presented by Foss and Gaul (1956) but are summarized here as presented by Mower et al. (1985). The Foss and Gaul model is not rigorous, because it assumes constant friction associated with plunger rise velocities of 1,000 ft/min, does not calculate reservoir inflow, assumes a value for gas slippage past the plunger, assumes an open unrestricted annulus, and assumes the user can determine unloaded gas and liquid rates independently of the model. Also, this model was originally designed for oil well operation that assumed the well would be shut-in on plunger arrival, so the average casing pressure, Pc_{avg}, is only an average during plunger travel. The net result of these assumptions is an overprediction of required casing pressure. If a well meets the Foss and Gaul (1956) criteria, it is almost certainly a candidate for plunger lift.

14.5.2.3 Plunger Lift Models
14.5.2.3.1 Basic Foss and Gaul Equations (modified by Mower et al) The required minimum casing pressure is expressed as

$$Pc_{\min} = [P_p + 14.7 + P_t + (P_{lh} + P_{lf}) \times V_{slug}] \times \left(1 + \frac{D}{K}\right), \tag{14.25}$$

where

$P_{c\min}$ = required minimum casing pressure, psia
$P_p = W_p/A_t$, psia
W_p = plunger weight, lb$_f$
A_t = tubing inner cross-sectional area, in.^{2}
P_{lh} = hydrostatic liquid gradient, psi/bbl slug
P_{lf} = flowing liquid gradient, psi/bbl slug
P_t = tubing head pressure, psia
V_{slug} = slug volume, bbl
D = depth to plunger, ft
K = characteristic length for gas flow in tubing, ft.

Foss and Gaul suggested an approximation where K and $P_{lh} + P_{lf}$ are constant for a given tubing size and a plunger velocity of 1,000 ft/min:

Tubing size (in.)	K (ft)	$P_{lh} + P_{lf}$ (psi/bbl)
2⅜	33,500	165
2⅞	45,000	102
3½	57,600	63

To successfully operate the plunger, casing pressure must build to Pc_{max} given by

$$Pc_{\max} = Pc_{\min}\left(\frac{A_a + A_t}{A_a}\right). \tag{14.26}$$

The average casing pressure can then be expressed as

$$Pc_{avg} = Pc_{\min}\left(1 + \frac{A_t}{2A_a}\right), \tag{14.27}$$

where A_a is annulus cross-sectional area in squared inch.

The gas required per cycle is formulated as

$$V_g = \frac{37.14 F_{gs} Pc_{avg} V_t}{Z(T_{avg} + 460)}, \tag{14.28}$$

where

V_g = required gas per cycle, Mscf
$F_{gs} = 1 + 0.02\,(D/1{,}000)$, modified Foss and Gaul slippage factor
$V_t = A_t(D - V_{slug}L)$, gas volume in tubing, Mcf
L = tubing inner capacity, ft/bbl
Z = gas compressibility factor in average tubing condition
T_{avg} = average temperature in tubing, °F.

The maximum number of cycles can be expressed as

$$N_{C\max} = \frac{1440}{\frac{D}{V_r} + \frac{D-V_{slug}L}{V_{fg}} + \frac{V_{slug}L}{V_{fl}}}, \quad (14.29)$$

where

$N_{C\max}$ = the maximum number of cycles per day
V_{fg} = plunger falling velocity in gas, ft/min
V_{fl} = plunger falling velocity in liquid, ft/min
V_r = plunger rising velocity, ft/min.

The maximum liquid production rate can be expressed as

$$q_{L\max} = N_{C\max} V_{slug}. \quad (14.30)$$

The required GLR can be expressed as

$$GLR_{\min} = \frac{V_g}{V_{slug}}. \quad (14.31)$$

Example Problem 14.3: Plunger Lift Calculations
Calculate required GLR, casing pressure, and plunger lift operating range for the following given well data:

Gas rate:	200 Mcfd expected when unloaded
Liquid rate:	10 bbl/day expected when unloaded
Liquid gradient:	0.45 psi/ft
Tubing, ID:	1.995 in.
Tubing, OD:	2.375 in.
Casing, ID:	4.56 in.
Depth to plunger:	7,000 ft
Line pressure:	100 psi
Available casing pressure:	800 psi
Reservoir pressure:	1200 psi
Average Z factor:	0.99
Average temperature:	140 °F
Plunger weight:	10 lb
Plunger fall in gas:	750 fpm
Plunger fall in liquid:	150 fpm
Plunger rise velocity:	1,000 fpm

Solution The minimum required GLR by a rule of thumb is

$$GLR_{\min} = 400\frac{D}{1,000} = 400\frac{7,000}{1,000} = 2,800 \text{ scf/bbl}.$$

The well's GLR of 2,857 scf/bbl is above 2,800 scf/bbl and is, therefore, considered adequate for plunger lift.

The minimum required casing pressure can be estimated using two rules of thumb. The simple rule of thumb gives

$$p_c = 1.5 p_{L\max} = (1.5)(100) = 150 \text{ psi}.$$

To calculate the minimum required casing pressure with the improved rule of thumb, the slug hydrostatic pressure needs to be known. For this case, assuming 10 cycles/day, equivalent to a plunger trip every 2.4 hours, and 10 bbls of liquid, the plunger will lift 1 bbl/cycle. The hydrostatic pressure of 1 bbl of liquid in 2⅜-in. tubing with a 0.45-psi/ft liquid gradient is about 120 psi. Then

$$p_c = p_{L\max} + \frac{p_{sh}}{f_{sl}} = 100 + \frac{120}{0.5 \text{ to } 0.6} = 300 \text{ to } 340 \text{ psi}.$$

Since the well has 800 psi of available casing pressure, it meets the pressure requirements for plunger lift.

The Foss and Gaul–type method can be used to determine plunger lift operating range. Basic parameters are given in Table 14.3.

Since the Foss and Gaul–type calculations involve determination of Z-factor values in Eq. (14.28) at different pressures, a spreadsheet program *PlungerLift.xls* was developed to speed up the calculation procedure. The solution is given in Table 14.4.

It was given that the estimated production when unloaded is 200 Mcfd with 10 bbl/day of liquid (GLR = 200/10 = 20 Mscf/bbl), and the maximum casing pressure buildup is 800 psi. From the Table 14.4, find casing pressure of about 800 psi, GLR of 20 Mscf/bbl, and production rates of 10 bbl/day. This occurs at slug sizes between about 0.25 and 3 bbl. The well will operate on plunger lift.

14.6 Hydraulic Jet Pumping

Figure 14.10 shows a hydraulic jet pump installation. The pump converts the energy from the injected power fluid (water or oil) to pressure that lifts production fluids. Because there is no moving parts involved, dirty and gassy fluids present no problem to the pump. The jet pumps can be set at any depth as long as the suction pressure is sufficient to prevent pump cavitation problem. The disadvantage of hydraulic jet pumps is their low efficiency (20–30%).

14.6.1 Working Principle
Figure 14.11 illustrates the working principle of a hydraulic jet pump. It is a dynamic-displacement pump that differs from a hydraulic piston pump in the manner in which it increases the pressure of the pumped fluid with a jet nozzle. The power fluid enters the top of the pump from an injection tubing. The power fluid is then accelerated through the nozzle and mixed with the produced fluid in the throat of the pump. As the fluids mix, the momentum of the power fluid is partially transferred to the produced fluid and increases its kinetic energy (velocity head).

Table 14.3 *Summary of Calculated Parameters*

Tubing inner cross-sectional area (A_t) =	3.12 in.²
Annulus cross-sectional area (A_a) =	11.90 in.²
Plunger-weight pressure (P_p) =	3.20 psi
Slippage factor (F_{gs}) =	1.14
Tubing inner capacity (L) =	258.80 ft/bbl
The average temperature (T_{avg}) =	600 °R

Table 14.4 *Solution Given by Spreadsheet Program* PlungerLift.xls

V_{slug} (bbl)	P_{Cmin} (psia)	P_{Cmax} (psia)	P_{Cavg} (psia)	Z	V_t (Mcf)	V_g (Mscf)	N_{Cmax} (cyc/day)	q_{Lmax} (bbl/day)	GLR_{min} (Mscf/bbl)
0.05	153	193	173	0.9602	0.1516	1.92	88	4.4	38.44
0.1	162	205	184	0.9624	0.1513	2.04	87	8.7	20.39
0.25	192	243	218	0.9744	0.1505	2.37	86	21.6	9.49
0.5	242	306	274	0.9689	0.1491	2.98	85	42.3	5.95
1	342	432	387	0.9499	0.1463	4.20	81	81.3	4.20
2	541	684	613	0.9194	0.1406	6.61	75	150.8	3.31
3	741	936	838	0.8929	0.1350	8.95	70	211.0	2.98
4	940	1,187	1,064	0.8666	0.1294	11.21	66	263.6	2.80
5	1,140	1,439	1,290	0.8457	0.1238	13.32	62	309.9	2.66

OTHER ARTIFICIAL LIFT METHODS 14/221

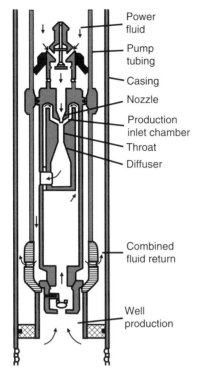

Figure 14.10 *Sketch of a hydraulic jet pump installation.*

Some of the kinetic energy of the mixed stream is converted to static pressure head in a carefully shaped diffuser section of expanding area. If the static pressure head is greater than the static column head in the annulus, the fluid mixture in the annulus is lifted to the surface.

14.6.2 Technical Parameters
The nomenclatures in Fig. 14.11 are defined as
p_1 = power fluid pressure, psia
q_1 = power fluid rate, bbl/day
p_2 = discharge pressure, psia
$q_2 = q_1 + q_3$, total fluid rate in return column, bbl/day
p_3 = intake pressure, psia
q_3 = intake (produced) fluid rate, bbl/day
A_j = jet nozzle area, in.^{2}
A_s = net throat area, in.^{2}
A_t = total throat area, in.^{2}.

The following dimensionless variables are also used in jet pump literature (Cholet, 2000):

$$R = \frac{A_j}{A_t} \qquad (14.32)$$

$$M = \frac{q_3}{q_1} \qquad (14.33)$$

$$H = \frac{p_2 - p_3}{p_1 - p_2} \qquad (14.34)$$

$$\eta = MH, \qquad (14.35)$$

where
  R = dimensionless nozzle area
  M = dimensionless flow rate
  H = dimensionless head
  η = pump efficiency.

14.6.3 Selection of Jet Pumps
Selection of jet pumps is made on the basis of manufacturer's literatures where pump performance charts are usually available. Figure 14.12 presents an example chart. It shows the effect of M on H and η. For a given jet pump specified by R value, there exists a peak efficiency η_p. It is good field practice to attempt to operate the pump at its peak efficiency. If M_p and H_p are used to denote M and H at the peak efficiency, respectively, pump parameters should be designed using

$$M_p = \frac{q_3}{q_1} \qquad (14.36)$$

and

$$H_p = \frac{p_2 - p_3}{p_1 - p_2}, \qquad (14.37)$$

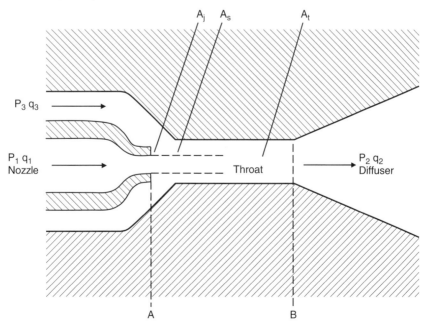

Figure 14.11 *Working principle of a hydraulic jet pump.*

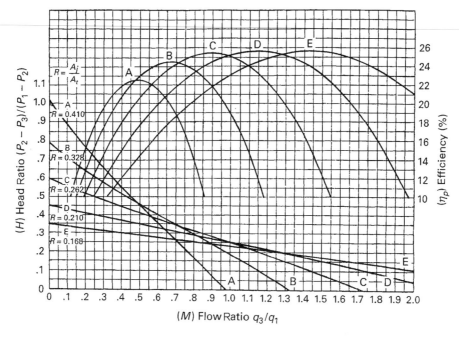

Figure 14.12 Example jet pump performance chart.

where M_p and H_p values can be determined from the given performance chart. If the H scale is not provided in the chart, H_p can be determined by

$$H_p = \frac{\eta_p}{M_p}. \tag{14.38}$$

The power fluid flow rate and pump pressure differential are related through jet nozzle size by

$$q_1 = 1214.5 A_j \sqrt{\frac{p_1 - p_3}{\gamma_1}}, \tag{14.39}$$

where γ_1 is the specific gravity of the power fluid, q_1 is in bbl/day, and p_1 and p_3 are both in psi.

The following procedure can be taken to select a jet pump:

1. Select a desired production rate of reservoir fluid q_3 based on well IPR. Determine the required bottom-hole pressure p_{wf}.
2. Design a pump setting depth D and estimate required pump intake pressure p_3 based on p_{wf} and flow gradient below the pump.
3. From manufacturer's literature, choose a pump with R value and determine M_p and H_p values for the pump based on pump performance curves.
4. Calculate power fluid rate q_1 by

$$q_1 = \frac{q_3}{M_p}.$$

5. Based on tubing flow performance, calculate the required discharge pressure $p_{2,r}$ using production rate $q_2 = q_1 + q_3$. This step can be performed with the spreadsheet program *HagedornBrownCorrelation.xls*.
6. Determine the power fluid pressure p_1 required to provide power fluid rate q_1 with Eq. (14.39), that is,

$$p_1 = p_3 + \gamma_1 \left(\frac{q_1}{1214.5 A_j}\right)^2.$$

7. Determine the available discharge pressure p_2 from the pump with Eq. (14.37), that is,

$$p_2 = \frac{p_3 + H_p p_1}{1 + H_p}.$$

8. If the p_2 value is greater than $p_{2,r}$ value with a reasonable safety factor, the chosen pump is okay to use, and go to Step 9. Otherwise, go to Step 3 to choose a different pump. If no pump meets the requirements for the desired production rate q_3 and/or lifting pressure $p_{2,r}$, go to Step 2 to change pump setting depth or reduce the value of the desired fluid production rate q_3.
9. Calculate the required surface operating pressure p_s based on the values of p_1 and q_1 and single-phase flow in tubing.
10. Calculate input power requirement by

$$HP = 1.7 \times 10^{-5} q_1 p_s,$$

where
HP = required input power, hp
p_s = required surface operating pressure, psia.

Summary

This chapter provides a brief introduction to the principles of electrical submersible pumping, hydraulic piston pumping, hydraulic jet pumping, progressive cavity pumping, and plunger lift systems. Design guidelines are also presented. Example calculations are illustrated with spreadsheet programs.

References

ABERCROMBIE, B. Plunger lift. In: *The Technology of Artificial Lift Methods* (Brown, K.E., ed.), Vol. 2b. Tulsa: PennWell Publishing Co., 1980, pp. 483–518.

BEESON, C.M., KNOX, D.G., and STODDARD, J.H. Plunger lift correlation equations and nomographs. Presented at AIME Petroleum Branch Annual meeting, 2–5 October 1955, New Orleans, Louisiana. Paper 501-G.

BROWN, K.E. *The Technology of Artificial Lift Methods*, Vol. 2b. Tulsa: PennWell Publishing Co., 1980.

Centrilift-Hughes, Inc. *Oilfield Centrilift-Hughes Submersible Pump Handbook*. Claremore, Oaklahoma, 1998.

CHOLET, H. *Well Production Practical Handbook*. Paris: Editions TECHNIP, 2000.

FOSS, D.L. and GAUL, R.B. Plunger lift performance criteria with operating experience: Ventura Avenue field. *Drilling Production Practices API* 1965:124–140.

GUO, B. and GHALAMBOR, A. *Natural Gas Engineering Handbook*. Houston, TX: Gulf Publishing Co., 2005.

GUO, B. and GHALAMBOR, A. *Gas Volume Requirements for Underbalanced Drilling Deviated Holes*. PennWell Books Tulsa, Oaklahoma, 2002.

LEA, J.F. Plunger lift vs velocity strings. Energy Sources Technology Conference & Exhibition (ETCE '99), 1–2 February 1999, Houston Sheraton Astrodome Hotel in Houston, Texas.

LEA, J.F. Dynamic analysis of plunger lift operations. Presented at the 56th Annual Fall Technical Conference and Exhibition, 5–7 October 1981, San Antonio, Texas. Paper SPE 10253.

LEBEAUX, J.M. and SUDDUTH, L.F. Theoretical and practical aspects of free piston operation. *JPT* September 1955:33–35.

LISTIAK, S.D. Plunger lift. In: *Petroleum Engineering Handbook* (Lake, L., ed.). Dallas: Society of Petroleum Engineers, 2006.

MOWER, L.N., LEA, J.F., BEAUREGARD, E., and FERGUSON, P.L. Defining the characteristics and performance of gas-lift plungers. Presented at the SPE Annual Technical Conference and Exhibition held in 22–26 September 1985, Las Vegas, Nevada. SPE Paper 14344.

ROSINA, L. A study of plunger lift dynamics [Masters Thesis], University of Tulsa, 1983.

TURNER, R.G., HUBBARD, M.G., and DUKLER, A.E. Analysis and prediction of minimum flow rate for the continuous removal of liquids from gas wells. J. *Petroleum Technol.* November 1969.

Problems

14.1 A 9,000-ft-deep well produces 26 °API oil with GOR 50 scf/stb and zero water cut through a 3-in. (2.992-in. ID) tubing in a 7-in. casing. The oil has a formation volume factor of 1.20 and average viscosity of 8 cp. Gas-specific gravity is 0.75. The surface and bottom-hole temperatures are 70 and 160 °F, respectively. The IPR of the well can be described by Vogel's model with a reservoir pressure 4,050 psia and AOF 12,000 stb/day. If the well is put in production with an ESP to produce liquid at 7,000 stb/day against a flowing well head pressure of 150 psia, determine the required specifications for an ESP for this application. Assume the minimum pump suction pressure is 220 psia.

14.2 A 9,000-ft-deep well has a potential to produce 35 °API oil with GOR 120 scf/stb and 10% water cut through a 2-in. (1.995-in. ID) tubing in a 7-in. casing with a pump installation. The oil has a formation volume factor of 1.25 and average viscosity of 5 cp. Gas- and water-specific gravities are 0.75 and 1.05, respectively. The surface and bottom-hole temperatures are 70 and 170 °F, respectively. The IPR of the well can be described by Vogel's model with a reservoir pressure 2,000 psia and AOF 400 stb/day. If the well is to put in production with a HPP at depth of 8,500 ft in an open power fluid system to produce liquid at 210 stb/day against a flowing well head pressure of 65 psia, determine the required specifications for the HPP for this application. Assume the overall efficiencies of the engine, HHP, and surface pump to be 0.90, 0.80, and 0.85, respectively.

14.3 Calculate required GLR, casing pressure, and plunger lift operating range for the following given well data:

Gas rate:	250 Mcfd expected when unloaded
Liquid rate:	12 bbl/day expected when unloaded
Liquid gradient:	0.40 psi/ft
Tubing, ID:	1.995 in.
Tubing, OD:	2.375 in.
Casing, ID:	4.56 in.
Depth to plunger:	7,000 ft
Line pressure:	120 psi
Available casing pressure:	850 psi
Reservoir pressure:	1250 psi
Average Z-factor:	0.99
Average temperature:	150 °F
Plunger weight:	10 lb
Plunger fall in gas:	750 fpm
Plunger fall in liquid:	150 fpm
Plunger rise velocity:	1,000 fpm

Part IV Production Enhancement

Good production engineers never stop looking for opportunities to improve the performance of their production systems. Performance enhancement ideas are from careful examinations and thorough analyses of production data to find the controlling factors affecting the performance. Part IV of this book presents procedures taken in the petroleum industry for identifying well problems and means of solving the problems. Materials are presented in the following four chapters.

Chapter 15:   Well Problem Identification
Chapter 16:   Matrix Acidizing
Chapter 17:   Hydraulic Fracturing
Chapter 18:   Production Optimization

15 Well Problem Identification

Contents
15.1 Introduction 15/228
15.2 Low Productivity 15/228
15.3 Excessive Gas Production 15/231
15.4 Excessive Water Production 15/231
15.5 Liquid Loading of Gas Wells 15/231
Summary 15/241
References 15/241
Problems 15/242

15.1 Introduction

The engineering work for sustaining and enhancing oil and gas production rates starts from identifying problems that cause low production rates of wells, quick decline of the desirable production fluid, or rapid increase in the undesirable fluids. For oil wells, these problems include

- Low productivity
- Excessive gas production
- Excessive water production
- Sand production

For gas wells, the problems include

- Low productivity
- Excessive water production
- Liquid loading
- Sand production

Although sand production is easy to identify, well testing and production logging are frequently needed to identify the causes of other well problems.

15.2 Low Productivity

The lower than expected productivity of oil or gas well is found on the basis of comparison of the well's actual production rate and the production rate that is predicted by Nodal analysis. If the reservoir inflow model used in the Nodal analysis is correct (which is often questionable), the lower than expected well productivity can be attributed to one or more of the following reasons:

- Overestimate of reservoir pressure
- Overestimate of reservoir permeability (absolute and relative permeabilities)
- Formation damage (mechanical and pseudo skins)
- Reservoir heterogeneity (faults, stratification, etc.)
- Completion ineffectiveness (limited entry, shallow perforations, low perforation density, etc.)
- Restrictions in wellbore (paraffin, asphaltane, scale, gas hydrates, sand, etc.)

The first five factors affect reservoir inflow performance, that is, deliverability of reservoir. They can be evaluated on the basis of pressure transient data analyses.

The true production profile from different zones can be obtained based on production logging such as temperature and spinner flow meter logs. An example is presented in Fig. 15.1, which shows that Zone A is producing less than 10% of the total flow, Zone B is producing almost 70% of the total rate, and Zone C is contributing about 25% of the total production.

The last factor controls well deliverability. It can be evaluated using data from production logging such as flowing gradient survey (FGS). The depth interval with high-pressure gradient is usually the interval where the depositions of paraffins, asphaltanes, scales, or gas hydrates are suspected.

15.2.1 Pressure Transient Data Analysis

Pressure transient testing plays a key role in evaluating exploration and development prospects. Properly designed well tests can provide reservoir engineers with reservoir pressure, reserves (minimum economic or total), and flow capacity, all of which are essential in the reservoir evaluation process. Some of the results one can obtain from pressure transient testing include the following:

- Initial reservoir pressure
- Average reservoir pressure
- Directional permeability
- Radial effective permeability changes from the wellbore
- Gas condensate fallout effect on flow

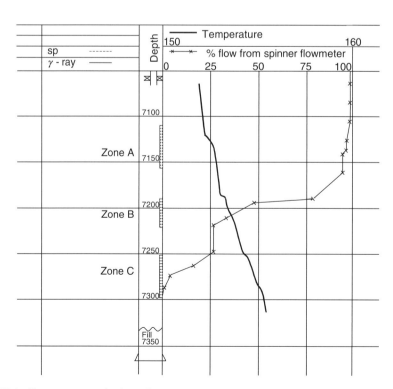

Figure 15.1 Temperature and spinner flowmeter-derived production profile (Economides et al., 1994).

- Near wellbore damage/stimulation
- Rate-dependent skin
- Boundary identification
- Partial penetration effect on flow
- Effective fracture length
- Effective fracture conductivity
- Dual-porosity characteristics (storativity and transmissivity ratios)

The theoretical basis of pressure transient data analysis is beyond the scope of this book. It can be found elsewhere (Chaudhry, 2004; Horne, 1995; Lee et al., 2003). Modern computer software packages are available for data analyses. These packages include PanSystem (EPS, 2004) and F.A.S.T. WellTest (Fekete, 2003). The following subsections briefly present some principles of data analyses that lead to deriving reservoir properties directly affecting well productivity.

Reservoir Pressure. Reservoir pressure is a key parameter controlling well deliverability. A simple way to determine the magnitude of initial reservoir pressure may be the Horner plot of data from pressure buildup test if the reservoir boundary was not reached during the test. If the boundary effects are seen, the average reservoir pressure can be estimated on the basis of the extrapolated initial reservoir pressure from Horner plot and the MBH plot (Dake, 2002).

Effective Permeability. The effective reservoir permeability that controls the well's deliverability should be derived from the flow regime that prevails in the reservoir for long-term production. To better understand the flow regimes, the commonly used equations describing flow in oil reservoirs are summarized first in this subsection. Similar equations for gas reservoirs can be found in Lee et al. (2003).

Horizontal Radial Flow. For vertical wells fully penetrating nonfractured reservoirs, the horizontal radial flow can be mathematically described in consistent units as

$$p_{wf} = p_i - \frac{qB\mu}{4\pi k_h h}\left[\ln\left(\frac{k_h t}{\phi \mu c_t r_w^2}\right) + 2S + 0.80907\right], \quad (15.1)$$

where
- p_{wf} = flowing bottom-hole pressure
- p_i = initial reservoir pressure
- q = volumetric liquid production rate
- B = formation volume factor
- μ = fluid viscosity
- k_h = the average horizontal permeability
- h = pay zone thickness
- t = flow time
- ϕ = initial reservoir pressure
- c_t = total reservoir compressibility
- r_w = wellbore radius
- S = total skin factor.

Horizontal Linear Flow. For hydraulically fractured wells, the horizontal linear flow can be mathematically described in consistent units as

$$p_{wf} = p_i - \frac{qB\mu}{2\pi k_y h}\left[\sqrt{\frac{\pi k_y t}{\phi \mu c_t x_f^2}} + S\right], \quad (15.2)$$

where x_f is fracture half-length and k_y is the permeability in the direction perpendicular to the fracture face.

Vertical Radial Flow. For horizontal wells as depicted in Fig. 15.2, the early-time vertical radial flow can be mathematically described in consistent units as

$$p_{wf} = p_i - \frac{qB\mu}{4\pi k_{yz} L}\left[\ln\left(\frac{k_{yz} t}{\phi \mu c_t r_w^2}\right) + 2S + 0.80907\right], \quad (15.3)$$

where L is the horizontal wellbore length and k_{yz} is the geometric mean of horizontal and vertical permeabilities, that is,

$$k_{yz} = \sqrt{k_y k_z}. \quad (15.4)$$

Horizontal Pseudo-Linear Flow. The pseudo-linear flow toward a horizontal wellbore can be mathematically described in consistent units as

$$p_{wf} = p_i - \frac{qB\mu}{2\pi k_y (h - Z_w)}\left[\sqrt{\frac{4\pi k_y t}{\phi \mu c_t L^2}} + S\right]. \quad (15.5)$$

Horizontal Pseudo-Radial Flow. The pseudo-radial flow toward a horizontal wellbore can be mathematically described in consistent units as

$$p_{wf} = p_i - \frac{qB\mu}{4\pi k_h h}\left[\ln\left(\frac{k_h t}{\phi \mu c_t r_w^2}\right) + 2S + 0.80907\right]. \quad (15.6)$$

For vertical wells fully penetrating nonfractured reservoirs, it is usually the average (geometric mean) of horizontal permeabilities, k_h, that dominates long-term production performance. This average horizontal permeability can be derived from the horizontal radial flow regime. For wells draining relatively small portions of hydraulically fractured reservoir segments, it is usually the permeability in the direction perpendicular to the fracture face that controls long-term production performance. This permeability can be derived from the horizontal linear flow regime. For horizontal wells draining relatively large portions of nonfractured reservoir segments, it is usually again the geometric mean of horizontal permeabilities

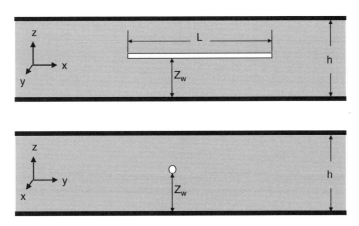

Figure 15.2 Notations for a horizontal wellbore.

that dominates long-term production performance. This average horizontal permeability can be derived from the pseudo-radial flow regime. For vertical wells partially penetrating nonfractured reservoirs, both horizontal and vertical permeabilities influence long-term production performance. These permeabilities can usually be derived from the hemispherical flow regime.

Flow regimes are usually identified using the diagnostic pressure derivative p' defined as

$$p' = \frac{d\Delta p}{d\ln(t)} = t\frac{d\Delta p}{dt}, \quad (15.7)$$

where t is time and Δp is defined as

$$\Delta p = p_i - p_{wf} \quad (15.8)$$

for drawdown tests, where p_i and p_{wf} are initial reservoir pressure and flowing bottom-hole pressure, respectively. For pressure buildup tests, the Δp is defined as

$$\Delta p = p_{sw} - p_{wfe}, \quad (15.9)$$

where p_{ws} and p_{wfe} are ship-in bottom-hole pressure and the flowing bottom-hole pressure at the end of flow (before shut-in), respectively.

For any type of radial flow (e.g., horizontal radial flow, vertical radial flow, horizontal pseudo-radial flow), the diagnostic derivative is derived from Eqs. (15.1), (15.3), and (15.6) as

$$p' = \frac{d\Delta p}{d\ln(t)} = \frac{qB\mu}{4\pi \bar{k} H_R}, \quad (15.10)$$

where $\bar{k}$ is the average permeability in the flow plane (k_h or k_{yz}) and

$$k_h = \sqrt{k_x k_y}$$

H_R is the thickness of the radial flow (h or L). Apparently, the diagnostic derivative is constant over the radial flow time regime. The plot of p' versus t data should show a trend of straight line parallel to the t-axis.

For linear flow (e.g., flow toward a hydraulic fracture), the diagnostic derivative is derived from Eq. (15.2) as

$$p' = \frac{d\Delta p}{d\ln(t)} = \frac{qB}{4hx_f}\sqrt{\frac{\mu t}{\pi \phi c_t k_y}}. \quad (15.12)$$

For pseudo-linear flow (e.g., flow toward a horizontal well), the diagnostic derivative is derived from Eq. (15.5) as

$$p' = \frac{d\Delta p}{d\ln(t)} = \frac{qB}{2L(h-z_w)}\sqrt{\frac{\mu t}{\pi \phi c_t k_y}}. \quad (15.13)$$

Taking logarithm of Eqs. (15.12) and (15.13) gives

$$\log(p') = \frac{1}{2}\log(t) + \log\left(\frac{qB}{4hx_f}\sqrt{\frac{\mu}{\pi \phi c_t k_y}}\right) \quad (15.14)$$

and

$$\log(p') = \frac{1}{2}\log(t) + \log\left(\frac{qB}{2L(h-z_w)}\sqrt{\frac{\mu}{\pi \phi c_t k_y}}\right). \quad (15.15)$$

Equations (15.13) and (15.14) indicate that the signature of the linear flow regime is the $\frac{1}{2}$ slope on the log-log plot of diagnostic derivative versus time.

Once the flow regimes are identified, permeabilities associated with the flow regime can be determined based on slope analyses. For any types of radial flow, Eqs. (15.1), (15.3), and (15.6) indicate that plotting of bottom-hole pressure versus time data on a semilog scale will show a trend with a constant slope m_R, where

$$m_R = -\frac{qB\mu}{4\pi \bar{k} H_R}. \quad (15.16)$$

Then the average permeability in the flow plane (k_h or k_{yz}) can be estimated by

$$\bar{k} = -\frac{qB\mu}{4\pi H_R m_R}. \quad (15.17)$$

For any types of linear flow, Eqs. (15.2) and (15.5) indicate that plotting of the bottom-hole pressure versus the square-root of time data will show a trend with a constant slope m_L, where

$$m_L = -\frac{qB}{H_L X_L}\sqrt{\frac{\mu}{\pi \phi c_t k_y}}, \quad (15.18)$$

where $H_L = h$ and $X_L = 2x_f$ for linear flow, and $H_L = h - Z_w$ and $X_L = L$ for pseudo-linear flow, respectively. Then the permeability in the flow plane can be estimated by

$$k_y = \frac{\mu}{\pi \phi c_t}\left(\frac{qB}{m_L H_L X_L}\right)^2. \quad (15.19)$$

If a horizontal well is tested for a time long enough to detect the pseudo-radial flow, then it is possible to estimate other directional permeabilities by

$$k_x = \frac{k_h^2}{k_y} \quad (15.20)$$

and

$$k_z = \frac{k_{yz}^2}{k_y}. \quad (15.21)$$

Although k_x and k_z are not used in well productivity analysis, they provide some insight about reservoir anisotropy.

Skin Factor. Skin factor is a constant that is used to adjust the flow equation derived from the ideal condition (homogeneous and isotropic porous media) to suit the applications in nonideal conditions. It is an empirical factor employed to consider the lumped effects of several aspects that are not considered in the theoretical basis when the flow equations were derived. The value of the skin factor can be derived from pressure transient test analysis with Eqs. (15.1), (15.2), (15.3), (15.5), and (15.6). But its value has different meanings depending on flow regime. A general expression of the skin factor is

$$S = S_D + S_{C+\theta} + S_P + \sum S_{PS}, \quad (15.22)$$

where S_D is damage skin during drilling, cementing, well completion, fluid injection, and even oil and gas production. Physically, it is due to plugging of pore space by external or internal solid particles and fluids. This component of skin factor can be removed or averted with well stimulation operations. The $S_{C+\theta}$ is a skin component due to partial completion and deviation angle, which make the flow pattern near the wellbore deviate from ideal radial flow pattern. This skin component is not removable in water coning and gas coning systems. The S_P is a skin component due to the nonideal flow condition around the perforations associated with cased-hole completion. It depends on a number of parameters including perforation density, phase angle, perforation depth, diameter, compacted zone, and others. This component can be minimized with optimized perorating technologies. The ΣS_{PS} represents pseudo-skin components due to non–Darcy flow effect, multiphase effect, and flow convergence near the wellbore. These components cannot be eliminated.

It is essential to know the magnitude of components of the skin factor S derived from the pressure transient test data analysis. Commercial software packages are available for decomposition of the skin factor on the basis of well completion method. One of the packages is WellFlo (EPS, 2005).

Example Problem 15.1 A horizontal wellbore was placed in a 100-ft thick oil reservoir of 0.23 porosity. Oil formation volume factor and viscosity are 1.25 rb/stb and 1 cp, respectively. The total reservoir compressibility factor is $10^{-5}\,\text{psi}^{-1}$. The well was tested following the schedule shown in Fig. 15.3. The measured flowing bottom-hole pressures are also presented in Fig. 15.3. Estimate directional permeabilities and skin factors from the test data.

Solution Figure 15.4 presents a log-log diagnostic plot of test data. It clearly indicates a vertical radial flow at early time, a pseudo-linear flow at mid-time, and the beginning of a pseudo-radial flow at late time.

The semi-log analysis for the vertical radial flow is shown in Fig. 15.5, which gives $k_{yz} = 0.9997$ md and near-wellbore skin factor $S = -0.0164$.

The square-root time plot analysis for the pseudo-linear flow is shown in Fig. 15.6, which gives the effective wellbore length of $L = 1{,}082.75$ ft and a skin factor due to convergence of $S = 3.41$.

The semi-log analysis for the horizontal pseudo-radial flow is shown in Fig. 15.7, which gives $k_h = 1.43$ md and pseudo-skin factor $S = -6.17$.

Figure 15.8 shows a match between the measured and model-calculated pressure responses given by an optimization technique. This match was obtained using the following parameter values:

$k_h = 1.29$ md
$k_z = 0.80$ md
$S = 0.06$
$L = 1{,}243$ ft.

To estimate the long-term productivity of this horizontal well, the $k_h = 1.29$ md and $S = 0.06$ should be used in the well inflow equation presented in Chapter 3.

15.3 Excessive Gas Production

Excessive gas production is usually due to channeling behind the casing (Fig. 15.9), preferential flow through high-permeability zones (Fig. 15.10), gas coning (Fig. 15.11), and casing leaks (Clark and Schultz, 1956).

The channeling behind the casing and gas coning problems can be identified based on production logging such as temperature and noise logs. An example is depicted in Fig. 15.12, where both logs indicate that gas is being produced from an upper gas sand and channeling down to some perforations in the oil zone.

Excessive gas production of an oil well could also be due to gas production from unexpected gas zones. This can be identified using production logging such as temperature and density logs. An example is presented in Fig. 15.13, where both logs indicate gas production from the thief zone B.

15.4 Excessive Water Production

Excessive water production is usually from water zones, not from the connate water in the pay zone. Water enters the wellbore due to channeling behind the casing (Fig. 15.14), preferential flow through high-permeability zones (Fig. 15.15), water coning (Fig. 15.16), hydraulic fracturing into water zones, and casing leaks.

Figure 15.17 shows how to identify fracture height using prefracture and postfracture temperature logs to tell whether the hydraulic fracture has extended into a water zone.

In addition to those production logging tools that are mentioned in the previous section, other production logging tools can be used for identifying water-producing zones. Fluid density logs are especially useful for identifying water entries. Comparison between water-cut data and spinner flowmeter log can sometimes give an idea of where the water is coming from. Figure 15.18 shows a spinner flowmeter log identifying a watered zone at the bottom of a well with a water-cut of nearly 50%.

15.5 Liquid Loading of Gas Wells

Gas wells usually produce natural gas-carrying liquid water and/or condensate in the form of mist. As the gas flow velocity in the well drops because of reservoir pressure depletion, the carrying capacity of the gas decreases.

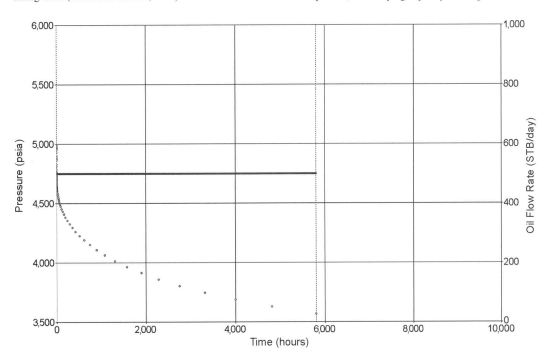

Figure 15.3 Measured bottom-hole pressures and oil production rates during a pressure drawdown test.

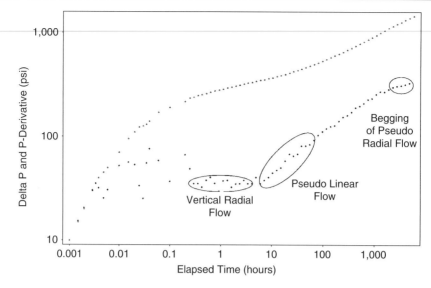

Figure 15.4 Log-log diagnostic plot of test data.

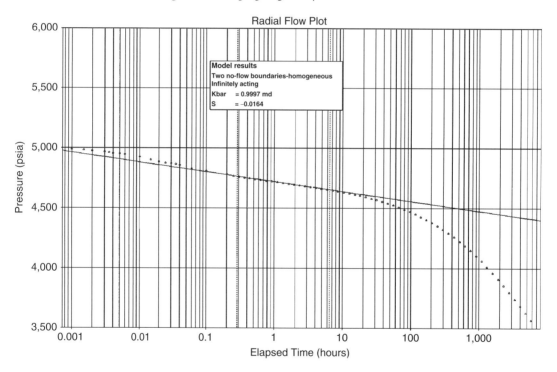

Figure 15.5 Semi-log plot for vertical radial flow analysis.

When the gas velocity drops to a critical level, liquids begin to accumulate in the well and the well flow can undergo an annular flow regime followed by a slug flow regime. The accumulation of liquids (liquid loading) increases the bottom-hole pressure, which reduces gas production rate. A low gas production rate will cause gas velocity to drop further. Eventually, the well will undergo a bubbly flow regime and cease producing.

Several measures can be taken to solve the liquid-loading problem. Foaming the liquid water can enable the gas to lift water from the well. Using smaller tubing or creating a lower wellhead pressure sometimes can keep mist flowing. The well can be unloaded by gas-lifting or pumping the liquids out of the well. Heating the wellbore can prevent oil condensation. Down-hole injection of water into an underlying disposal zone is another option. However, liquid-loading is not always obvious and recognizing the liquid-loading problem is not an easy task. A thorough diagnostic analysis of well data needs to be performed. The symptoms to look for include onset of liquid slugs at the surface of well, increasing difference between the tubing and casing pressures with time, sharp changes in gradient on a flowing pressure survey, and sharp drops in production decline curve.

15.5.1 The Turner et al. Method

Turner et al. (1969) were the pioneer investigators who analyzed and predicted the minimum gas flow rate to prevent liquid-loading. They presented two mathematical

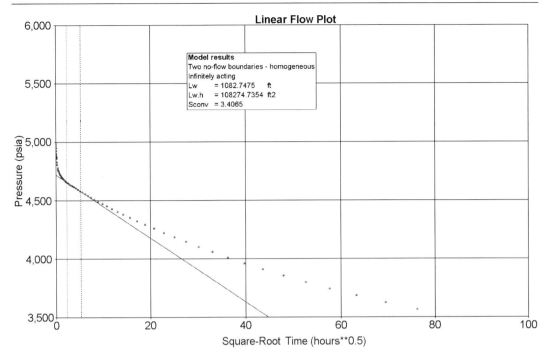

Figure 15.6 Square-root time plot for pseudo-linear flow analysis.

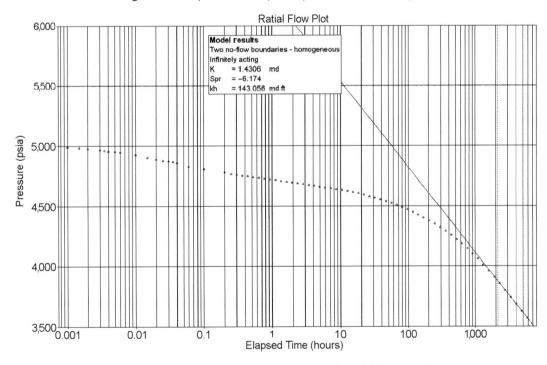

Figure 15.7 Semi-log plot for horizontal pseudo-radial flow analysis.

models to describe the liquid-loading problem: the film-movement model and the entrained drop movement model. On the basis of analyses on field data, they concluded that the film-movement model does not represent the controlling liquid transport mechanism.

Turner et al.'s entrained drop movement model was derived on the basis of the terminal free settling velocity of liquid drops and the maximum drop diameter corresponding to the critical Weber number of 30. Turner et al.'s terminal slip velocity equation is expressed in U.S. field units as

$$v_{sl} = \frac{1.3\sigma^{1/4}\left(\rho_L - \rho_g\right)^{1/4}}{C_d^{1/4}\rho_g^{1/2}}. \quad (15.23)$$

According to Turner et al., gas will continuously remove liquids from the well until its velocity drops to below the

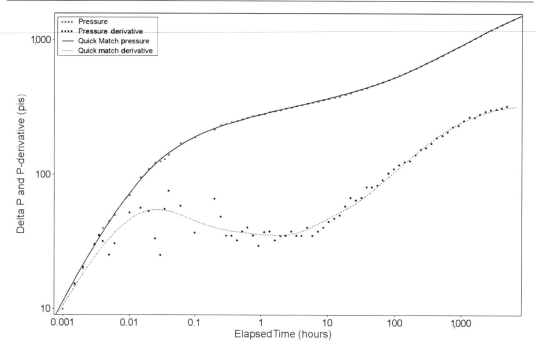

Figure 15.8 Match between measured and model calculated pressure data.

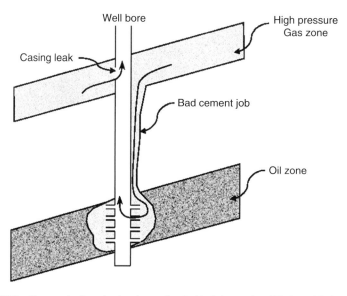

Figure 15.9 Gas production due to channeling behind the casing (Clark and Schultz, 1956).

terminal slip velocity. The minimum gas flow rate (in MMcf/D) for a particular set of conditions (pressure and conduit geometry) can be calculated using Eqs. (15.23) and (15.24):

$$Q_{gslMM} = \frac{3.06 p v_{sl} A}{Tz} \qquad (15.24)$$

Figure 15.19 shows a comparison between the results of Turner et al.'s entrained drop movement model. The map shows many loaded points in the unloaded region. Turner et al. recommended the equation-derived values be adjusted upward by approximately 20% to ensure removal of all drops. Turner et al. believed that the discrepancy was attributed to several facts including the use of drag coefficients for solid spheres, the assumption of stagnation velocity, and the critical Weber number established for drops falling in air, not in compressed gas.

The main problem that hinders the application of Turner et al.'s entrained drop model to gas wells comes from the difficulties of estimating the values of fluid density and pressure. Using an average value of gas-specific gravity (0.6) and gas temperature (120 °F), Turner et al. derived an expression for gas density as 0.0031 times the pressure. However, they did not present a method for calculating the gas pressure in a multiphase flow wellbore.

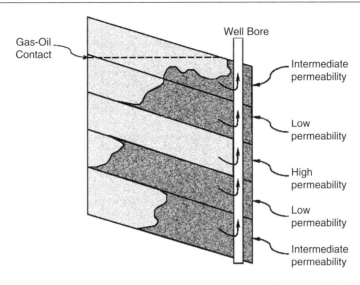

Figure 15.10 Gas production due to preferential flow through high-permeability zones (Clark and Schultz, 1956).

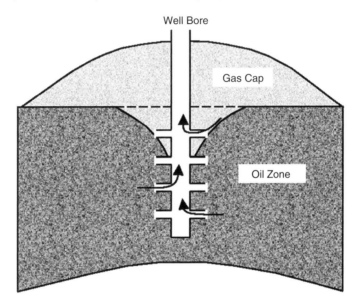

Figure 15.11 Gas production due to gas coning (Clark and Schultz, 1956).

The spreadsheet program *TurnerLoading.xls* has been developed for quick calculation associated with this book.

Turner et al.'s entrained drop movement model was later modified by a number of authors. Coleman et al. (1991) suggested to use Eq. (15.23) with a lower constant value. Nosseir et al. (2000) expanded Turner et al.'s entrained drop model to more than one flow regime in a well. Lea and Nickens (2004) made some corrections to Turner et al.'s simplified equations. However, the original drawbacks (neglected transport velocity and multiphase flow pressure) with Turner et al.'s approach still remain unsolved.

15.5.2 The Guo et al. Method

Starting from Turner et al.'s entrained drop model, Guo et al. (2006) determined the minimum kinetic energy of gas that is required to lift liquids. A four-phase (gas, oil, water, and solid particles) mist-flow model was developed. Applying the minimum kinetic energy criterion to the four-phase flow model resulted in a closed-form analytical equation for predicting the minimum gas flow rate.

15.5.2.1 Minimum Kinetic Energy

Kinetic energy per unit volume of gas can be expressed as

$$E_k = \frac{\rho_g v_g^2}{2g_c}. \qquad (15.25)$$

Substituting Eq. (15.23) into Eq. (15.25) gives an expression for the minimum kinetic energy required to keep liquid droplets from falling:

$$E_{ksl} = 0.026\sqrt{\frac{\sigma(\rho_L - \rho_g)}{C_d}} \qquad (15.26)$$

If the value of drag coefficient $C_d = 0.44$ (recommended by Turner et al.) is used and the effect of gas density is neglected (a conservative assumption), Eq. (15.26) becomes

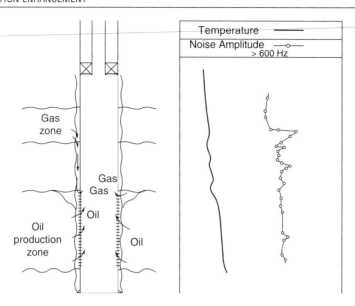

Figure 15.12 Temperature and noise logs identifying gas channeling behind casing (Economides et al., 1994).

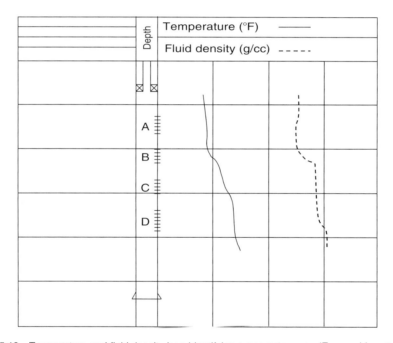

Figure 15.13 Temperature and fluid density logs identifying a gas entry zone (Economides et al., 1994).

$$E_{ksl} = 0.04\sqrt{\sigma\rho_L}. \qquad (15.27)$$

In gas wells producing water, typical values for water–gas interfacial tension and water density are 60 dynes/cm and 65 lb$_m$/ft^3, respectively. This yields the minimum kinetic energy value of 2.5 lb$_f$-ft/ft^3. In gas wells producing condensate, typical values for condensate–gas interfacial tension and condensate density are 20 dynes/cm and 45 lb$_m$/ft^3, respectively. This yields the minimum kinetic energy value of 1.2 lb$_f$-ft/ft^3.

The minimum gas velocity required for transporting the liquid droplets upward is equal to the minimum gas velocity required for floating the liquid droplets (keeping the droplets from falling) plus the transport velocity of the droplets, that is,

$$v_{gm} = v_{sl} + v_{tr}. \qquad (15.28)$$

The transport velocity v_{tr} may be calculated on the basis of liquid production rate, geometry of the conduit, and liquid volume fraction, which is difficult to quantify. Instead of trying to formulate an expression for the transport velocity v_{tr}, Guo et al. used v_{tr} as an empirical constant to lump the effects of nonstagnation velocity, drag coefficients for solid spheres, and the critical Weber number established for drops falling in air. On the

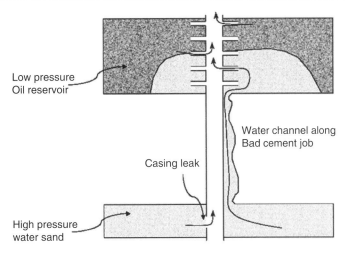

Figure 15.14 Water production due to channeling behind the casing.

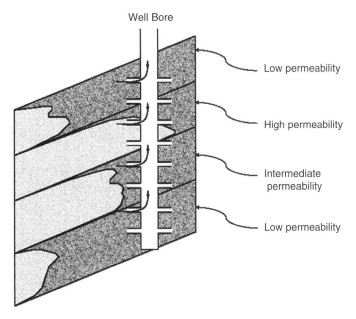

Figure 15.15 Preferential water flow through high-permeability zones.

basis of the work by Turner et al., the value of v_{tr} was taken as 20% of v_{sl} in this study. Use of this value results in

$$v_{gm} \approx 1.2 v_{sl}. \quad (15.29)$$

Substituting Eqs. (15.23) and (15.29) into Eq. (15.25) results in the expression for the minimum kinetic energy required for transporting the liquid droplets as

$$E_{km} = 0.0576\sqrt{\sigma \rho_L}. \quad (15.30)$$

For typical gas wells producing water, this equation yields the minimum kinetic energy value of $3.6\,lb_f\text{-}ft/ft^3$. For typical gas wells producing condensate, this equation gives the minimum kinetic energy value of $1.73\,lb_f\text{-}ft/ft^3$. These numbers imply that the required minimum gas production rate in water-producing gas wells is approximately twice that in condensate-producing gas wells.

To evaluate the gas kinetic energy E_k in Eq. (15.25) at a given gas flow rate and compare it with the minimum required kinetic energy E_{km} in Eq. (15.30), the values of gas density ρ_g and gas velocity v_g need to be determined.

Expressions for ρ_g and v_g can be obtained from ideal gas law:

$$\rho_g = \frac{2.7 S_g p}{T} \quad (15.31)$$

$$v_g = 4.71 \times 10^{-2} \frac{T Q_G}{A_i p} \quad (15.32)$$

Substituting Eqs. (15.31) and (15.32) into Eq. (15.25) yields

$$E_k = 9.3 \times 10^{-5} \frac{S_g T Q_G^2}{A_i^2 p}. \quad (15.33)$$

Equation (15.33) indicates that the gas kinetic energy decreases with increased pressure, which means that the controlling conditions are bottom-hole conditions where gas has higher pressure and lower kinetic energy. This analysis is consistent with the observations from air-drilling operations where solid particles accumulate at the bottom-hole rather than at the top-hole. However, this analysis is in contradiction with the results by Turner et al., which indicated that the wellhead conditions are in most instances, controlling.

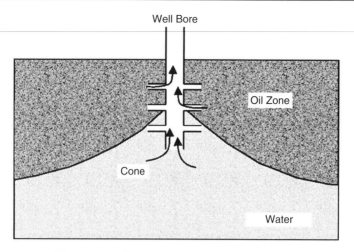

Figure 15.16 Water production due to water coning.

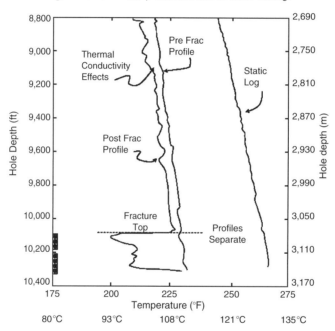

Figure 15.17 Prefracture and postfracture temperature logs identifying fracture height (Dobkins, 1981).

15.5.2.2 Four-Phase Flow Model

To accurately predict the bottom-hole pressure p in Eq. (15.33), a gas-oil-water-solid four-phase mist-flow model was developed by Guo et al. (2006). According to the four-phase flow model, the flowing pressure p at depth L can be solved numerically from the following equation:

$$144b(p - p_{hf}) + \frac{1-2bm}{2} \ln\left|\frac{(144p+m)^2 + n}{(144p_{hf}+m)^2 + n}\right|$$

$$b(P - P_{hf}) + \frac{1-2bm}{2} \ln\left|\frac{(P+m)^2 + n}{(P_{hf}+m)^2 + n}\right|$$

$$- \frac{m + \frac{b}{c}n - bm^2}{\sqrt{n}} \left[\tan^{-1}\left(\frac{144p+m}{\sqrt{n}}\right)\right.$$

$$\left. - \tan^{-1}\left(\frac{144p_{hf}+m}{\sqrt{n}}\right)\right]$$

$$= a(1 + d^2 e)L, \quad (15.34)$$

where

$$a = \frac{15.33 S_s Q_s + 86.07 S_w Q_w + 86.07 S_o Q_o + 18.79 S_g Q_G}{10^3 T_{av} Q_G}$$

$$\times \cos(\theta), \quad (15.35)$$

$$b = \frac{0.2456 Q_s + 1.379 Q_w + 1.379 Q_o}{10^3 T_{av} Q_G}, \quad (15.36)$$

$$c = \frac{6.785 \times 10^{-6} T_{av} Q_G}{A_i}, \quad (15.37)$$

$$d = \frac{Q_s + 5.615(Q_w + Q_o)}{600 A_i}, \quad (15.38)$$

$$e = \frac{6f}{g D_h \cos(\theta)}, \quad (15.39)$$

$$f_M = \left[\frac{1}{1.74 - 2\log\left(\frac{2\varepsilon'}{D_h}\right)}\right]^2, \quad (15.40)$$

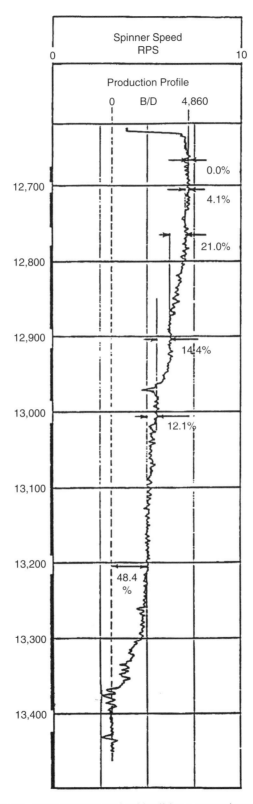

Figure 15.18 Spinner flowmeter log identifying a watered zone at bottom.

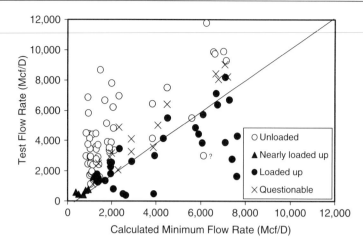

Figure 15.19 Calculated minimum flow rates with the Turner et al. model and test flow rates.

$$m = \frac{cde}{1+d^2e}, \quad (15.41)$$

and

$$n = \frac{c^2e}{(1+d^2e)^2}, \quad (15.42)$$

where

A = cross-sectional area of conduit, ft^2
D_h = hydraulic diameter, in.
f_M = Moody friction factor
g = gravitational acceleration, 32.17 ft/s^2
L = conduit length, ft
p = pressure, psia
p_{hf} = wellhead flowing pressure, psia
Q_G = gas production rate, Mscf/day
Q_o = oil production rate, bbl/day
Q_s = solid production rate, ft^3/day
Q_w = water production rate, bbl/day
S_g = specific gravity of gas, air = 1
S_o = specific gravity of produced oil, freshwater = 1
S_w = specific gravity of produced water, freshwater = 1
S_s = specific gravity of produced solid, freshwater = 1
T_{av} = the average temperature in the butting, °R
ε' = pipe wall roughness, in.
θ = inclination angle, degrees.

15.5.2.3 Minimum Required Gas Production Rate

A logical procedure for predicting the minimum required gas flow rate Q_{gm} involves calculating gas density ρ_g, gas velocity v_g, and gas kinetic energy E_k at bottom-hole condition using an assumed gas flow rate Q_G, and compare the E_k with E_{km}. If the E_k is greater than E_{km}, the Q_G is higher than the Q_{gm}. The value of Q_G should be reduced and the calculation should be repeated until the E_k is very close to E_{km}. Because this procedure is tedious, a simple equation was derived by Guo et al. for predicting the minimum required gas flow rate in this section. Under the minimum unloaded condition (the last point of the mist flow regime), Eq. (15.33) becomes

$$E_{km} = 9.3 \times 10^{-5} \frac{S_g T_{bh} Q_{gm}^2}{A_i^2 p}, \quad (15.43)$$

which gives

$$p = 9.3 \times 10^{-5} \frac{S_g T_{bh} Q_{gm}^2}{A_i^2 E_{km}}. \quad (15.44)$$

Substituting Eq. (15.44) into Eq. (15.34) results in

$$144 b\alpha_1 + \frac{1-2bm}{2} \ln \alpha_2 - \frac{m + \frac{b}{c}n - bm^2}{\sqrt{n}} \quad (15.45)$$
$$\times \left[\tan^{-1} \beta_1 - \tan^{-1} \beta_2\right] = \gamma,$$

where

$$\alpha_1 = 9.3 \times 10^{-5} \frac{S_g T_{bh} Q_{gm}^2}{A_i^2 E_{km}} - p_{hf}, \quad (15.46)$$

$$\alpha_2 = \frac{\left(1.34 \times 10^{-2} \frac{S_g T_{bh} Q_{gm}^2}{A_i^2 E_{km}} + m\right)^2 + n}{(144 p_{hf} + m)^2 + n}, \quad (15.47)$$

$$\beta_1 = \frac{1.34 \times 10^{-2} \frac{S_g T_{bh} Q_{gm}^2}{A_i^2 E_{km}} + m}{\sqrt{n}}, \quad (15.48)$$

$$\beta_2 = \frac{144 p_{hf} + m}{\sqrt{n}}, \quad (15.49)$$

and

$$\gamma = a(1 + d^2 e)L. \quad (15.50)$$

All the parameter values should be evaluated at Q_{gm}. The minimum required gas flow rate Q_{gm} can be solved from Eq. (15.45) with a trial-and-error or numerical method such as the Bisection method. It can be shown that Eq. (15.45) is a one-to-one function of Q_{gm} for Q_{gm} values greater than zero. Therefore, the Newton–Raphson iteration technique can also be used for solving Q_{gm}. Commercial software packages such as MS Excel can be used as solvers. In fact, the Goal Seek function built into MS Excel was used for generating solutions presented in this chapter. The spreadsheet program is named *GasWellLoading.xls*.

Example Problem 15.2 To demonstrate how to use Eq. (15.45) for predicting the minimum unloading gas flow rate, consider a vertical gas well producing 0.70 specific gravity gas and 50 bbl/day condensate through a 2.441-in. inside diameter (ID) tubing against a wellhead pressure of 900 psia. Suppose the tubing string is set at a depth of 10,000 ft, and other data are given in Table 15.1.

Solution The solution given by the spreadsheet program *GasWellLoading.xls* is shown in Table 15.2.

Table 15.1 Basic Parameter Values for Example Problem 15.1

Gas-specific gravity	0.7 (air = 1)
Hole inclination	0 degrees
Wellhead temperature	60°
Geothermal gradient	0.01 °F/ft
Condensate gravity	60 °API
Water-specific gravity	1.05 (water = 1)
Solid-specific gravity	2.65 (water = 1)
Interfacial tension	20 dyne/cm
Tubing wall roughness	0.000015 in.

Table 15.2 Result Given by the Spreadsheet Program GasWellLoading.xls

Calculated Parameters

Hydraulic diameter	0.2034 ft
Conduit cross-sectional area	0.0325 ft^2
Average temperature	570 °R
Minimum kinetic energy	1.6019 lb-ft/ft^3
$a =$	2.77547E-05
$b =$	1.20965E-07
$c =$	875999.8117
$d =$	0.10598146
$e =$	0.000571676
$f_M =$	0.007481992
$m =$	53.07387106
$n =$	438684299.6

Solution

Critical gas production rate	1,059 Mscf/day
Pressure (p) =	1,189 psia
Objective function f(Q_{gm}) =	−1.78615E-05

15.5.3 Comparison of the Turner et al. and the Guo et al. Methods

Figure 15.20 illustrates Eq. (15.45)–calculated minimum flow rates mapped against the test flow rates for the same wells used in Fig. 15.19. This map shows six loaded points in the unloaded region, but they are very close to the boundary. This means the Guo et al. method is more accurate than the Turner et al. method in estimating the minimum flow rates.

Summary

This chapter presents a guideline to identifying problems commonly encountered in oil and gas wells. Well test analysis provides a means of estimating properties of individual pay zones. Production logging analysis identifies fluid entries to the wellbore from different zones. The Guo et al. method is more accurate than the Turner et al. method for predicting liquid-loading problems in gas production wells.

References

CHAUDHRY, A.C. *Oil Well Testing Handbook*. Burlington: Gulf Professional Publishing, 2004.

CLARK, N.J. and SCHULTZ, W.P. The analysis of problem wells. *Petroleum Engineer* September 1956;28:B30–B38.

COLEMAN, S.B., CLAY, H.B., MCCURDY, D.G., and NORRIS, L.H., III. A new look at predicting gas well loading-up. *JPT* (March 1991), *Trans. AIME* 1991;291:329.

DAKE, L.P. *Fundamentals of Reservoir Engineering*. Amsterdam: Elsevier, 2002.

DOBKINS, T.A. Improved method to determine hydraulic fracture height. *JPT* April 1981:719–726.

ECONOMIDES, M.J., HILL, A.D., and EHLIG-ECONOMIDES, C. *Petroleum Production Systems*. New Jersey: Prentice Hall PTR, 1994.

E-Production Services, Inc. *FloSystem User Manual*. Edinburgh: E-Production Services, Inc., 2005.

E-Production Services, Inc. *PanSystem User Manual*. Edinburgh: E-Production Services, Inc., 2004.

FEKETE., F.A.S.T. *WellTest User Manual*. Calgary: Fekete Associates, Inc., 2003.

GUO, B., GHALAMBOR, A., and XU, C. A systematic approach to predicting liquid loading in gas well. *SPE Production Operations* J. February 2006.

HORNE, R.N. *Modern Well Test Analysis: A Computer-Aided Approach*. New York: Petroway Publishing, 1995.

LEA, J.F. and NICKENS, H.V. Solving gas-well liquid-loading problems. *SPE Prod. Facilities* April 2004:30.

LEE, J.W., ROLLINS, J.B., and SPIVEY, J.P. *Pressure Transient Testing*. Richardson: Society of Petroleum Engineers, 2003.

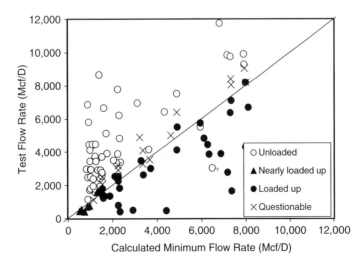

Figure 15.20 The minimum flow rates given by the Guo et al. model and the test flow rates.

NOSSEIR, M.A., DARWICH, T.A., SAYYOUH, M.H., and SALLALY, M.E. A new approach for accurate prediction of loading in gas wells under different flowing conditions. *SPE Prod. Facilities* November 2000;15(4):245.

TURNER, R.G., HUBBARD, M.G., and DUKLER, A.E. Analysis and prediction of minimum flow rate for the continuous removal of liquids from gas wells. *JPT* November 1969, *Trans. AIME* 1969;246:1475.

Problems

15.1 Consider a gas well producing 50 bbl/d of condensate and 0.1 cubic foot of sand through a 2.441-in. I.D. tubing against a wellhead pressure of 500 psia. Suppose the tubing string is set at a depth of 8,000 ft, use the following data and estimate the minimum gas production rate before the gas well gets loaded.

Gas-specific gravity:	0.75 (air = 1)
Hole inclination:	0 degrees
Wellhead temperature:	60 °F
Geothermal gradient:	0.01 °F/ft
Condensate gravity:	60 °API
Water-specific gravity:	1.07 (water = 1)
Solid-specific gravity:	2.65 (water = 1)
Oil–gas interface tension:	20 dyne/cm
Tubing wall roughness:	0.000015 in.

15.2 Consider a gas well producing 50 bbl/day of water and 0.2 ft^3 of sand through a 2.441-in. ID tubing against a wellhead pressure of 600 psia and temperature of 80 °F. Suppose the tubing string is set at a depth of 9,000 ft and geothermal gradient is 0.01 °F/ft, estimate the minimum gas production rate before the gas well gets loaded.

15.3 Consider a gas well producing 80 bbl/day of water and 0.1 ft^3 of sand through a 1.995-in. ID tubing against a wellhead pressure of 400 psia and temperature of 70 °F. Suppose the tubing string is set at a depth of 7,000 ft and geothermal gradient is 0.01 °F/ft, estimate the minimum gas production rate before the gas well gets loaded.

15.4 Consider a gas well producing 70 bbl/day of oil and 0.1 ft^3 of sand through a 1.995-in. ID tubing against a wellhead pressure of 600 psia and temperature of 80 °F. Suppose the tubing string is set at a depth of 6,000 ft and geothermal gradient is 0.01 °F/ft, estimate the minimum gas production rate before the gas well gets loaded.

16 Matrix Acidizing

Contents
16.1 Introduction 16/244
16.2 Acid–Rock Interaction 16/244
16.3 Sandstone Acidizing Design 16/244
16.4 Carbonate Acidizing Design 16/247
Summary 16/248
References 16/248
Problems 16/249

16.1 Introduction

Matrix acidizing is also called *acid matrix treatment*. It is a technique to stimulate wells for improving well inflow performance. In the treatment, acid solution is injected into the formation to dissolve some of the minerals to recover permeability of sandstones (removing skin) or increase permeability of carbonates near the wellbore. After a brief introduction to acid–rock interaction, this chapter focuses on important issues on sandstone acidizing design and carbonate acidizing design. More in-depth information can be found from Economides and Nolte (2000).

16.2 Acid–Rock Interaction

Minerals that are present in sandstone pores include montmorillonite (bentonite), kaolinite, calcite, dolomite, siderite, quartz, albite (sodium feldspar), orthoclase, and others. These minerals can be either from invasion of external fluid during drilling, cementing, and well completion or from host materials that exist in the naturally occurring rock formations. The most commonly used acids for dissolving these minerals are hydrochloric acid (HCl) and hydrofluoric acid (HF).

16.2.1 Primary Chemical Reactions

Silicate minerals such as clays and feldspars in sandstone pores are normally removed using mixtures of HF and HCl, whereas carbonate minerals are usually attacked with HCl. The chemical reactions are summarized in Table 16.1. The amount of acid required to dissolve a given amount of mineral is determined by the stoichiometry of the chemical reaction. For example, the simple reaction between HCl and $CaCO_3$ requires that 2 mol of HCl is needed to dissolve 1 mol of $CaCO_3$.

16.2.2 Dissolving Power of Acids

A more convenient way to express reaction stoichiometry is the dissolving power. The dissolving power on a mass basis is called *gravimetric dissolving power* and is defined as

$$\beta = C_a \frac{\nu_m MW_m}{\nu_a MW_a}, \qquad (16.1)$$

where

β = gravimetric dissolving power of acid solution, lb_m mineral/lb_m solution
C_a = weight fraction of acid in the acid solution
ν_m = stoichiometry number of mineral
ν_a = stoichiometry number of acid
MW_m = molecular weight of mineral
MW_a = molecular weight of acid.

For the reaction between 15 wt% HCl solution and $CaCO_3$, $C_a = 0.15$, $\nu_m = 1$, $\nu_a = 2$, $MW_m = 100.1$, and $MW_a = 36.5$. Thus,

$$\beta_{15} = (0.15)\frac{(1)(100.1)}{(2)(36.5)}$$

$$= 0.21\, lb_m\ CaCO_3/lb_m\ 15\,wt\%\ HCl\ solution.$$

The dissolving power on a volume basis is called *volumetric dissolving power* and is related to the gravimetric dissolving power through material densities:

$$X = \beta \frac{\rho_a}{\rho_m}, \qquad (16.2)$$

where

X = volumetric dissolving power of acid solution, ft^3 mineral/ft^3 solution
ρ_a = density of acid, lb_m/ft^3
ρ_m = density of mineral, lb_m/ft^3

16.2.3 Reaction Kinetics

The acid–mineral reaction takes place slowly in the rock matrix being acidized. The reaction rate can be evaluated experimentally and described by kinetics models. Research work in this area has been presented by many investigators including Fogler et al. (1976), Lund et al. (1973, 1975), Hill et al. (1981), Kline and Fogler (1981), and Schechter (1992). Generally, the reaction rate is affected by the characteristics of mineral, properties of acid, reservoir temperature, and rates of acid transport to the mineral surface and removal of product from the surface. Detailed discussion of reaction kinetics is beyond the scope of this book.

16.3 Sandstone Acidizing Design

The purpose of sandstone acidizing is to remove the damage to the sandstone near the wellbore that occurred during drilling and well completion processes. The acid treatment is only necessary when it is sure that formation damage is significant to affect well productivity. A major formation damage is usually indicated by a large positive skin factor derived from pressure transit test analysis in a flow regime of early time (see Chapter 15).

16.3.1 Selection of Acid

The acid type and acid concentration in acid solution used in acidizing is selected on the basis of minerals in the formation and field experience. For sandstones, the typical treatments usually consist of a mixture of 3 wt% HF and 12 wt% HCl, preceded by a 15 wt% HCl preflush. McLeod (1984) presented a guideline to the selection of acid on the basis of extensive field experience. His recommendations for sandstone treatments are shown in Table 16.2. McLeod's recommendation should serve only as a starting point. When many wells are treated in a particular formation, it is worthwhile to conduct laboratory tests of the responses of cores to different acid strengths. Figure 16.1 shows typical acid–response curves.

Table 16.1 Primary Chemical Reactions in Acid Treatments

Montmorillonite (Bentonite)-HF/HCl:	$Al_4Si_8O_{20}(OH)_4 + 40HF + 4H^+ \leftrightarrow 4AlF_2^+ + 8SiF_4 + 24H_2O$
Kaolinite-HF/HCl:	$Al_4Si_8O_{10}(OH)_8 + 40HF + 4H^+ \leftrightarrow 4AlF_2^+ + 8SiF_4 + 18H_2O$
Albite-HF/HCl:	$NaAlSi_3O_8 + 14HF + 2H^+ \leftrightarrow Na^+ + AlF_2^+ + 3SiF_4 + 8H_2O$
Orthoclase-HF/HCl:	$KAlSi_3O_8 + 14HF + 2H^+ \leftrightarrow K^+ + AlF_2^+ + 3SiF_4 + 8H_2O$
Quartz-HF/HCl:	$SiO_2 + 4HF \leftrightarrow SiF_4 + 2H_2O$
	$SiF_4 + 2HF \leftrightarrow H_2SiF_6$
Calcite-HCl:	$CaCO_3 + 2HCl \rightarrow CaCl_2 + CO_2 + H_2O$
Dolomite-HCl:	$CaMg(CO_3)_2 + 4HCl \rightarrow CaCl_2 + MgCl_2 + 2CO_2 + 2H_2O$
Siderite-HCl:	$FeCO_3 + 2HCl \rightarrow FeCl_2 + CO_2 + H_2O$

16.3.2 Acid Volume Requirement

The acid volume should be high enough to remove near-wellbore formation damage and low enough to reduce cost of treatment. Selection of an optimum acid volume is complicated by the competing effects. The volume of acid needed depends strongly on the depth of the damaged zone, which is seldom known. Also, the acid will never be distributed equally to all parts of the damaged formation. The efficiency of acid treatment and, therefore, acid volume also depends on acid injection rate. To ensure that an adequate amount of acid contacts most of the damaged formation, a larger amount of acid is necessary.

The acid preflush volume is usually determined on the basis of void volume calculations. The required minimum acid volume is expressed as

$$V_a = \frac{V_m}{X} + V_P + V_m, \tag{16.3}$$

where

V_a = the required minimum acid volume, ft^3
V_m = volume of minerals to be removed, ft^3
V_P = initial pore volume, ft^3

and

$$V_m = \pi(r_a^2 - r_w^2)(1 - \phi)C_m, \tag{16.4}$$
$$V_P = \pi(r_a^2 - r_w^2)\phi, \tag{16.5}$$

where

r_a = radius of acid treatment, ft
r_w = radius of wellbore, ft
ϕ = porosity, fraction
C_m = mineral content, volume fraction.

Example Problem 16.1 A sandstone with a porosity of 0.2 containing 10 v% calcite (CaCO$_3$) is to be acidized with HF/HCl mixture solution. A preflush of 15 wt% HCl solution is to be injected ahead of the mixture to dissolve the carbonate minerals and establish a low pH environment. If the HCl preflush is to remove all carbonates in a region within 1 ft beyond a 0.328-ft radius wellbore before the HF/HCl stage enters the formation, what minimum preflush volume is required in terms of gallon per foot of pay zone?

Table 16.2 Recommended Acid Type and Strength for Sandstone Acidizing

HCl Solubility > 20%	Use HCl Only
High-perm sand (k > 100 md)	
High quartz (80%), low clay (<5%)	10% HCl-3% HF^a
High feldspar (>20%)	13.5% HCl-1.5% HF^a
High clay (>10%)	6.5% HCl-1% HF^b
High iron chlorite clay	3% HCl-0.5% HF^b
Low-perm sand (k < 10 md)	
Low clay (<5%)	6% HCl-1.5% HF^c
High chlorite	3% HCl-0.5% HF^d

^{a} Preflush with 15% HCl.
^{b} Preflush with sequestered 5% HCl.
^{c} Preflush with 7.5% HCl or 10% acetic acid.
^{d} Preflush with 5% acetic acid.

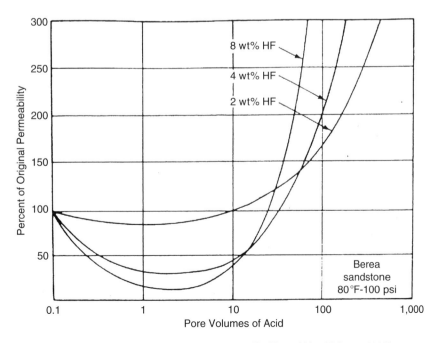

Figure 16.1 Typical acid response curves (Smith and Hendrickson, 1965).

Solution

Volume of $CaCO_3$ to be removed:

$$V_m = \pi(r_a^2 - r_w^2)(1-\phi)C_m$$
$$= \pi(1.328^2 - 0.328^2)(1-0.2)(0.1)$$
$$= 0.42\,\text{ft}^3\ CaCO_3/\text{ft pay zone}$$

Initial pore volume:

$$V_P = \pi(r_a^2 - r_w^2)\phi$$
$$= \pi(1.328^2 - 0.328^2)(0.2) = 1.05\,\text{ft}^3/\text{ft pay zone}$$

Gravimetric dissolving power of the 15 wt% HCl solution:

$$\beta = C_a \frac{v_m MW_m}{v_a MW_a}$$
$$= (0.15)\frac{(1)(100.1)}{(2)(36.5)}$$
$$= 0.21\,\text{lb}_m\ CaCO_3/\text{lb}_m\ 15\,\text{wt\% HCl solution}$$

Volumetric dissolving power of the 15 wt% HCl solution:

$$X = \beta \frac{\rho_a}{\rho_m}$$
$$= (0.21)\frac{(1.07)(62.4)}{(169)}$$
$$= 0.082\,\text{ft}^3\ CaCO_3/\text{ft}^3\ 15\,\text{wt\% HCl solution}$$

The required minimum HCl volume

$$V_a = \frac{V_m}{X} + V_P + V_m$$
$$= \frac{0.42}{0.082} + 1.05 + 0.42$$
$$= 6.48\,\text{ft}^3\ 15\,\text{wt\% HCl solution/ft pay zone}$$
$$= (6.48)(7.48)$$
$$= 48\,\text{gal}\ 15\,\text{wt\% HCl solution/ft pay zone}$$

The acid volume requirement for the main stage in a mud acid treatment depends on mineralogy and acid type and strength. Economides and Nolte (2000) provide a listing of typical stage sequences and volumes for sandstone acidizing treatments. For HCl acid, the volume requirement increases from 50 to 200 gal/ft pay zone with HCl solubility of HF changing from less than 5% to 20%. For HF acid, the volume requirement is in the range of 75–100 gal/ft pay zone with 3.0–13.5% HCl and 0.5–3.0% HF depending on mineralogy.

Numerous efforts have been made to develop a rigorous method for calculating the minimum required acid volume in the past 2 decades. The most commonly used method is the two-mineral model (Hekim et al., 1982; Hill et al., 1981; Taha et al., 1989). This model requires a numerical technique to obtain a general solution. Schechter (1992) presented an approximate solution that is valid for Damkohler number being greater than 10. This solution approximates the HF fast-reacting mineral front as a sharp front. Readers are referred to Schechter (1992) for more information.

Because mud acid treatments do not dissolve much of the formation minerals but dissolve the materials clogging the pore throats, Economides and Nolte (2000) suggest taking the initial pour volume (Eq. [16.5]) within the radius of treatment as the minimum required acid volume for the main stage of acidizing treatment. Additional acid volume should be considered for the losses in the injection tubing string.

16.3.3 Acid Injection Rate

Acid injection rate should be selected on the basis of mineral dissolution and removal and depth of damaged zone. Selecting an optimum injection rate is a difficult process because the damaged zone is seldom known with any accuracy and the competing effects of mineral dissolution and reaction product precipitation. Fortunately, research results have shown that acidizing efficiency is relatively insensitive to acid injection rate and that the highest rate possible yields the best results. McLeod (1984) recommends relatively low injection rates based on the observation that acid contact time with the formation of 2–4 hours appears to give good results. da Motta (1993) shows that with shallow damage, acid injection rate has little effect on the residual skin after 100 gal/ft of injection rate; and with deeper damage, the higher the injection rate, the lower the residual skin. Paccaloni et al. (1988) and Paccaloni and Tambini (1990) also report high success rates in numerous field treatments using the highest injection rates possible.

There is always an upper limit on the acid injection rate that is imposed by formation breakdown (fracture) pressure p_{bd}. Assuming pseudo–steady-state flow, the maximum injection rate limited by the breakdown pressure is expressed as

$$q_{i,\max} = \frac{4.917 \times 10^{-6} kh(p_{bd} - \bar{p} - \Delta p_{sf})}{\mu_a \left(\ln \frac{0.472 r_e}{r_w} + S\right)}, \quad (16.6)$$

where

q_i = maximum injection rate, bbl/min
k = permeability of undamaged formation, md
h = thickness of pay zone to be treated, ft
p_{bd} = formation breakdown pressure, psia
$\bar{p}$ = reservoir pressure, psia
Δp_{sf} = safety margin, 200 to 500 psi
μ_a = viscosity of acid solution, cp
r_e = drainage radius, ft
r_w = wellbore radius, ft
S = skin factor, ft.

The acid injection rate can also be limited by surface injection pressure at the pump available to the treatment. This effect is described in the next section.

16.3.4 Acid Injection Pressure

In most acid treatment operations, only the surface tubing pressure is monitored. It is necessary to predict the surface injection pressure at the design stage for pump selection. The surface tubing pressure is related to the bottom-hole flowing pressure by

$$p_{si} = p_{wf} - \Delta p_h + \Delta p_f, \quad (16.7)$$

where

p_{si} = surface injection pressure, psia
p_{wf} = flowing bottom-hole pressure, psia
Δp_h = hydrostatic pressure drop, psia
Δp_f = frictional pressure drop, psia.

The second and the third term in the right-hand side of Eq. (16.7) can be calculated using Eq. (11.93). However, to avert the procedure of friction factor determination, the following approximation may be used for the frictional pressure drop calculation (Economides and Nolte, 2000):

$$\Delta p_f = \frac{518 \rho^{0.79} q^{1.79} \mu^{0.207}}{1{,}000 D^{4.79}} L, \quad (16.8)$$

where

ρ = density of fluid, g/cm^3
q = injection rate, bbl/min
μ = fluid viscosity, cp
D = tubing diameter, in.
L = tubing length, ft.

Equation (16.8) is relatively accurate for estimating frictional pressures for newtonian fluids at flow rates less than 9 bbl/min.

Example Problem 16.2 A 60-ft thick, 50-md sandstone pay zone at a depth of 9,500 ft is to be acidized with an acid solution having a specific gravity of 1.07 and a viscosity of 1.5 cp down a 2-in. inside diameter (ID) coil tubing. The formation fracture gradient is 0.7 psi/ft. The wellbore radius is 0.328 ft. Assuming a reservoir pressure of 4,000 psia, drainage area radius of 1,000 ft, and a skin factor of 15, calculate

(a) the maximum acid injection rate using safety margin 300 psi.

(b) the maximum expected surface injection pressure at the maximum injection rate.

Solution

(a) The maximum acid injection rate:

$$q_{i,\max} = \frac{4.917 \times 10^{-6} \, kh(p_{bd} - \bar{p} - \Delta p_{sf})}{\mu_a \left(\ln \frac{0.472 r_e}{r_w} + S \right)}$$

$$= \frac{4.917 \times 10^{-6}(50)(60)((0.7)(9,500) - 4,000 - 300)}{(1.5)\left(\ln \frac{0.472(1,000)}{(0.328)} + 15 \right)}$$

$$= 1.04 \, \text{bbl/min}$$

(b) The maximum expected surface injection pressure:

$$p_{wf} = p_{bd} - \Delta p_{sf} = (0.7)(9,500) - 300 = 6,350 \, \text{psia}$$
$$\Delta p_h = (0.433)(1.07)(9,500) = 4,401 \, \text{psi}$$
$$\Delta p_f = \frac{518 \rho^{0.79} q^{1.79} \mu^{0.207}}{1,000 D^{4.79}} L$$
$$= \frac{518(1.07)^{0.79}(1.04)^{1.79}(1.5)^{0.207}}{1,000(2)^{4.79}}(9,500)$$
$$= 218 \, \text{psi}$$
$$p_{si} = p_{wf} - \Delta p_h + \Delta p_f$$
$$= 6,350 - 4,401 + 218 = 2,167 \, \text{psia}$$

16.4 Carbonate Acidizing Design

The purpose of carbonate acidizing is not to remove the damage to the formation near the wellbore, but to create wormholes through which oil or gas will flow after stimulation. Figure 16.2 shows wormholes created by acid dissolution of limestone in a laboratory (Hoefner and Fogler, 1988).

Carbonate acidizing is a more difficult process to predict than sandstone acidizing because the physics is much more complex. Because the surface reaction rates are very high and mass transfer often plays the role of limiting step locally, highly nonuniform dissolution patterns are usually created. The structure of the wormholes depends on many factors including flow geometry, injection rate, reaction kinetics, and mass transfer rates. Acidizing design relies on mathematical models calibrated by laboratory data.

16.4.1 Selection of Acid

HCl is the most widely used acid for carbonate matrix acidizing. Weak acids are suggested for perforating fluid and perforation cleanup, and strong acids are recommended for other treatments. Table 16.3 lists recommended acid type and strength for carbonate acidizing (McLeod, 1984).

All theoretical models of wormhole propagation predict deeper penetration for higher acid strengths, so a high concentration of acid is always preferable.

Figure 16.2 Wormholes created by acid dissolution of limestone (Hoefner and Fogler, 1988; courtesy AIChE).

16.4.2 Acidizing Parameters

Acidizing parameters include acid volume, injection rate, and injection pressure. The acid volume can be calculated with two methods: (1) Daccord's wormhole propagation model and (2) the volumetric model, on the basis of desired penetration of wormholes. The former is optimistic, whereas the latter is more realistic (Economides et al., 1994).

Based on the wormhole propagation model presented by Daccord et al. (1989), the required acid volume per unit thickness of formation can be estimated using the following equation:

$$V_h = \frac{\pi \phi D^{2/3} q_h^{1/3} r_{wh}^{d_f}}{b N_{Ac}} \tag{16.9}$$

where

V_h = required acid volume per unit thickness of formation, m³/m
ϕ = porosity, fraction
D = molecular diffusion coefficient, m²/s
q_h = injection rate per unit thickness of formation, m³/sec-m
r_{wh} = desired radius of wormhole penetration, m
d_f = 1.6, fractal dimension
b = 105×10^{-5} in SI units
N_{Ac} = acid capillary number, dimensionless,

where the acid capillary number is defined as

$$N_{Ac} = \frac{\phi \beta \gamma_a}{(1 - \phi) \gamma_m}, \tag{16.10}$$

Table 16.3 Recommended Acid Type and Strength for Carbonate Acidizing

Perforating fluid:	5% acetic acid
Damaged perforations:	9% formic acid
	10% acetic acid
	15% HCl
Deep wellbore damage:	15% HCl
	28% HCl
	Emulsified HCl

where

γ_a = acid specific gravity, water = 1.0
γ_m = mineral specific gravity, water = 1.0.

Based on the volumetric model, the required acid volume per unit thickness of formation can be estimated using the following equation:

$$V_h = \pi\phi(r_{wh}^2 - r_w^2)(PV)_{bt}, \tag{16.11}$$

where $(PV)_{bt}$ is the number of pore volumes of acid injected at the time of wormhole breakthrough at the end of the core. Apparently, the volumetric model requires data from laboratory tests.

Example Problem 16.3 A 28 wt% HCl is needed to propagate wormholes 3 ft from a 0.328-ft radius wellbore in a limestone formation (specific gravity 2.71) with a porosity of 0.15. The designed injection rate is 0.1 bbl/min-ft, the diffusion coefficient is 10^{-9} m^2/sec, and the density of the 28% HCl is 1.14 g/cm^3. In linear core floods, 1.5 pore volume is needed for wormhole breakthrough at the end of the core. Calculate the acid volume requirement using (a) Daccord's model and (b) the volumetric model.

Solution

(a) Daccord's model:

$$\beta = C_a \frac{v_m MW_m}{v_a MW_a} = (0.28)\frac{(1)(100.1)}{(2)(36.5)}$$
$$= 0.3836\, \text{lb}_m\, \text{CaCO}_3/\text{lb}_m\, 28\,\text{wt\%}\,\text{HCl solution}.$$
$$N_{Ac} = \frac{\phi\beta\gamma_a}{(1-\phi)\gamma_m} = \frac{(0.15)(0.3836)(1.14)}{(1-0.15)(2.71)} = 0.0285$$
$$q_h = 0.1\, \text{bbl/min-ft} = 8.69 \times 10^{-4}\, \text{m}^3/\text{sec-m}$$
$$r_{wh} = 0.328 + 3 = 3.328\,\text{ft} = 1.01\,\text{m}$$
$$V_h = \frac{\pi\phi D^{2/3} q_h^{1/3} r_{wh}^{d_f}}{bN_{Ac}}$$
$$= \frac{\pi(0.15)(10^{-9})^{2/3}(8.69\times 10^{-4})^{1/3}(1.01)^{1.6}}{(1.5\times 10^{-5})(0.0285)}$$
$$= 0.107\, \text{m}^3/\text{m} = 8.6\,\text{gal/ft}$$

(b) Volumetric model:

$$V_h = \pi\phi(r_{wh}^2 - r_w^2)(PV)_{bt}$$
$$= \pi(0.15)(3.328^2 - 0.328^2)(1.5)$$
$$= 7.75\, \text{ft}^3/\text{ft} = 58\,\text{gal/ft}.$$

This example shows that the Daccord model gives optimistic results and the volumetric model gives more realistic results.

The maximum injection rate and pressure for carbonate acidizing can be calculated the same way as that for sandstone acidizing. Models of wormhole propagation predict that wormhole velocity increases with injection rate to the power of $1/2$ to 1. Therefore, the maximum injection rate is preferable. However, this approach may require more acid volume. If the acid volume is constrained, a slower injection rate may be preferable. If a sufficient acid volume is available, the maximum injection rate is recommended for limestone formations. However, a lower injection rate may be preferable for dolomites. This allows the temperature of the acid entering the formation to increase, and thus, the reaction rate increases.

The designed acid volume and injection rate should be adjusted based on the real-time monitoring of pressure during the treatment.

Summary

This chapter briefly presents chemistry of matrix acidizing and a guideline to acidizing design for both sandstone and carbonate formations. More in-depth materials can be found in McLeod (1984), Economides et al. (1994), and Economides and Nolte (2000).

References

DACCORD, G., TOUBOUL, E., and LENORMAND, R. Carbonate acidizing: toward a quantitative model of the wormholing phenomenon. *SPEPE* Feb. 1989:63–68.

DA MOTTA, E.P. Matrix Acidizing of Horizontal Wells, Ph.D. Dissertation. Austin: University of Texas at Austin, 1993.

ECONOMIDES, M.J., HILL, A.D., and EHLIG-ECONOMIDES, C. *Petroleum Production Systems*. Englewood Cliffs, NJ: Prentice Hall, 1994.

ECONOMIDES, M.J. and NOLTE, K.G. *Reservoir Stimulation*, 3rd edition. New York: John Wiley & Sons, 2000.

FOGLER, H.S., LUND, K., and MCCUNE, C.C. Predicting the flow and reaction of HCl/HF mixtures in porous sandstone cores. *SPEJ* Oct. 1976, *Trans. AIME*, 1976;234:248–260.

HEKIM, Y., FOGLER, H.S., and MCCUNE, C.C. The radial movement of permeability fronts and multiple reaction zones in porous media. *SPEJ* Feb. 1982:99–107.

HILL, A.D. and GALLOWAY, P.J. Laboratory and theoretical modeling of diverting agent behavior. *JPT* June 1984:1157–1163.

HILL, A.D., LINDSAY, D.M., SILBERBERG, I.H., and SCHECHTER, R.S. Theoretical and experimental studies of sandstone acidizing. *SPEJ* Feb. 1981;21:30–42.

HOEFNER, M.L. and FOGLER, H.S. Pore evolution and channel formation during flow and reaction in porous media. *AIChE J.* Jan. 1988;34:45–54.

LUND, K., FOGLER, H.S., and MCCUNE, C.C. Acidization I: the dissolution of dolomite in hydrochloric acid. *Chem. Eng. Sci.* 1973;28:691.

LUND, K., FOGLER, H.S., MCCUNE, C.C., and AULT, J.W. Acidization II: the dissolution of calcite in hydrochloric acid. *Chem. Eng. Sci.* 1975;30:825.

MCLEOD, H.O., JR. Matrix acidizing. *JPT* 1984;36:2055–2069.

PACCALONI, G. and TAMBINI, M. Advances in matrix stimulation technology. *JPT* 1993;45:256–263.

PACCALONI, G., TAMBINI, M., and GALOPPINI, M. Key factors for enhanced results of matrix stimulation treatment. Presented at the SPE Formation Damage Control Symposium held in Bakersfield, California on February 8–9, 1988. SPE Paper 17154.

SCHECHTER, R.S. *Oil Well Stimulation*. Englewood Cliffs, NJ: Prentice Hall, 1992.

SMITH, C.F., and HENDRICKSON, A.R. Hydrofluoric acid stimulation of sandstone reservoirs. *JPT* Feb. 1965, *Trans. AIME* 1965;234:215–222.

TAHA, R., HILL, A.D., and SEPEHRNOORI, K. Sandstone acidizing design with a generalized model. *SPEPE* Feb. 1989:49–55.

Problems

16.1 For the reaction between 20 wt% HCl solution and calcite, calculate the gravimetric and volumetric dissolving power of the acid solution.

16.2 For the reaction between 20 wt% HCl solution and dolomite, calculate the gravimetric and volumetric dissolving power of the acid solution.

16.3 A sandstone with a porosity of 0.18 containing 8 v% calcite is to be acidized with HF/HCl mixture solution. A preflush of 15 wt% HCl solution is to be injected ahead of the mixture to dissolve the carbonate minerals and establish a low-pH environment. If the HCl preflush is to remove all carbonates in a region within 1.5 ft beyond a 0.328-ft-radius wellbore before the HF/HCl stage enters the formation, what minimum preflush volume is required in terms of gallon per foot of pay zone?

16.4 A sandstone with a porosity of 0.15 containing 12 v% dolomite is to be acidized with HF/HCl mixture solution. A preflush of 15 wt% HCl solution is to be injected ahead of the mixture to dissolve the carbonate minerals and establish a low-pH environment. If the HCl preflush is to remove all carbonates in a region within 1.2 feet beyond a 0.328-ft-radius wellbore before the HF/HCl stage enters the formation, what minimum preflush volume is required in terms of gallon per foot of pay zone?

16.5 A 30-ft thick, 40-md sandstone pay zone at a depth of 9,000 ft is to be acidized with an acid solution having a specific gravity of 1.07 and a viscosity of 1.2 cp down a 2-in. ID coil tubing. The formation fracture gradient is 0.7 psi/ft. The wellbore radius is 0.328 ft. Assuming a reservoir pressure of 4,000 psia, drainage area radius of 1,500 ft and skin factor of 10, calculate

   (a) the maximum acid injection rate using safety margin 200 psi.

   (b) the maximum expected surface injection pressure at the maximum injection rate.

16.6 A 40-ft thick, 20-md sandstone pay zone at a depth of 8,000 ft is to be acidized with an acid solution having a specific gravity of 1.07 and a viscosity of 1.5 cp down a 2-in. ID coil tubing. The formation fracture gradient is 0.65 psi/ft. The wellbore radius is 0.328 ft. Assuming a reservoir pressure of 3,500 psia, drainage area radius of 1,200 ft, and skin factor of 15, calculate

   (a) the maximum acid injection rate using a safety margin of 400 psi.

   (b) the maximum expected surface injection pressure at the maximum injection rate.

16.7 A 20 wt% HCl is needed to propagate wormholes 2 ft from a 0.328-ft radius wellbore in a limestone formation (specific gravity 2.71) with a porosity of 0.12. The designed injection rate is 0.12 bbl/min-ft, the diffusion coefficient is 10^{-9} m^2/sec, and the density of the 20% HCl is 1.11 g/cm^3. In linear core floods, 1.2 pore volume is needed for wormhole breakthrough at the end of the core. Calculate the acid volume requirement using (a) Daccord's model and (b) the volumetric model.

16.8 A 25 wt% HCl is needed to propagate wormholes 3 ft from a 0.328-ft radius wellbore in a dolomite formation (specific gravity 2.87) with a porosity of 0.16. The designed injection rate is 0.15 bbl/min-ft, the diffusion coefficient is 10^{-9} m^2/sec, and the density of the 25% HCl is 1.15 g/cm^3. In linear core floods, 4 pore volumes is needed for wormhole breakthrough at the end of the core. Calculate the acid volume requirement using (a) Daccord's model and (b) the volumetric model.

17 Hydraulic Fracturing

Contents
17.1 Introduction   17/252
17.2 Formation Fracturing Pressure   17/252
17.3 Fracture Geometry   17/254
17.4 Productivity of Fractured Wells   17/256
17.5 Hydraulic Fracturing Design   17/258
17.6 Post-Frac Evaluation   17/262
Summary   17/264
References   17/264
Problems   17/265

17.1 Introduction

Hydraulic fracturing is a well-stimulation technique that is most suitable to wells in low- and moderate-permeability reservoirs that do not provide commercial production rates even though formation damages are removed by acidizing treatments.

Hydraulic fracturing jobs are carried out at well sites using heavy equipment including truck-mounted pumps, blenders, fluid tanks, and proppant tanks. Figure 17.1 illustrates a simplified equipment layout in hydraulic fracturing treatments of oil and gas wells. A hydraulic fracturing job is divided into two stages: the pad stage and the slurry stage (Fig. 17.2). In the pad stage, fracturing fluid only is injected into the well to break down the formation and create a pad. The pad is created because the fracturing fluid injection rate is higher than the flow rate at which the fluid can escape into the formation. After the pad grows to a desirable size, the slurry stage is started. During the slurry stage, the fracturing fluid is mixed with sand/proppant in a blender and the mixture is injected into the pad/fracture. After filling the fracture with sand/proppant, the fracturing job is over and the pump is shut down. Apparently, to reduce the injection rate requirement, a low leaf-off fracturing fluid is essential. Also, to prop the fracture, the sand/proppant should have a compressive strength that is high enough to resist the stress from the formation.

This chapter concisely describes hydraulic fracturing treatments. For detailed information on this subject, see Economides and Nolte (2000). This chapter focuses on the following topics:

- Formation fracturing pressure
- Fracture geometry
- Productivity of fractured wells
- Hydraulic fracturing design
- Post-frac evaluation

17.2 Formation Fracturing Pressure

Formation fracturing pressure is also called *breakdown pressure*. It is one of the key parameters used in hydraulic fracturing design. The magnitude of the parameter depends on formation depth and properties. Estimation of the parameter value begins with *in situ* stress analysis.

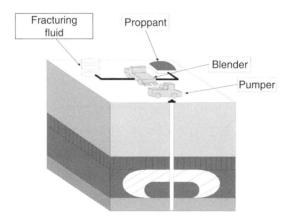

Figure 17.1 Schematic to show the equipment layout in hydraulic fracturing treatments of oil and gas wells.

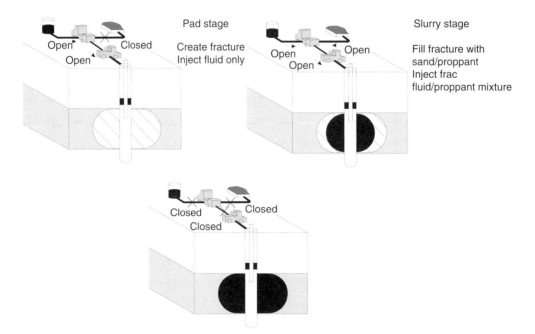

Figure 17.2 A schematic to show the procedure of hydraulic fracturing treatments of oil and gas wells.

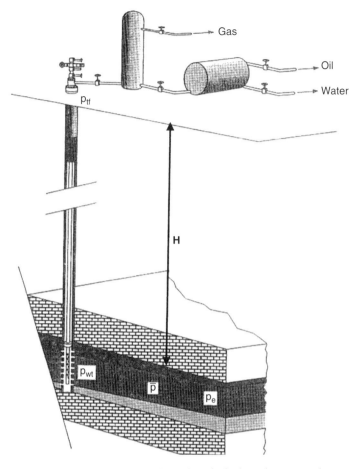

Figure 17.3 Overburden formation of a hydrocarbon reservoir.

Consider a reservoir rock at depth H as shown in Fig. 17.3. The *in situ* stress caused by the weight of the overburden formation in the vertical direction is expressed as

$$\sigma_v = \frac{\rho H}{144}, \quad (17.1)$$

where

σ_v = overburden stress, psi
ρ = the average density of overburden formation, lb/ft^3
H = depth, ft.

The overburden stress is carried by both the rock grains and the fluid within the pore space between the grains. The contact stress between grains is called *effective stress* (Fig. 17.4):

$$\sigma'_v = \sigma_v - \alpha p_p, \quad (17.2)$$

where

σ'_v = effective vertical stress, psi
α = Biot's poro-elastic constant, approximately 0.7
p_p = pore pressure, psi.

The effective horizontal stress is expressed as

$$\sigma'_h = \frac{\nu}{1-\nu}\sigma'_v, \quad (17.3)$$

where ν is Poison's ratio. The total horizontal stress is expressed as

$$\sigma_h = \sigma'_h + \alpha p_p. \quad (17.4)$$

Because of the tectonic effect, the magnitude of the horizontal stress may vary with direction. The maximum horizontal stress may be $\sigma_{h,\max} = \sigma_{h,\min} + \sigma_{tect}$, where σ_{tect} is called *tectonic stress*.

Based on a failure criterion, Terzaghi presented the following expression for the breakdown pressure:

$$p_{bd} = 3\sigma_{h,\min} - \sigma_{h,\max} + T_0 - p_p, \quad (17.5)$$

where T_0 is the tensile strength of the rock.

Example Problem 17.1 A sandstone at a depth of 10,000 ft has a Poison's ratio of 0.25 and a poro-elastic constant of 0.72. The average density of the overburden formation is 165 lb/ft^3. The pore pressure gradient in the sandstone is 0.38 psi/ft. Assuming a tectonic stress of 2,000 psi and a tensile strength of the sandstone of 1,000 psi, predict the breakdown pressure for the sandstone.

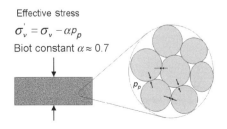

Figure 17.4 Concept of effective stress between grains.

Solution

Overburden stress:

$$\sigma_v = \frac{\rho H}{144} = \frac{(165)(10,000)}{144} = 11,500 \text{ psi}$$

Pore pressure:

$$p_p = (0.38)(10,000) = 3,800 \text{ psi}$$

The effective vertical stress:

$$\sigma'_v = \sigma_v - \alpha p_p = 11,500 - (0.72)(3,800) = 8,800 \text{ psi}$$

The effective horizontal stress:

$$\sigma'_h = \frac{\nu}{1-\nu}\sigma'_v = \frac{0.25}{1-0.25}(8,800) = 2,900 \text{ psi}$$

The minimum horizontal stress:

$$\sigma_{h,\min} = \sigma'_h + \alpha p_p = 2,900 + (0.72)(3,800) = 5,700 \text{ psi}$$

The maximum horizontal stress:

$$\sigma_{h,\max} = \sigma_{h,\min} + \sigma_{tect} = 5,700 + 2,000 = 7,700 \text{ psi}$$

Breakdown pressure:

$$\begin{aligned}p_{bd} &= 3\sigma_{h,\min} - \sigma_{h,\max} + T_0 - p_p \\ &= 3(5,700) - 7,700 + 1,000 - 3,800 = 6,600 \text{ psi}\end{aligned}$$

17.3 Fracture Geometry

It is still controversial about whether a single fracture or multiple fractures are created in a hydraulic fracturing job. Whereas both cases have been evidenced based on the information collected from tiltmeters and microseismic data, it is commonly accepted that each individual fracture is sheet-like. However, the shape of the fracture varies as predicted by different models.

17.3.1 Radial Fracture Model

A simple radial (penny-shaped) crack/fracture was first presented by Sneddon and Elliot (1946). This occurs when there are no barriers constraining height growth or when a horizontal fracture is created. Geertsma and de Klerk (1969) presented a radial fracture model showing that the fracture width at wellbore is given by

$$w_w = 2.56\left[\frac{\mu q_i(1-\nu)R}{E}\right]^{\frac{1}{4}}, \qquad (17.6)$$

where

w_w = fracture width at wellbore, in.
μ = fluid viscosity, cp
q_i = pumping rate, bpm
R = the radius of the fracture, ft
E = Young's modulus, psi.

Assuming the fracture width drops linearly in the radial direction, the average fracture width may be expressed as

$$\bar{w} = 0.85\left[\frac{\mu q_i(1-\nu)R}{E}\right]^{\frac{1}{4}}. \qquad (17.7)$$

17.3.2 The KGD Model

Assuming that a fixed-height vertical fracture is propagated in a well-confined pay zone (i.e., the stresses in the layers above and below the pay zone are large enough to prevent fracture growth out of the pay zone), Khristianovich and Zheltov (1955) presented a fracture model as shown in Fig. 17.5. The model assumes that the width of the crack at any distance from the well is independent of vertical position, which is a reasonable approximation for a fracture with height much greater than its length. Their solution included the fracture mechanics aspects of the fracture tip. They assumed that the flow rate in the fracture was constant, and that the pressure in the fracture could be approximated by a constant pressure in the majority of the fracture body, except for a small region near the tip with no fluid penetration, and hence, no fluid pressure. This concept of fluid lag has remained an element of the mechanics of the fracture tip. Geertsma and de Klerk (1969) gave a much simpler solution to the same problem. The solution is now referred to as the *KGD model*. The average width of the KGD fracture is expressed as

$$\bar{w} = 0.29\left[\frac{q_i\mu(1-\nu)x_f^2}{Gh_f}\right]^{1/4}\left(\frac{\pi}{4}\right), \qquad (17.8)$$

where

$\bar{w}$ = average width, in.
q_i = pumping rate, bpm

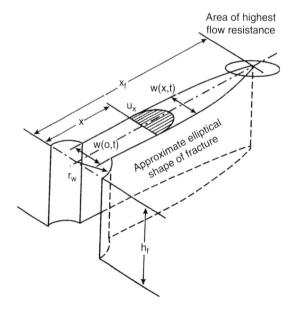

Figure 17.5 The KGD fracture geometry.

μ = fluid viscosity, cp
G = $E/2(1+\nu)$, shear modulus, psia
h_f = fracture height, ft

17.3.3 The PKN model
Perkins and Kern (1961) also derived a solution for a fixed-height vertical fracture as illustrated in Fig. 17.6. Nordgren (1972) added leakoff and storage within the fracture (due to increasing width) to the Perkins and Kern model, deriving what is now known as the *PKN model*. The average width of the PKN fracture is expressed as

$$\overline{w} = 0.3 \left[\frac{q_i \mu (1-\nu) x_f}{G} \right]^{1/4} \left(\frac{\pi}{4} \gamma \right), \qquad (17.9)$$

where $\gamma \approx 0.75$. It is important to emphasize that even for contained fractures, the PKN solution is only valid when the fracture length is at least three times the height.

The three models discussed in this section all assume that the fracture is planar, that is, fracture propagates in a particular direction (perpendicular to the minimum stress), fluid flow is one-dimensional along the length (or radius) of the fracture, and leakoff behavior is governed by a simple expression derived from filtration theory. The rock in which the fracture propagates is assumed to be a continuous, homogeneous, isotropic linear elastic solid, and the fracture is considered to be of fixed height (PKN and KGD) or completely confined in a given layer (radial). The KGD and PKN models assume respectively that the fracture height is large or small relative to length, while the radial model assumes a circular shape. Since these models were developed, numerous extensions have been made, which have relaxed these assumptions.

17.3.4 Three-Dimensional and Pseudo-3D Models
The planar 2D models discussed in the previous section are deviated with significant simplifying assumptions. Although their accuracies are limited, they are useful for understanding the growth of hydraulic fractures. The power of modern computer allows routine treatment designs to be made with more complex models, which are solved numerically. The biggest limitation of the simple models is the requirement to specify the fracture height or to assume that a radial fracture will develop. It is not always obvious from data such as logs where, or whether, the fracture will be contained. In addition, the fracture height will usually vary from the well to the tip of the fracture, as the pressure varies.

There are two major types of *pseudo–three-dimensional* (P3D) *models: lumped* and *cell based*. In the lumped (or elliptical) models, the fracture shape is assumed to consist of two half-ellipses joined at the center. The horizontal length and wellbore vertical tip extensions are calculated at each time-step, and the assumed shape is made to match these positions. Fluid flow is assumed to occur along streamlines from the perforations to the edge of the ellipse, with the shape of the streamlines derived from simple analytical solutions. In cell-based models, the fracture shape is not prescribed. The fracture is treated as a series of connected cells, which are linked only via the fluid flow from cell to cell. The height at any cross-section is calculated from the pressure in that cell, and fluid flow in the vertical direction is generally approximated.

Lumped models were first introduced by Cleary (1980), and numerous papers have since been presented on their use (e.g., Cleary et al., 1994). As stated in the 1980 paper, "The heart of the formulae can be extracted very simply by a non-dimensionalization of the governing equations; the remainder just involves a good physics-mathematical choice of the undetermined coefficients." The lumped models implicitly require the assumption of a self-similar fracture shape (i.e., one that is the same as time evolves, except for length scale). The shape is generally assumed to consist of two half-ellipses of equal lateral extent, but with different vertical extent.

In cell-based P3D models, the fracture length is discretized into cells along the length of the fracture. Because only one direction is discretized and fluid flow is assumed to be essentially horizontal along the length of the fracture, the model can be solved much more easily than planar 3D models. Although these models allow the calculation of fracture height growth, the assumptions make them primarily suitable for reasonably contained fractures, with length much greater than height.

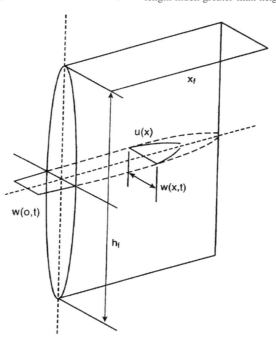

Figure 17.6 The PKN fracture geometry.

Planar 3D models: The geometry of a hydraulic fracture is defined by its width and the shape of its periphery (i.e., height at any distance from the well and length). The width distribution and the overall shape change as the treatment is pumped, and during closure. They depend on the pressure distribution, which itself is determined by the pressure gradients caused by the fluid flow within the fracture. The relation between pressure gradient and flow rate is very sensitive to fracture width, resulting in a tightly coupled calculation. Although the mechanics of these processes can be described separately, this close coupling complicates the solution of any fracture model. The nonlinear relation between width and pressure and the complexity of a moving-boundary problem further complicate numerical solutions. Clifton and Abou-Sayed (1979) reported the first numerical implementation of a planar model. The solution starts with a small fracture, initiated at the perforations, divided into a number of equal elements (typically 16 squares). The elements then distort to fit the evolving shape. The elements can develop large aspect ratios and very small angles, which are not well handled by the numerical schemes typically used to solve the model. Barree (1983) developed a model that does not show grid distortion. The layered reservoir is divided into a grid of equal-size rectangular elements, over the entire region that the fracture may cover.

Simulators based on such models are much more computationally demanding than P3D-based simulators, because they solve the fully 2D fluid-flow equations and couple this solution rigorously to the elastic-deformation equations. The elasticity equations are also solved more rigorously, using a 3D solution rather than 2D slices. Computational power and numerical methods have improved to the point that these models are starting to be used for routine designs. They should be used whenever a significant portion of the fracture volume is outside the zone where the fracture initiates or where there is significant vertical fluid flow. Such cases typically arise when the stress in the layers around the pay zone is similar to or lower than that within the pay.

Regardless of which type of model is used to calculate the fracture geometry, limited data are available on typical treatments to validate the model used. On commercial treatments, the pressure history during the treatment is usually the only data available to validate the model. Even in these cases, the quality of the data is questionable if the bottom-hole pressure must be inferred from the surface pressure. The bottom-hole pressure is also not sufficient to uniquely determine the fracture geometry in the absence of other information, such as that derived from tiltmeters and microseismic data. If a simulator incorporates the correct model, it should match both treating pressure and fracture geometry.

Table 17.1 summarizes main features of fracture models in different categories. Commercial packages are listed in Table 17.2.

17.4 Productivity of Fractured Wells

Hydraulically created fractures gather fluids from reservoir matrix and provide channels for the fluid to flow into wellbores. Apparently, the productivity of fractured wells depends on two steps: (1) receiving fluids from formation and (2) transporting the received fluid to the wellbore. Usually one of the steps is a limiting step that controls the well-production rate. The efficiency of the first step depends on fracture dimension (length and height), and the efficiency of the second step depends on fracture permeability. The relative importance of each of the steps can be analyzed using the concept of fracture conductivity defined as (Argawal et al., 1979; Cinco-Ley and Samaniego, 1981):

$$F_{CD} = \frac{k_f w}{k x_f}, \tag{17.10}$$

where

F_{CD} = fracture conductivity, dimensionless
k_f = fracture permeability, md
w = fracture width, ft
x_f = fracture half-length, ft.

Table 17.1 *Features of Fracture Geometry Models*

A. 2D models
    Constant height
    Plain strain/stress
    Homogeneous stress/elastic properties
    Engineering oriented: quick look
    Limited computing requirements
B. Pseudo-3D (2D × 2D) models
    Limited height growth
    Planar frac properties of layers/adjacent zones
    State of stress
    Specialized field application
    Moderate computer requirements
C. Fully 3D models
    Three-dimensional propagation
    Nonideal geometry/growth regimes
    Research orientated
    Large database and computer requirements
    Calibration of similar smaller models in conjunction with laboratory experiments

Table 17.2 *Summary of Some Commercial Fracturing Models*

Software name	Model type	Company	Owner
PROP	Classic 2D	Halliburton	
Chevron 2D	Classic 2D	ChevronTexaco	
CONOCO 2D	Classic 2D	CONOCO	
Shell 2D	Classic 2D	Shell	
TerraFrac	Planar 3D	Terra Tek	ARCO
HYRAC 3D	Planar 3D	Lehigh U.	S.H. Advani
GOHFER	Planar 3D	Marathon	R. Barree
STIMPLAN	Pseudo–3D "cell"	NSI Technologies	M. Smith
ENERFRAC	Pseudo–3D "cell"	Shell	
TRIFRAC	Pseudo–3D "cell"	S.A. Holditch & Association	
FracCADE	Pseudo–3D "cell"	Schlumberger	EAD sugar-land
PRACPRO	Pseudo–3D "parametric"	RES, Inc.	GTI
PRACPROPT	Pseudo–3D "parametric"	Pinnacle Technologies	GTI
MFRAC-III	Pseudo–3D "parametric"	Meyer & Associates	Bruce Meyer
Fracanal	Pseudo–3D "parametric"	Simtech	A. Settari

In the situations in which the fracture dimension is much less than the drainage area of the well, the long-term productivity of the fractured well can be estimated assuming pseudo-radial flow in the reservoir. Then the inflow equation can be written as

$$q = \frac{kh(p_e - p_{wf})}{141.2B\mu\left(\ln\frac{r_e}{r_w} + S_f\right)}, \quad (17.11)$$

where S_f is the equivalent skin factor. The fold of increase can be expressed as

$$\frac{J}{J_o} = \frac{\ln\frac{r_e}{r_w}}{\ln\frac{r_e}{r_w} + S_f}, \quad (17.12)$$

where

J = productivity of fractured well, stb/day-psi
J_o = productivity of nonfractured well, stb/day-psi.

The effective skin factor S_f can be determined based on fracture conductivity and Fig. 17.7.

It is seen from Fig. 17.7 that the parameter $S_f + \ln(x_f/r_w)$ approaches a constant value in the range of $F_{CD} > 100$, that is, which gives

$$S_f \approx 0.7 - \ln(x_f/r_w), \quad (17.13)$$

meaning that the equivalent skin factor of fractured wells depends only on fracture length for high-conductivity fractures, not fracture permeability and width. This is the situation in which the first step is the limiting step. On the other hand, Fig. 17.7 indicates that the parameter $S_f + \ln(x_f/r_w)$ declines linearly with log (F_{CD}) in the range of $F_{CD} < 1$, that is,

$$S_f \approx 1.52 + 2.31\log(r_w) - 1.545\log\left(\frac{k_f w}{k}\right)$$
$$- 0.765\log(x_f). \quad (17.14)$$

Comparing the coefficients of the last two terms in this relation indicates that the equivalent skin factor of fractured well is more sensitive to the fracture permeability and width than to fracture length for low-conductivity fractures. This is the situation in which the second step is the limiting step.

The previous analyses reveal that low-permeability reservoirs, leading to high-conductivity fractures, would benefit greatly from fracture length, whereas high-permeability reservoirs, naturally leading to low-conductivity fractures, require good fracture permeability and width.

Valko et al. (1997) converted the data in Fig. 17.7 into the following correlation:

$$s_f + \ln\left(\frac{x_f}{r_w}\right) = \frac{1.65 - 0.328u + 0.116u^2}{1 + 0.180u + 0.064u^2 + 0.05u^3} \quad (17.15)$$

where

$$u = \ln(F_{CD}) \quad (17.16)$$

Example Problem 17.2 A gas reservoir has a permeability of 1 md. A vertical well of 0.328-ft radius draws the reservoir from the center of an area of 160 acres. If the well is hydraulically fractured to create a 2,000-ft long, 0.12-in. wide fracture of 200,000 md permeability around the center of the drainage area, what would be the fold of increase in well productivity?

Solution Radius of the drainage area:

$$r_e = \sqrt{\frac{A}{\pi}} = \sqrt{\frac{(43,560)(160)}{\pi}} = 1,490\,\text{ft}$$

Fracture conductivity:

$$F_{CD} = \frac{k_f w}{k x_f} = \frac{(200,000)(0.12/12)}{(1)(2,000/2)} = 2$$

Figure 17.7 reads

$$S_f + \ln(x_f/r_w) \approx 1.2,$$

which gives

$$S_f \approx 1.2 - \ln(x_f/r_w) = 1.2 - \ln(1,000/0.328) = -6.82.$$

The fold of increase is

$$\frac{J}{J_o} = \frac{\ln\frac{r_e}{r_w}}{\ln\frac{r_e}{r_w} + S_f} = \frac{\ln\frac{1,490}{0.328}}{\ln\frac{1,490}{0.328} - 6.82} = 5.27.$$

In the situations in which the fracture dimension is comparable to the drainage area of the well, significant error may result from using Eq. (17.12), which was derived based

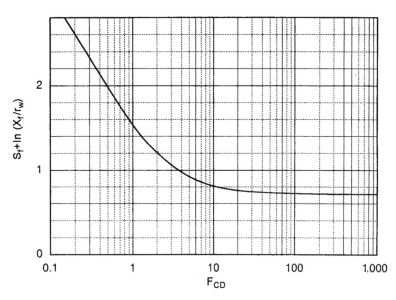

Figure 17.7 Relationship between fracture conductivity and equivalent skin factor (Cinco-Ley and Samaniego, 1981).

on radial flow. In these cases, the long-term productivity of the well may be estimated assuming bilinear flow in the reservoir. Pressure distribution in a linear flow reservoir and a linear flow in a finite conductivity fracture is illustrated in Fig. 17.8. An analytical solution for estimating fold of increase in well productivity was presented by Guo and Schechter (1999) as follows:

$$\frac{J}{J_o} = \frac{0.72\left(\ln\frac{r_e}{r_w} - \frac{3}{4} + S_o\right)}{(z_e\sqrt{c} + S)\left(\frac{1}{1-e^{-\sqrt{c}x_f}} - \frac{1}{2x_f\sqrt{c}}\right)}, \tag{17.17}$$

where $c = \frac{2k}{z_e w k_f}$ and z_e are distance between the fracture and the boundary of the drainage area.

17.5 Hydraulic Fracturing Design

Hydraulic fracturing designs are performed on the basis of parametric studies to maximize net present values (NPVs) of the fractured wells. A hydraulic fracturing design should follow the following procedure:

1. Select a fracturing fluid
2. Select a proppant
3. Determine the maximum allowable treatment pressure
4. Select a fracture propagation model
5. Select treatment size (fracture length and proppant concentration)
6. Perform production forecast analyses
7. Perform NPV analysis

A complete design must include the following components to direct field operations:

- Specifications of fracturing fluid and proppant
- Fluid volume and proppant weight requirements
- Fluid injection schedule and proppant mixing schedule
- Predicted injection pressure profile

17.5.1 Selection of Fracturing Fluid

Fracturing fluid plays a vital role in hydraulic fracture treatment because it controls the efficiencies of carrying proppant and filling in the fracture pad. Fluid loss is a major fracture design variable characterized by a fluid-loss coefficient C_L and a spurt-loss coefficient S_p. Spurt loss occurs only for wall-building fluids and only until the filter cake is established. Fluid loss into the formation is a more steady process than spurt loss. It occurs after the filter cake is developed. Excessive fluid loss prevents fracture propagation because of insufficient fluid volume accumulation in the fracture. Therefore, a fracture fluid with the lowest possible value of fluid-loss (leak-off) coefficient C_L should be selected.

The second major variable is fluid viscosity. It affects transporting, suspending, and deposition of proppants, as well as back-flowing after treatment. The viscosity should be controlled in a range suitable for the treatment. A fluid viscosity being too high can result in excessive injection pressure during the treatment.

However, other considerations may also be major for particular cases. They are compatibility with reservoir fluids and rock, compatibility with other materials (e.g., resin-coated proppant), compatibility with operating pressure and temperature, and safety and environmental concerns.

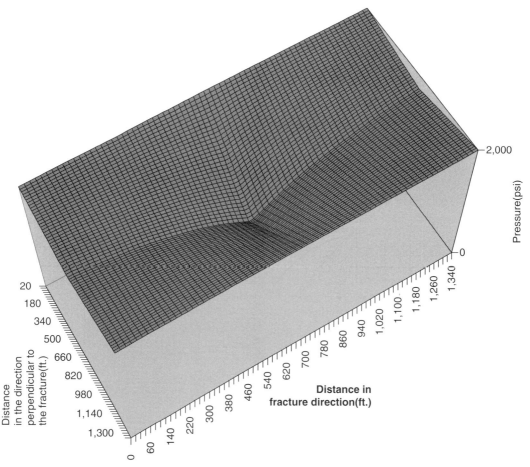

Figure 17.8 Relationship between fracture conductivity and equivalent skin factor.

17.5.2 Selection of Proppant

Proppant must be selected on the basis of *in situ* stress conditions. Major concerns are compressive strength and the effect of stress on proppant permeability. For a vertical fracture, the compressive strength of the proppant should be greater than the effective horizontal stress. In general, bigger proppant yields better permeability, but proppant size must be checked against proppant admittance criteria through the perforations and inside the fracture. Figure 17.9 shows permeabilities of various types of proppants under fracture closure stress.

Example Problem 17.3 For the following situation, estimate the minimum required compressive strength of 20/40 proppant. If intermediate-strength proppant is used, estimate the permeability of the proppant pack:

Formation depth: 10,000 ft
Overburden density: 165 lb_m/ft^3
Poison's ratio: 0.25
Biot constant: 0.7
Reservoir pressure: 6,500 psi
Production drawdown: 2,000 and 4,000 psi

Solution

The initial effective horizontal stress:

$$\sigma'_h = \frac{\nu}{1-\nu}\left(\frac{\rho H}{144} - \alpha p_p\right)$$

$$= \frac{0.25}{1-0.25}\left[\frac{(165)(10,000)}{144} - (0.7)(6500)\right] = 2,303\,\text{psi}$$

The effective horizontal stress under 2,000-psi pressure drawdown:

$$\sigma'_h = \frac{\nu}{1-\nu}\left(\frac{\rho H}{144} - \alpha p_p\right)$$

$$= \frac{0.25}{1-0.25}\left[\frac{(165)(10,000)}{144} - (0.7)(4500)\right] = 2,770\,\text{psi}$$

The effective horizontal stress under 4,000-psi pressure drawdown:

$$\sigma'_h = \frac{\nu}{1-\nu}\left(\frac{\rho H}{144} - \alpha p_p\right)$$

$$= \frac{0.25}{1-0.25}\left[\frac{(165)(10,000)}{144} - (0.7)(2500)\right] = 3,236\,\text{psi}$$

Therefore, the minimum required proppant compressive strength is 3,236 psi. Figure 17.9 indicates that the pack of the intermediate-strength proppants will have a permeability of about $k_f = 500$ darcies.

17.5.3 The maximum Treatment Pressure

The maximum treatment pressure is expected to occur when the formation is broken down. The bottom-hole pressure is equal to the formation breakdown pressure p_{bd} and the expected surface pressure can be calculated by

$$p_{si} = p_{bd} - \Delta p_h + \Delta p_f, \tag{17.18}$$

where

p_{si} = surface injection pressure, psia
p_{bd} = formation breakdown pressure, psia
Δp_h = hydrostatic pressure drop, psia
Δp_f = frictional pressure drop, psia.

The second and the third term in the right-hand side of Eq. (17.18) can be calculated using Eq. (11.93) (see Chapter 11). However, to avert the procedure of friction factor determination, the following approximation may be used for the frictional pressure drop calculation (Economides and Nolte, 2000):

$$\Delta p_f = \frac{518\rho^{0.79}q^{1.79}\mu^{0.207}}{1,000D^{4.79}}L, \tag{17.19}$$

where

ρ = density of fluid, g/cm^3
q = injection rate, bbl/min
μ = fluid viscosity, cp
D = tubing diameter, in.
L = tubing length, ft.

Equation (17.19) is relatively accurate for estimating frictional pressures for newtonian fluids at low flow rates.

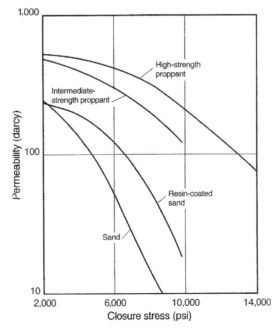

Figure 17.9 *Effect of fracture closure stress on proppant pack permeability (Economides and Nolte, 2000).*

Example Problem 17.4 For Example Problem 17.1, predict the maximum expected surface injection pressure using the following additional data:

Specific gravity of fracturing fluid: 1.2
Viscosity of fracturing fluid: 20 cp
Tubing inner diameter: 3.0 in.
Fluid injection rate: 10 bpm

Solution

Hydrostatic pressure drop:
$$\Delta p_h = (0.433)(1.2)(10,000) = 5,196 \text{ psi}$$

Frictional pressure drop:
$$\Delta p_f = \frac{518\rho^{0.79} q^{1.79} \mu^{0.207}}{1,000 D^{4.79}} L$$
$$= \frac{518(1.2)^{0.79}(10)^{1.79}(20)^{0.207}}{1,000(3)^{4.79}} (10,000) = 3,555 \text{ psi}$$

Expected surface pressure:
$$p_{si} = p_{bd} - \Delta p_h + \Delta p_f = 6,600 - 5,196 + 3,555$$
$$= 4,959 \text{ psia}$$

17.5.4 Selection of Fracture Model

An appropriate fracture propagation model is selected for the formation characteristics and pressure behavior on the basis of *in situ* stresses and laboratory tests. Generally, the model should be selected to match the level of complexity required for the specific application, quality and quantity of data, allocated time to perform a design, and desired level of output. Modeling with a planar 3D model can be time consuming, whereas the results from a 2D model can be simplistic. Pseudo-3D models provide a compromise and are most often used in the industry. However, 2D models are still attractive in situations in which the reservoir conditions are simple and well understood. For instance, to simulate a short fracture to be created in a thick sandstone, the KGD model may be beneficial. To simulate a long fracture to be created in a sandstone tightly bonded by strong overlaying and underlaying shales, the PKN model is more appropriate. To simulate frac-packing in a thick sandstone, the radial fracture model may be adequate. It is always important to consider the availability and quality of input data in model selection: garbage-in garbage-out (GIGO).

17.5.5 Selection of Treatment Size

Treatment size is primarily defined by the fracture length. Fluid and proppant volumes are controlled by fracture length, injection rate, and leak-off properties. A general statement can be made that the greater the propped fracture length and greater the proppant volume, the greater the production rate of the fractured well. Limiting effects are imposed by technical and economical factors such as available pumping rate and costs of fluid and proppant. Within these constraints, the optimum scale of treatment should be ideally determined based on the maximum NPV. This section demonstrates how to design treatment size using the KGD fracture model for simplicity. Calculation procedure is summarized as follows:

1. Assume a fracture half-length x_f and injection rate q_i, calculate the average fracture width $\bar{w}$ using a selected fracture model.
2. Based on material balance, solve injection fluid volume V_{inj} from the following equation:

$$V_{inj} = V_{frac} + V_{Leakoff}, \quad (17.20)$$

where
$$V_{inj} = q_i t_i \quad (17.21)$$
$$V_{frac} = A_f \bar{w} \quad (17.22)$$
$$V_{Leakoff} = 2 K_L C_L A_f r_p \sqrt{t_i} \quad (17.23)$$
$$K_L = \frac{1}{2}\left[\frac{8}{3}\eta + \pi(1-\eta)\right] \quad (17.24)$$

$$r_p = \frac{h}{h_f} \quad (17.25)$$
$$A_f = 2 x_f h_f \quad (17.26)$$
$$\eta = \frac{V_{frac}}{V_{inj}} \quad (17.27)$$
$$V_{pad} = V_{inj} \frac{1-\eta}{1+\eta} \quad (17.28)$$

Since K_L depends on fluid efficiency η, which is not known in the beginning, a numerical iteration procedure is required. The procedure is illustrated in Fig. 17.10.

3. Generate proppant concentration schedule using:

$$c_p(t) = c_f \left(\frac{t - t_{pad}}{t_{inj} - t_{pad}}\right)^{\varepsilon}, \quad (17.29)$$

where c_f is the final concentration in ppg. The proppant concentration in pound per gallon of added fluid (ppga) is expressed as

$$c'_p = \frac{c_p}{1 - c_p/\rho_p} \quad (17.30)$$

and
$$\varepsilon = \frac{1-\eta}{1+\eta}. \quad (17.31)$$

4. Predict propped fracture width using

$$w = \frac{C_p}{(1-\phi_p)\rho_p}, \quad (17.32)$$

where
$$C_p = \frac{M_p}{2 x_f h_f} \quad (17.33)$$
$$M_p = \bar{c}_p (V_{inj} - V_{pad}) \quad (17.34)$$
$$\bar{c}_p = \frac{c_f}{1+\varepsilon} \quad (17.35)$$

Example Problem 17.5 The following data are given for a hydraulic fracturing treatment design:

Pay zone thickness: 70 ft
Young's modulus of rock: 3×10^6 psi
Poison's ratio: 0.25
Fluid viscosity: 1.5 cp
Leak-off coefficient: $0.002 \text{ ft}/\min^{1/2}$
Proppant density: 165 lb/ft^3
Proppant porosity: 0.4
Fracture half-length: 1,000 ft

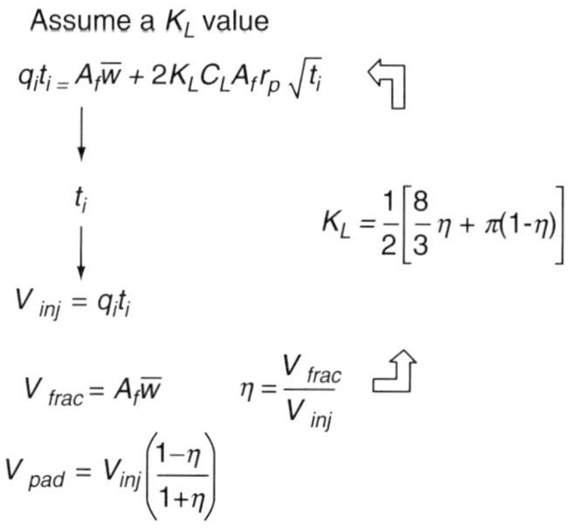

Assume a K_L value

$$q_i t_i = A_f \bar{w} + 2 K_L C_L A_f r_p \sqrt{t_i}$$

↓

t_i

$$K_L = \frac{1}{2}\left[\frac{8}{3}\eta + \pi(1-\eta)\right]$$

↓

$V_{inj} = q_i t_i$

$V_{frac} = A_f \bar{w}$ $\eta = \dfrac{V_{frac}}{V_{inj}}$ ↰

$V_{pad} = V_{inj}\left(\dfrac{1-\eta}{1+\eta}\right)$

Figure 17.10 Iteration procedure for injection time calculation.

Fracture height: 100 ft
Fluid injection rate: 40 bpm
Final proppant concentration: 3 ppg

Assuming KGD fracture, estimate

a. Fluid volume requirement
b. Proppant mixing schedule
c. Proppant weight requirement
d. Propped fracture width

Solution

a. Fluid volume requirements:

The average fracture width:

$$\bar{w} = 0.29 \left[\frac{q_i \mu (1-\nu) x_f^2}{G h_f}\right]^{1/4} \left(\frac{\pi}{4}\right)$$

$$= 0.29 \left[\frac{(40)(1.5)(1-0.25)(1,000)^2}{\frac{(3 \times 10^6)}{2(1+0.25)}(70)}\right]^{1/4} \left(\frac{\pi}{4}\right) = 0.195 \text{ in.}$$

Fracture area:

$$A_f = 2 x_f h_f = 2(1,000)(100) = 2 \times 10^5 \text{ ft}^2$$

Fluid volume based on volume balance:

$$q_i t_i = A_f \bar{w} + 2 K_L C_L A_f r_p \sqrt{t_i}.$$

Assuming $K_L = 1.5$,

$$(40)(5.615) t_i = (2 \times 10^5)\left(\frac{0.195}{12}\right) + 2(1.5)(2 \times 10^{-3})$$

$$\times (2 \times 10^5)\left(\frac{70}{100}\right)\sqrt{t_i}$$

gives $t_i = 37$ min.
Check K_L value:

$$V_{inj} = q_i t = (40)(42)(37) = 6.26 \times 10^4 \text{ gal}$$

$$V_{frac} = A_f \bar{w} = (2 \times 10^5)\left(\frac{0.195}{12}\right)(7.48) = 2.43 \times 10^4 \text{ gal}$$

$$\eta = \frac{V_{frac}}{V_{inj}} = \frac{2.43 \times 10^4}{6.26 \times 10^4} = 0.3875$$

$$K_L = \frac{1}{2}\left[\frac{3}{8}\eta + \pi(1-\eta)\right] = \frac{1}{2}\left[\frac{3}{8}(0.3875) + \pi(1-0.3875)\right]$$

$$= 1.48 \text{ OK}$$

Pad volume:

$$\varepsilon = \frac{1-\eta}{1+\eta} = \frac{1-0.3875}{1+0.3875} = 0.44$$

$$V_{pad} = V_{inj}\varepsilon = (6.26 \times 10^4)(0.44) = 2.76 \times 10^4 \text{ gal}$$

It will take 17 min to pump the pad volume at an injection rate of 40 bpm.

b. Proppant mixing schedule:

$$c_p(t) = (3)\left(\frac{t-17}{37-17}\right)^{0.44}$$

gives proppant concentration schedule shown in Table 17.3. Slurry concentration schedule is plotted in Fig. 17.11.

c. Proppant weight requirement:

$$\bar{c}_p = \frac{c_f}{1+\varepsilon} = \frac{3}{1+0.44} = 2.08 \text{ ppg}$$

$$M_p = \bar{c}_p(V_{inj} - V_{pad}) = (2.08)(6.26 \times 10^4 - 2.76 \times 10^4)$$

$$= 72,910 \text{ lb}$$

d. Propped fracture width:

$$C_p = \frac{M_p}{2 x_f h_f} = \frac{72,910}{2(1,000)(100)} = 0.3645 \text{ lb/ft}^3$$

$$w = \frac{C_p}{(1-\phi_p)\rho_p} = \frac{0.3645}{(1-0.4)(165)} = 0.00368 \text{ ft} = 0.04 \text{ in.}$$

17.5.6 Production forecast and NPV Analyses

The hydraulic fracturing design is finalized on the basis of production forecast and NPV analyses. The information

Table 17.3 Calculated Slurry Concentration

t (min)	c_p (ppg)
0	0
17	0.00
20	1.30
23	1.77
26	2.11
29	2.40
32	2.64
35	2.86
37	3.00

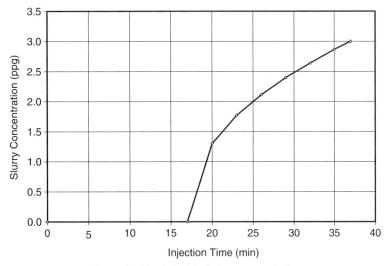

Figure 17.11 Calculated slurry concentration.

of the selected fracture half-length x_f and the calculated fracture width w, together with formation permeability (k) and fracture permeability (k_f), can be used to predict the dimensionless fracture conductivity F_{CD} with Eq. (17.10). The equivalent skin factor S_f can be estimated based on Fig. 17.7. Then the productivity index of the fractured well can be calculated using Eq. (17.11). Production forecast can be performed using the method presented in Chapter 7.

Comparison of the production forecast for the fractured well and the predicted production decline for the unstimulated well allows for calculations of the annual incremental cumulative production for year n for an oil well:

$$\Delta N_{p,n} = N_{p,n}^f - N_{p,n}^{nf}, \quad (17.36)$$

where

$\Delta N_{p,n}$ = predicted annual incremental cumulative production for year n
$N_{p,n}^f$ = forecasted annual cumulative production of fractured well for year n
$N_{p,n}^{nf}$ = predicted annual cumulative production of nonfractured well for year n.

If Eq. (17.36) is used for a gas well, the notations $\Delta N_{p,n}$, $N_{p,n}^f$, and $N_{p,n}^{nf}$ should be replaced by $\Delta N_{p,n}$, $N_{p,n}^f$, and $N_{p,n}^{nf}$, respectively.

The annual incremental revenue above the one that the unstimulated well would deliver is expressed as

$$\Delta R_n = (\$)\Delta N_{p,n}, \quad (17.37)$$

where ($\$$) is oil price. The present value of the future revenue is then

$$NPV_R = \sum_{n=1}^{m} \frac{\Delta R_n}{(1+i)^n}, \quad (17.38)$$

where m is the remaining life of the well in years and i is the discount rate. The NPV of the hydraulic fracture project is

$$NPV = NPV_R - \text{cost}. \quad (17.39)$$

The cost should include the expenses for fracturing fluid, proppant, pumping, and the fixed cost for the treatment job. To predict the pumping cost, the required hydraulic horsepower needs to be calculated by

$$HHP = \frac{q_i p_{si}}{40.8}. \quad (17.40)$$

17.6 Post-Frac Evaluation

Post-frac evaluation can be performed by pressure matching, pressure transient data analysis, and other techniques including pumping radioactive materials stages and running tracer logs, running production logging tools, and conducting back-pressure and performing Nodal analysis.

17.6.1 Pressure Matching

Pressure matching with a computer software is the first step to evaluate the fracturing job. It is understood that the more refined the design model is, the more optional parameters we have available for pressure matching and the more possible solutions we will get. The importance of capturing the main trend with the simplest model possible can only be beneficial. Attention should be paid to those critical issues in pressure matching such as fracture confinement. Therefore, all the lumped pseudo-3D models developed for processing speed of pressure-matching applications are widely used.

The final result of the net pressure-matching process should ideally be an exact superposition of the simulation on the pumping record. A perfect match is obtainable by adjusting controlling parameter of a fracture simulator, but this operation is quite time consuming and is not the goal of the exercise. Perfect matches are sometimes proposed by manually changing the number of fractures during the propagation. Unfortunately, there is no independent source that can be used to correlate a variation of the number of fractures. The option of multiple fractures is not available to all simulators. Nevertheless, much pressure adjustment can be obtained by changing parameters controlling the near-wellbore effect. Example parameters are the number of perforations, the relative erosion rate of perforation with proppant, and the characteristics of fracture tortuosity. These parameters have a major impact on the bottom-hole response but have nothing to do with the net pressure to be matched for fracture geometry estimate.

Matching the Net Pressure during Calibration Treatment and the Pad. The calibration treatment match is part of the set of analysis performed on-site for redesign of the injection schedule. This match should be reviewed before proceeding with the analysis of the main treatment itself. Consistency between the parameters obtained from both matches should be maintained and deviation recognized.

The first part of the treatment-match process focusing on the pad is identical to a match performed on the calibration treatment. The shut-in net pressure obtained from a minifrac (calibration treatment decline) gives the magnitude of the net pressure. The pad net pressure history (and low prop concentration in the first few stages) is adjusted by changing either the compliance or the tip pressure. The Nolte–Smith Plot (Nolte and Smith, 1981) provides indication of the degree of confinement of the fracture. A positive slope is an indication of confinement, a negative slope an indication of height growth, and a zero slope an indication of toughness-dominated short fracture or moderate height growth.

Using 2D Models. In general, when the fracture is confined (PKN model) and viscous dominated, we either decrease the height of the zone or increase the Young's modulus to obtain higher net pressure (compliance is $\sim h/E$). For a radial fracture (KGD model), we adjust the tip pressure effect to achieve net pressure match. If the fractured formation is a clean sand section and the fracture is confined or with moderate height growth, the fracture height should be fixed to the pay zone. In a layered formation/dirty sandstone, the fracture height could be adjusted because any of the intercalated layers may or may not have been broken down. The fracture could still be confined, but the height cannot *a priori* be set as easily as in the case of a clean sand zone section. Unconsolidated sands show low Young's modulus ($\sim 5 \times 10^5$ psi), this should not be changed to match the pressure. A low Young modulus value often gives insufficient order of magnitude of net pressure because the viscous force is not the dominating factor. The best way to adjust a fracture elastic model to match the behavior of a loosely consolidated sand is to increase the "apparent toughness" that controls the tip effect propagating pressure.

Using Pseudo-3D Models. Height constraint is adjusted by increasing the stress difference between the pay-zone and the bounding layer. Stiffness can be increased by an increase of the Young modulus of all the layers that are fractured or to some extent by adding a small shale layer with high stress in the middle of the zone (pinch-point effect). Very few commercial fracturing simulators actually use a layer description of the modulus. All of the lumped 3D models use an average value. Tip effect can also be adjusted by changing toughness (Meyer et al., 1990). For some simulators, the users have no direct control of this effect, as an apparent toughness is recalculated from the rock toughness and fluid-lag effect.

Simulating controlled height growth with a pseudo-3D model can be tricky. Height growth is characterized by a slower rate of pressure increase than in the case of a confined fracture. To capture the big picture, a simplification to a three-layer model can help by reducing the number of possible inputs. Pressure-matching slow height growth of a fracture is tedious and lengthy. In the first phase, we should adjust the magnitude of the simulated net pressure. The match can be considered excellent if the difference between the recorded pressure and the simulated pressure is less than 15% over the length of the pad.

The pressure matching can be performed using data from real-time measurements (Wright et al., 1996; Burton et al., 2002). Computer simulation of fracturing operations with recorded job parameters can yield the following fracture dimensions:

- Fracture height
- Fracture half-length
- Fracture width

A typical pressure matching with a pseudo-3D fracturing model is shown in Fig. 17.12 (Burton et al., 2002).

Efficiency and Leakoff. The first estimate of efficiency and leakoff is obtained from the calibration treatment decline analysis. The calibration treatment provides a direct measurement of the efficiency using the graphical G-plot analysis and the $\frac{3}{4}$ rules or by using time to closure with a fracturing simulator. Then calibration with a model that estimates the geometry of the fracture provides the corresponding leakoff coefficient (Meyer and Jacot, 2000). This leakoff coefficient determination is model dependent.

Propped Fracture Geometry. Once we have obtained both a reasonable net pressure match, we have an estimate of length and height. We can then directly calculate the average width expressed in mass/area of the propped fracture from mass balance. The propped geometry given by any simulator after closure should not be any different.

Post-propped Frac Decline. The simulator-generated pressure decline is affected by the model of extension recession that is implemented and by the amount of surface area that still have leakoff when the simulator cells are packed with proppant. It is very unlikely that the simulator matches any of those extreme cases. The lumped solution used in FracProPT does a good job of matching pressure decline. The analysis methodology was indeed developed around pressure matching the time to closure. The time to closure always relates to the efficiency of the fluid regardless of models (Nolte and Smith, 1981).

17.6.2 Pressure Buildup Test Analysis

Fracture and reservoir parameters can be estimated using data from pressure transient well tests (Cinco-Ley and Samaniego, 1981; Lee and Holditch, 1981). In the pressure transient well-test analysis, the log-log plot of pressure derivative versus time is called a *diagnostic plot*. Special slope values of the derivative curve usually are used for identification of reservoir and boundary models. The transient behavior of a well with a finite-conductivity fracture includes several flow periods. Initially, there is a fracture linear flow characterized by a half-slope straight line; after a transition flow period, the system may or may not exhibit a bilinear flow period, indicated by a one-fourth–slope straight line. As time increases, a formation linear flow period might develop. Eventually, the system reaches a pseudo-radial flow period if the drainage area is significantly larger than the fracture dimension (Fig. 17.13).

During the fracture linear flow period, most of the fluid entering the wellbore comes from the expansion of the system within the fracture. The behavior in the period occurs at very small amounts of time, normally a few seconds for the fractures created during frac-packing operations. Thus, the data in this period, even if not distorted by wellbore storage effect, are still not of practical use.

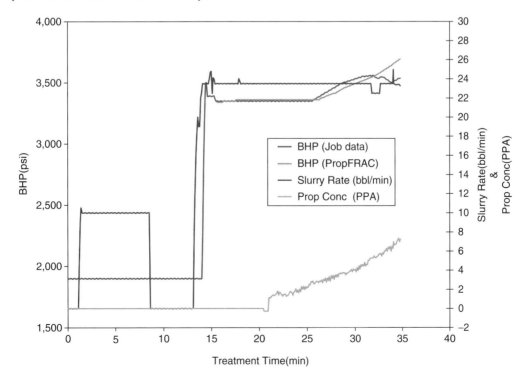

Figure 17.12 Bottom-hole pressure match with three-dimensional fracturing model PropFRAC.

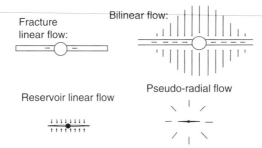

Figure 17.13 Four flow regimes that can occur in hydraulically fractured reservoirs.

The bilinear flow regime means two linear flows occur simultaneously. One flow is a linear flow within the fracture and the other is a linear flow in the formation toward the fracture. Bilinear flow analysis gives an estimate of fracture length and conductivity. A calculated pressure distribution during a bilinear flow is illustrated in Fig. 17.2 (Guo and Schechter, 1999).

The formation linear flow toward the fracture occurs after the bilinear flow. Linear flow analysis yields an estimate of formation permeability in the direction perpendicular to the fracture face. If the test time is long enough and there is no boundary effect, a system pseudo-radial flow will eventually occur. Pseudo-radial flow analysis provides an estimate of formation permeability in the radial direction. The reader is referred to Chapter 15 for analysis and interpretation of pressure transient data.

It is important to note that by no means does the pressure-match procedure and the pressure transient data analysis give details of the fracture geometry such as fracture width near the wellbore, which frequently dominates the post-treatment well performance. The fracture width near the wellbore can be significantly lower than that in the region away from the wellbore. This can occur because of a number of mishaps. Overdisplacement of proppant leads to the fracture unsupported near the wellbore, resulting in fracture closure. Fluid backflow reduces the amount of proppant near the wellbore, which results in less fracture width supported. If the proppant grains do not have compressive strength to withstand the stress concentration in the near-wellbore region, they will be crushed during fracture closure, resulting in tight fracture near the wellbore. The reduced fracture width near the wellbore affects well productivity because of the fracture choking effect. Post-treatment flow tests should be run to verify well performance.

The effect of near-wellbore fracture geometry on post-treatment well production is of special significance in deviated and horizontal wells (Chen and Economides, 1999). This is because a fracture from an arbitrarily oriented well "cuts" the wellbore at an angle, thereby limiting the communication between the wellbore and the reservoir. This feature of fluid entry to the wellbore itself causes the fracture-choking effect, even though the near-wellbore fracture is perfectly propped. Certainly, a horizontal well in the longitudinal to the fracture direction and using 180-degree perforation phasing that can be oriented will eliminate the problem. However, to align the horizontal wellbore in the longitudinal to the fracture direction, the horizontal wellbore has to be drilled in the direction parallel to the maximum horizontal stress direction. The orientation of the stress can be obtained by running tests in a vertical pilot hole of the horizontal well. Special log imaging (e.g., FMI and FMS) can be run in combination with an injection test at small-rate MDT or large-scale minifrac to fracture the formation and read directly the image in the wellbore after the fracture has been created.

17.6.3 Other evaluation techniques

In addition to the pressure-matching and pressure buildup data analyses, other techniques can be used to verify the fracture profile created during a fracpack operation. These techniques include (1) pumping radioactive materials in the proppant stages and running tracer logs to verify the fracture heights, (2) running production logging tools to determine the production profiles, and (3) conducting back-pressure and performing Nodal analysis to verify the well deliverability.

Summary

This chapter presents a brief description of hydraulic fracturing treatments covering formation fracturing pressure, fracture geometry, productivity of fractured wells, hydraulic fracturing design, and post-frac evaluation. More in-depth discussions can be found from Economides et al. (1994) and Economides and Nolte (2000).

References

ARGAWAL, R.G., CARTER, R.D., and POLLOCK, C.B. Evaluation and prediction of performance of low-permeability gas wells stimulated by massive hydraulic fracturing. *JPT* March 1979, *Trans. AIME* 1979;267:362–372.

BARREE, R.D. A practical numerical simulator for three dimensional fracture propagation in heterogeneous media. Proceedings of the Reservoir Simulation Symposium, San Francisco, CA, 403-411 Nov. 1983. SPE 12273.

BURTON, R.C., DAVIS, E.R., HODGE, R.M., STOMP, R.J., PALTHE, P.W., and SALDUNGARAY, P. Innovative completion design and well performance evaluation for effective Frac-packing of long intervals: a case study from the West Natuna Sea, Indonesia. Presented at the SPE International Petroleum Conference and Exhibition held 10–12 February 2002, in Villahermosa, Mexico. Paper SPE 74351.

CHEN, Z. and ECONOMIDES, M.J. Effect of near-wellbore fracture geometry on fracture execution and post-treatment well production of deviated and horizontal wells. *SPE Prod. Facilities* August 1999.

CINCO-LEY, H. and SAMANIEGO, F. Transient pressure analysis for fractured wells. *J. Petroleum Technol.* September 1981.

CLEARY, M.P. Comprehensive design formulae for hydraulic fracturing. Presented at the SPE Annual Technology Conference held in Dallas, Texas, September 1980. SPE 9259.

CLEARY, M.P., COYLE, R.S., TENG, E.Y., CIPOLLA, C.L., MEEHAN, D.N., MASSARAS, L.V., and WRIGHT, T.B. Major new developments in hydraulic fracturing, with documented reductions in job costs and increases in normalized production. Presented at the 69th Annual Technical Conference and Exhibition of the Society of Petroleum Engineers, held in New Orleans, Louisiana, 25 28 September 1994. SPE 28565.

CLIFTON, R.J. and ABOU-SAYED, A.S. On the computation of the three-dimensional geometry of hydraulic fractures. Presented at the SPE/DOE Low Perm. Gas Res. Symposium, heid in Denver, Colorado, May 1979. SPE 7943.

ECONOMIDES, M.J., HILL, A.D., AND EHLIG-ECONOMIDES, C. Petroleum Production Systems, Upper Saddle River, New Jersey, Prentice Hall PTR, 1994.

ECONOMIDES, M.J. and NOLTE, K.G. *Reservoir Stimulation*, 3rd edition. New York: John Wiley & Sons, 2000.

GEERTSMA, J. and de KLERK, F. A rapid method of predicting width and extent of hydraulic induced fractures. *J. Petroleum Technol.* Dec. 1969;21:1571–1581.

GUO, B. and SCHECHTER, D.S. A simple and rigorous IPR equation for vertical and horizontal wells intersecting long fractures. *J. Can. Petroleum Technol.* July 1999.

KHRISTIANOVICH, S.A. and ZHELTOV, Y.P. Formation of vertical fractures by means of highly viscous liquid. In: *Proceedings of the SPE Fourth World Petroleum Congress held in Rome*, Section II. 1955, pp. 579–586.

LEE, W.J. and HOLDITCH, S.A. Fracture evaluation with pressure transient testing in low-permeability gas reservoirs. *J. Petroleum Technol.* September 1981.

MEYER, B.R., COOPER, G.D., and NELSON, S.G. Real-time 3-D hydraulic fracturing simulation: theory and field case studies. Presented at the 65th Annual Technical Conference and Exhibition of the Society of Petroleum Engineers, held in New Orleans, Louisiana, 23–26 September 1990. Paper SPE 20658.

MEYER, B.R. and JACOT, R.H. Implementation of fracture calibration equations for pressure dependent leakoff. Presented at the 2000 SPE/AAPG Western Regional Meeting, held in Long Beach, California, 19–23 June 2000. Paper SPE 62545.

NOLTE, K.G. and SMITH, M.B. Interpretation of fracturing pressures. *J. Petroleum Technol.* September 1981.

NORDGREN, R.P. Propagation of vertical hydraulic fracture. *SPEJ* Aug. 1972:306–314.

PERKINS, T.K. and KERN, L.R. Width of Hydraulic Fracture. *J. Petroleum Technol.* Sept. 1961:937–949.

SNEDDON, I.N. and ELLIOTT, A.A. The opening of a GRIFFITH crack under internal pressure. *Quart. Appl. Math.* 1946;IV:262.

VALKO, P., OLIGNEY, R.E., ECONOMIDES, M.J. *High permeability fracturing of gas wells.* Gas TIPS. October 1997;3:31–40.

WRIGHT, C.A., WEIJERS, L., GERMANI, G.A., MACLVOR, K.H., WILSON, M.K., and WHITMAN, B.A. Fracture treatment design and evaluation in the Pakenham field: a real-data approach. Presented at the SPE Annual Technical Conference and Exhibition, held in Denver, Colorado, 6–9 October 1996. Paper SPE 36471.

Problems

17.1 A sandstone at a depth of 8,000 ft has a Poison's ratio of 0.275 and a poro-elastic constant of 0.70. The average density of the overburden formation is 162 lb/ft^3. The pore–pressure gradient in the sandstone is 0.36 psi/ft. Assuming a tectonic stress of 1,000 psi and a tensile strength of the sandstone of 800 psi, predict the breakdown pressure for the sandstone.

17.2 A carbonate at a depth of 12,000 ft has a Poison's ratio of 0.3 and a poro-elastic constant of 0.75. The average density of the overburden formation is 178 lb/ft^3. The pore–pressure gradient in the sandstone is 0.35 psi/ft. Assuming a tectonic stress of 2,000 psi and a tensile strength of the sandstone of 1,500 psi, predict the breakdown pressure for the sandstone.

17.3 A gas reservoir has a permeability of 5 md. A vertical well of 0.328-ft radius draws the reservoir from the center of an area of 320 acres. If the well is hydraulically fractured to create a 2,000-ft long, 0.15-in. wide fracture of 200,000-md permeability around the center of the drainage area, what would be the fold of increase in well productivity?

17.4 A reservoir has a permeability of 100 md. A vertical well of 0.328-ft radius draws the reservoir from the center of an area of 160 acres. If the well is hydraulically fractured to create a 2,800-ft long, 0.12-in. wide fracture of 250,000-md permeability around the center of the drainage area, what would be the fold of increase in well productivity?

17.5 For the following situation, estimate the minimum required compressive strength of 20/40 proppant. If high-strength proppant is used, estimate the permeability of the proppant pack:

Formation depth:	12,000 ft
Overburden density:	165 lb$_m$/ft^3
Poison's ratio:	0.25
Biot constant:	0.72
Reservoir pressure:	6,800 psi
Production drawdown:	3,000 psi

17.6 For the Problem 17.5, predict the maximum expected surface injection pressure using the following additional data:

Specific gravity of fracturing fluid:	1.1
Viscosity of fracturing fluid:	10 cp
Tubing inner diameter:	3.0 in.
Fluid injection rate:	20 bpm

17.7 The following data are given for a hydraulic fracturing treatment design:

Pay zone thickness:	50 ft
Young's modulus of rock:	4×10^6 psi
Poison's ratio:	0.25
Fluid viscosity:	1.25 cp
Leakoff coefficient:	0.003 ft/min$^{1/2}$
Proppant density:	185 lb/ft^3
Proppant porosity:	0.4
Fracture half length:	1,200 ft
Fracture height:	70 ft
Fluid injection rate:	35 bpm
Final proppant concentration:	5 ppg

Assuming KGD fracture, estimate

a. Fluid volume requirement

b. Proppant mixing schedule

c. Proppant weight requirement

d. Propped fracture width

17.8 Predict the productivity index of the fractured well described in Problem 17.7.

18 Production Optimization

Contents
18.1 Introduction 18/268
18.2 Naturally Flowing Well 18/268
18.3 Gas-Lifted Well 18/268
18.4 Sucker Rod–Pumped Well 18/269
18.5 Separator 18/270
18.6 Pipeline Network 18/272
18.7 Gas-Lift Facility 18/275
18.8 Oil and Gas Production Fields 18/276
18.9 Discounted Revenue 18/279
Summary 18/279
References 18/279
Problems 18/280

18.1 Introduction

The term "production optimization" has been used to describe different processes in the oil and gas industry. A rigorous definition of the term has not been found from the literature. The book by Beggs (2003) "Production Optimization Using NODAL Analysis" presents a systems analysis approach (called NODAL analysis, or Nodal analysis) to analyze performance of production systems. Although the entire production system is analyzed as a total unit, interacting components, electrical circuits, complex pipeline networks, pumps, and compressors are evaluated individually using this method. Locations of excessive flow resistance or pressure drop in any part of the network are identified.

To the best of our understanding, production optimization means determination and implementation of the optimum values of parameters in the production system to maximize hydrocarbon production rate (or discounted revenue) or to minimize operating cost under various technical and economical constraints. Because a system can be defined differently, the production optimization can be performed at different levels such as well level, platform/facility level, and field level. This chapter describes production optimization of systems defined as

- Naturally flowing well
- Gas-lifted well
- Sucker rod–pumped well
- Separator
- Pipeline network
- Gas lift facility
- Oil and gas production fields

In the upstream oil and gas production, various approaches and technologies are used to address different aspects of hydrocarbon production optimization. They serve to address various business objectives. For example, online facility optimizer addresses the problem of maximizing the value of feedstock throughput in real time. This chapter presents principals of production optimization with the aids of computer programs when necessary.

18.2 Naturally Flowing Well

A naturally flowing well may be the simplest system in production optimization. The production rate from a single flowing well is dominated by inflow performance, tubing size, and wellhead pressure controlled by choke size. Because the wellhead pressure is usually constrained by surface facility requirements, there is normally not much room to play with the choke size.

Well inflow performance is usually improved with well-stimulation techniques including matrix acidizing and hydraulic fracturing. While matrix-acidizing treatment is effective for high-permeability reservoirs with significant well skins, hydraulic-fracturing treatment is more beneficial for low-permeability reservoirs. Inflow equations derived from radial flow can be used for predicting inflow performance of acidized wells, and equations derived from both linear flow and radial flow may be employed for forecasting deliverability of hydraulically fractured wells. These equations are found in Chapter 15.

Figure 18.1 illustrates inflow performance relationship (IPR) curves for a well before and after stimulation. It shows that the benefit of the stimulation reduces as bottom-hole pressure increases. Therefore, after predicting inflow performance of the stimulated well, single-well Nodal analysis needs to be carried out. The operating points of stimulated well and nonstimulated wells are compared. This comparison provides an indication of

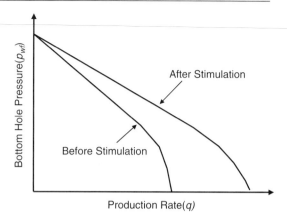

Figure 18.1 Comparison of oil well inflow performance relationship (IPR) curves before and after stimulation.

whether the well inflow is the limiting step that controls well deliverability. If yes, treatment design may proceed (Chapters 16 and 17) and economic evaluation should be performed (see Section 18.9). If no, optimization of tubing size should be investigated.

It is not true that the larger the tubing size is, the higher the well deliverability is. This is because large tubing reduces the gas-lift effect in oil wells. Large tubing also results in liquid loading of gas wells due to the inadequate kinetic energy of gas flow required to lift liquid. The optimal tubing size yields the lowest frictional pressure drop and the maximum production rate. Nodal analysis can be used to generate tubing performance curve (plot of operating rate vs tubing size) from which the optimum tubing size can be identified. Figure 18.2 shows a typical tubing performance curve. It indicates that a 3.5-in. inner diameter (ID) tubing will give a maximum oil production rate of 600 stb/day. However, this tubing size may not be considered optimal because a 3.0-in. ID tubing will also deliver a similar oil production rate and this tubing may be cheaper to run. An economics evaluation should be performed (see Section 18.9).

18.3 Gas-Lifted Well

The optimization of individual gas-lift wells mainly focuses on determining and using the optimal gas-lift gas injection rate. Overinjection of gas-lift gas is costly and results in lower oil production rate. The optimal gas

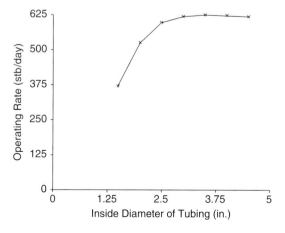

Figure 18.2 A typical tubing performance curve.

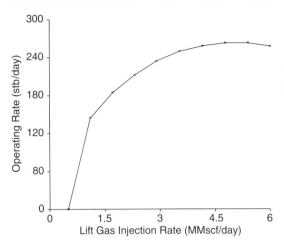

Figure 18.3 A typical gas lift performance curve of a low-productivity well.

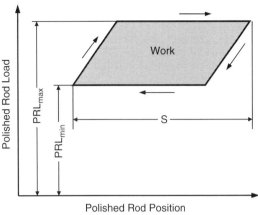

Figure 18.4 Theoretical load cycle for elastic sucker rods.

injection rate can be identified from a gas-lift performance curve, which can be generated using Nodal analysis software such as WellFlo (1997). Figure 18.3 presents a typical gas-lift performance curve. It shows that a 5.0-MMscf/day gas injection rate will give a maximum oil production rate of 260 stb/day. However, this gas injection rate may not be the optimum rate because slightly lower gas injection rates will also deliver a similar oil production rate with lower high-pressure gas consumption. An economics evaluation should be performed on a scale of a batch of similar wells (see Section 18.9).

18.4 Sucker Rod–Pumped Well

The potential of increasing oil production rate of a normal sucker rod–pumped well is usually low. Optimization of this type of well mainly focuses on two areas:

- Improving the volumetric efficiency of the plunger pump
- Improving the energy efficiency of the pumping unit

Estimating the volumetric efficiency of plunger pump and improving the energy efficiency of the pumping unit require the use of the information from a dynamometer card that records polished rod load. Figure 18.4 demonstrates a theoretical load cycle for elastic sucker rods. However, because of the effects of acceleration and friction, the actual load cycles are very different from the theoretical cycles. Figure 18.5 demonstrates an actual load cycle of a sucker rod under normal working conditions. It illustrates that the peak polished rod load can be significantly higher than the theoretical maximum polished rod load.

Much information can be obtained from the dynamometer card. The procedure is illustrated with the parameters shown in Fig. 18.6. The nomenclature is as follows:

- C = calibration constant of the dynamometer, lb/in.
- D_1 = maximum deflection, in.
- D_2 = minimum deflection, in.
- D_3 = load at the counterbalance line (CB) drawn on the dynamometer card by stopping the pumping unit at the position of maximum counterbalance effect (crank arm is horizontal on the upstroke), in.
- A_1 = lower area of card, in.^{2}
- A_2 = upper area of card, in.^{2}.

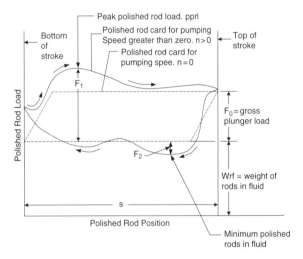

Figure 18.5 Actual load cycle of a normal sucker rod.

The following information can be obtained from the card parameter values:

Peak polished rod load: $\quad$ PPRL $= CD_1$
Minimum polished rod load: $\quad$ MPRL $= CD_2$
Range of load: $\quad$ ROL $= C(D_1 - D_2)$
Average upstroke load: $\quad$ AUL $= \dfrac{C(A_1 + A2)}{L}$
Average downstroke load: $\quad$ ADL $= \dfrac{CA_1}{L}$
Work for rod elevation: $\quad$ WRE $= A_1$ converted to ft-lb
Work for fluid elevation and friction: $\quad$ WFEF $= A_2$ converted to ft-lb
Approximate "ideal" counterbalance: $\quad$ AICB $= \dfrac{PPRL + MPRL}{2}$
Actual counterbalance effect: $\quad$ ACBE $= CD_3$
Correct counterbalance: $\quad$ CCB $= (AUL + ADL)/2$
$\quad = \dfrac{C\left(A_1 + \frac{A_2}{2}\right)}{L}$
Polished rod horsepower: $\quad$ PRHP $= \dfrac{CSNA_2}{33,000(12)L}$

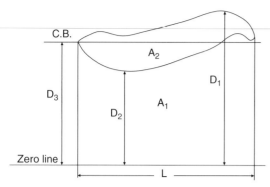

Figure 18.6 Dimensional parameters of a dynamometer card.

Example Problem 18.1 Analyze the dynamometer card shown in Fig. 18.6 assuming the following parameter values:

$S = 45$ in.
$N = 18.5$ spm
$C = 12,800$ lb/in.
$D_1 = 1.2$ in.
$D_2 = 0.63$ in.
$L = 2.97$ in.
$A_1 = 2.1$ in.^{2}
$A_2 = 1.14$ in.^{2}

Solution

Peak polished rod load:
$$\text{PPRL} = (12,800)(1.20) = 15,400 \, \text{lb}$$

Minimum polished rod load:
$$\text{MPRL} = (12,800)(0.63) = 8,100 \, \text{lb}$$

Average upstroke load:
$$\text{AUL} = \frac{(12,800)(1.14 + 2.10)}{2.97} = 14,000 \, \text{lb}$$

Average downstroke load:
$$\text{ADL} = \frac{(12,800)(2.10)}{2.97} = 9,100 \, \text{lb}$$

Correct counterbalance:
$$\text{CCB} = \frac{(12,800)(2.10 + \frac{1.14}{2})}{2.97} = 11,500 \, \text{lb}$$

Polished rod horsepower:
$$\text{PRHP} = \frac{(12,800)(45)(18.5)(1.14)}{33,000(12)(2.97)} = 10.3 \, \text{hp}$$

The information of the CCB can be used for adjusting the positions of counterweights to save energy.

In addition to the dimensional parameter values taken from the dynamometer card, the shape of the card can be used for identifying the working condition of the plunger pump. The shape of the dynamometer cards are influenced by

- Speed and pumping depth
- Pumping unit geometry
- Pump condition
- Fluid condition
- Friction factor

Brown (1980) listed 13 abnormal conditions that can be identified from the shape of the dynamometer cards. For example, the dynamometer card shown in Fig. 18.7 indicates synchronous pumping speeds, and the dynamometer card depicted in Fig. 18.8 reveals a gas-lock problem.

18.5 Separator

Optimization of the separation process mainly focuses on recovering more oil by adjusting separator temperature and pressure. Field experience proves that lowering the operating temperature of a separator increases the liquid recovery. It is also an efficient means of handling high-pressure gas and condensate at the wellhead. A low-temperature separation unit consists of a high-pressure separator, pressure-reducing chokes, and various pieces of heat exchange equipment. When the pressure is reduced by the use of a choke, the fluid temperature decreases because of the Joule–Thomson or the throttling effect. This is an irreversible adiabatic process whereby the heat content of the gas remains the same across the choke but the pressure and temperature of the gas stream are reduced.

Generally at least 2,500–3,000 psi pressure drop is required from wellhead flowing pressure to pipeline pressure for a low-temperature separation unit to pay out in increased liquid recovery. The lower the operating temperature of the separator, the lighter the liquid recovered will be. The lowest operating temperature recommended for low-temperature units is usually around $-20\,°\text{F}$. This is constrained by carbon steel embitterment, and high-alloy steels for lower temperatures are usually not economical for field installations. Low-temperature separation units are normally operated in the range of 0–20 °F. The actual temperature drop per unit pressure drop is affected by several factors including composition of gas stream, gas and liquid flow rates, bath temperature, and ambient temperature. Temperature reduction in the process can be estimated using the equations presented in Chapter 5.

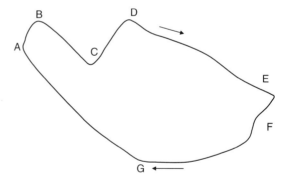

Figure 18.7 A dynamometer card indicating synchronous pumping speeds.

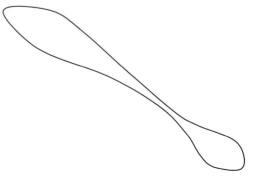

Figure 18.8 A dynamometer card indicating gas lock.

Gas expansion pressures for hydrate formation can be found from the chart prepared by Katz (1945) or Guo and Ghalambor (2005). Liquid and vapor phase densities can be predicted by flash calculation.

Following the special requirement for construction of low-temperature separation units, the pressure-reducing choke is usually mounted directly on the inlet of the high-pressure separator. Hydrates form in the downstream of the choke because of the low gas temperature and fall to the bottom settling section of the separator. They are heated and melted by liquid heating coils located in the bottom of the separator.

Optimization of separation pressure is performed with flash calculations. Based on the composition of wellstream fluid, the quality of products from each stage of separation can be predicted, assuming phase equilibriums are reached in the separators. This requires the knowledge of the equilibrium ratio defined as

$$k_i = \frac{y_i}{x_i}, \quad (18.1)$$

where

k_i = liquid/vapor equilibrium ratio of compound i
y_i = mole fraction of compound i in the vapor phase
x_i = mole fraction of compound i in the liquid phase.

Accurate determination of k_i values requires computer simulators solving the Equation of State (EoS) for hydrocarbon systems. Ahmed (1989) presented a detailed procedure for solving the EoS. For pressures lower than 1,000 psia, a set of equations presented by Standing (1979) provide an easy and accurate means of determining k_i values. According to Standing, k_i can be calculated by

$$k_i = \frac{1}{p} 10^{a+cF_i}, \quad (18.2)$$

where

$$a = 1.2 + 4.5 \times 10^{-4} p + 1.5 \times 10^{-9} p^2 \quad (18.3)$$

$$c = 0.89 - 1.7 \times 10^{-4} p - 3.5 \times 10^{-8} p^2 \quad (18.4)$$

$$F_i = b_i \left(\frac{1}{T_{bi}} - \frac{1}{T} \right) \quad (18.5)$$

$$b_i = \frac{\log \left(\frac{p_{ci}}{14.7} \right)}{\frac{1}{T_{bi}} - \frac{1}{T_{ci}}}, \quad (18.6)$$

where

p_c = critical pressure, psia
T_b = boiling point, °R
T_c = critical temperature, °R.

Consider 1 mol of fed-in fluid and the following equation holds true on the basis of mass balance:

$$n_L + n_V = 1, \quad (18.7)$$

where

n_L = number of mole of fluid in the liquid phase
n_V = number of mole of fluid in the vapor phase.

For compound i,

$$z_i = x_i n_L + y_i n_V, \quad (18.8)$$

where z_i is the mole fraction of compound i in the fed-in fluid. Combining Eqs. (18.1) and (18.8) gives

$$z_i = x_i n_L + k_i x_i n_V, \quad (18.9)$$

which yields

$$x_i = \frac{z_i}{n_L + k_i n_V}. \quad (18.10)$$

Mass balance applied to Eq. (18.10) requires

$$\sum_{i=1}^{N_c} x_i = \sum_{i=1}^{N_c} \frac{z_i}{n_L + k_i n_V} = 1, \quad (18.11)$$

where N_c is the number of compounds in the fluid. Combining Eqs. (18.1) and (18.8) also gives

$$z_i = \frac{y_i}{k_i} n_L + y_i n_V, \quad (18.12)$$

which yields

$$y_i = \frac{z_i k_i}{n_L + k_i n_V}. \quad (18.13)$$

Mass balance applied to Eq. (18.13) requires

$$\sum_{i=1}^{N_c} y_i = \sum_{i=1}^{N_c} \frac{z_i k_i}{n_L + k_i n_V} = 1. \quad (18.14)$$

Subtracting Eq. (18.14) from Eq. (18.11) gives

$$\sum_{i=1}^{N_c} \frac{z_i}{n_L + k_i n_V} - \sum_{i=1}^{N_c} \frac{z_i k_i}{n_L + k_i n_V} = 0, \quad (18.15)$$

which can be rearranged to obtain

$$\sum_{i=1}^{N_c} \frac{z_i(1 - k_i)}{n_L + k_i n_V} = 0. \quad (18.16)$$

Combining Eqs. (18.16) and (18.7) results in

$$\sum_{i=1}^{N_c} \frac{z_i(1 - k_i)}{n_V(k_i - 1) + 1} = 0. \quad (18.17)$$

This equation can be used to solve for the number of mole of fluid in the vapor phase n_v. Then, x_i and y_i can be calculated with Eqs. (18.10) and (18.13), respectively. The apparent molecular weights of liquid phase (MW) and vapor phase (MW) can be calculated by

$$MW_a^L = \sum_{i=1}^{N_c} x_i MW_i \quad (18.18)$$

$$MW_a^V = \sum_{i=1}^{N_c} y_i MW_i, \quad (18.19)$$

where MW_i is the molecular weight of compound i. With the apparent molecular weight of the vapor phase known, the specific gravity of the vapor phase can be determined, and the density of the vapor phase in lb_m/ft^3 can be calculated by

$$\rho_V = \frac{MW_a^V p}{zRT}. \quad (18.20)$$

The liquid phase density in lb_m/ft^3 can be estimated by the Standing method (1981), that is,

$$\rho_L = \frac{62.4 \gamma_{oST} + 0.0136 R_s \gamma_g}{0.972 + 0.000147 \left[R_s \sqrt{\frac{\gamma_g}{\gamma_o}} + 1.25(T - 460) \right]^{1.175}}, \quad (18.21)$$

where

γ_{oST} = specific gravity of stock-tank oil, water
γ_g = specific gravity of solution gas, air = 1
R_s = gas solubility of the oil, scf/stb.

Then the specific volumes of vapor and liquid phases can be calculated by

$$V_{Vsc} = \frac{z n_V R T_{sc}}{p_{sc}} \quad (18.22)$$

$$V_L = \frac{n_L MW_a^L}{\rho_L}, \quad (18.23)$$

18/272 PRODUCTION ENHANCEMENT

where

V_{Vsc} = specific volume of vapor phase under standard condition, scf/mol-lb
R = gas constant, 10.73 ft^3-psia/lb mol-R
T_{sc} = standard temperature, 520 °R
p_{sc} = standard pressure, 14.7 psia
V_L = specific volume of liquid phase, ft^3/mol-lb.

Finally, the gas–oil ratio (GOR) in the separator can be calculated by

$$GOR = \frac{V_{Vsc}}{V_L}. \tag{18.24}$$

Specific gravity and the American Petroleum Institute (API) gravity of oil at the separation pressure can be calculated based on liquid density from Eq. (18.21). The lower the GOR, the higher the API gravity, and the higher the liquid production rate. For gas condensates, there exists an optimum separation pressure that yields the lower GOR at a given temperature.

Example Problem 18.2 Perform flash calculation under the following separator conditions:

Pressure:	600 psia
Temperature:	200 °F
Specific gravity of stock-tank oil:	0.90 water = 1
Specific gravity of solution gas:	0.70 air = 1
Gas solubility (R_s):	500 scf/stb

Solution The flash calculation can be carried out using the spreadsheet program *LP-Flash.xls*. The results are shown in Table 18.1.

18.6 Pipeline Network

Optimization of pipelines mainly focuses on de-bottlenecking of the pipeline network, that is, finding the most restrictive pipeline segments and replacing/adding larger segments to remove the restriction effect. This requires the knowledge of flow of fluids in the pipe. This section presents mathematical models for gas pipelines. The same principle applies to oil flow. Equations for oil flow are presented in Chapter 11.

18.6.1 Pipelines in Series

Consider a three-segment gas pipeline in a series of total length L depicted in Fig. 18.9a. Applying the Weymouth equation to each of the three segments gives

$$p_1^2 - p_2^2 = \frac{\gamma_g \overline{T}\overline{z}L_1}{D_1^{16/3}} \left(\frac{q_h p_b}{18.062 T_b}\right)^2 \tag{18.25}$$

$$p_2^2 - p_3^2 = \frac{\gamma_g \overline{T}\overline{z}L_2}{D_2^{16/3}} \left(\frac{q_h p_b}{18.062 T_b}\right)^2 \tag{18.26}$$

$$p_3^2 - p_4^2 = \frac{\gamma_g \overline{T}\overline{z}L_3}{D_3^{16/3}} \left(\frac{q_h p_b}{18.062 T_b}\right)^2. \tag{18.27}$$

Adding these three equations gives

$$p_1^2 - p_4^2 = \gamma_g \overline{T}\overline{z} \left(\frac{L_1}{D_1^{16/3}} + \frac{L_2}{D_2^{16/3}} + \frac{L_3}{D_3^{16/3}}\right)$$

$$\times \left(\frac{q_h p_b}{18.062 T_b}\right)^2 \tag{18.28}$$

or

$$q_h = \frac{18.062 T_b}{p_b} \sqrt{\frac{p_1^2 - p_4^2}{\gamma_g \overline{T}\overline{z} \left(\frac{L_1}{D_1^{16/3}} + \frac{L_2}{D_2^{16/3}} + \frac{L_3}{D_3^{16/3}}\right)}}. \tag{18.29}$$

Gas composition

Compound	Mole fraction
C_1	0.6599
C_2	0.0869
C_3	0.0591
i-C_4	0.0239
n-C_4	0.0278
i-C_5	0.0157
n-C_5	0.0112
C_6	0.0181
C_{7+}	0.0601
N_2	0.0194
CO_2	0.0121
H_2S	0.0058

Capacity of a single-diameter (D_1) pipeline is expressed as

$$q_1 = \frac{18.062 T_b}{p_b} \sqrt{\frac{p_1^2 - p_4^2}{\gamma_g \overline{T}\overline{z} \left(\frac{L}{D_1^{16/3}}\right)}}. \tag{18.30}$$

Dividing Eq. (18.29) by Eq. (18.30) yields

$$\frac{q_t}{q_1} = \sqrt{\frac{\left(\frac{L}{D_1^{16/3}}\right)}{\left(\frac{L_1}{D_1^{16/3}} + \frac{L_2}{D_2^{16/3}} + \frac{L_3}{D_3^{16/3}}\right)}}. \tag{18.31}$$

18.6.2 Pipelines in Parallel

Consider a three-segment gas pipeline in parallel as depicted in Fig. 18.9b. Applying the Weymouth equation to each of the three segments gives

$$q_1 = 18.062 \frac{T_b}{p_b} \sqrt{\frac{(p_1^2 - p_2^2)D_1^{16/3}}{\gamma_g \overline{T}\overline{z}L}} \tag{18.32}$$

$$q_2 = 18.062 \frac{T_b}{p_b} \sqrt{\frac{(p_1^2 - p_2^2)D_2^{16/3}}{\gamma_g \overline{T}\overline{z}L}} \tag{18.33}$$

$$q_3 = 18.062 \frac{T_b}{p_b} \sqrt{\frac{(p_1^2 - p_2^2)D_3^{16/3}}{\gamma_g \overline{T}\overline{z}L}}. \tag{18.34}$$

Adding these three equations gives

$$q_t = q_1 + q_2 + q_3$$

$$= 18.062 \frac{T_b}{p_b} \sqrt{\frac{(p_1^2 - p_2^2)}{\gamma_g \overline{T}\overline{z}L}}$$

$$\times \left(\sqrt{D_1^{16/3}} + \sqrt{D_2^{16/3}} + \sqrt{D_3^{16/3}}\right). \tag{18.35}$$

Dividing Eq. (18.35) by Eq. (18.32) yields

$$\frac{q_t}{q_1} = \frac{\sqrt{D_1^{16/3}} + \sqrt{D_2^{16/3}} + \sqrt{D_3^{16/3}}}{\sqrt{D_1^{16/3}}}. \tag{18.36}$$

18.6.3 Looped Pipelines

Consider a three-segment looped gas pipeline depicted in Fig. 18.10. Applying Eq. (18.35) to the first two (parallel) segments gives

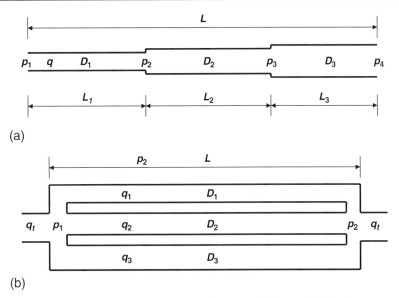

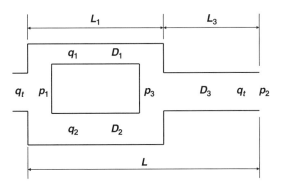

Figure 18.9 Sketch of (a) a series pipeline and (b) a parallel pipeline.

Figure 18.10 Sketch of a looped pipeline.

$q_t = q_1 + q_2$

$$= 18.062 \frac{T_b}{p_b} \sqrt{\frac{(p_1^2 - p_2^2)}{\gamma_g \overline{T}\overline{z}L}} \left(\sqrt{D_1^{16/3}} + \sqrt{D_2^{16/3}} \right) \quad (18.37)$$

or

$$p_1^2 - p_2^2 = \frac{\gamma_g \overline{T}\overline{z}L_1}{\left(\sqrt{D_1^{16/3}} + \sqrt{D_2^{16/3}}\right)^2} \left(\frac{q_t p_b}{18.062 T_b}\right)^2. \quad (18.38)$$

Applying the Weymouth equation to the third segment (with diameter D_3) yields

$$p_3^2 - p_2^2 = \frac{\gamma_g \overline{T}\overline{z}L_3}{D_3^{16/3}} \left(\frac{q_t p_b}{18.062 T_b}\right)^2. \quad (18.39)$$

Adding Eqs. (18.38) and (18.39) results in

$$p_1^2 - p_2^2 = \gamma_g \overline{T}\overline{z} \left(\frac{q_t p_b}{18.062 T_b}\right)^2$$

$$\times \left(\frac{L_1}{\left(\sqrt{D_1^{16/3}} + \sqrt{D_2^{16/3}}\right)^2} + \frac{L_3}{D_3^{16/3}} \right) \quad (18.40)$$

or

$$q_t = \frac{18.062 T_b}{p_b}$$

$$\times \sqrt{\frac{(p_1^2 - p_2^2)}{\gamma_g \overline{T}\overline{z} \left(\frac{L_1}{\left(\sqrt{D_1^{16/3}} + \sqrt{D_2^{16/3}}\right)^2} + \frac{L_3}{D_3^{16/3}} \right)}}.$$

(18.41)

Capacity of a single-diameter (D_3) pipeline is expressed as

$$q_3 = \frac{18.062 T_b}{p_b} \sqrt{\frac{p_1^2 - p_2^2}{\gamma_g \overline{T}\overline{z} \left(\frac{L}{D_3^{16/3}}\right)}}. \quad (18.42)$$

Dividing Eq. (18.41) by Eq. (18.42) yields

$$\frac{q_t}{q_3} = \sqrt{\frac{\left(\frac{L}{D_3^{16/3}}\right)}{\left(\frac{L_1}{\left(\sqrt{D_1^{16/3}} + \sqrt{D_2^{16/3}}\right)^2} + \frac{L_3}{D_3^{16/3}} \right)}}. \quad (18.43)$$

Let Y be the fraction of looped pipeline and X be the increase in gas capacity, that is,

Table 18.1 Flash Calculation with Standing's Method for k_i Values

Flash calculation
$n_v = 0.8791$

Compound	z_i	k_i	$z_i(k_i - 1)/[n_v(k_i - 1) + 1]$
C_1	0.6599	6.5255	0.6225
C_2	0.0869	1.8938	0.0435
C_3	0.0591	0.8552	−0.0098
i-C_4	0.0239	0.4495	−0.0255
n-C_4	0.0278	0.3656	−0.0399
i-C_5	0.0157	0.1986	−0.0426
n-C_5	0.0112	0.1703	−0.0343
C_6	0.0181	0.0904	−0.0822
C_{7+}	0.0601	0.0089	−0.4626
N_2	0.0194	30.4563	0.0212
CO_2	0.0121	3.4070	0.0093
H_2S	0.0058	1.0446	0.0002
		Sum:	0.0000

$n_L = 0.1209$

Compound	x_i	y_i	$x_i MW_i$	$y_i MW_i$
C_1	0.1127	0.7352	1.8071	11.7920
C_2	0.0487	0.0922	1.4633	2.7712
C_3	0.0677	0.0579	2.9865	2.5540
i-C_4	0.0463	0.0208	2.6918	1.2099
n-C_4	0.0629	0.0230	3.6530	1.3356
i-C_5	0.0531	0.0106	3.8330	0.7614
n-C_5	0.0414	0.0070	2.9863	0.5085
C_6	0.0903	0.0082	7.7857	0.7036
C_{7+}	0.4668	0.0042	53.3193	0.4766
N_2	0.0007	0.0220	0.0202	0.6156
CO_2	0.0039	0.0132	0.1709	0.5823
H_2S	0.0056	0.0058	0.1902	0.1987

Apparent molecular weight of liquid phase:	23.51			80.91
Apparent molecular weight of vapor phase:	0.76			
Specific gravity of liquid phase:		water = 1		
Specific gravity of vapor phase:	0.81	air = 1		
Input vapor phase z factor:	0.958			
Density of liquid phase:	47.19	lb_m/ft^3		
Density of vapor phase:	2.08	lb_m/ft^3		
Volume of liquid phase:	0.04	bbl		
Volume of vapor phase:	319.66	scf		
GOR:	8,659	scf/bbl		
API gravity of liquid phase:	56			

$$Y = \frac{L_1}{L}, \quad X = \frac{q_t - q_3}{q_3}. \tag{18.44}$$

If, $D_1 = D_3$, Eq. (18.43) can be rearranged as

$$Y = \frac{1 - \dfrac{1}{(1 + X)^2}}{1 - \dfrac{1}{(1 + R_D^{2.31})^2}}, \tag{18.45}$$

where R_D is the ratio of the looping pipe diameter to the original pipe diameter, that is, $R_D = D_2/D_3$. Equation (18.45) can be rearranged to solve for X explicitly

$$X = \frac{1}{\sqrt{1 - Y\left(1 - \dfrac{1}{(1 + R_D^{2.31})^2}\right)}} - 1. \tag{18.46}$$

The effects of looped line on the increase of gas flow rate for various pipe diameter ratios are shown in Fig. 18.11. This figure indicates an interesting behavior of looping: The increase in gas capacity is not directly proportional to the fraction of looped pipeline. For example, looping of 40% of pipe with a new pipe of the same diameter will increase only 20% of the gas flow capacity. It also shows that the benefit of looping increases with the fraction of looping. For example, looping of 80% of the pipe with a new pipe of the same diameter will increase 60%, not 40%, of gas flow capacity.

Example Problem 18.3 Consider a 4-in. pipeline that is 10 miles long. Assuming that the compression and delivery pressures will maintain unchanged, calculate gas capacity increases by using the following measures of improvement: (a) replace 3 miles of the 4-in. pipeline by a 6-in. pipeline segment; (b) place a 6-in. parallel pipeline to share gas

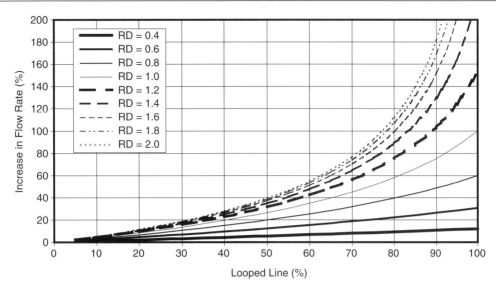

Figure 18.11 Effects of looped line and pipe diameter ratio on the increase of gas flow rate.

transmission; and (c) loop 3 miles of the 4-in. pipeline with a 6-in. pipeline segment.

Solution

(a) Replace a portion of pipeline:

$L = 10\,\text{mi}$
$L_1 = 7\,\text{mi}$
$L_2 = 3\,\text{mi}$
$D_1 = 4\,\text{in.}$
$D_2 = 6\,\text{in.}$

$$\frac{q_t}{q_1} = \sqrt{\frac{\left(\frac{10}{4^{16/3}}\right)}{\left(\frac{7}{4^{16/3}} + \frac{3}{6^{16/3}}\right)}}$$

$= 1.1668$, or 16.68% increase in flow capacity.

(b) Place a parallel pipeline:

$D_1 = 4\,\text{in.}$
$D_2 = 6\,\text{in.}$

$$\frac{q_t}{q_1} = \frac{\sqrt{4^{16/3}} + \sqrt{6^{16/3}}}{\sqrt{4^{16/3}}}$$

$= 3.9483$, or 294.83% increase in flow capacity.

(c) Loop a portion of the pipeline:

$L = 10\,\text{mi}$
$L_1 = 7\,\text{mi}$
$L_2 = 3\,\text{mi}$
$D_1 = 4\,\text{in.}$
$D_2 = 6\,\text{in.}$

$$\frac{q_t}{q_3} = \sqrt{\frac{\left(\frac{10}{4^{16/3}}\right)}{\left(\frac{L_1}{\left(\sqrt{4^{16/3}} + \sqrt{6^{16/3}}\right)^2} + \frac{L_3}{4^{16/3}}\right)}}$$

$= 1.1791$, or 17.91% increase in flow capacity.

Similar problems can also be solved using the spreadsheet program *LoopedLines.xls*. Table 18.2 shows the solution to Example Problem 18.3 given by the spreadsheet.

18.7 Gas-Lift Facility

Optimization of gas lift at the facility level mainly focuses on determination of the optimum lift-gas distribution among the gas-lifted wells. If lift-gas volume is not limited by the capacity of the compression station, every well should get the lift-gas injection rate being equal to its optimal gas injection rate (see Section 18.3). If limited lift-gas volume is available from the compression station, the lift gas should be assigned first to those wells that will produce more incrementals of oil production for a given incremental of lift-gas injection rate. This can be done by calculating and comparing the slopes of the gas-lift performance curves of individual wells at the points of adding more lift-gas injection rate. This principle can be illustrated by the following example problem.

Example Problem 18.4 The gas-lift performance curves of two oil wells are known based on Nodal analyses at well level. The performance curve of Well A is presented in Fig. 18.3 and that of Well B is in Fig. 18.12. If a total lift-gas injection rate of 1.2 to 6.0 MMscf/day is available to the two wells, what lift-gas flow rates should be assigned to each well?

Solution Data used for plotting the two gas-lift performance curves are shown in Table 18.3. Numerical derivatives (slope of the curves) are also included.

At each level of given total gas injection rate, the incremental gas injection rate (0.6 MMscf/day) is assigned to one of the wells on the basis of their performance curve slope at the present gas injection rate of the well. The procedure and results are summarized in Table 18.4. The results indicate that the share of total gas injection rate by wells depends on the total gas rate availability and performance of individual wells. If only 2.4 MMscf/day of gas is available, no gas should be assigned to Well A. If only 3.6 MMscf/day of gas is available, Well A should share one-third of the total gas rate. If only 6.0 MMscf/day of gas is available, each well should share 50% of the total gas rate.

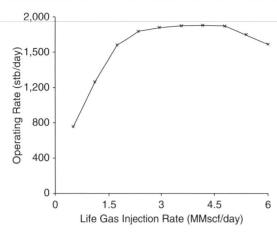

Figure 18.12 *A typical gas lift performance curve of a high-productivity well.*

18.8 Oil and Gas Production Fields

An oil or gas field integrates wells, flowlines, separation facilities, pump stations, compressor stations, and transportation pipelines as a whole system. Single-phase and multiphase flow may exist in different portions in the system. Depending on system complexity and the objective of optimization task, field level production optimization can be performed using different approaches.

18.8.1 Types of Flow Networks

Field-level production optimization deals with complex flow systems of two types: (1) hierarchical networks and (2) nonhierarchical networks. A hierarchical network is defined as a treelike converging system with multiple inflow points (sources) and one outlet (sink). Figure 18.13 illustrates two hierarchical networks. Flow directions in this type of network are known. Fluid flow in this type of network can be simulated using sequential solving approach. Commercial software to perform this type of computation are those system analysis (Nodal analysis) programs such as FieldFlo and PipeSim, among others.

A nonhierarchical network is defined as a general system with multiple inflow points (sources) and multiple outlets (sinks). Loops may exist, so the flow directions in some portions of the network are not certain. Figure 18.14 presents a nonhierarchical network. Arrows in this figure represent flow directions determined by a computer program. Fluid flow in this type of network can be simulated using simultaneous solving approaches. Commercial software to perform this type of computations include ReO, GAP, HYSYS, FAST Piper, and others.

18.8.2 Optimization Approaches

Field-level production optimizations are carried out with two distinct approaches: (a) the simulation approach and (b) the optimization approach.

18.8.2.1 Simulation Approach

The simulation approach is a kind of trial-and-error approach. A computer program simulates flow conditions (pressures and flow rates) with fixed values of variables in each run. All parameter values are input manually before each run. Different scenarios are investigated with different sets of input data. Optimal solution to a given problem is selected on the basis of results of many simulation runs with various parameter values. Thus, this approach is more time consuming.

18.8.2.2 Optimization Approach

The optimization approach is a kind of intelligence-based approach. It allows some values of parameters to be determined by the computer program in one run. The parameter values are optimized to ensure the objective function is either maximized (production rate as the objective function) or minimized (cost as the objective function) under given technical or economical constraints. Apparently, the optimization approach is more efficient than the simulation approach.

18.8.3 Procedure for Production Optimization

The following procedure may be followed in production optimization:

1. Define the main objective of the optimization study. The objectives can be maximizing the total oil/gas production rate or minimizing the total cost of operation.
2. Define the scope (boundary) of the flow network.
3. Based on the characteristics of the network and fluid type, select a computer program.
4. Gather the values of component/equipment parameters in the network such as well-inflow performance, tubing sizes, choke sizes, flowline sizes, pump capacity, compressor horsepower, and others.
5. Gather fluid information including fluid compositions and properties at various points in the network.
6. Gather the fluid-flow information that reflects the current operating point, including pressures, flow rates, and temperatures at all the points with measurements.

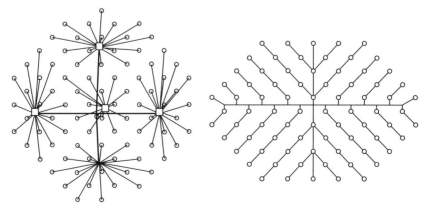

Figure 18.13 *Schematics of two hierarchical networks.*

Table 18.2 Solution to Example Problem 18.3 Given by the Spreadsheet LoopedLines.xls

LoopedLines.xls
This spreadsheet computes capacities of series, parallel, and looped pipelines.

Input data
Original pipe ID:				4	in.
Total pipeline length:				10	mi
Series pipe ID:	4		6	4	in.
Segment lengths:	7		3	0	mi
Parallel pipe ID:	4		6	0	in.
Looped pipe ID:	4		6	4	in.
Segment lengths:	3			7	mi

Solution

Capacity improvement by series pipelines: $= 1.1668$

$$q_h = \frac{3.23 T_b}{p_b} \sqrt{\frac{1}{f}} \sqrt{\frac{(p_1^2 - p_2^2) D^5}{\gamma_g \bar{T} \bar{z} L}}$$

Capacity improvement by parallel pipelines: $= 3.9483$

Capacity improvement by looped pipelines: $= 1.1791$

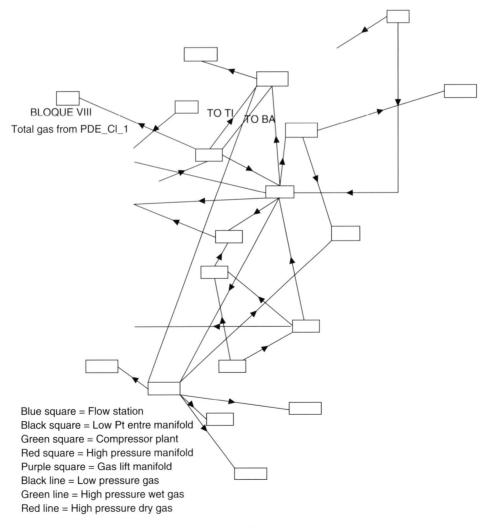

Blue square = Flow station
Black square = Low Pt entre manifold
Green square = Compressor plant
Red square = High pressure manifold
Purple square = Gas lift manifold
Black line = Low pressure gas
Green line = High pressure wet gas
Red line = High pressure dry gas

Figure 18.14 An example of a nonhierarchical network.

Table 18.3 Gas Lift Performance Data for Well A and Well B

Lift gas injection rate (MMscf/day)	Oil production rate (stb/day)		Slope of performance curve (stb/MMscf)	
	Well A	Well B	Well A	Well B
0.6	0	740	242	850
1.2	145	1,250	150	775
1.8	180	1,670	54	483
2.4	210	1,830	46	142
3	235	1,840	33	13
3.6	250	1,845	17	6
4.2	255	1,847	8	0
4.8	259	1,845	4	−56
5.4	260	1,780	−3	−146
6	255	1,670		

Table 18.4 Assignments of Different Available Lift Gas Injection Rates to Well A and Well B

Total lift gas (MMscf/day)	Gas injection rate before assignment (stb/day)		Slope of performance curve (stb/MMscf)		Lift gas assignment (MMscf/day)		Gas injection rate after assignment (stb/day)	
	Well A	Well B	Well A	Well B	Well A	Well B	Well A	Well B
1.2	0	0	242	850	0	1.2	0	1.2
1.8	0	1.2	242	775	0	0.6	0	1.8
2.4	0	1.8	242	483	0	0.6	0	2.4
3	0	2.4	242	142	0.6	0	0.6	2.4
3.6	0.6	2.4	242	142	0.6	0	1.2	2.4
4.2	1.2	2.4	150	142	0.6	0	1.8	2.4
4.8	1.8	2.4	54	142	0	0.6	1.8	3
5.4	1.8	3	54	13	0.6	0	2.4	3
6	2.4	3	46	13	0.6	0	3	3

7. Construct a computer model for the flow network.
8. Validate equipment models for each well/equipment in the network by simulating and matching the current operating point of the well/equipment.
9. Validate the computer model at facility level by simulating and matching the current operating point of the facility.
10. Validate the computer model at field level by simulating and matching the current operating point of the field.
11. Run simulations for scenario investigations with the computer model if a simulation-type program is used.
12. Run optimizations with the computer model if an optimization-type program is used.
13. Implement the result of optimization with an open-loop or closed-loop method.

18.8.4 Production Optimization Software

Commercial software packages are available for petroleum production optimization at all levels. Field-level optimization can be performed with ReO, GAP, HYSYS, FAST Piper, among others. This section makes a brief introduction to these packages.

18.8.4.1 ReO

The software ReO (EPS, 2004) is a compositional production simulator that can simulate and optimize highly nonhierarchical networks of multiphase flow. Its optimizer technology is based on sequential linear programming techniques. Because the network is solved simultaneously rather than sequentially, as is the case for nodal analysis techniques, the system can optimize and simulate accounting for targets, objectives, and constraints anywhere in the network.

A key feature of ReO is that it is both a production simulation and an optimization tool. Simulation determines the pressures, temperatures, and fluid flow rates within the production system, whereas optimization determines the most economical production strategy subject to engineering or economic constraints. The economic modeling capability inherent within ReO takes account of the revenues from hydrocarbon sales in conjunction with the production costs, to optimize the net revenue from the field. The ReO Simulation option generates distributions of pressure, temperature, and flow rates of water, oil, and gas in a well-defined network. The ReO Optimization option determines optimum parameter values that will lead to the maximum hydrocarbon production rate or the minimum operating cost under given technical and economical constraints. ReO addresses the need to optimize production operations, that is, between reservoir and facilities, in three main areas:

- To aid in the design of new production capacity, both conceptual and in detail
- To optimize production systems either off-line or in real time
- To forecast performance and create production profiles for alternative development scenarios

ReO integrates complex engineering calculations, practical constraints, and economic parameters to determine the optimal configuration of production network. It can be employed in all phases of field life, from planning through development and operations, and to enable petroleum, production, facility, and other engineers to share the same integrated model of the field and perform critical analysis and design activities such as the following:

- Conceptual design in new developments

- Equipment sizing, evaluation and selection
- Daily production optimization, on-line or off-line
- Problem and bottleneck detection/diagnosis
- Production forecasting
- Reservoir management
- Data management

Target and penalty functions are used in ReO within a valid region. This type of "target" is required to find the best compromise among conflicting objectives in a system. An example might be ensuring maximum production by driving down wellhead pressure in a gas field while maintaining optimum intake pressures to a compressor train.

One of the most important aspects of modeling production systems is the correct calculation of fluid PVT properties. Variable detail and quality often characterizes the PVT data available to the engineer, and ReO is designed to accommodate this. If complete compositional analysis has been performed, this can be used directly. If only Black Oil data are available, ReO will use a splitting technique to define a set of components to use in the compositional description. This approach means that different fluids, with different levels of detailed description can be combined into the same base set of components. Where wells are producing fluids of different composition, the mixing of these fluids is accurately modeled in the system. The composition is reported at all the nodes in the network. This is highly valuable in fields with differing wells compositions.

The facility models available in ReO for gas networks include pipeline, chokes (both variable and fixed diameter), block valves, standard compressors (polytropic model), heat exchangers (intercoolers), gas and gas condensate wells, sinks (separators, gas export and delivery points, flares, or vents), manifolds, links (no pressure loss pipelines), and flanges (no flow constraint).

Production constraints may be defined at any point within the production system in terms of pressure and/or flow rate along with objective functions for maximizing and minimizing flow rate or pressure in terms of sales revenues and costs.

ReO is seamlessly integrated with the program WellFlo application. WellFlo may be run from within ReO and new well models may be defined or existing well models used to simulate inflow and tubing performance.

The most complex application of ReO has been in Latin America where a network system including several hundred wells is optimized on a daily basis through a SCADA system. This system includes a low-pressure gas-gathering network integrated with a number of compressor trains and a high-pressure gas injection and distribution network.

18.8.4.2 HYSYS

HYSYS is an integrated steady-state and dynamic process simulator (AspenTech, 2005). HYSYS creates simulation models for the following:

- Plant design
- Performance monitoring
- Troubleshooting
- Operational improvement
- Business planning
- Asset management

HYSYS offers an integrated set of intuitive and interactive simulation and analysis tools and real-time applications. It provides rapid evaluations of safe and reliable designs through quick creation of interactive models for "what if" studies and sensitivity analysis.

HYSYS Upstream is for handling petroleum fluids and RefSYS is for handling multiunit modeling and simulation of refinery systems. HYSYS interfaces with applications such as Microsoft Excel and Visual Basic and features ActiveX compliance.

18.8.4.3 FAST Piper

FAST Piper (Fekete, 2001) is a gas pipeline, wellbore, and reservoir deliverability model that enables the user to optimize both existing and proposed gas-gathering systems. FAST Piper is designed to be a "quick and simple looking tool" that can solve very complicated gathering system designs and operating scenarios.

Developed and supported under Microsoft Windows 2000 and Windows XP, FAST Piper deals with critical issues such as multiphase flow, compressors, contracts, rate limitations, multiple wells, multiple pools, gas composition tracking, among others. The Key Features FAST Piper include the following:

- Allows matching of current production conditions
- Analyzes "what-if" scenarios (additional wells, compression, contracts, etc.)
- Integrated the coal bed methane (CBM) reservoir model allowing the user to predict the total gas and water production of an interconnected network of CBM wells, while incorporating compressor capacity curves, facility losses, and pipeline friction losses.

18.9 Discounted Revenue

The economics of production optimization projects is evaluated on the basis of discounted revenue to be generated by the projects. The most widely used method for calculating the discounted revenue is to predict the net present value (NPV) defined as

$$NPV = NPV_R - \text{cost}, \quad (18.47)$$

where

$$NPV_R = \sum_{n=1}^{m} \frac{\Delta R_n}{(1+i)^n}, \quad (18.48)$$

where m is the remaining life of the system in years, and i is the discount rate. The annual incremental revenue after optimization is expressed as

$$\Delta R_n = (\$)\Delta N_{p,n}, \quad (18.49)$$

where ($\$$) is oil or gas price and the $\Delta N_{p,n}$ is the predicted annual incremental cumulative production for year n, which is expressed as

$$\Delta N_{p,n} = N_{p,n}^{op} - N_{p,n}^{no}, \quad (18.50)$$

where

$N_{p,n}^{op}$ = forcasted annual cumulative production of optimized system for year n

$N_{p,n}^{no}$ = predicted annual cumulative production of non-optimized well for year n.

Summary

This chapter presents principles of production optimization of well, facility, and field levels. While well- and facility-level optimization computations can be carried out using Nodal analysis approach, field-level computations frequently require simulators with simultaneous solvers. Production optimization is driven by production economics.

References

AHMED, T. *Hydrocarbon Phase Behavior*. Houston: Gulf Publishing Company, 1989.

AspenTech. Aspen HYSYS. Aspen Technology, Inc., 2005.

BEGGS, H.D. *Production Optimization Using NODAL Analysis*, 2nd edition. Tulsa: OGCI, Inc., Petroskils, LLC., and H. Dale Beggs, 2003.

BROWN, K.E. *The Technology of Artificial Lift Methods*, Vol. 2a. Tulsa, OK: Petroleum Publishing Co., 1980.

EDINBURGH Petroleum Services. *FloSystem User Documentation*. Edinburgh: Edinburgh Petroleum Services, Ltd., 1997.

E-Production Solutions. *ReO User Documentation*. Edinburgh: E-Production Solutions, 2004.

Fekete Associates. *Fekete Production Optimization*. Fekete Associates, Inc., Calgary, Canada, 2001.

GUO, B. and GHALAMBOR, A. *Natural Gas Engineering Handbook*. Houston, TX: Gulf Publishing Company, 2005.

STANDING, M.B. A set of equations for computing equilibrium ratios of a crude oil/natural gas system at pressures below 1,000 psia. *J. Petroleum Technol. Trans. AIME* 1979;31(Sept):1193.

STANDING, M.B. *Volume and Phase Behavior of Oil Field Hydrocarbon Systems*, 9th edition. Dallas: Society of Petroleum Engineers, 1981.

Problems

18.1 Analyze the dynamometer card shown in Figure 18.7 (scale = 1:1.5) assuming the following parameter values:
$S = 40$ in.
$N = 20$ spm
$C = 12,500$ lb/in.

18.2 Perform flash calculation under the following separator conditions:
Pressure: 500 psia
Temperature: 150 °F
Specific gravity of stock-tank oil: 0.85 (water = 1)
Specific gravity of solution gas: 0.65 (air = 1)
Gas solubility (R_s): 800 scf/stb

Gas Composition	
Compound	Mole fraction
C_1	0.6899
C_2	0.0969
C_3	0.0591
i-C_4	0.0439
n-C_4	0.0378
i-C_5	0.0157
n-C_5	0.0112
C_6	0.0081
C_{7+}	0.0101
N_2	0.0094
CO_2	0.0021
H_2S	0.0058

18.3 Consider a 6-in. pipeline that is 20 miles long. Assuming that the compression and delivery pressures will remain unchanged, calculate gas-capacity increases using the following measures of improvement: (a) replace 10 miles of the pipeline by a 8-in. pipeline segment; (b) place an 8-in. parallel pipeline to share gas transmission; and (c) loop 10 miles of the pipeline with an 8-in. pipeline segment.

18.4 The gas lift performance data of four oil wells are as follows: If a total lift gas injection rate of 12 MMscf/day is available to the four wells, what lift gas flow rates should be assigned to each well?

Lift gas injection rate (MMscf/day)	Oil production rate		(stb/day)	
	Well A	Well B	Well C	Well D
0.6	80	740	870	600
1.2	145	1,250	1,450	1,145
1.8	180	1,670	1,800	1,180
2.4	210	1,830	2,100	1,210
3	235	1,840	2,350	1,235
3.6	250	1,845	2,500	1,250
4.2	255	1,847	2,550	1,255
4.8	259	1,845	2,590	1,259
5.4	260	1,780	2,600	1,260
6	255	1,670	2,550	1,255

Appendices

Contents
Appendix A Unit Conversion Factors   282
Appendix B The Minimum Performance
                    Properties of API Tubing   283

Appendix A: Unit Conversion Factors

Quantity	U.S. Field unit	To SI unit	To U.S. Field unit	SI unit
Length (L)	feet (ft)	0.3084	3.2808	meter (m)
	mile (mi)	1.609	0.6214	kilometer (km)
	inch (in.)	25.4	0.03937	millimeter (mm)
Mass (M)	ounce (oz)	28.3495	0.03527	gram (g)
	pound (lb)	0.4536	2.205	kilogram (kg)
	lbm	0.0311	32.17	slug
Volume (V)	gallon (gal)	0.003785	264.172	meter^3 (m^3)
	cu. ft. (ft^3)	0.028317	35.3147	meter^3 (m^3)
	barrel (bbl)	0.15899	6.2898	meter^3 (m^3)
	Mcf (1,000 ft^3, 60 °F, 14.7 psia)	28.317	0.0353	Nm^3 (15 °C, 101.325 kPa)
	sq. ft (ft^2)	9.29×10^{-2}	10.764	meter^2 (m^2)
Area (A)	acre	4.0469×10^3	2.471×10^{-4}	meter^2 (m^2)
	sq. mile	2.59	0.386	(km)^2
Pressure (P)	lb/in.^{2} (psi)	6.8948	0.145	kPa (1000 Pa)
	psi	0.0680	14.696	atm
	psi/ft	22.62	0.0442	kPa/m
	inch Hg	3.3864×10^3	0.2953×10^{-3}	Pa
Temperature (t)	F	0.5556(F-32)	1.8C+32	C
	Rankine (°R)	0.5556	1.8	Kelvin (K)
Energy/work (w)	Btu	252.16	3.966×10^{-3}	cal
	Btu	1.0551	0.9478	kilojoule (kJ)
	ft-lbf	1.3558	0.73766	joule (J)
	hp-hr	0.7457	1.341	kW-hr
Viscosity (μ)	cp	0.001	1,000	Pa·s
	lb/ft·sec	1.4882	0.672	kg/(m-sec) or (Pa·s)
	lbf-s/ft^2	479	0.0021	dyne-s/cm^2 (poise)
Thermal conductivity (k)	Btu-ft/hr-ft^2-F	1.7307	0.578	W/(m·K)
Specific heat (C_p)	Btu/(lbm·°F)	1	1	cal/(g·°C)
	Btu/(lbm·°F)	4.184×10^3	2.39×10^{-4}	J.(kg·K)
Density (P)	lbm/ft^3	16.02	0.0624	kg/m^3
Permeability (k)	md	0.9862	1.0133	mD ($= 10^{-15}$ m^2)
	md ($= 10^{-3}$ darcy)	9.8692×10^{-16}	1.0133×10^{15}	m^2

Appendix B: The Minimum Performance Properties of API Tubing

Nom. (in.)	O.D. (in.)	Grade	Wt per ft with couplings (lb) Non-Upset	Wt per ft with couplings (lb) Upset	Inside diameter (in.)	Drift diameter (in.)	O.D. of upset (in.)	O.D. of Cplg. (in.) Non-Upset	O.D. of Cplg. (in.) Upset	Collapse resistance (psi)	Internal yield pressure (psi)	Joint yield strength (lb) Non-Upset	Joint yield strength (lb) Upset
¾	1.050	F-25		1.20	0.824	0.730	1.315		1.660	5,960	4,710		8,320
		H-40		1.20	0.824	0.730	1.315		1.660	7,680	7,530		13,300
		J-55		1.20	0.824	0.730	1.315		1.660	10,560	10,360		18,290
		C-75	1.14	1.20	0.824	0.730	1.315	1.313	1.660	14,410	14,120	11,920	24,950
		N-80		1.20	0.824	0.730	1.315		1.660	15,370	15,070		26,610
1	1.315	F-25		1.80	1.049	0.955	1.469		1.900	5,540	4,430		12,350
		H-40		1.80	1.049	0.955	1.469		1.900	7,270	7,080		19,760
		J-55		1.80	1.049	0.955	1.469		1.900	10,000	9,730		27,160
		C-75	1.70	1.80	1.049	0.955	1.469	1.660	1.900	13,640	13,270	20,540	37,040
		N-80		1.80	1.049	0.955	1.469		1.900	14,650	14,160		39,510
1¼	1.660	F-25		2.40	1.380	1.286	1.812		2.200	4,400	3,690		16,710
		H-40		2.40	1.380	1.286	1.812		2.200	6,180	5,910		26,740
		J-55		2.40	1.380	1.286	1.812		2.200	8,490	8,120		36,770
		C-75	2.30	2.40	1.380	1.286	1.812	2.054	2.200	11,580	11,070	29,120	50,140
		N-80		2.40	1.380	1.286	1.812		2.200	12,360	11,800		53,480
1½	1.900	F-25	2.75	2.90	1.610	1.516	2.094	2.200	2.500	3,920	3,340	11,930	19,900
		H-40	2.75	2.90	1.610	1.516	2.094	2.200	2.500	5,640	5,350	19,090	31,980
		J-55	2.75	2.90	1.610	1.516	2.094	2.200	2.500	7,750	7,350	26,250	43,970
		C-75	2.75	2.90	1.610	1.516	2.094	2.200	2.500	10,570	10,020	35,800	59,960
		N-80	2.75	2.90	1.610	1.516	2.094	2.200	2.500	11,280	10,680	38,180	63,960
2	2.375	F-25	4.00		2.041	1.947		2.875		3,530	3,080	18,830	
		F-25	4.60	4.70	1.995	1.901	2.594	2.875	3.063	4,160	3,500	22,480	32,600
		H-40	4.00		2.041	1.947		2.875		5,230	4,930	30,130	
		H-40	4.60	4.70	1.995	1.901	2.594	2.875	3.063	5,890	5,600	35,960	52,170
		J-55	4.00		2.041	1.947		2.875		7,190	6,770	41,430	
		J-55	4.60	4.70	1.995	1.901	2.594	2.875	3.063	8,100	7,700	49,440	71,730
		C-75	4.00		2.041	1.947		2.875		9,520	9,230	56,500	
		C-75	4.60	4.70	1.995	1.901	2.594	2.875	3.063	11,040	10,500	67,430	97,820
		C-75	5.80	5.95	1.867	1.773	2.594	2.875	3.063	14,330	14,040	96,560	126,940
		N-80	4.00		2.041	1.947		2.875		9,980	9,840	60,260	
		N-80	4.60	4.70	1.995	1.901	2.594	2.875	3.063	11,780	11,200	71,920	104,340
		N-80	5.80	5.95	1.867	1.773	2.594	2.875	3.063	15,280	14,970	102,980	135,400
		P-105	4.60	4.70	1.995	1.901	2.594	2.875	3.063	15,460	14,700	94,400	136,940
		P-105	5.80	5.95	1.867	1.773	2.594	2.875	3.063	20,060	19,650	135,170	177,710
2½	2.875	F-25	6.40	6.50	2.441	2.347	3.094	3.500	3.668	3,870	3,300	32,990	45,300
		H-40	6.40	6.50	2.441	2.347	3.094	3.500	3.668	5,580	5,280	52,780	72,480
		J-55	6.40	6.50	2.441	2.347	3.094	3.500	3.668	7,680	7,260	72,570	99,660
		C-75	6.40	6.50	2.441	2.347	3.094	3.500	3.668	10,470	9,910	98,970	135,900
		C-75	8.60	8.70	2.259	2.165	3.094	3.500	3.668	14,350	14,060	149,360	186,290

(*Continued*)

Appendix B: *(Continued)*

Nom. (in.)	O.D. (in.)	Grade	Wt per ft with couplings (lb) Non-Upset	Wt per ft with couplings (lb) Upset	Inside diameter (in.)	Drift diameter (in.)	O.D. of upset (in.)	O.D. of Cplg. (in.) Non-Upset	O.D. of Cplg. (in.) Upset	Collapse resistance (psi)	Internal yield pressure (psi)	Joint yield strength (lb) Non-Upset	Joint yield strength (lb) Upset
		N-80	6.40	6.50	2.441	2.347	3.094	3.500	3.668	11,160	10,570	105,560	144,960
		N-80	8.60	8.70	2.259	2.165	3.094	3.500	3.668	15,300	15,000	159,310	198,710
		P-105	6.40	6.50	2.441	2.347	3.094	3.500	3.668	14,010	13,870	138,550	190,260
		P-105	8.60	8.70	2.259	2.165	3.094	3.500	3.668	20,090	19,690	209,100	260,810
3	3.500	F-25	7.70		3.068	2.943		4.250		2,970	2,700	40,670	
		F-25	9.20	9.3	2.992	2.867	3.750	4.250	4.500	3,680	3,180	49,710	64,760
		F-25	10.20		2.922	2.797		4.250		4,330	3,610	57,840	
		H-40	7.70		3.068	2.943		4.250		4,630	4,320	65,070	
		H-40	9.20	9.3	2.992	2.867	3.750	4.250	4.500	5,380	5,080	79,540	103,610
		H-40	10.20		2.922	2.797		4.250		6,060	5,780	92,550	
		J-55	7.70		3.068	2.943		4.250		5,970	5,940	89,470	
		J-55	9.20	9.3	2.992	2.867	3.750	4.250	4.500	7,400	6,980	109,370	142,460
		J-55	10.20		2.922	2.797		4.250		8,330	7,940	127,250	
		C-75	7.70		3.068	2.943		4.250		7,540	8,100	122,010	
		C-75	9.20	9.3	2.992	2.867	3.750	4.250	4.500	10,040	9,520	149,140	194,260
		C-75	10.20		2.922	2.797		4.250		11,360	10,840	173,530	
		C-75	12.70	12.95	2.750	2.625	3.750	4.250	4.500	14,350	14,060	230,990	276,120
		N-80	7.70		3.068	2.943		4.250		7,870	8,640	130,140	
		N-80	9.20	9.3	2.992	2.867	3.750	4.250	4.500	10,530	10,160	159,080	207,220
		N-80	10.20		2.922	2.797		4.250		12,120	11,560	185,100	
		N-80	12.70	12.95	2.750	2.625	3.750	4.250	4.500	15,310	15,000	246,390	294,530
		P-105	9.20	9.3	2.992	2.867	3.750	4.250	4.500	13,050	13,340	208,790	271,970
		P-105	12.70	12.95	2.750	2.625	3.750	4.250	4.500	20,090	19,690	323,390	386,570
3½	4.000	F-25	9.50		3.548	3.423		4.750		2,630	2,470	15,000	
		F-25		11.00	3.476	3.351	4.250		5.000	3,220	2,870		76,920
		H-40	9.50		3.548	3.423		4.750		4,060	3,960	72,000	
		H-40		11.00	3.476	3.351	4.250		5.000	4,900	4,580		123,070
		J-55	9.50		3.548	3.423		4.750		5,110	5,440	99,010	
		J-55		11.00	3.476	3.351	4.250		5.000	6,590	6,300		169,220
		C-75	9.50		3.548	3.423		4.750		6,350	7,420	135,010	
		C-75		11.00	3.476	3.351	4.250		5.000	8,410	8,600		230,760
		N-80	9.50		3.548	3.423		4.750		6,590	7,910	144,010	
		N-80		11.00	3.476	3.351	4.250		5.000	8,800	9,170		246,140
4	4.500	F-25	12.60	12.75	3.958	3.833	4.750	5.200	5.563	2,870	2,630	65,230	90,010
		H-40	12.60	12.75	3.958	3.833	4.750	5.200	5.563	4,500	4,220	104,360	144,020
		J-55	12.60	12.75	3.958	3.833	4.750	5.200	5.563	5,720	5,790	143,500	198,030
		C-75	12.60	12.75	3.958	3.833	4.750	5.200	5.563	7,200	7,900	195,680	270,030
		N-80	12.60	12.75	3.958	3.833	4.750	5.200	5.563	7,500	8,440	208,730	288,040

Index

A

Acid, 10/129, 16/244–249
  volume, 16/245–249
Acidizing, 16/243
  design, 16/243–244, 16/247–248
  models, 16/248
Acid mineral reaction, 16/244
  kinetics, 16/244, 16/247
  stoichiometry, 16/244
American Gas Association, 11/157
American Petroleum Institute, 1/6, 1/17, 2/26 9/110, 12/163, 18/272
Annular flow, 4/46, 4/48, 14/216, 15/232
API gravity, 2/20, 2/26, 5/65, 11/145, 11/157, 12/170, 12/179–180, 18/272, 18/274
Artificial lift, 3/30, 4/57, 5/66, 12/159, 12/162, 12/164, 12/179, 13/182, 13/184, 13/205, 14/207, 14/208, 14/209, 14/216, 14/222, 18/279
  method, 3/30, 4/57, 5/66, 12/159, 13/182–206, 14/207, 14/208–223, 18/279

B

Boiling point, 18/271
Buckling, 9/110, 9/112–115, 11/151
Buoyancy, 9/111–112, 12/179, 14/215

C

Capacity, 1/11, 10/121–122, 10/124, 10/126–131, 11/137, 11/142–143, 11/148, 11/153, 12/162, 13/188–189, 13/191–192, 13/203, 14/210–211, 14/216, 14/219–220, 15/228, 15/231, 18/272–280
Carbon dioxide, 2/23, 2/25, 10/118, 11/151
Carbonate acidizing, 16/243–244, 16/247–248
  design, 16/243–244, 16/247
Casings, 1/5
Cavitation, 14/209, 14/220
Centrifugal, 1/10–11, 10/118–119, 11/137, 11/142–143, 11/157, 13/188, 13/190–193, 13/206, 14/208–209
  efficiency, 10/120, 10/125, 10/127, 10/129, 11/135, 11/137, 11/139–140, 11/142–143, 11/148, 11/150, 11/152–153, 11/157, 12/162, 12/169–170, 12/172–174, 12/177, 12/180, 13/183, 13/188–192, 13/206
  volumetric, 1/5, 1/10, 2/20, 4/48, 4/52, 5/65, 7/88, 7/92, 8/98, 11/135–137, 11/139–141, 11/148, 12/159, 12/170, 12/172–174, 12/177, 12/180, 13/188–189, 14/209, 14/211, 15/229, 16/244, 16/246–249, 18/269
  horsepower, 11/135–137, 11/139–140, 11/142–143, 11/157, 12/173, 12/173, 12/177, 13/189–192, 13/205–206, 14/208–210, 14/213, 17/262, 18/270, 18/274
  actual, 3/42, 5/65, 11/140, 11/142, 11/145–148, 11/152, 12/168, 13/190–191, 13/196, 13/201, 14/218, 15/228, 18/269–270
  brake, 11/136, 11/157, 12/173, 13/190, 13/206
  isentropic, 5/60, 5/62, 5/64, 11/138, 11/142, 13/187, 13/189–191
Channeling, 15/231, 15/234, 15/236–237
Chokes, 1/5, 1/7, 1/17, 5/60–62, 5/64–66, 13/182, 13/187, 18/270, 18/279
Coating, 4/48, 11/148–149, 11/153
Collapse, 4/48, 9/110–112, 9/114, 11/150–151, 11/157
Completion, 3/30, 9/111–112, 9/115, 12/162, 12/177, 14/208, 14/211, 15/228, 15/230, 16/244, 17/264
Compressibility, 2/20–23, 2/25–27, 3/30, 3/33–35, 3/43, 4/50, 4/53–56, 4/58, 5/66, 6/82–84, 6/86, 7/88–89, 7/93, 7/95–96, 8/98, 10/121, 11/142–143, 11/146, 13/186, 13/190–192, 13/200, 14/219, 15/229, 15/231
Compressor, 1/3–4, 1/10–11, 10/118, 10/126, 11/133–134, 11/136–140, 11/142–143, 11/146, 11/156–157, 13/182–183, 13/185, 13/187–193, 13/197, 13/205–206, 18/268, 18/276–277, 18/279
Conductivity, 11/152–153, 15/229, 15/238, 17/256–258, 17/262–264
Corrosion, 1/12, 10/126, 10/129, 11/148–149, 11/151–152, 11/157
Critical point, 1/5
Cylinders, 12/162, 13/189

D

Damage characterization
Decline curve analysis, 14/218
  constant fractional decline, 8/98
  harmonic decline, 8/98, 8/100–103
  hyperbolic decline, 8/98, 8/100–101, 8/103
Dehydration, 10/117, 10/118, 10/121, 10/125–129, 10/132
  cooling, 10/125–126, 10/128
  glycol, 10/126–132,
  stripping still, 10/127–128, 10/131–132
Density of gas, 2/24, 10/121
Dewatering, 14/214
Downhole, 12/162, 12/179
Drilling, 1/6, 4/57, 5/66, 10/127, 11/157, 14/216, 14/223, 15/230, 16/244
  mud, 10/127
Drums, 11/139
Drying, 10/126
Dynamometer cards, 12/174, 12/177, 18/270

E

Economics, 1/4, 7/88, 14/219, 18/268–269, 18/279
Enthalpy, 11/157, 13/189
Entropy, 11/157, 13/189
Equation of state, 2/26, 18/217
Exploration well, 3/39

F

Fittings, 1/7, 5/66
Flow metering, 5/66
Flow efficiency, 10/125, 11/153
Flow regime, 3/30–31, 3/42, 4/48–49, 4/51, 4/53, 5/60, 5/63–65, 7/88, 11/144, 14/216, 15/229–230, 15/232, 15/235, 15/240, 16/244, 17/264
Flowline, 1/4, 1/7, 1/11, 1/13, 1/15, 5/63–64, 5/67, 6/75–76, 6/85, 10/124, 10/132, 11/143, 11/150–151, 11/153, 13/183, 14/217, 14/218, 18/276
Fluid, 1/1, 1/4–5, 1/7–8, 1/11–12, 2/20, 2/22, 2/26, 3/30, 3/33–35, 3/37, 3/40, 3/43, 4/46–48, 4/51, 4/57, 5/60, 5/63–64, 5/66, 6/70, 6/72, 6/79, 6/82, 6/84, 7/88–92, 7/94, 9/110–115, 10/118, 10/120, 10/132, 11/134–136, 11/138, 11/143–146, 11/150–154, 11/156–159, 12/162, 12/164–165, 12–169–173, 12/177–179, 13/182–184, 13/186, 13/192–193, 13/196–206, 14/208–215, 14/219–223, 15/228–231, 15/234, 15/236, 15/241, 16/244, 16/247, 17/252–256, 17/258–266, 18/269–272, 18/276–279
  loss, 17/258
  volume, 7/89, 14/214, 17/258, 17/260–261, 17/265
Formation damage, 15/228, 16/244–245, 17/252
Formation volume factor, 2/20, 2/22, 2/25–26, 3/30, 3/33–35, 3/40, 3/43, 4/50–51, 6/72, 6/76, 6/78, 6/85–86, 7/88, 7/93, 7/95, 12/170, 12/173, 12/179, 14/210, 14/211–213, 14/223, 25/339
Forming, 1/7, 6/74, 6/84, 17/262, 17/264

INDEX

Fracture direction, 17/258, 17/264
Friction, 1/6, 1/11, 4/46–48, 4/50–51, 4/53, 4/55, 4/57, 5/60, 6/72, 6/81, 11/137, 11/142–148, 11/157, 12/162, 12/168–169, 12/172–173, 13/183–188, 13/191, 14/210, 14/212–213, 14/216, 14/218–219, 15/240, 16/246, 17/259–160, 18/268–270, 18/279
    factor, 4/46–57, 6/72, 6/81, 11/144–148, 11/157, 13/184, 15/240, 16/246, 17/259, 18/270
    pressure drop, 4/57, 16/246, 17/259–260, 18/268

G

Gas, 2/22–27, 4/50, 6/82–86
    compressibility, 2/22–27, 4/50, 6/82–86, 10/121, 11/142–143, 13/186, 13/191–192, 13/200, 14/219, 13/200, 14/219
    compressors, 10/126, 11/156
    condensate, 1/4, 4/56, 5/64–66, 10/120, 10/124, 15/228, 18/272, 18/279
    flow, 1/4, 1/9–10, 3/42, 4/48, 4/53, 5/60–67, 6/70, 6/72–75, 6/82, 10/118, 10/121, 10/126–132, 11/136–137, 11/139–140, 11/142–148, 11/157, 13/182, 13/185, 13/187–188, 13/190–196, 13/205–206, 14/216, 14/219, 15/231–232, 15/234–235, 15/237, 15/240, 18/268, 18/274–275, 18/280
    formation volume factor, 2/22, 2/25–26
    gravity, 11/146, 11/148, 13/201, 13/206
    injection rate, 13/183–187, 18/268–269, 18/275–276, 18/278, 18/280
    lift, 1/10, 4/46, 5/66, 12/159, 13/181–185, 13/187, 13/189, 13/191–195, 13/197–201, 13/203–206, 14/208, 14/216, 14/219, 14/223, 15/232, 18/268–269, 18/275–278, 18/280
    lines, 11/149–151
    pipelines, 10/125, 11/143, 11/149, 11/152, 18/272
    transmission, 11/147, 11/157, 18/280
    viscosity, 2/21, 2/23–24, 2/26, 4/52, 5/61, 6/75–76, 6/82–83, 6/85–86, 7/93, 11/144, 13/187, 13/200, 13/206
    well deliverability, 6/71, 6/82–83
    well performance, 6/85
Gathering lines, 11/150–151
GOR, 6/82–84, 6/86, 10/118, 10/125, 11/143, 11/148, 11/153, 13/182, 13/185, 13/188–189, 13/203, 14/210, 14/213, 14/219, 14/223, 16/246, 17/256, 17/265, 18/268, 18/272, 18/274, 18/280
Gravel pack, 9/112

H

Harmonic decline, 8/97, 8/98, 8/100–103
HCl preflush, 16/244, 16/245, 16/249
Head, 1/4–1/10, 1/13, 1/15, 1/17, 3/30, 3/37, 4/46, 4/50–55, 4/57–58, 5/60, 5/63–64, 5/66–67, 6/70–82, 6/84–86, 7/88, 7/93, 7/95–96, 10/120, 10/127, 10/129, 11/136, 11/142–143, 11/148, 11/150–151, 11/157–158, 2/162, 12/168, 12/172–173, 12/177, 12/180, 13/184–185, 13/187–188, 13/191–192, 13/198–199, 3/201–202, 13/205–206, 14/208–216, 14/218–221, 14/223, 15/232, 15/237, 15/240–242, 16/245, 16/249
    rating, 14/214–215
Heavy oil, 14/213
Hoop stress, 11/150–152
Horizontal well, 3/31–33, 3/42, 9/111, 14/214, 15/229–230, 16/248, 17/264–265
Hydrates, 1/7, 5/60, 5/62, 10/125–126, 11/152, 15/228
Hydraulically fractured well, 15/229
Hydraulic fracturing, 9/112, 15/225, 15/231, 17/251–255, 17/257–261, 17/263–265
Hydraulic piston pumping, 12/159, 14/207–208, 14/211, 14/222
Hyperbolic decline, 8/97–98, 8/100–101, 8/103
Hydrogen sulfide, 2/23, 10/118, 10/126, 10/129, 11/151
Hydrostatic pressure, 11/149–151, 13/199, 14/212, 14/217, 14/219–220, 16/246, 17/259–260

I

Inflow performance relationship (IPR), 3/29–30, 3/32, 3/42–43, 14/210
Injected gas, 13/183

Insulation, 11/148, 11/152–154, 11/156, 11/158, 14/209
Interest, 2/20, 11/145
IPR curve, 3/29, 3/32–43, 7/88, 7/92, 13/183, 14/218

J

Jet pumping, 12/159, 14/207–208, 14/220, 14/222
Jet pumps, 14/220–221

K

KGD model, 17/254, 17/262

L

Laminar flow, 4/46–47, 9/112, 11/144
Leak, 1/7, 1/12, 10/129, 12/177, 12/179, 13/185, 13/189, 14/214, 15/231, 15/234, 15/237, 17/255, 17/258, 17/260, 17/263, 17/265
Lifting, 12/172, 13/182, 13/184, 14/208, 14/213, 14/218, 14/222, 15/232
Line pipe, 1/11, 11/145
Line size, 1/11
Liquid holdup, 4/48, 4/51–53
Liquid phase, 4/48, 4/50, 4/52, 10/120, 10/124

M

Maintenance, 1/5, 1/11, 7/95, 10/127, 11/137, 13/183, 13/188
Manifolds, 1/11, 1/11, 11/143, 13/187, 13/205, 13/206
Matrix acidizing, 16/224, 16/247, 16/248
Measurement, 1/7, 2/20, 2/23, 4/48, 4/50, 4/57, 5/66, 7/94, 17/263
Meters, 14/210
Methane, 14/214
Mist flow, 4/56, 14/216, 15/232, 15/235, 15/238, 15/239
Molecular weight, 2/22, 2/24, 2/26–27, 11/143, 13/192, 16/244
Mole fraction, 2/22, 2/24–27, 4/51, 6/72, 6/85
Mollier diagram, 13/189, 13/190
Motor, 10/128, 11/137, 11/140, 11/142, 12/162, 12/174, 13/189, 4/208–211, 14/215, 14/217
Multilateral, 3/37, 6/79, 6/80–85
Multiphase flow 3/41, 4/45–46, 4/48, 4/53, 4/57, 5/59, 5/63, 5/66, 13/184, 14/216, 15/234–235 3/41, 4/46, 4/48, 4/53, 5/63, 13/184, 14/216, 15/234, 15/235
Multiphase fluid, 4/46, 5/64, 11/145

N

Natural gas, 1/10–11, 2/20–26, 4/53, 4/57, 5/60, 5/62, 5/67, 10/125–128, 10/132, 11/134, 11/136–137, 11/140–142, 11/144, 11/146, 11/148, 11/152, 11/157, 13/186–187, 13/189–200, 13/205–206, 15/231
    composition of, 10/118, 15/230
    water content, 10/125–126, 10/129–130, 10/132
Nodal analysis, 6/70–71, 6/74–75, 6/77, 6/80, 6/84, 7/88–90, 7/92–94, 15/228, 17/262, 17/264

O

Offshore, 1/9, 1/11, 9/114, 10/118, 10/125, 11/148, 11/151–152, 12/162, 13/182, 14/208–14/209, 14/211
    Operations, 1/11, 12/162, 13/182, 14/208, 14/211
Oil properties, 2/20, 5/66
Operating costs, 10/127, 14/214
Operating pressure, 6/71, 6/76, 10/120, 10/122, 10/124, 10/128–132, 11/147, 11/150, 13/190, 13/192, 13/197–199, 14/212–213, 14/222, 17/258
Operators, 10/120, 10/132, 11/140, 11/150
Orifice, 5/61–62, 5/64, 5/67, 13/182, 13/184–185, 13/187, 13/192, 13/194, 13/198, 13/200
    Charts, 2/21–23, 4/52, 4/54, 5/64, 6/74, 6/78, 6/80, 10/121, 11/145, 13/187, 14/221
    expansion factor, 2/26
Outflow performance curve, 6/70, 6/74–76, 13/183, 6/185

P

Panhandle equation 11/146
Packer, 1/5–7, 9/112–115, 13/183, 13/184, 13/203–205, 14/218
Paraffin, 1/7, 14/209, 14/215, 15/228
Parametric study, 17/256, 17/258

Pay thickness, 6/82, 6/82, 6/86
Perforating, 16/247
Performance curve, 5/60, 6/70, 6/74–76, 13/183–185, 14/218, 14/222, 18/268, 18/269, 18/275, 18/276, 18/278
Permeability, 3/30, 3/32–35, 3/37, 3/39–41, 3/43, 6/82, 6/83, 6/86, 7/88, 7/93, 7/95, 15/228–231, 15/235, 15/237, 16/244–246, 16/248, 17/252, 17/256, 17/257, 17/259, 17/262, 17/264, 17/265, 18/268
Petroleum hydrocarbons, 1/4, 1/11, 10/126, 10/127, 10/129, 10/131, 11/157
Phase behavior, 10/120
Pipeline, 1/4, 1/9–1/11, 1/16–1/17, 2/20, 9/114, 10/120, 10/125, 10/126, 11/133, 11/134, 11/136, 11/143–158, 18/267, 18/268, 18/270, 18/272–277, 18/278, 18/280
PKN model, 17/255, 17/260, 17/261
Platform, 1/9, 1/11, 1/15, 1/17, 10/188, 10/132, 11/150, 11/151, 18/268
Polymer, 11/152
Pore space, 3/34, 15/230, 17/253
Porosity, 3/30, 3/33–3/35, 3/43, 7/88, 7/93, 7/95, 7/96, 15/229, 15/230, 16/245–249, 17/260, 17/265
Positive-displacement pump, 14/209
Precipitation, 16/246
Preflush/postflush, 16/244, 16/245, 16/249
Pressure, 1/5, 1/7
 drop, 4/46, 4/48, 4/53, 4/57, 4/60, 5/66, 7/88–90, 7/95, 10/128, 11/114, 12/172, 13/183, 13/197, 16/246, 17/259, 17/260, 18/268, 18/270
 gauge, 1/6–1/8, 5/60, 10/118
 traverse, 4/46, 4/48, 4/53, 4/55, 4/58, 13/198
Processing plant, 1/1
Produced water, 4/50, 15/240
Production, 1/11
 facility, 1/11, 11/143, 18/278
 injection well, 3/31
 logging, 15/228, 15/231, 15/241, 17/262, 17/264
Progressing cavity pump (PCP), 14/213–215, 14/241
Proppant, 17/252, 17/258–265
Pseudo–steady-state flow, 3/31–3/34, 3/37/, 3/43, 7/87, 7/88, 7/90, 7/92, 7/94, 7/95, 8/98, 16/246
Pump, 1/4, 1/9–10, 1/16–17, 10/128, 10/129, 10/131, 11/133–136, 11/156, 12/162, 12/164, 12/165, 12/172, 14/208, 14/209, 14/220, 14/221, 17/252, 18/268
 intake pressure, 14/212–214, 14/222
 Jet, 12/159, 14/207–208, 14/220–222
PVT, 2/20–2/21, 2/23, 2/26–2/27, 7/94, 18/279

R
Range, 1/10–11, 2/20, 2/26–27, 3/36, 6/70, 6/73, 6/75, 6/78, 10/125, 11/136, 11/142, 11/144, 11/145, 11/147, 12/126, 12/172, 13/188–191, 13/196, 14/209, 14/217, 14/220, 14/223, 16/246, 17/257, 17/258, 18/269, 18/270
Real gas, 2/23–24, 2/26, 3/30–31, 7/94, 10/121, 11/142, 11/143, 11/147, 13/189
Regulation, 5/60
Relative permeability, 3/34, 3/39–41, 3/43,
Relative roughness, 4/46, 4/47, 4/55, 4/57, 4/58, 6/70, 6/71, 6/75, 6/76, 6/85, 11/144–147
Reserves, 2/26, 15/228
Reservoir, 1/4–7, 1/17, 2/20–21, 2/25, 3/30–43, 4/46, 4/55, 4/58, 5/60, 6/70–86, 7/88–95, 8/98, 9/110, 11/153, 11/157, 13/182–184, 13/186, 13/192, 13/199, 13/201, 14/208, 14/210–213, 14/216, 14/218–220, 14/222, 14/223, 15/228–231, 15/237, 15/241, 15/242, 16/244, 16/246–249, 17/252, 17/253, 17/256–260, 17/263–265, 18/268, 18/278, 18/279
 engineering, 2/20, 3/42, 7/94, 15/241
 hydrocarbons, 1/4,
 thickness, 3/30
Reynolds number, 4/46, 4/47, 4/50, 4/53, 4/55, 5/60–62, 6/81, 11/144–148, 11/157, 13/187
Riser, 1/11, 11/151, 11/152, 11/157

S
Safety, 1/3, 1/7, 1/11–13, 1/15–16, 9/111, 10/118, 11/150–151, 12/170, 12/172–174, 13/185, 13/187–188, 14/215, 14/222, 16/246–247, 17/258
Sandstone, 16/243–248, 17/253, 17/260, 17/262
 Acidizing, 16/243–248
Saturated oil, 1/4, 3/34, 3/36, 3/38, 7/88
Saturations, 3/37, 7/89
Scales, 11/148, 15/228
Separator, 1/3–4, 1/8–9, 10/118–129, 10/131–132, 13/183, 14/208, 18/267–268, 18/270–272, 18/279
SI units, 2/21, 2/24–25, 4/51–52, 4/54, 6/72–74, 6/78–80, 13/188, 16/247
Single-phase flow, 3/31, 3/34, 3/37, 7/88–89, 14/212, 14/222
Slug, 1/9, 4/48–49, 10/118, 10/125, 10/127, 13/192, 13/201–203, 14/216–220, 15/232
Specific gravity, 2/20–26, 4/47, 4/49–55, 5/61–65, 6/70–80, 6/82–85, 7/88–89, 7/93, 10/122, 10/128–131, 11/143–146, 11/148, 12/168, 12/172–173, 13/187–188, 13/190–191, 13/193, 13/200, 14/210, 14/212, 14/216, 14/222, 15/234, 15/240–241, 16/247–248, 17/260, 18/271–272, 18/274
Spread, 13/193–195
Stability, 11/148–149, 11/153
Stabilization, 11/149
Steady-state, 7/90, 7/92, 7/94, 8/98, 11/146, 11/152, 13/182, 13/192, 16/246, 18/279
Storage, 1/16, 10/120, 10/124, 11/136, 11/151, 13/204, 17/255, 17/263
Stress, 9/110–112, 11/149–152, 12/165, 12/170, 12/173–174, 12/177–179, 14/215, 17/252–256, 17/259–260, 17/262, 17/264
Subsea, 1/11, 11/152
Sucker rod pumping, 12/159, 12/161–163, 12/165, 12/167, 12/169, 12/171, 12/173–175, 12/177, 12/179
Surface equipment, 1/6–7, 5/60, 12/174
System analysis, 6/70, 6/84, 13/184–186, 18/276

T
Temperature, 1/4–5, 1/12, 2/20–25, 3/30, 4/50–56, 5/60–65, 6/70–80, 6/82–84, 7/88, 7/92–93, 9/112–114, 10/120–122, 10/124–127, 10/129–130, 10/132, 11/138–140, 11/142–144, 11/146, 11/148, 11/150, 11/152–156, 12/162, 13/186–193, 13/200–201, 13/203, 14/208–210, 14/213, 14/215–216, 14/219–220, 15/228, 15/231, 15/234, 15/236, 15/238, 15/240–241, 16/244, 16/248, 17/258, 18/270–272, 18/276, 18/278
Thermal conductivity, 11/152–153, 15/238
Thermodynamic, 4/46, 4/53
Torque, 12/162–163, 12/166–171, 12/173, 12/177, 12/179, 14/214–215
Transient flow, 3/30–31, 3/33, 3/39, 7/87–88, 7/90, 7/92–93, 7/95, 11/153
Transportation, 1/4, 1/9, 1/11, 9/107, 10/118, 11/133–137, 11/139, 11/141, 11/143, 11/145, 11/147, 11/149, 11/151, 11/153, 11/155–156, 18/276
Transmission lines, 11/136, 11/147
Tubing movement, 9/114
Turbulent flow, 4/46–48, 4/55, 6/81, 9/112, 11/144–146
Two-phase flow, 1/5, 3/32, 3/34, 3/38, 3/41, 3/48, 5/63–64, 5/66, 7/88, 7/90–91, 7/94, 11/149, 11/151, 13/198
Two-phase reservoirs, 3/34–35, 3/39

U
Unsaturated oil, 3/35
Undersaturated oil, 1/4, 3/33, 3/35–37, 7/88–89
 Reservoirs, 3/33, 7/88
Units, 1/17, 2/21, 2/24–25, 4/46, 4/51–54, 4/56, 5/60, 6/72–74, 6/78–81, 8/99, 10/118, 10/121, 10/128, 10/132, 11/140, 11/144–146, 11/150–151, 12/162–163, 12/165, 12/168–169, 12/171, 12/174–175, 13/188, 14/208–209, 15/229, 15/233, 16/247, 18/270–271

V

Valves, 1/6–8, 1/10, 10/119–120, 10/126, 11/136, 13/181–183, 13/188–201, 13/203–205, 18/279
Velocity, 1/8, 1/11, 4/46–47, 4/49, 4/52–53, 5/60–63, 10/121, 10/126, 10/137–138, 11/144–145, 11/154, 12/165, 13/188, 13/203, 14/216, 14/218–220, 15/231–237, 15/240, 16/248
Vertical lift performance (VLP), 4/46, 13/183
Viscosity, 2/20–21, 2/23–24, 2/26, 3/30, 3/33–35, 3/39–40, 4/46, 4/48, 4/52–54, 4/58, 5/61, 6/74–76, 6/78, 6/80, 6/82–83, 7/88, 7/93, 7/95, 10/127, 11/144, 11/147, 12/172, 13/187, 13/200, 14/209–210, 14/212–213, 14/215, 15/229, 15/231, 16/246–247, 17/254–255, 17/258–260

W

Wall thickness, 1/11, 9/110, 11/148–152, 11/154
Water, 1/4
  coning, 15/230–231, 15/238
  flow, 13/183, 15/237
  production, 4/50–52, 4/57, 9/112, 13/202, 15/227–228, 15/231, 15/237–238, 15/240, 18/279
Well, 1/5–9, 2/20–21, 3/30–43, 4/46, 4/55, 5/60, 6/69–71, 7/88–90, 8/98–99, 8/101, 9/107, 10/118–122, 11/153, 12/159, 13/182–185, 14/208–214, 15/225–235, 16/244, 17/252, 18/267–269
  deliverability, 1/1, 3/30, 6/69–71, 6/73, 6/76–77, 6/79, 6/81–84, 13/205, 15/228–229, 17/264, 18/268
  operation, 14/219
  productivity, 1/4, 3/42, 12/172, 15/228–230, 16/244, 17/257–258, 17/264,
  test, 15/241, 17/263
Wellbore flow, 4/46, 6/70
Wellhead, 1/4–7, 1/9, 1/13, 1/15, 3/30, 3/37, 4/46, 4/50–51, 5/60, 5/63–64, 6/70–72, 6/74–82, 6/84, 7/93, 10/127, 11/150–151, 12/173, 13/184–185, 13/198–199, 13/201, 13/205, 14/210, 14/212–214, 14/216, 14/219, 15/232, 15/237, 15/240–241, 18/268, 18/270, 18/279
Weymouth equation, 11/146–148, 13/187, 18/272–273
Wormhole, 16/247–248

Y

Yield stress, 9/110–112

Z

Z-factor, 2/23–25, 5/65, 7/92–93, 11/146, 13/188, 14/220